Statistics with *Mathematica*®

LIMITED WARRANTY AND DISCLAIMER OF LIABILITY

ACADEMIC PRESS ("AP") AND ANYONE ELSE WHO HAS BEEN INVOLVED IN THE CREATION OR PRODUCTION OF THE ACCOMPANYING CODE ("THE PRODUCT") CANNOT AND DO NOT WARRANT THE PERFORMANCE OR RESULTS THAT MAY BE OBTAINED BY USING THE PRODUCT. THE PRODUCT IS SOLD "AS IS" WITHOUT WARRANTY OF ANY KIND (EXCEPT AS HEREAFTER DESCRIBED), EITHER EXPRESSED OR IMPLIED, INCLUDING, BUT NOT LIMITED TO, ANY WARRANTY OF PERFORMANCE OR ANY IMPLIED WARRANTY OF MERCHANTABILITY OR FITNESS FOR ANY PARTICULAR PURPOSE. AP WARRANTS ONLY THAT THE CD-ROM ON WHICH THE CODE IS RECORDED IS FREE FROM DEFECTS IN MATERIAL AND FAULTY WORKMANSHIP UNDER THE NORMAL USE AND SERVICE FOR A PERIOD OF NINETY (90) DAYS FROM THE DATE THE PRODUCT IS DELIVERED. THE PURCHASER'S SOLE AND EXCLUSIVE REMEDY IN THE EVENT OF A DEFECT IS EXPRESSLY LIMITED TO EITHER REPLACEMENT OF THE CD-ROM OR REFUND OF THE PURCHASE PRICE, AT AP'S SOLE DISCRETION.

IN NO EVENT, WHETHER AS A RESULT OF BREACH OF CONTRACT, WARRANTY, OR TORT (INCLUDING NEGLIGENCE), WILL AP OR ANYONE WHO HAS BEEN INVOLVED IN THE CREATION OR PRODUCTION OR THE PRODUCT BE LIABLE TO PURCHASER FOR ANY DAMAGES, INCLUDING ANY LOST PROFITS, LOST SAVINGS OR OTHER INCIDENTAL OR CONSEQUENTIAL DAMAGES ARISING OUT OF THE USE OR INABILITY TO USE THE PRODUCT OR ANY MODIFICATIONS THEREOF, OR DUE TO THE CONTENTS OF THE CODE, EVEN IF AP HAS BEEN ADVISED OF THE POSSIBILITY OF SUCH DAMAGES, OR FOR ANY CLAIM BY ANY OTHER PARTY.

Any request for replacement of a defective CD-ROM must be postage prepaid and must be accompanied by the original defective CD-ROM, your mailing address and telephone number, and proof of date of purchase and purchase price. Send such requests, stating the nature of the problem, to Academic Press Customer Service, 6277 Sea Harbor Drive, Orlando, FL 32887, 1-800-321-5068. AP shall have no obligation to refund the purchase price or to replace a CD-ROM based on claims of defects in the nature or operation of the Product.

Some states do not allow limitation on how long an implied warranty lasts, nor exclusions or limitations of incidental or consequential damages, so the above limitations and exclusions may not apply to you. This warranty gives you specific legal rights, and you may also have other rights which vary from jurisdiction to jurisdiction.

THE RE-EXPORT OF UNITED STATES ORIGIN SOFTWARE IS SUBJECT TO THE UNITED STATES LAWS UNDER THE EXPORT ADMINISTRATION ACT OF 1969 AS AMENDED. ANY FURTHER SALE OF THE PRODUCT SHALL BE IN COMPLIANCE WITH THE UNITED STATES DEPARTMENT OF COMMERCE ADMINISTRATION REGULATIONS. COMPLIANCE WITH SUCH REGULATIONS IS YOUR RESPONSIBILITY AND NOT THE RESPONSIBILITY OF AP.

Statistics with *Mathematica*®

Martha L. Abell
James P. Braselton
John A. Rafter
Department of Mathematics and Computer Science
Georgia Southern University
Statesboro, Georgia

ACADEMIC PRESS
San Diego London Boston
New York Sydney Tokyo Toronto

This book is printed on acid-free paper. ♾

Academic Press
A Division of Harcourt Brace & Company
525 B Street, Suite 1900, San Diego, California 92101-4495, USA
http://www.apnet.com

Academic Press
24–28 Oval Road, London NW1 7DX, UK
http://www.hbuk.co.uk/ap/

Library of Congress Cataloging-in-Publication Data
Abell, Martha L.,
Statistics with Mathematica® / Martha L. Abell, James P. Braselton, John A. Rafter.
p. cm.
Includes bibliographical references and index.
ISBN 0-12-041554-2 (book : alk. paper). — ISBN 0-12-041555-0 (CD-ROM)
1. Mathematical statistics—Data processing. 2. Mathematica (Computer file) I. , James P. II. Rafter, John A. (John Arthur).
III. Title.
QA276.4.A24 1998
519.5′0285′53—dc21 98-27879
CIP

Printed in the United States of America
98 99 00 01 02 CP 9 8 7 6 5 4 3 2 1

Contents

Preface

Statistics with Mathematica® is an appropriate reference for all users of *Mathematica*, in particular for users such as students, instructors, engineers, business people, and other professionals using statistics and *Mathematica*. Readers from the most elementary to advanced levels will find that the range of topics covered addresses their needs. The text can be easily used and understood by anyone with a background in statistics, such as that typically taught in a two-semester introductory statistics course. In fact, all of the information discussed in such a course is presented, so the text is particularly useful as a supplement for instructors of the course and students enrolled in the course. Because the text is a reference book, pertinent information is provided without the detail and discussion that is common in textbooks and is especially terse for procedures or methods that are readily available in most introductory statistical textbooks. For those topics that are not typically covered in these sorts of texts, we have included more detail and discussion. For one who is learning or relearning the vocabulary, the methodology of the book is sequential in nature. Information is often restated to help the reader who needs to refer to specific sections for specific information. We hope that the extensive index will also prove useful.

In addition to using virtually all of *Mathematica*'s built-in statistical commands as well as those contained in various packages, we have written several procedures, which are contained on the CD, to extend *Mathematica*'s capabilities. We discuss how to access and use these commands when they are introduced.

Other features to help make *Statistics with Mathematica* as easy to use and as useful as possible include the following.

1. **Version 3 CoStibility.** All examples illustrated in *Statistics with Mathematica*, were completed using Version 3 of *Mathematica*. Although most computations can continue to be carried out with earlier versions of *Mathematica*, such as Version 2.2, we have taken advantage of the new features in Version 3, such as two-dimensional typset input and output, as much as possible.
2. **Detailed Table of Contents.** The table of contents includes all chapter, section, and subsection headings. Along with the comprehensive index, we hope that users will be able to locate information quickly and easily.
3. **Comprehensive Index.** In the index, mathematical examples and applications are listed by topic, or name, as well as commands and frequently used options. Particular mathematical examples, as well as examples illustrating how to use frequently used commands, are easy to locate. In addition, commands in the index are cross-referenced with frequently used options. Functions available in the various packages are cross-referenced both by package and alphabetically.
4. **CD-ROM.** All *Mathematica* input that appears in *Statistics with Mathematica* is included on the disk packaged with the text.

Of course, we must express our appreciation to those who assisted in this project, especially our colleague, Dr. Patricia Humphrey, who provided numerous suggestions in addition to substantial technical advice regarding many of the topics. In addition, we would like to express appreciation to our editor, Charles B. Glaser, editorial assistant, Della Grayson, and our assistant production manager, Diane Grossman, at Academic Press for providing a pleasant environment in which to work. Also, Wolfram Research has been most helpful in providing us up-to-date information about *Mathematica*. Finally, we thank those close to us, especially Imogene Abell, Carolyn Rafter, Lori Braselton, Ada Braselton, and Martha Braselton, for enduring with us the pressures of meeting a deadline and for graciously accepting our demanding work schedules. We certainly could not have completed this task without their care and understanding.

M. L. Abell
J. P. Braselton
Statesboro, Georgia

CHAPTER 1

Data and Data File Manipulation with Mathematica

1.1 Introduction to Mathematica

The objectives of *Statistics with Mathematica®* are many. One, of course, is to provide the user of *Mathematica* with the information necessary to generate a wide range of descriptive statistics, to perform inferential procedures, and to employ a variety of other related techniques. We introduce the topics covered, provide a brief discussion of their usefulness, point out details required for correct application and interpretation, provide formulas when appropriate, and give examples of the use of the statistical functions in *Mathematica*. In addition, if there is no *Mathematica* function available in one of the standard packages to perform a needed operation, we provide the necessary files written in the *Mathematica* language.

Mathematica, first released in 1988 by Wolfram Research, Inc., is a system for doing mathematics on a computer. It combines symbolic manipulation, numerical mathematics, outstanding graphics, and a sophisticated programming language. Because of its versatility, *Mathematica* has established itself as the computer algebra system of choice for many computer users. Among the over 1 million users of *Mathematica*, 28% are engineers, 21% are computer scientists, 20% are physical scientists, 12% are mathematical scientists, and 12% are business, social, and life scientists. Two-thirds of the users are in industry and government, and there is a small (8%) but growing number of student users.

Mathematica Version 3

With the release of Version 3, many new functions and features have been added to *Mathematica*. We encourage users of earlier versions of *Mathematica* to update to Version 3 as soon as they can. All examples in *Statistics with Mathematica* were completed with Version 3. In most cases, the same results will be obtained if you are using Version 2.0 or later, although the appearance of the output will almost certainly differ from that presented here. Occasionally, however, particular features of Version 3 are utilized (such as editable two-dimensional input and output) and in those cases, of course, these features are not available in earlier versions. If you are using an earlier or later version of *Mathematica*, your results may not appear in a form identical to those found in this book: some commands found in Version 3 are not available in earlier versions of *Mathematica*, and in later versions some commands will certainly be changed, new commands added, and obsolete commands removed. For details regarding these changes, please see *The Mathematica Book*.* You can determine the version of *Mathematica* you are using during a given *Mathematica* session by entering either the command `$Version` or the command `$VersionNumber`. In this text, we assume that *Mathematica* has been correctly installed on the computer you are using. If you need to install *Mathematica* on your computer, please refer to the documentation that came with the *Mathematica* software package.

1.2 Getting Started with Mathematica

After the *Mathematica* program has been properly installed, a user can access *Mathematica*. The following are brief descriptions of methods for starting *Mathematica* on several platforms.

Macintosh: In the same manner that folders are opened, the *Mathematica* application can be started by selecting the *Mathematica* icon, going to **File,** and selecting **Open**, or by simply clicking twice on the *Mathematica* icon. Of course, opening an existing *Mathematica* document also opens the *Mathematica* program.

Windows: Double-click on the *Mathematica* for Windows application icon. This operation opens *Mathematica* for Windows with a blank worksheet.

DOS: Type the command **MATH** at the DOS prompt. After a few seconds, the screen will clear, the *Mathematica* logo will appear briefly, and a prompt will appear at the bottom of the screen above the status line.

UNIX: Type **math** at the UNIX prompt. A new window will open, containing *Mathematica*.

---

* Stephen Wolfram, *The Mathematica Book*, Third Edition, Wolfram Media, Inc. (1996).

Once *Mathematica* has been started, computations can be carried out immediately. *Mathematica* commands are evaluated by pressing **SHIFT-RETURN** or by pressing **ENTER**. Note that on systems where both **RETURN** and **ENTER** keys are available, **ENTER** alone is typically equivalent to **SHIFT-RETURN**. On these systems **RETURN** allows input to be continued on the next line. For systems which do not have both keys, *Mathematica* will automatically continue on to successive lines until it receives a complete command. We will use **ENTER** to mean "begin processing." Generally, when a semicolon is placed at the end of the command, the resulting output is *not* displayed. When it is displayed, output is located below input. Some of the typical steps involved in working with *Mathematica* are illustrated in the following calculations. In each case, type the command and press **SHIFT-RETURN**. *Mathematica* evaluates the command, displays the result, and inserts a new horizontal black line. For example, entering

```
N[π, 50]
```

```
3.1415926535897932384626433832795028841971693993751l
```

returns a 50-digit approximation of π. In *Mathematica*, both π and `Pi` represent the mathematical constant π. Thus, entering `N[Pi,50]` returns the same result. Subsequent calculations are entered in the same way. For example, entering

```
Plot[{Sin[x], 2 Cos[2 x]}, {x, 0, 3 π},
  PlotStyle→{GrayLevel[0], GrayLevel[0.5]}]
```

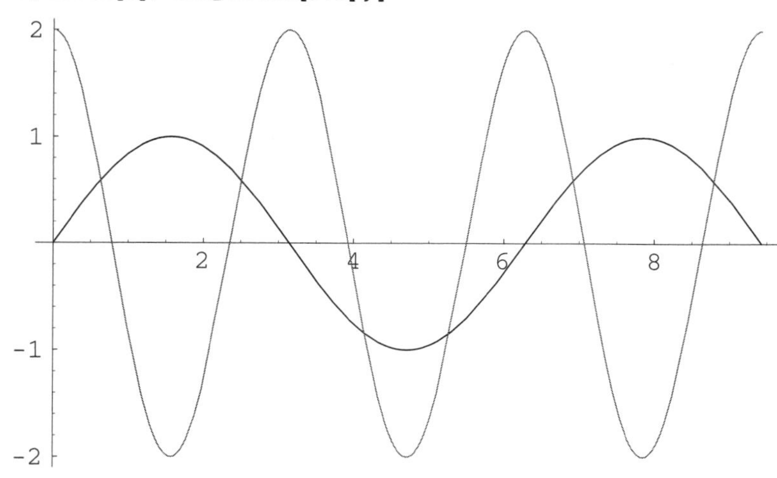

graphs the functions $\sin x$ (in black) and $2\cos 2x$ (in gray) on the interval $[0, 3\pi]$. Similarly, entering

```
Plot3D[Sin[x + Cos[y]], {x, 0, 4 π}, {y, 0, 4 π},
 Ticks → None, Boxed → False, Axes → None, PlotPoints→30]
```

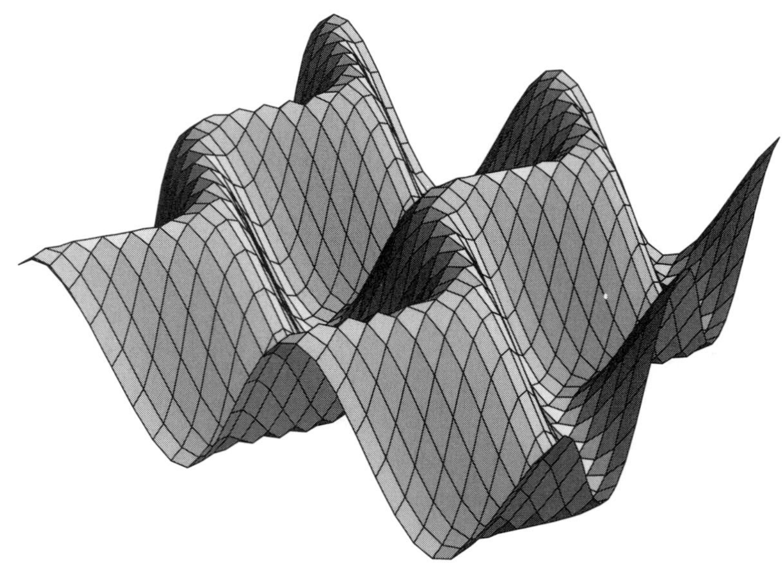

graphs the function $\sin(x + \cos y)$ on the rectangle $[0, 4\pi] \times [0, 4\pi]$. (If you are using a color monitor, the result will be displayed in color.) Notice that *every Mathematica* command begins with a *capital letter* and the argument is enclosed by *square brackets* "[...]."

Input and output can be displayed in a variety of ways: **InputForm**, **OutputForm**, **StandardForm**, and **TraditionalForm**. For example, entering

```
Solve[x^3 - 2 x + 1 == 0]
```

$$\left\{\{x \to 1\}, \left\{x \to \frac{1}{2}(-1-\sqrt{5})\right\}, \left\{x \to \frac{1}{2}(-1+\sqrt{5})\right\}\right\}$$

in **InputForm** or

```
Solve[x^3 - 2 x + 1 == 0]
```

$$\left\{\{x \to 1\}, \left\{x \to \frac{1}{2}(-1-\sqrt{5})\right\}, \left\{x \to \frac{1}{2}(-1+\sqrt{5})\right\}\right\}$$

in **StandardForm** solves the equation $x^3 - 2x + 1 = 0$ for x.

To choose the form for input/output, move the cursor to the *Mathematica* menu, select **Cell**, and then **Convert To**, as illustrated in the following figure.

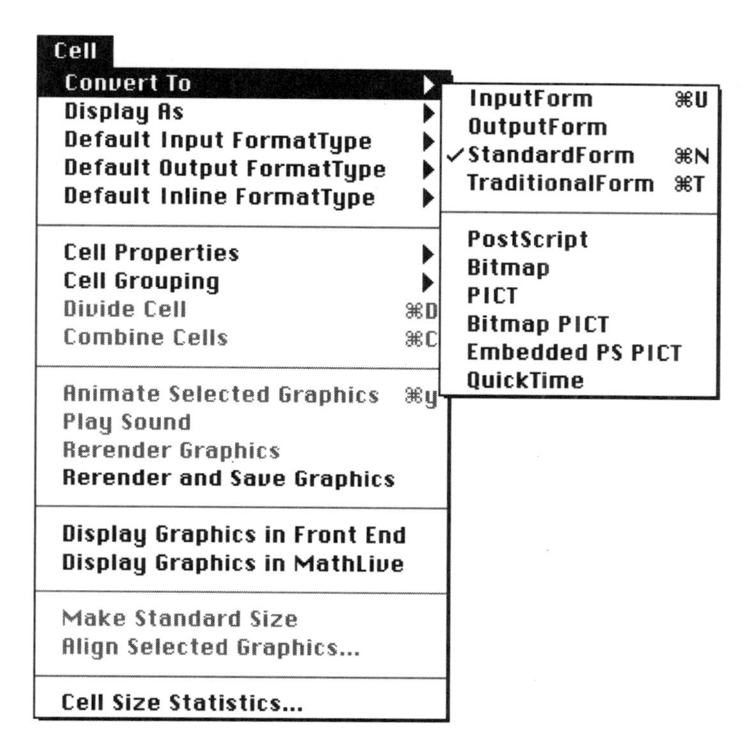

Here, the input is in **InputForm** and the output in **OutputForm**.

```
Solve[x^3 - 2*x + 1 == 0]
```

```
                        1                               1
{{x -> 1}, {x -> - (-1 - Sqrt[5])}, {x -> - (-1 + Sqrt[5])}}
                        2                               2
```

Here both the input and output appear in **TraditionalForm**.

Solve$(x^3 - 2\ x + 1 == 0)$

$$\left\{\{x \to 1\}, \left\{x \to \frac{1}{2}\left(-1-\sqrt{5}\right)\right\}, \left\{x \to \frac{1}{2}\left(-1+\sqrt{5}\right)\right\}\right\}$$

You can determine the form of input/output by looking at the cell bracket that contains the input/output. For example, even though all three of the following commands look different, all three evaluate $\int_0^{2\pi} x^3 \sin x\ dx$.

```
Integrate[x^3*Sin[x], {x, 0, 2*Pi}]
```

$$\int_0^{2\pi} x^3 \, \mathrm{Sin}[x] \, dx$$

$$\int_0^{2\pi} x^3 \sin(x) \, dx$$

A cell bracket like this] means the input is in **InputForm** and the output is in **OutputForm**. A cell bracket like this] means the input and output of the cell are in **StandardForm**. A cell bracket like this] means the input and output of the cell are in **TraditionalForm**. Throughout *Statistics with Mathematica* we display input and output using **StandardForm**, unless otherwise stated. To enter code in **StandardForm**, we often take advantage of the BasicTypesetting palette which is accessed by going to **File** under the *Mathematica* menu and then selecting **Palettes**

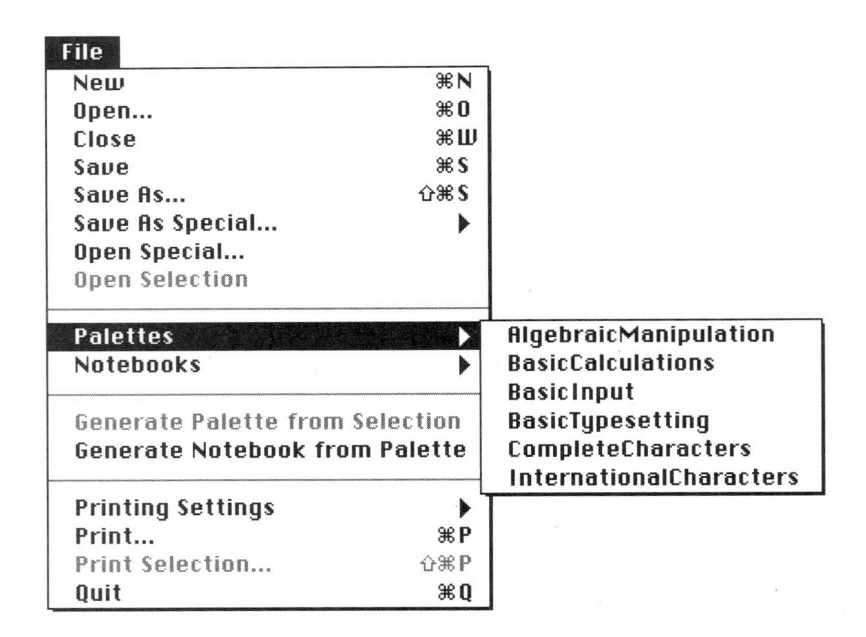

followed by **BasicTypesetting**.

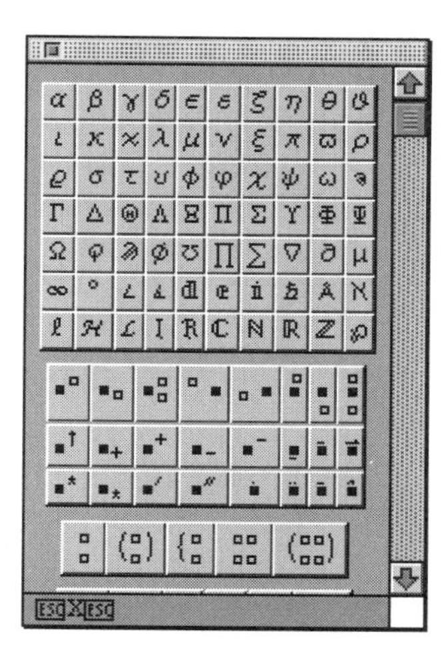

Alternatively, you can find a complete list of typesetting shortcuts in *The Mathematica Book* (see appendix **Listing of Named Characters**).

Mathematica sessions are terminated by entering `Quit[]`. On several platforms with notebook interfaces (such as Macintosh, Windows, and NeXT), *Mathematica* sessions are ended by selecting **Quit** from the **File** menu, or by using the appropriate keyboard shortcut. *Mathematica* notebooks can be saved by referring to the **File** menu. *Mathematica* allows you to save notebooks in a variety of formats, in addition to the standard *Mathematica* format.

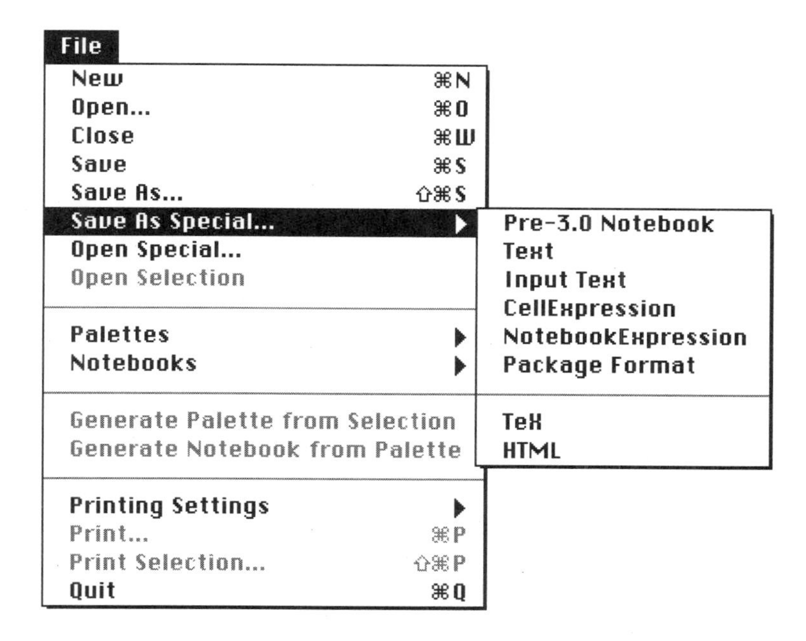

On platforms with a notebook interface, input and text regions in notebook interfaces can be edited. Editing input can create a notebook in which the mathematical output does not make sense in the sequence it appears. It is also possible to simply go into a notebook and alter input without doing any recalculation. This also creates misleading notebooks. Hence, common sense and caution should be used when editing the input regions of notebooks. Recalculating all commands in the notebook will clarify any confusion.

Preview

In order for the *Mathematica* user to take full advantage of the capabilities of this powerful software, an understanding of its syntax is imperative. Although the rules of *Mathematica* syntax are far too numerous to list here, knowledge of the following five rules equips the beginner with the necessary tools to start using the *Mathematica* program with little trouble.

Five Basic Rules

1. The arguments of *all* functions (both built-in and ones that you define) must be contained in brackets `[...]`. Parentheses `(...)` are used for grouping operations; vectors, matrices, and lists are given in braces `{...}`; and double square brackets `[[...]]` are used for indexing lists and tables.
2. The names of built-in functions have their first letters capitalized; if a name consists of two or more words, the first letter of each word is capitalized.
3. Multiplication is represented by a space or an asterisk `(*)`.
4. Powers are denoted by a caret `(^)`.
5. If you get no response or an incorrect response, you may have entered or executed the command incorrectly. In some cases, the amount of memory allocated to *Mathematica* can cause a crash.

1.3 Loading Packages

Although *Mathematica* contains many built-in functions, a tremendous number of additional commands are available in various **packages** that are shipped with each version of *Mathematica*. Experienced users can create their own packages; other packages are available from user groups and **MathSource**, which electronically distributes *Mathematica*-related products. For information about **MathSource**, send the message "help" to mathsource@wri.com. If desired, you can purchase **MathSource** on a CD directly from Wolfram

Research, Inc. or you can access **MathSource** from the Wolfram Research World Wide Web site (http://www.wri.com).

On a computer with a notebook interface, the folder containing the packages shipped with *Mathematica* is shown in the following figure. Descriptions of the various packages are contained in the *Technical Report: Guide to Standard Mathematica Packages* published by and available from Wolfram Research, Inc. Descriptions are also available via the **Help Browser**, discussed in the next section.

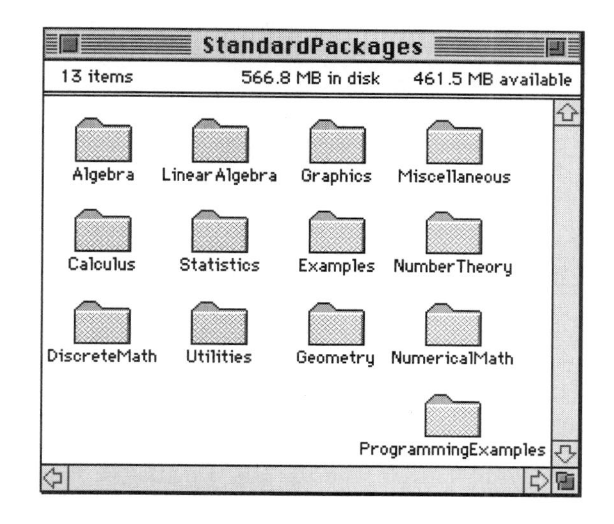

A package is loaded by entering the command `<<directory`packagename`` where `directory` is the location of the package, `packagename`. Entering the command `<<directory`Master`` makes all the functions contained in all the packages in `directory` available. In this case, each package need not be loaded individually. For example, to load the package **Shapes** contained in the **Graphics** folder (or directory), enter `<<Graphics`Shapes``.

```
<<Graphics`Shapes`
```

After the **Shapes** package has been loaded, entering

```
Show[Graphics3D[Torus[1, 0.5, 30, 30]], Boxed→ False]
```

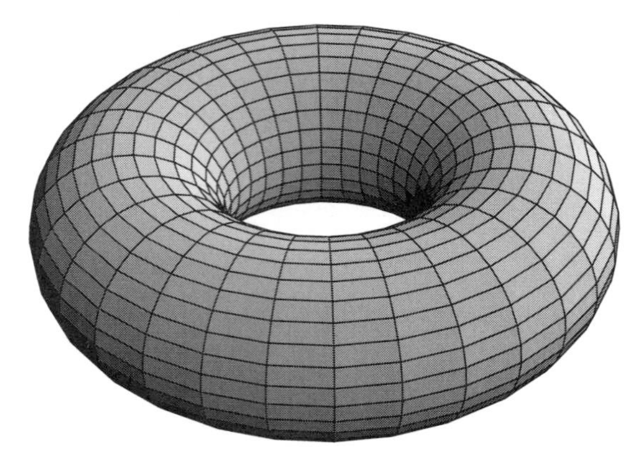

generates the graph of a torus. A Möbius strip and a sphere can be generated and displayed side-by-side using the built-in function GraphicsArray and the shapes from the Shapes package.

```
mstrip= Graphics3D[MoebiusStrip[1, 0.5, 40], Boxed→ False];
sph = Graphics3D[Sphere[1, 25, 25], Boxed→False];
Show[GraphicsArray[{mstrip, sph}]]
```

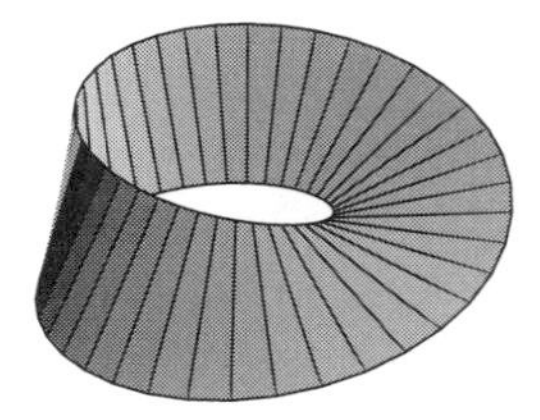

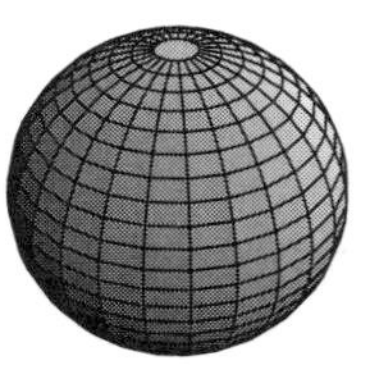

The **Shapes** package contains definitions of familiar three-dimensional shapes, including the cone, cylinder, helix, and double helix. In addition, it allows us to perform transformations such as rotations and translations on three-dimensional graphics.

A Word of Caution

The package must be loaded before using a command contained in the package; otherwise, the command cannot be processed. For example, the command `GramSchmidt[{v1,v2,...,vn}]` returns an orthonormal set of vectors with the same span as the vectors $v_1, v_2, \ldots, v_n$. It is contained in the **Orthogonalization** package located in the

LinearAlgebra folder. Before the package has been loaded, *Mathematica* does not know the meaning of `GramSchmidt`, so the input is returned.

```
GramSchmidt[{{1, 1, 0}, {0, 2, 1}, {1, 0, 3}}]
```

```
GramSchmidt[{{1, 1, 0}, {0, 2, 1}, {1, 0, 3}}]
```

If you do forget to load the package, then the original command must be removed in order for the command to work properly once the package is loaded. If, at this point in the example, we load the **Orthogonalization** package, several error messages result.

```
<< LinearAlgebra'Orthogonalization'
```

```
GramSchmidt::shdw: Symbol GramSchmidt appears in multiple
contexts{LinearAlgebra'Orthogonalization',Global'};
definitions in context LinearAlgebra'Orthogonalization' may
shadow or be shadowed by other definitions.
```

In fact, when we reenter the command, we obtain the same result as that obtained previously.

```
GramSchmidt[{{1, 1, 0}, {0, 2, 1}, {1, 0, 3}}]
```

```
GramSchmidt[{{1, 1, 0}, {0, 2, 1}, {1, 0, 3}}]
```

However, after using the command `Remove`, the command `GramSchmidt` works as expected.

```
Remove[GramSchmidt]
```

```
GramSchmidt[{{1, 1, 0}, {0, 2, 1}, {1, 0, 3}}]
```

$$\{\{\frac{1}{\sqrt{2}}, \frac{1}{\sqrt{2}}, 0\}, \{-\frac{1}{\sqrt{3}}, \frac{1}{\sqrt{3}}, \frac{1}{\sqrt{3}}\}, \{\frac{1}{\sqrt{6}}, -\frac{1}{\sqrt{6}}, \sqrt{\frac{2}{3}}\}\}$$

Alternatively, quit *Mathematica*, restart, load the package, and then execute the command.

1.4 Getting Help from Mathematica

Becoming competent with *Mathematica* can take a serious investment of time. One way to obtain information about commands and functions, including user-defined functions, is the command `?`. `?object` gives information on the *Mathematica* object `object`.

EXAMPLE 1: Use ? to obtain information about the command PolynomialDivision.

SOLUTION:

```
?PolynomialDivision
```

```
PolynomialDivision[p, q, x] gives a list of the quotient and remainder obtained by
  division of the polynomials p and q in x.
```

We illustrate PolynomialDivision by computing the quotient and remainder obtained by dividing $x^3 + 1$ by $x - 1$.

```
PolynomialDivision[x^3 + 1, x - 1, x]
```

```
{1 + x + x^2, 2}
```

The result means that $(x-1)(x^2+x+1)+2 = x^3+1$, which is verified with Expand.

```
Expand[(1 + x + x^2) (x - 1) + 2]
```

```
1 + x^3
```

■

Another way to obtain information on *Mathematica* commands is the command Options. Options[object] returns a list of the available options associated with object along with their current settings. Notice that the default value (the value automatically assumed by *Mathematica*) for each option is given in the output.

EXAMPLE 2: Use Options to obtain a list of the options and their current settings for the command ParametricPlot.

SOLUTION: The command Options[ParametricPlot] lists all the options and their current settings for the command ParametricPlot.

```
Options[ParametricPlot]
```

$$\left\{\text{AspectRatio} \to \frac{1}{\text{GoldenRatio}}, \text{Axes} \to \text{Automatic},\right.$$

```
AxesLabel→None, AxesOrigin→Automatic, AxesStyle→Automatic,
Background→Automatic, ColorOutput→Automatic,
Compiled→True, DefaultColor→Automatic, Epilog→{},
Frame→False, FrameLabel→None, FrameStyle→Automatic,
FrameTicks→Automatic, GridLines→None, ImageSize→Automatic,
MaxBend→10., PlotDivision→30., PlotLabel→None,
PlotPoints→25, PlotRange→Automatic, PlotRegion→Automatic,
PlotStyle→Automatic, Prolog→{}, RotateLabel→True,
Ticks→Automatic, DefaultFont:→$DefaultFont,
DisplayFunction:→$DisplayFunction, FormatType:→$FormatType,
TextStyle:→$TextStyle}
```

■

`??object` or, equivalently, `Information[object]` yields the information on the *Mathematica* object `object` returned by both ?object and `Options[object]` in addition to a list of attributes of object. Note that `object` may either be a user-defined object or a built-in *Mathematica* object.

EXAMPLE 3: Use `??` to obtain information about the commands `Solve` and `Map`. Use `Information` to obtain information about the command `PolynomialLCM`.

SOLUTION: We use `??` to obtain information about the commands `Solve` and `Map`, including a list of options and their current settings.

```
?? Solve

Solve[eqns, vars] attempts to solve an equation or set of equations
  for the variables vars. Solve[eqns, vars, elims] attempts
  to solve the equations for vars, eliminating the variables elims.

Attributes[Solve] = {Protected}

Options[Solve] = {InverseFunctions-> Automatic,
   MakeRules -> False, Method -> 3, Mode -> Generic, Sort -> True,
   VerifySolutions -> Automatic, WorkingPrecision -> Infinity}

?? Map

Map[f, expr] or f /@ expr applies f to each element on the
  first level in expr. Map[f, expr, levelspec] applies f to
  parts of expr specified by levelspec.

Attributes[Map] = {Protected}

Options[Map] = {Heads -> False}
```

Similarly, we use `Information` to obtain information about the command `PolynomialLCM`, including a list of options and their current settings.

```
Information[PolynomialLCM]

PolynomialLCM[poly1, poly2, ... ] gives the least common
  multiple of the polynomials polyi. PolynomialLCM
  [poly1, poly2, ... , Modulus->p] evaluates the LCM
  modulo the prime p.

Attributes[PolynomialLCM] = {Listable, Protected}

Options[PolynomialLCM] =
   {Extension -> None, Modulus -> 0, Trig -> False}
```

■

The command `Names["form"]` lists all objects that match the pattern defined in `form`. For example, `Names["Plot"]` returns `Plot`; `Names["*Plot"]` returns all objects that end with the string `Plot`; `Names["Plot*"]` lists all objects that begin with the string `Plot`; `Names["*Plot*"]` lists all objects that contain the string `Plot`. `Names["form",SpellingCorrection->True]` finds those symbols that match the pattern defined in `form` after a spelling correction.

EXAMPLE 4: Create a list of all built-in functions beginning with the string `Plot`.

SOLUTION: We use `Names` to find all object that match the pattern `Plot`.

```
Names["Plot"]
```

```
{Plot}
```

Next, we use `Names` to create a list of all built-in functions beginning with the string `Plot`.

```
Names["Plot*"]
```

```
{Plot, Plot3D, Plot3Matrix, PlotDivision, PlotJoined,
 PlotLabel, PlotPoints, PlotRange, PlotRegion, PlotStyle}
```

■

The command `?` can be used in several other ways. Entering `?letters*` gives all *Mathematica* commands that begin with the string `letters`; `?*letters*` gives all *Mathematica* commands that contain the string `letters`; and `?*letters` gives all *Mathematica* commands that end in the string `letters`.

EXAMPLE 5: What are the *Mathematica* functions that (a) end in the string `Cos`; (b) contain the string `Sin`; and (c) begin with the string `Polynomial`?

SOLUTION: Entering

```
? * Cos
```

```
ArcCos Cos
```

yields all functions that end with the string `Cos`, entering

```
? * Sin*
```

```
ArcSin                 SingleLetterItalics  Sinh
ArcSinh                SingularityDepth     SinhIntegral
IncludeSingularTerm    SingularValues       SinIntegral
Sin
```

returns all functions containing the string `Sin`, and entering

```
?Polynomial*

PolynomialDivision PolynomialMod       PolynomialReduce
PolynomialForm     PolynomialQ         PolynomialRemainder
PolynomialGCD      PolynomialQuotient  Polynomials
PolynomialLCM
```

returns all functions that begin with the string `Polynomial`.

■

Mathematica Help

On platforms with a notebook interface (such as Macintosh, Windows, and NeXT), additional help features are accessed from the *Mathematica* menu under **Help**. For basic information about *Mathematica*, go to **Help** and select **Help...**

If you are a beginning *Mathematica* user, you may choose to select **Welcome Screen...**

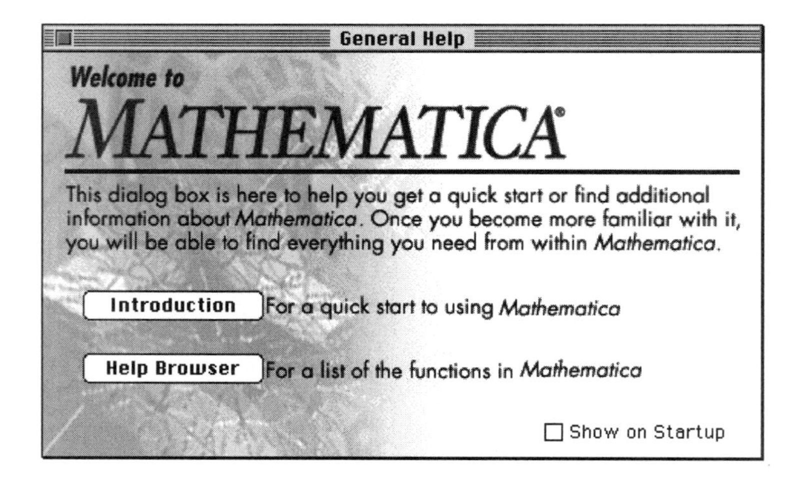

and then select **Introduction** or **Help Browser**.

On the other hand, if you click on `Getting Started` and then select `Using the Help Browser`, you obtain the following basic information about *Mathematica* help.

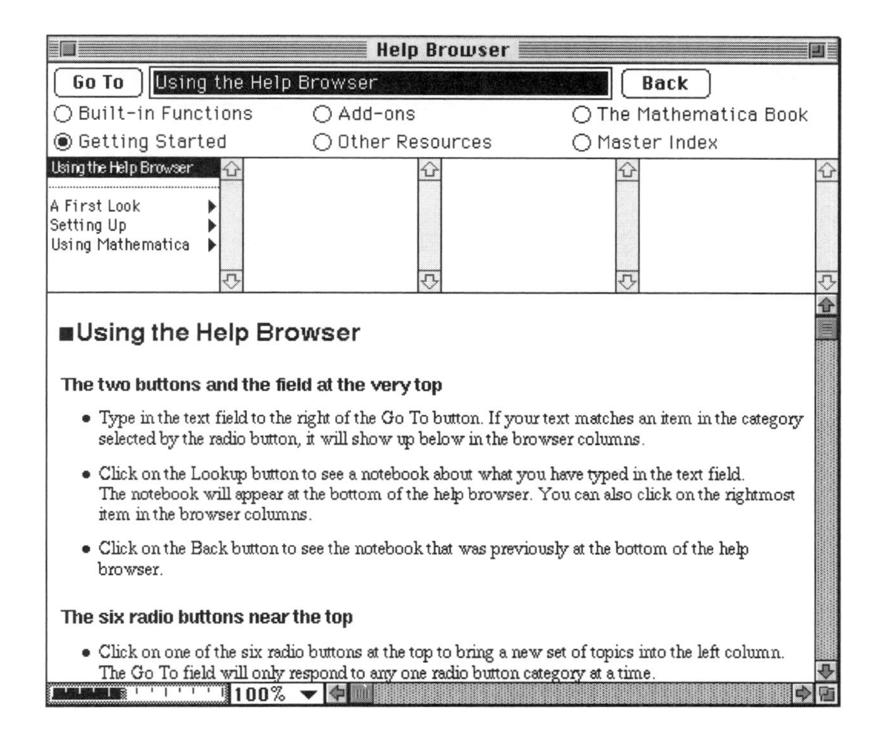

To obtain information about a particular *Mathematica* object or function, open the **Help Browser**, type the name of the object, function, or topic, and press the Go To button. Alternatively, you can type the name of a function that you wish to obtain help about, select it, go to **Help**, and then select **Find in Help...**, as we do here with the command `FullSimplify`.

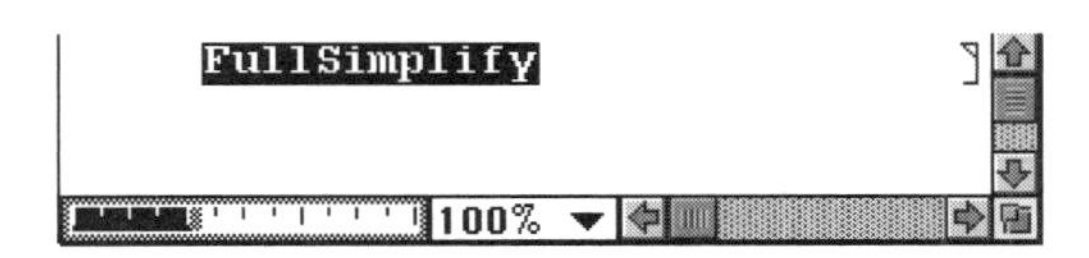

Mathematica presents the following information about `FullSimplify`. Notice that a typical help window not only contains a detailed description of the command and its options,

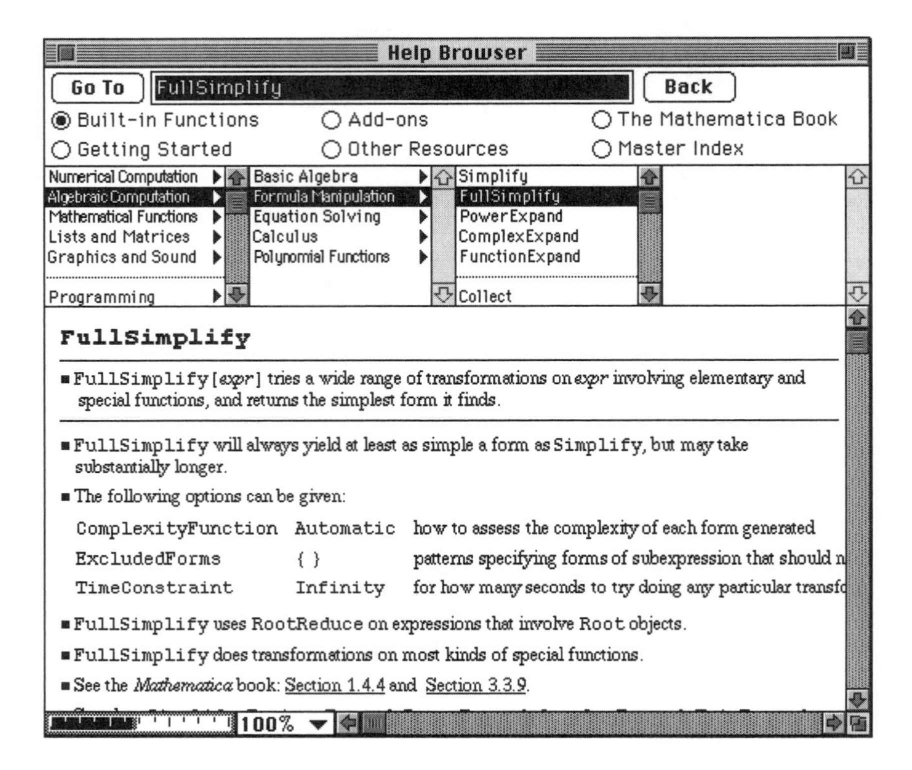

but also several examples that illustrate the command,

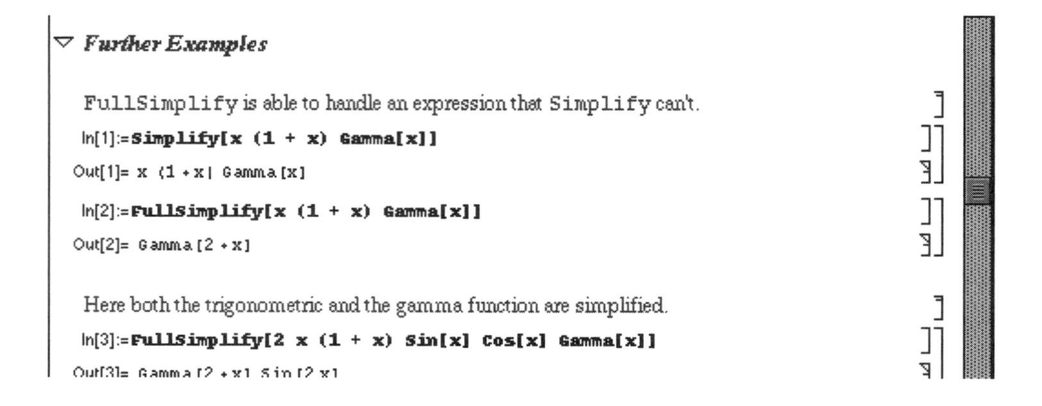

as well as hyperlinked cross-references to related commands and *The Mathematica Book*, which can be accessed by clicking.

TimeConstraint Infinity for how many seconds to try doing any particular transfo

- FullSimplify uses RootReduce on expressions that involve Root objects.
- FullSimplify does transformations on most kinds of special functions.
- See the *Mathematica* book: Section 1.4.4 and Section 3.3.9.
- See also: Simplify, Factor, Expand, PowerExpand, ComplexExpand, TrigExpand, FunctionExpand.

You can also use the **Help Browser** to access an on-line version of *The Mathematica Book*. Here is a portion of Section **3.2.14, Statistical Distributions and Related Functions**.

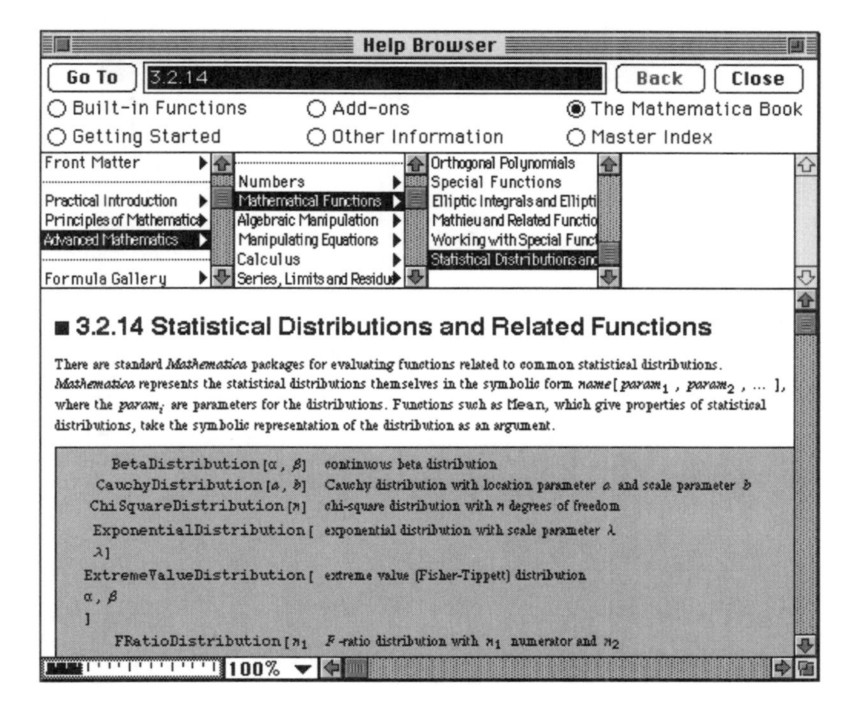

If you click on Add-ons, you can obtain detailed information about the standard packages that are included with *Mathematica*.

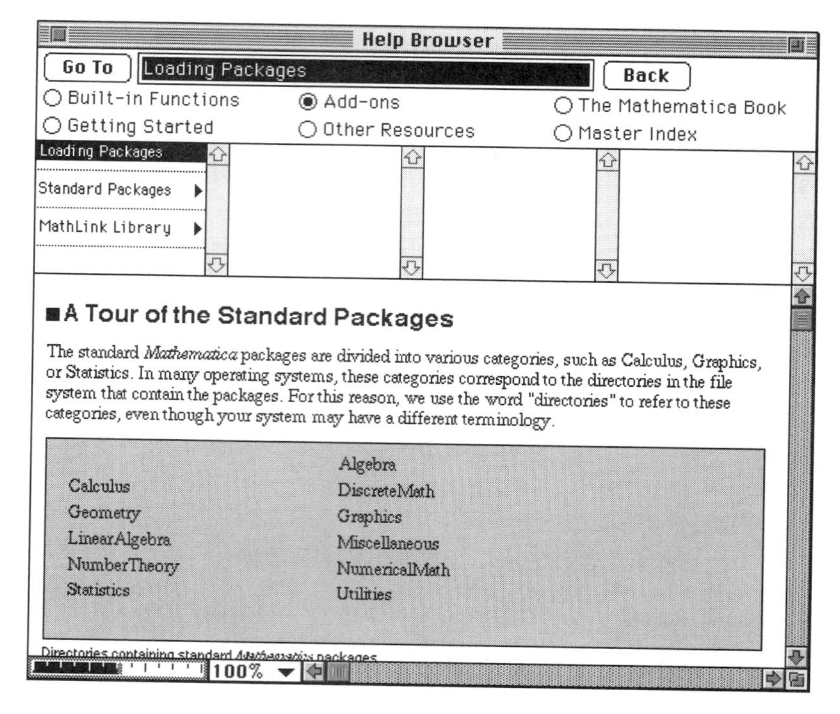

The `Master Index` contains hyperlinks to all portions of *Mathematica* help.

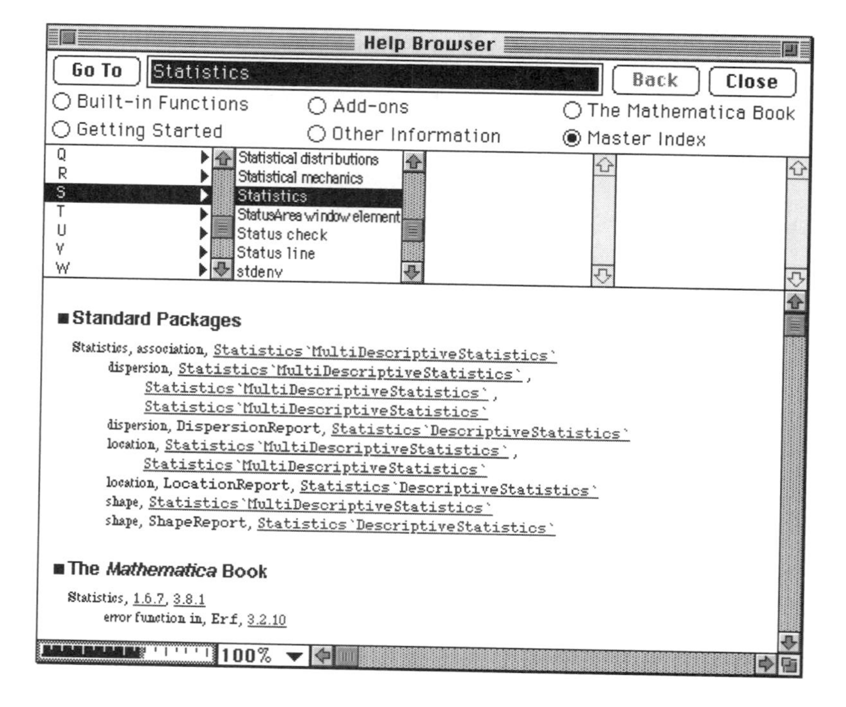

1.5 Data Manipulation

Types of Data

The measurements and other observations which are subjected to analysis are, in general, called **data**. It is important to have a clear understanding of the various properties of your data, because the methods used to summarize and analyze the data will often depend on these properties. Measurements for which arithmetic operations such as differences and averages make sense are called **quantitative data**. These measurements may be **discrete**. That is, the possible values are separate and distinct. An example of this is the number of runs scored in the first inning by a baseball team. Or the measurements may be **continuous**. That is, there is a possible value between any other two possible values. This is true of measurements of time and distance, for example. When the observations are of some quality or attribute rather than a numerical measure, they are called **qualitative data**. These observations may have an inherent order, such as position in the batting lineup or military rank. Or there may be no order to the observations, such as right or left handedness or favorite soft drink. We will refer to qualitative data as **categorical data** to avoid possible confusion between the terms qualitative and quantitative, which are spelled so much alike.

Types of Statistics

The term statistics is often used to refer to the procedures used in presenting, analyzing, and interpreting the data. The word "statistic" is derived from the Latin for "state." Historically, the ideas and practices of statistics grew out of the desire of rulers to count the number of inhabitants or measure the value of the taxable land in their domains.

Statistical methods are often discussed in two parts. **Descriptive statistics** refers to methods for organizing and summarizing data with the goal of their orderly presentation. These include tables, graphs, and numerical measures of location, variability, shape, and association. The primary concern of **inferential statistics** is finding out something about a **population**. A population consists of all individuals, objects, transactions, measurements, etc., of interest. To infer something about the population, we look at a subset of the population called a **sample**. When a sample is taken from a population and a numerical characteristic such as an average is calculated, the resulting value is called a **statistic**. The same calculation done on the population would result in a value which is called a **parameter**. A statistic is used to make an inference about a corresponding parameter. A data set in *Mathematica* can represent a population or a sample at the discretion of the user.

Accuracy and Significant Figures

Accuracy is the nearness of a measurement to the actual value of the variable being measured. In the statistical context, **precision** is not a synonymous term but refers to the closeness to each other of repeated measurements of the same quantity. If you step on a scale twice, and it gives two readings which are close together, but both are 10 pounds below your actual weight, you can say the scale is precise but not accurate. To be successful, a major league baseball umpire need not call balls and strikes accurately (per the rule book) as long as he is precise in his calls (i.e. he "calls it the same" for all batters).

It is important to designate the accuracy with which a measurement has been made. On a continuous scale, the value 8 cm represents a measurement between 7.5 and 8.49. This implies a range of accuracy of 1 cm. Likewise, the value 8.3 cm represents a measurement between 8.25 and 8.349, which implies a range of accuracy of 0.1 cm. Note that 8.0 cm represents a measurement between 7.95 and 8.049, with an implied range of accuracy of 0.1 cm. Those digits in a number that denote the accuracy of the measurement are referred to as **significant figures**. Thus, 8 has one significant figure, 8.0 and 8.3 each have two significant figures, and 8.00 and 8.32 each have three.

It is equally important not to overstate the accuracy by reporting too many significant figures after completing a calculation. *Mathematica* can yield results with more significant figures than are justified by the data. It is good practice to retain many significant figures for all steps until the last in the sequence. When reporting the result of the calculations, the value(s) should be rounded off to the appropriate number of significant figures. A good rule of thumb is to round off to one more significant figure than is meaningful for the original data values.

Entering Data into Mathematica

Data is often stored in rectangular arrays using a computer spreadsheet or editor. The rows of the array are called **records** and represent the units being studied, which are called **experimental units**. The columns of the array are called **variables** and represent the various observations or measurements taken. In general, a variable may be referred to as quantitative or categorical, discrete or continuous, depending on the type of data it represents.

With *Mathematica*, records are contained in braces (`{...}`) and separated by commas (`,`) and values of variables within a record are separated by commas (`,`). Moreover, you must place a left brace (`{`) at the beginning of the array and a right brace (`}`) at the end of the array.

For small data sets, we generally enter the data directly into *Mathematica*.

EXAMPLE 1: ENTERING DATA DIRECTLY INTO *MATHEMATICA*. The following table shows the number of vehicles exported from Brazil and Mexico for selected years. Enter the data into *Mathematica*.

Year	Brazil (number)	Brazil (as a percent of vehicle production)	Mexico (number)	Mexico (as a percent of vehicle production)
1985	207640	21.5	58423	12.7
1986	182134	17.2	72429	21.2
1987	345555	37.6	163073	41.3
1988	320310	30	173147	33.8
1989	254086	25.1	195999	30.6
1990	187314	20.5	276859	33.7
1991	193148	20.1	358666	36.3

Source: Rhys Jenkins, "The political economy of industrial policy: automobile manufacture in the newly industrialising countries," *Cambridge Journal of Economics* 1995, Volume 19, Academic Press Limited, pp. 625–645.

SOLUTION: We type the data directly into *Mathematica* and evaluate it by pressing **Enter**. Including `autodata=` at the beginning of the array instructs *Mathematica* to name the result `autodata`. *Mathematica* does not return a result because we include a semicolon at the end of the command.

```
autodata= {{1985, 207640, 21.5, 58423, 12.7},
  {1986, 182134, 17.2, 72429, 21.2},
  {1987, 345555, 37.6, 163073, 41.3},
  {1988, 320310, 30, 173147, 33.8},
  {1989, 254086, 25.1, 195999, 30.6},
  {1990, 187314, 20.5, 276859, 33.7},
  {1991, 193148, 20.1, 358666, 36.3}};
```

We view `autodata` in traditional row-and-column form using `TableForm`.

```
TableForm[autodata]

1985    207640    21.5    58423     12.7
1986    182134    17.2    72429     21.2
1987    345555    37.6    163073    41.3
1988    320310    30      173147    33.8
1989    254086    25.1    195999    30.6
1990    187314    20.5    276859    33.7
1991    193148    20.1    358666    36.3
```

Selecting Rows

Rows of an array are extracted with `Part`, which is abbreviated with double square brackets (`[[...]]`). First we obtain information about `Part` using the command `?`.

```
?Part
```

```
expr[[i]] or Part[expr, i] gives the ith part of
  expr. expr[[-i]] counts from the end. expr[[0]]
  gives the head of expr. expr[[i, j, ... ]] or
  Part[expr, i, j, ... ] is equivalent to expr[[i]]
  [[j]] .... expr[[ {i1, i2, ... } ]] gives a
  list of the parts i1, i2, ... of expr.
```

For example, entering

```
Part[autodata, 2]
```

```
{1986, 182134, 17.2, 72429, 21.2}
```

or

```
autodata[[2]]
```

```
{1986, 182134, 17.2, 72429, 21.2}
```

returns the second row of `autodata`; entering either of the commands

```
Part[autodata, {2, 4, 5}]
```

```
{{1986, 182134, 17.2, 72429, 21.2},
 {1988, 320310, 30, 173147, 33.8},
 {1989, 254086, 25.1, 195999, 30.6}}
```

or

```
autodata[[{2, 4, 5}]]
```

```
{{1986, 182134, 17.2, 72429, 21.2},
 {1988, 320310, 30, 173147, 33.8},
 {1989, 254086, 25.1, 195999, 30.6}}
```

returns the second fourth and fifth rows of autodata. Several rows of an array can be extracted using Take.

For example, entering

```
Take[autodata, 3]
```

```
{{1985, 207640, 21.5, 58423, 12.7},
 {1986, 182134, 17.2, 72429, 21.2},
 {1987, 345555, 37.6, 163073, 41.3}}
```

returns the first three rows of autodata; entering

```
Take[autodata, -3]
```

```
{{1989, 254086, 25.1, 195999, 30.6},
 {1990, 187314, 20.5, 276859, 33.7},
 {1991, 193148, 20.1, 358666, 36.3}}
```

returns the last three rows; and entering

```
Take[autodata, {3, 6}]
```

```
{{1987, 345555, 37.6, 163073, 41.3},
 {1988, 320310, 30, 173147, 33.8},
 {1989, 254086, 25.1, 195999, 30.6},
 {1990, 187314, 20.5, 276859, 33.7}}
```

returns the third through sixth rows.

Selecting Columns

To extract columns of an array, we use the commands Column and ColumnTake, which are both contained in the **DataManipulation** package that is located in the **Statistics** folder (or directory). We use **Mathematica Help** to obtain information about the commands contained in the **DataManipulation** package.

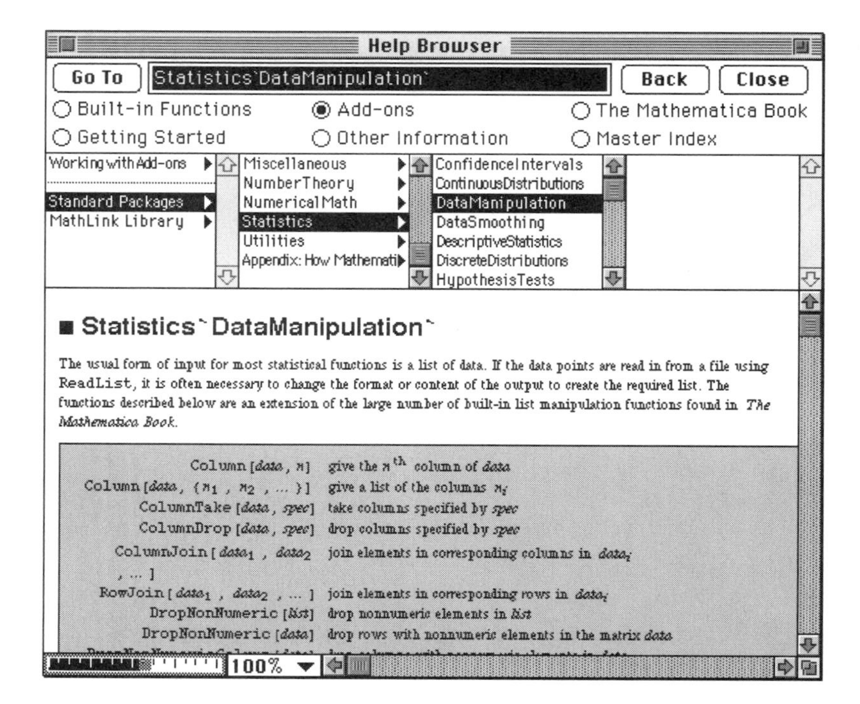

Next, we load the **DataManipulation** package.

```
<< Statistics'DataManipulation'
```

`Column` and `ColumnTake` extract columns from an array in the same way that `Part` and `Take` extract rows of an array. Thus, entering

```
Column[autodata, 3]
```

```
{21.5, 17.2, 37.6, 30, 25.1, 20.5, 20.1}
```

returns the third column of `autodata`; entering

```
Column[autodata, {3, 5}]
```

```
{{21.5, 12.7}, {17.2, 21.2}, {37.6, 41.3}, {30, 33.8},
 {25.1, 30.6}, {20.5, 33.7}, {20.1, 36.3}}
```

returns the third and fifth columns; and entering

```
ColumnTake[autodata, {2, 5}]
```

```
{{207640, 21.5, 58423, 12.7},
 {182134, 17.2, 72429, 21.2},
 {345555, 37.6, 163073, 41.3},
 {320310, 30, 173147, 33.8},
 {254086, 25.1, 195999, 30.6},
 {187314, 20.5, 276859, 33.7},
 {193148, 20.1, 358666, 36.3}}
```

returns the second through fifth columns.

■

Often, as with large data sets or data sets that have been manipulated by other programs, you may wish to import your data into *Mathematica* without manually reentering it. In some cases, you may find it easiest to manipulate your data with an editor, such as a text editor, to arrange it in a form usable by *Mathematica*.

Sample Data Set: The Hominoid Molar Data Set

The Hominoid Molar data set contains various types of data relating to the molars of humans, gorillas, and orangutans. The 67 rows in the Hominoid Molar data set each contain information about a distinct molar. The molars are the experimental units. Each cell of the Hominoid Molar data set contains a value that is (typically) a number or character string. The following table identifies each of the columns (variables) and the type of data that it contains. Note how **codes** are used to save space. For example, Hs is used for *Homo sapiens, Gg for Gorilla gorilla,* and Pp for *Pongo pygmaeus*. Often, numbers, such as 0 representing male and 1 representing female, are used as codes. When this is done, care should be taken to avoid doing arithmetic with codes. In addition, note that several entries are incomplete. Leaving a cell blank is sometimes appropriate, but not with *Mathematica*. We adopt the convention to use the symbol `null` to represent missing values.

Column Name	Description	Data Type
ID	Hs=Human, Gg=Gorilla, Pp=Orangutan	Categorical
Sex	0=male, 1=female	Categorical
Tooth Number (TN)	1=upper right, 2=upper left, 3=lower left, 4=lower right, 6=first molar, 7=second molar, 8=third molar	Categorical (ordinal)
Length (L)	Crown length	Quantitative

Column Name	Description	Data Type
Width (W)	Crown width	Quantitative
Thickness (T)	Average enamel thickness	Quantitative

ID	Sex	TN	L	W	T
Gg1	0	26	14.6	16.1	0.90
Gg1	0	27	16.2	17.3	0.93
Gg1	0	18	17.3	16.4	
Gg2	0	16	15.5	16.1	
Gg3	0	17	15.6	16.8	0.83
Gg4	0	46	16.2	13.6	0.89
Gg5	0	47	18.8	14.7	1.14
Gg3	0	38	18.1	14.8	1.05
Gg2	0	36	16.3	14.4	0.62
Gg3	0	47	18.2	15.5	1.01
Gg5	0	48	15.0	12.6	
Gg6	1	16	14.8	15.2	0.75
Gg6	1	27	14.1	15.1	
Gg7	1	28	15.1	14.9	0.83
Gg8	1	26	14.9	15.5	0.77
Gg9	1	17	14.9	15.8	0.81
Gg9	1	18	13.9	14.3	0.84
Gg6	1	36	16.4	13.5	
Gg6	1	47	16.3	14.6	0.96
Gg10	1	48	18.3	14.7	0.93
Gg9	1	36	15.3	13.3	
Gg9	1	37	16.5	14.6	0.73
Gg9	1	38	16.4	14.1	
Pp1	0	26	13.1	13.2	0.83
Pp1	0	27	13.2	14.4	
Pp1	0	28	12.4	13.1	
Pp2	0	27	12.2	13.9	1.13
Pp2	0	28	12.7	14.1	
Pp3	0	36	12.9	11.5	

ID	Sex	TN	L	W	T
Pp4	1	17	11.7	13.5	1.19
Pp5	1	16	11.5	12.9	1.02
Pp5	1	17	11.2	13.1	1.07
Pp6	1	28	11.7	13.8	1.19
Pp7	1	36	13.5	12.3	0.93
Pp5	1	38	13.1	11.1	1.17
Pp4	1	36	14.0	11.8	0.89
Pp4	1	47	14.2	13.2	
Pp4	1	48	14.8	12.9	1.36
Hs1	0	26	09.4	10.9	0.67
Hs1	0	27	08.7	10.7	
Hs2	0	28	08.9	11.7	
Hs2	0	26	09.9	11.5	
Hs3	0	27	08.8	10.9	
Hs3	0	28	08.9	11.1	1.01
Hs4	0	46	11.0	10.9	0.78
Hs4	0	47	11.7	10.1	1.47
Hs5	0	38	10.8	09.9	
Hs6	0	46	09.8	09.6	1.63
Hs6	0	47	10.4	09.6	
Hs7	0	48	10.8	09.7	1.21
Hs8	1	16	10.7	11.8	1.45
Hs8	1	17	10.2	12.5	1.80
Hs9	1	28	09.9	12.6	1.15
Hs10	1	26	10.9	0.98	
Hs10	1	27	08.6	10.9	0.82
Hs10	1	28	08.7	11.0	1.12
Hs11	1	46	10.5	09.9	
Hs12	1	37	10.3	09.7	

ID	Sex	TN	L	W	T
Pp2	0	37	14.8	12.7	1.08
Pp2	0	48	15.6	13.5	0.98
Pp1	0	46	13.0	12.3	0.94
Pp1	0	48	13.3	12.6	
Pp4	1	16	12.8	13.1	1.05

ID	Sex	TN	L	W	T
Hs12	1	48	09.5	09.0	
Hs13	1	46	10.3	09.6	0.92
Hs13	1	47	09.9	09.1	1.40
Hs13	1	48	11.4	09.7	1.30

EXAMPLE 2: IMPORTING DATA. Enter the Hominoid Molar data, which is contained in the text (ASCII) file HominoidData.txt, contained in the **StatsData** folder, into *Mathematica*.

SOLUTION: We show how (a portion of) the Hominoid Molar data set looks in a typical text editor. In this case, we are using Microsoft Word. Notice how the records occur on successive rows; the values in the records are separated by tab marks. Remember that with *Mathematica*, records are contained in braces (`{ . . . }`) and separated by commas (`,`); values within a record are separated by commas (`,`). Moreover, you must place a left brace (`{`) at the beginning of the array and a right brace (`}`) at the end of the array.

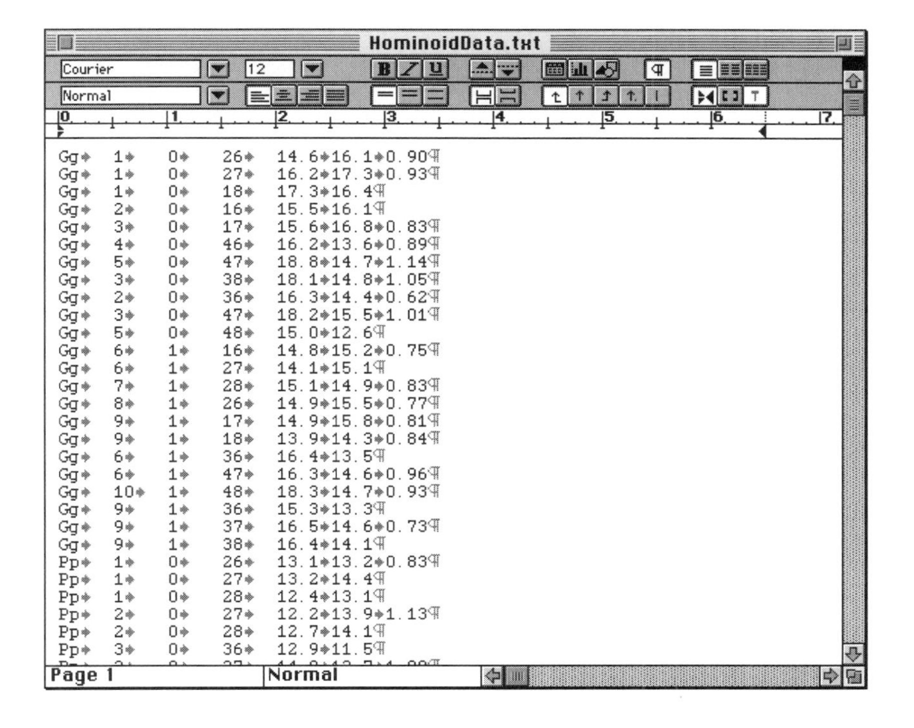

Because we have adopted the convention to use the symbol `null` to represent missing values, we first use Word's editing capabilities to replace all missing values with the symbol `null` (where the symbols for tab and end of paragraph are incerted using the button marked Special).

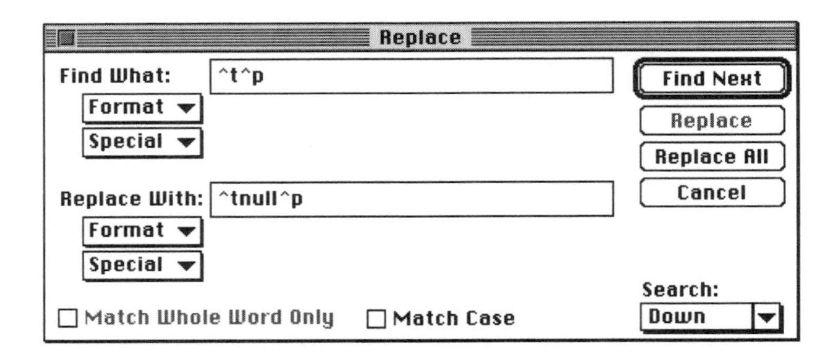

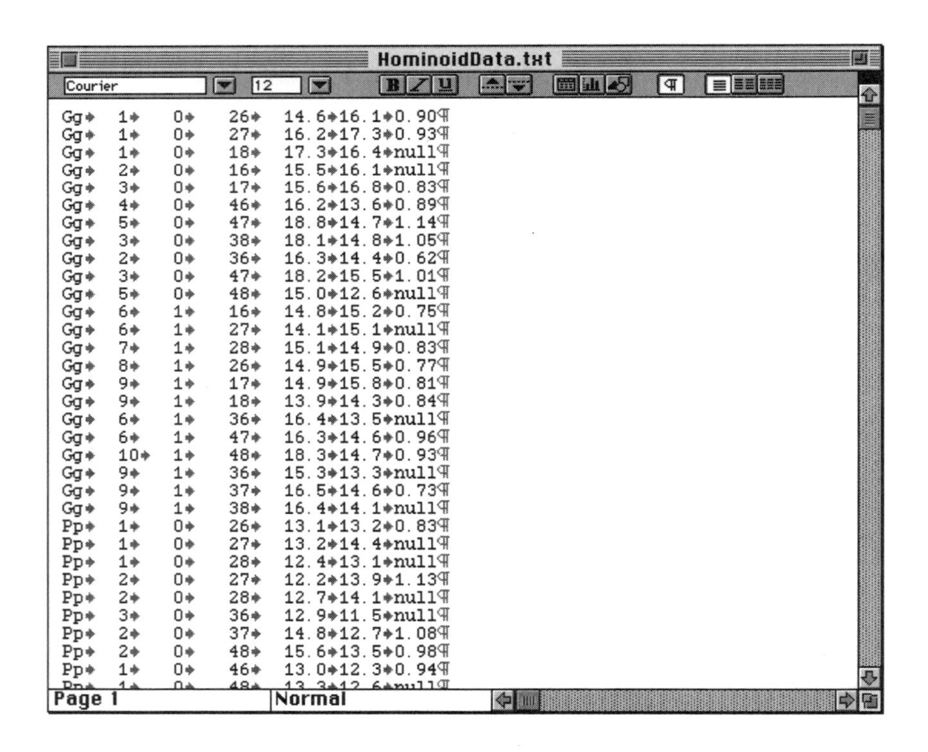


```
Gg	1	0	26	14.6	16.1	0.90
Gg	1	0	27	16.2	17.3	0.93
Gg	1	0	18	17.3	16.4	null
Gg	2	0	16	15.5	16.1	null
Gg	3	0	17	15.6	16.8	0.83
Gg	4	0	46	16.2	13.6	0.89
Gg	5	0	47	18.8	14.7	1.14
Gg	3	0	38	18.1	14.8	1.05
Gg	2	0	36	16.3	14.4	0.62
Gg	3	0	47	18.2	15.5	1.01
Gg	5	0	48	15.0	12.6	null
Gg	6	1	16	14.8	15.2	0.75
Gg	6	1	27	14.1	15.1	null
Gg	7	1	28	15.1	14.9	0.83
Gg	8	1	26	14.9	15.5	0.77
Gg	9	1	17	14.9	15.8	0.81
Gg	9	1	18	13.9	14.3	0.84
Gg	6	1	36	16.4	13.5	null
Gg	6	1	47	16.3	14.6	0.96
Gg	10	1	48	18.3	14.7	0.93
Gg	9	1	36	15.3	13.3	null
Gg	9	1	37	16.5	14.6	0.73
Gg	9	1	38	16.4	14.1	null
Pp	1	0	26	13.1	13.2	0.83
Pp	1	0	27	13.2	14.4	null
Pp	1	0	28	12.4	13.1	null
Pp	2	0	27	12.2	13.9	1.13
Pp	2	0	28	12.7	14.1	null
Pp	3	0	36	12.9	11.5	null
Pp	2	0	37	14.8	12.7	1.08
Pp	2	0	48	15.6	13.5	0.98
Pp	1	0	46	13.0	12.3	0.94
```

Because values are separated by commas, we next replace all tab marks with commas.

Replace
Find What: ^t
Format
Special
Replace With: ,
Format
Special
Find Next
Replace
Replace All
Cancel
Match Whole Word Only
Match Case
Search: Down

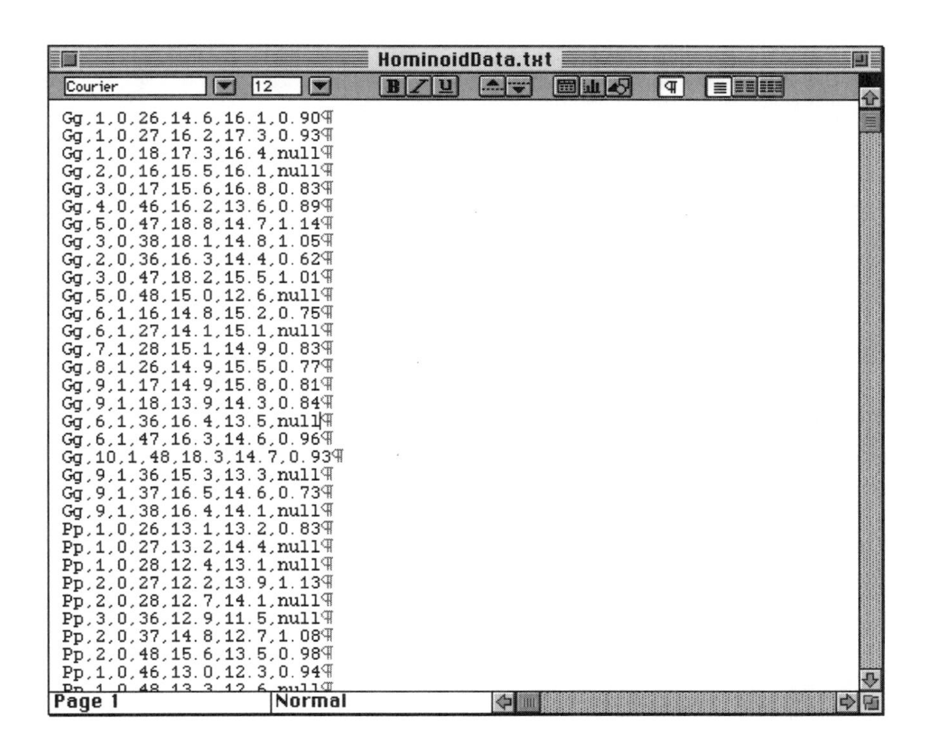

```
Gg,1,0,26,14.6,16.1,0.90¶
Gg,1,0,27,16.2,17.3,0.93¶
Gg,1,0,18,17.3,16.4,null¶
Gg,2,0,16,15.5,16.1,null¶
Gg,3,0,17,15.6,16.8,0.83¶
Gg,4,0,46,16.2,13.6,0.89¶
Gg,5,0,47,18.8,14.7,1.14¶
Gg,3,0,38,18.1,14.8,1.05¶
Gg,2,0,36,16.3,14.4,0.62¶
Gg,3,0,47,18.2,15.5,1.01¶
Gg,5,0,48,15.0,12.6,null¶
Gg,6,1,16,14.8,15.2,0.75¶
Gg,6,1,27,14.1,15.1,null¶
Gg,7,1,28,15.1,14.9,0.83¶
Gg,8,1,26,14.9,15.5,0.77¶
Gg,9,1,17,14.9,15.8,0.81¶
Gg,9,1,18,13.9,14.3,0.84¶
Gg,6,1,36,16.4,13.5,null¶
Gg,6,1,47,16.3,14.6,0.96¶
Gg,10,1,48,18.3,14.7,0.93¶
Gg,9,1,36,15.3,13.3,null¶
Gg,9,1,37,16.5,14.6,0.73¶
Gg,9,1,38,16.4,14.1,null¶
Pp,1,0,26,13.1,13.2,0.83¶
Pp,1,0,27,13.2,14.4,null¶
Pp,1,0,28,12.4,13.1,null¶
Pp,2,0,27,12.2,13.9,1.13¶
Pp,2,0,28,12.7,14.1,null¶
Pp,3,0,36,12.9,11.5,null¶
Pp,2,0,37,14.8,12.7,1.08¶
Pp,2,0,48,15.6,13.5,0.98¶
Pp,1,0,46,13.0,12.3,0.94¶
```

With *Mathematica*, records are contained in braces ({...}) and separated by commas, so we place a brace at the end and beginning of each record and a comma in between records.

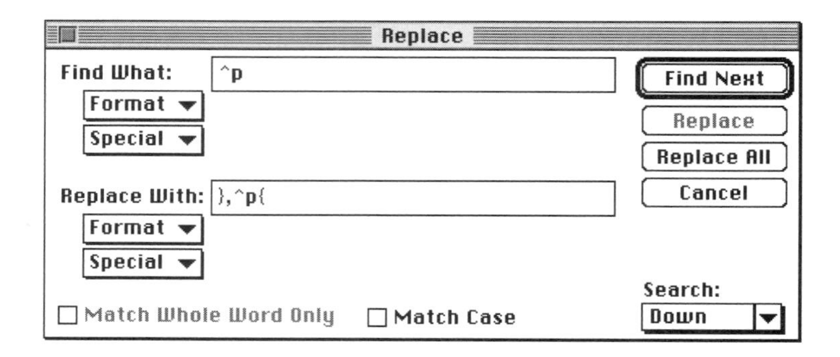

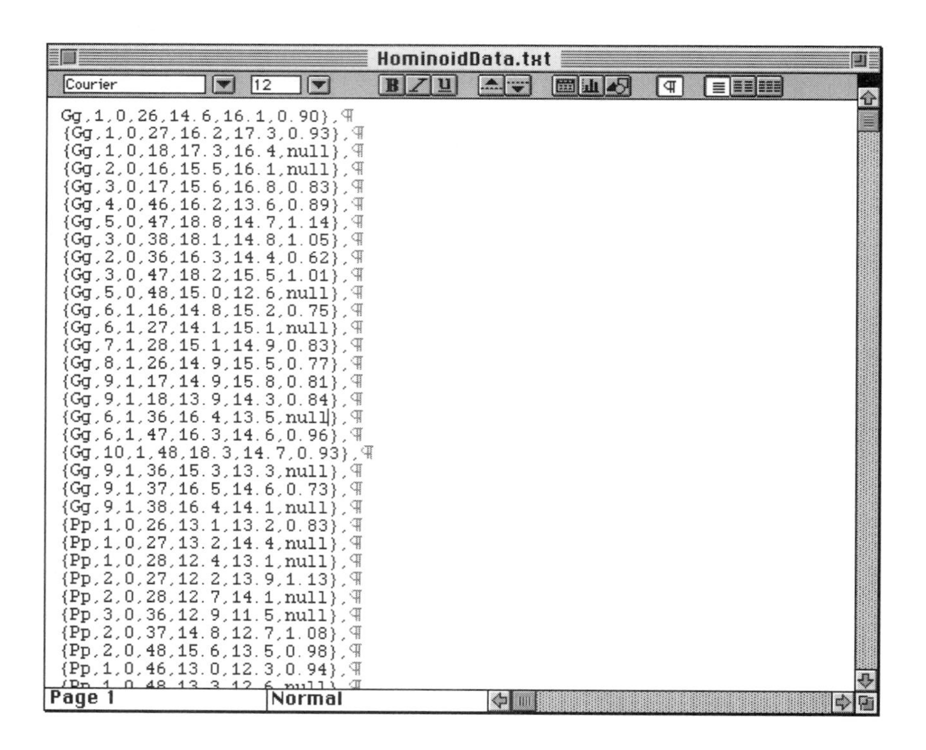

Last, we manually insert `hominoidmolar={{` at the beginning of the array

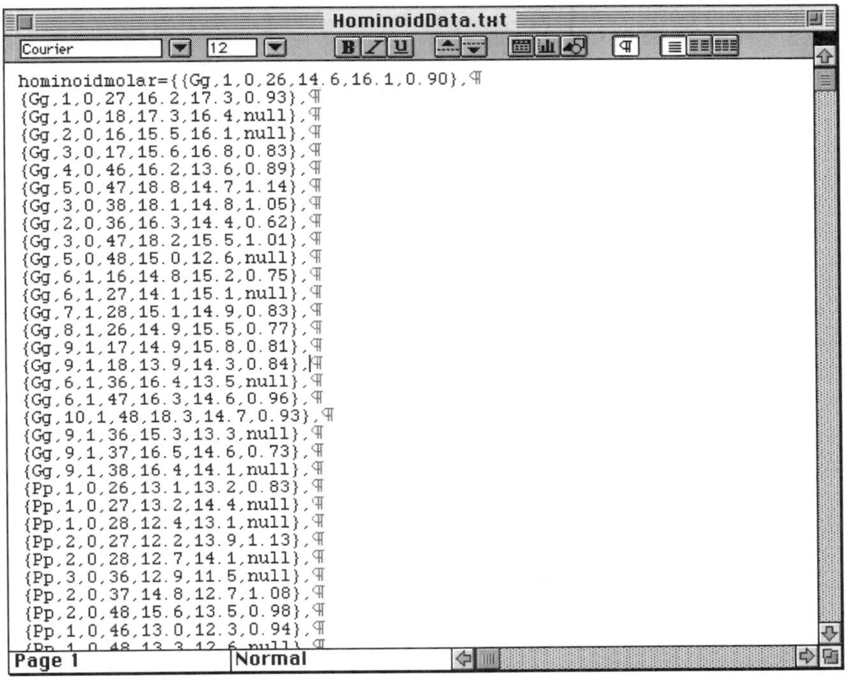

HominoidData.txt

```
hominoidmolar={{Gg,1,0,26,14.6,16.1,0.90},
{Gg,1,0,27,16.2,17.3,0.93},
{Gg,1,0,18,17.3,16.4,null},
{Gg,2,0,16,15.5,16.1,null},
{Gg,3,0,17,15.6,16.8,0.83},
{Gg,4,0,46,16.2,13.6,0.89},
{Gg,5,0,47,18.8,14.7,1.14},
{Gg,3,0,38,18.1,14.8,1.05},
{Gg,2,0,36,16.3,14.4,0.62},
{Gg,3,0,47,18.2,15.5,1.01},
{Gg,5,0,48,15.0,12.6,null},
{Gg,6,1,16,14.8,15.2,0.75},
{Gg,6,1,27,14.1,15.1,null},
{Gg,7,1,28,15.1,14.9,0.83},
{Gg,8,1,26,14.9,15.5,0.77},
{Gg,9,1,17,14.9,15.8,0.81},
{Gg,9,1,18,13.9,14.3,0.84},
{Gg,6,1,36,16.4,13.5,null},
{Gg,6,1,47,16.3,14.6,0.96},
{Gg,10,1,48,18.3,14.7,0.93},
{Gg,9,1,36,15.3,13.3,null},
{Gg,9,1,37,16.5,14.6,0.73},
{Gg,9,1,38,16.4,14.1,null},
{Pp,1,0,26,13.1,13.2,0.83},
{Pp,1,0,27,13.2,14.4,null},
{Pp,1,0,28,12.4,13.1,null},
{Pp,2,0,27,12.2,13.9,1.13},
{Pp,2,0,28,12.7,14.1,null},
{Pp,3,0,36,12.9,11.5,null},
{Pp,2,0,37,14.8,12.7,1.08},
{Pp,2,0,48,15.6,13.5,0.98},
{Pp,1,0,46,13.0,12.3,0.94},
```

Page 1 Normal

and `}};` at the end of the array.

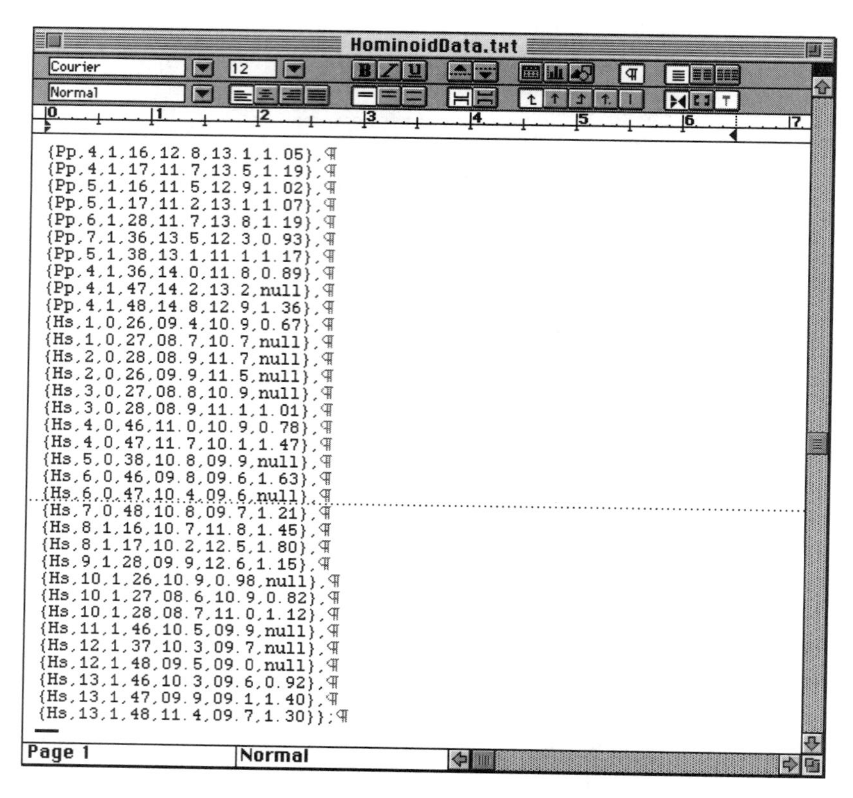


```
{Pp,4,1,16,12.8,13.1,1.05},
{Pp,4,1,17,11.7,13.5,1.19},
{Pp,5,1,16,11.5,12.9,1.02},
{Pp,5,1,17,11.2,13.1,1.07},
{Pp,6,1,28,11.7,13.8,1.19},
{Pp,7,1,36,13.5,12.3,0.93},
{Pp,5,1,38,13.1,11.1,1.17},
{Pp,4,1,36,14.0,11.8,0.89},
{Pp,4,1,47,14.2,13.2,null},
{Pp,4,1,48,14.8,12.9,1.36},
{Hs,1,0,26,09.4,10.9,0.67},
{Hs,1,0,27,08.7,10.7,null},
{Hs,2,0,28,08.9,11.7,null},
{Hs,2,0,26,09.9,11.5,null},
{Hs,3,0,27,08.8,10.9,null},
{Hs,3,0,28,08.9,11.1,1.01},
{Hs,4,0,46,11.0,10.9,0.78},
{Hs,4,0,47,11.7,10.1,1.47},
{Hs,5,0,38,10.8,09.9,null},
{Hs,6,0,46,09.8,09.6,1.63},
{Hs,6,0,47,10.4,09.6,null},
{Hs,7,0,48,10.8,09.7,1.21},
{Hs,8,1,16,10.7,11.8,1.45},
{Hs,8,1,17,10.2,12.5,1.80},
{Hs,9,1,28,09.9,12.6,1.15},
{Hs,10,1,26,10.9,0.98,null},
{Hs,10,1,27,08.6,10.9,0.82},
{Hs,10,1,28,08.7,11.0,1.12},
{Hs,11,1,46,10.5,09.9,null},
{Hs,12,1,37,10.3,09.7,null},
{Hs,12,1,48,09.5,09.0,null},
{Hs,13,1,46,10.3,09.6,0.92},
{Hs,13,1,47,09.9,09.1,1.40},
{Hs,13,1,48,11.4,09.7,1.30}};
```

After saving the result in a *text* (ASCII) format, we can import the file directly into *Mathematica* and manipulate the result in the same way that we manipulated the array in Example 1. To open **HominoidData.txt**, which is contained in the **StatsData** folder, in *Mathematica*, go to the *Mathematica* menu, select **File** and then **Open Special**.

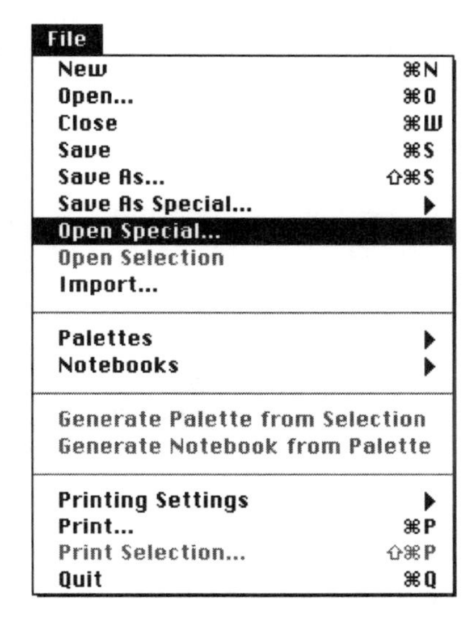

Click the **OK** button when the following pop-up window appears.

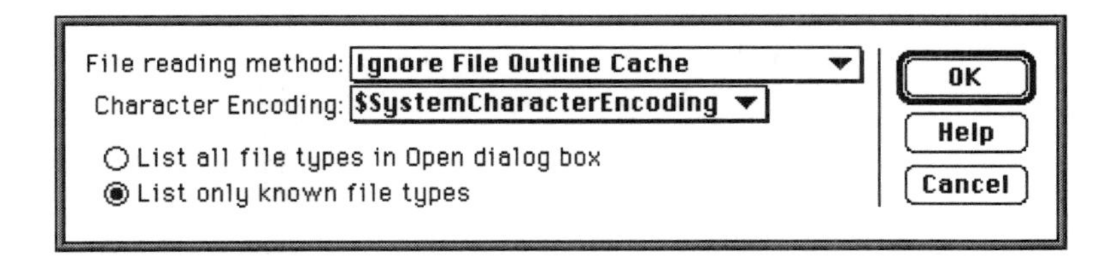

Then, locate the **StatsData** folder, select **HominoidData.txt**, and click the **OPEN** button. The **HominoidData.txt** file is opened as a (new) *Mathematica* notebook.

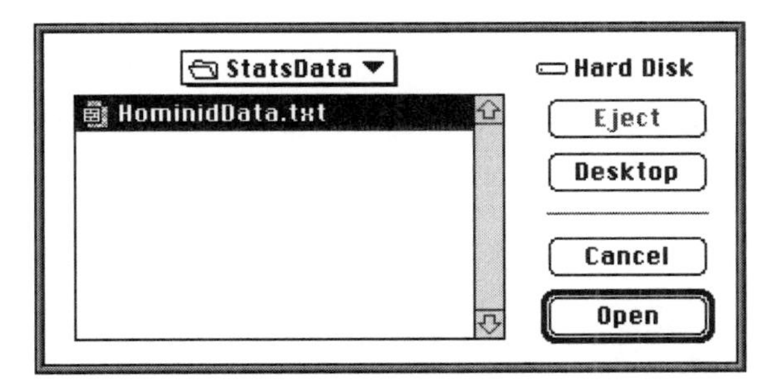

Note that if you had selected **Import...** instead of **Open Special...** at the beginning of this procedure, the data contained in the **HominoidData.txt** file would have been inserted at the position of the cursor in your active (as opposed to its own) *Mathematica* notebook.

Selecting Records

Once the **HominoidData.txt** file is open in *Mathematica*, we can use *Mathematica* to manipulate the data contained in the file. First, we enter the cell containing the Hominoid Molar data. Note that entering `hominoiddata` returns the entire Hominoid Molar data set. For length considerations, we view an abbreviated version with `Short`.

```
hominidmolar// Short
```

```
{{Gg, 1, 0, 26, <<20>>, 16.1, 0.9}, <<65>>, <<1>>}
```

We can extract records from the Hominoid Molar data set that meet particular criteria using the `Select` function.

```
?Select
```

```
Select[list, crit] picks out all elements e1 of list for
  which crit[e1] is True. Select[list, crit, n] picks
  out the first n elements for which crit[e1] is True.
```

For example, the first column of hominoidmolar identifies the species of the individual from which the molar was taken. Thus, entering

```
Select[hominidmolar, #[[1]] == Hs&]
```

```
{{Hs, 1, 0, 26, 9.4, 10.9, 0.67},
 {Hs, 1, 0, 27, 8.7, 10.7, null},
 {Hs, 2, 0, 28, 8.9, 11.7, null},
 {Hs, 2, 0, 26, 9.9, 11.5, null},
 {Hs, 3, 0, 27, 8.8, 10.9, null},
 {Hs, 3, 0, 28, 8.9, 11.1, 1.01},
 {Hs, 4, 0, 46, 11., 10.9, 0.78},
 {Hs, 4, 0, 47, 11.7, 10.1, 1.47},
 {Hs, 5, 0, 38, 10.8, 9.9, null},
 {Hs, 6, 0, 46, 9.8, 9.6, 1.63},
 {Hs, 6, 0, 47, 10.4, 9.6, null},
 {Hs, 7, 0, 48, 10.8, 9.7, 1.21},
 {Hs, 8, 1, 16, 10.7, 11.8, 1.45},
 {Hs, 8, 1, 17, 10.2, 12.5, 1.8},
 {Hs, 9, 1, 28, 9.9, 12.6, 1.15},
 {Hs, 10, 1, 26, Null, Null, 10.9, 0.98},
 {Hs, 10, 1, 27, 8.6, 10.9, 0.82},
 {Hs, 10, 1, 28, 8.7, 11., 1.12},
 {Hs, 11, 1, 46, 10.5, 9.9, null},
 {Hs, 12, 1, 37, 10.3, 9.7, null},
 {Hs, 12, 1, 48, 9.5, 9., null},
 {Hs, 13, 1, 46, 10.3, 9.6, 0.92},
 {Hs, 13, 1, 47, 9.9, 9.1, 1.4},
 {Hs, 13, 1, 48, 11.4, 9.7, 1.3}}
```

extracts the records from `hominoidmolar` for human subjects. Similarly, the seventh column of `hominoidmolar` indicates the average thickness of the molar. Entering

```
Select[hominidmolar, #[[7]] ≥ 1&]
```

or

```
Select[hominidmolar, #[[7]] >= 1&]
```

```
{{Gg, 5, 0, 47, 18.8, 14.7, 1.14},
 {Gg, 3, 0, 38, 18.1, 14.8, 1.05},
 {Gg, 3, 0, 47, 18.2, 15.5, 1.01},
 {Pp, 2, 0, 27, 12.2, 13.9, 1.13},
 {Pp, 2, 0, 37, 14.8, 12.7, 1.08},
 {Pp, 4, 1, 16, 12.8, 13.1, 1.05},
 {Pp, 4, 1, 17, 11.7, 13.5, 1.19},
 {Pp, 5, 1, 16, 11.5, 12.9, 1.02},
 {Pp, 5, 1, 17, 11.2, 13.1, 1.07},
 {Pp, 6, 1, 28, 11.7, 13.8, 1.19},
 {Pp, 5, 1, 38, 13.1, 11.1, 1.17},
 {Pp, 4, 1, 48, 14.8, 12.9, 1.36},
 {Hs, 3, 0, 28, 8.9, 11.1, 1.01},
 {Hs, 4, 0, 47, 11.7, 10.1, 1.47},
 {Hs, 6, 0, 46, 9.8, 9.6, 1.63},
 {Hs, 7, 0, 48, 10.8, 9.7, 1.21},
 {Hs, 8, 1, 16, 10.7, 11.8, 1.45},
 {Hs, 8, 1, 17, 10.2, 12.5, 1.8},
 {Hs, 9, 1, 28, 9.9, 12.6, 1.15},
 {Hs, 10, 1, 26, Null, Null, 10.9, 0.98},
 {Hs, 10, 1, 28, 8.7, 11., 1.12},
 {Hs, 13, 1, 47, 9.9, 9.1, 1.4},
 {Hs, 13, 1, 48, 11.4, 9.7, 1.3}}
```

extracts those records for which the average thickness is greater than or equal to 1. To extract records that meet multiple criteria, we use the `Or` and `And` functions.

```
?Or
```

```
e1 || e2 || ... is the logical OR function. It
  evaluates its arguments in order, giving True
  immediately if any of them are True, and False
  if they are all False.
```

```
?And
```

```
e1 && e2 && ... is the logical AND function. It
  evaluates its arguments in order, giving False
  immediately if any of them are False, and True
  if they are all True.
```

For example, entering

```
Select[hominidmolar, Or[#[[5]] >= 16, #[[6]] >= 16]&]

{{Gg, 1, 0, 26, 14.6, 16.1, 0.9},
 {Gg, 1, 0, 27, 16.2, 17.3, 0.93},
 {Gg, 1, 0, 18, 17.3, 16.4, null},
 {Gg, 2, 0, 16, 15.5, 16.1, null},
 {Gg, 3, 0, 17, 15.6, 16.8, 0.83},
 {Gg, 4, 0, 46, 16.2, 13.6, 0.89},
 {Gg, 5, 0, 47, 18.8, 14.7, 1.14},
 {Gg, 3, 0, 38, 18.1, 14.8, 1.05},
 {Gg, 2, 0, 36, 16.3, 14.4, 0.62},
 {Gg, 3, 0, 47, 18.2, 15.5, 1.01},
 {Gg, 6, 1, 36, 16.4, 13.5, null},
 {Gg, 6, 1, 47, 16.3, 14.6, 0.96},
 {Gg, 10, 1, 48, 18.3, 14.7, 0.93},
 {Gg, 9, 1, 37, 16.5, 14.6, 0.73},
 {Gg, 9, 1, 38, 16.4, 14.1, null}}
```

returns those records for which the crown length or the crown width of the molar is greater than or equal to 16. On the other hand, entering

```
Select[hominidmolar, And[#[[1]] == Pp, #[[3]] == 1]&]

{{Pp, 4, 1, 16, 12.8, 13.1, 1.05},
 {Pp, 4, 1, 17, 11.7, 13.5, 1.19},
 {Pp, 5, 1, 16, 11.5, 12.9, 1.02},
 {Pp, 5, 1, 17, 11.2, 13.1, 1.07},
 {Pp, 6, 1, 28, 11.7, 13.8, 1.19},
 {Pp, 7, 1, 36, 13.5, 12.3, 0.93},
 {Pp, 5, 1, 38, 13.1, 11.1, 1.17},
 {Pp, 4, 1, 36, 14., 11.8, 0.89},
 {Pp, 4, 1, 47, 14.2, 13.2, null},
 {Pp, 4, 1, 48, 14.8, 12.9, 1.36}}
```

returns the records of female orangutans. Note that we can use the `Or` and `And` functions together as well. For example, entering

```
Select[hominidmolar, And[#[[3]] == 0, Or[#[[5]] < 10, #[[6]] < 10]]&]
```

```
{{Hs, 1, 0, 26, 9.4, 10.9, 0.67},
 {Hs, 1, 0, 27, 8.7, 10.7, null},
 {Hs, 2, 0, 28, 8.9, 11.7, null},
 {Hs, 2, 0, 26, 9.9, 11.5, null},
 {Hs, 3, 0, 27, 8.8, 10.9, null},
 {Hs, 3, 0, 28, 8.9, 11.1, 1.01},
 {Hs, 5, 0, 38, 10.8, 9.9, null},
 {Hs, 6, 0, 46, 9.8, 9.6, 1.63},
 {Hs, 6, 0, 47, 10.4, 9.6, null},
 {Hs, 7, 0, 48, 10.8, 9.7, 1.21}}
```

returns the records of the male subjects for which either the crown length or the crown width is less than 10.

In Example 1, we saw that we can use the `Column` command, which is contained in the **DataManipulation** package, to extract columns of an array. First, we load the **DataManipulation** package and then use `Column` to extract the seventh column of `hominidmolar`, corresponding to the average enamel thickness of each molar collected. We name the resulting list `thicknesses`.

```
thicknesses = Column[hominidmolar, 7]
```

```
{0.9, 0.93, null, null, 0.83, 0.89, 1.14, 1.05, 0.62,
 1.01, null, 0.75, null, 0.83, 0.77, 0.81, 0.84,
 null, 0.96, 0.93, null, 0.73, null, 0.83, null,
 null, 1.13, null, null, 1.08, 0.98, 0.94, null,
 1.05, 1.19, 1.02, 1.07, 1.19, 0.93, 1.17, 0.89,
 null, 1.36, 0.67, null, null, null, null, 1.01,
 0.78, 1.47, null, 1.63, null, 1.21, 1.45, 1.8, 1.15,
 10.9, 0.82, 1.12, null, null, null, 0.92, 1.4, 1.3}
```

We will see that, in general, *Mathematica* can only perform arithmetic on numeric elements of an array or list. Note that thicknesses contains several nonnumeric entries, the symbol `null`, representing missing values. To remove nonnumeric elements of an array or list, we use `DropNonNumeric`, which is also contained in the **DataManipulation** package.

```
?DropNonNumeric
```

```
DropNonNumeric[data] drops elements or rows that
 contain non-numeric elements in data.
```

Thus, entering

```
thicknesses = DropNonNumeric[thicknesses]

{0.9, 0.93, 0.83, 0.89, 1.14, 1.05, 0.62, 1.01, 0.75, 0.83,
 0.77, 0.81, 0.84, 0.96, 0.93, 0.73, 0.83, 1.13, 1.08, 0.98,
 0.94, 1.05, 1.19, 1.02, 1.07, 1.19, 0.93, 1.17, 0.89,
 1.36, 0.67, 1.01, 0.78, 1.47, 1.63, 1.21, 1.45, 1.8,
 1.15, 10.9, 0.82, 1.12, 0.92, 1.4, 1.3}
```

removes the nonnumeric entries from `thicknesses` and names the resulting list `thicknesses`. We now sort the list `thicknesses` in ascending order with `Sort` and name the result `lowtohigh`.

```
?Sort

Sort[list] sorts the elements of list into canonical
  order. Sort[list, p] sorts using the ordering function p.

lowtohigh = Sort[thicknesses]

{0.62, 0.67, 0.73, 0.75, 0.77, 0.78, 0.81, 0.82, 0.83,
 0.83, 0.83, 0.84, 0.89, 0.89, 0.9, 0.92, 0.93, 0.93,
 0.93, 0.94, 0.96, 0.98, 1.01, 1.01, 1.02, 1.05, 1.05,
 1.07, 1.08, 1.12, 1.13, 1.14, 1.15, 1.17, 1.19, 1.19,
 1.21, 1.3, 1.36, 1.4, 1.45, 1.47, 1.63, 1.8, 10.9}
```

To arrange the list in descending order, we use `Reverse`, naming the result `hightolow`.

```
?Reverse

Reverse[expr] reverses the order of the elements in expr.

hightolow = Reverse[lowtohigh]

{10.9, 1.8, 1.63, 1.47, 1.45, 1.4, 1.36, 1.3, 1.21, 1.19,
 1.19, 1.17, 1.15, 1.14, 1.13, 1.12, 1.08, 1.07, 1.05,
 1.05, 1.02, 1.01, 1.01, 0.98, 0.96, 0.94, 0.93, 0.93,
 0.93, 0.92, 0.9, 0.89, 0.89, 0.84, 0.83, 0.83, 0.83,
 0.82, 0.81, 0.78, 0.77, 0.75, 0.73, 0.67, 0.62}
```

■

If a data set in a text file consists *entirely* of numerical entries and the records occur on successive lines, we can import it directly into *Mathematica* using the command `ReadList["file name",Number,RecordLists->True]`.

EXAMPLE 3: IMPORTING DATA. The following table shows the number of home runs hit by Babe Ruth in each of 21 seasons. The data is contained in the text file **BabeRuth.txt**. Import the data into *Mathematica*.

Season #	Home Runs
1	4
2	3
3	2
4	11
5	29
6	54
7	59
8	35
9	41
10	46
11	25

Season #	Home Runs
12	47
13	60
14	54
15	46
16	49
17	46
18	41
19	34
20	22
21	6

SOLUTION: The following screen shot shows how the data looks in a typical text editor. Notice that records occur on consecutive lines; values are separated by tab marks.

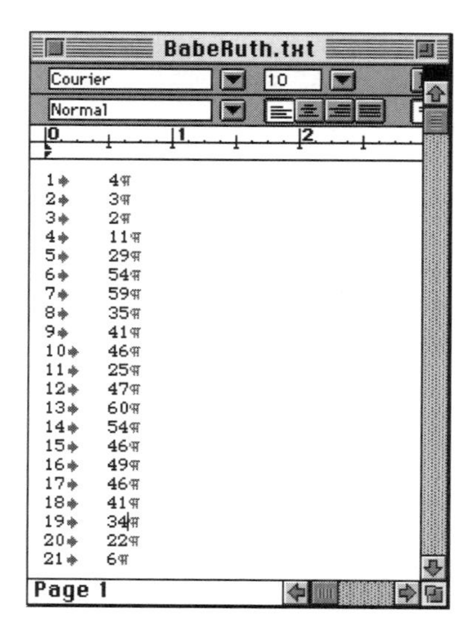

We use !! to display this text file in *Mathematica*,

```
!! "BabeRuth.txt"

1   4
2   3
3   2
4   11
5   29
6   54
7   59
8   35
9   41
10  46
11  25
12  47
13  60
14  54
15  46
16  49
17  46
18  41
19  34
20  22
21  6
```

and then use ReadList to load the data contained in the **BabeRuth.txt** text file and convert it to an array, which we name ruthruns.

```
ruthruns =
 ReadList["BabeRuth.txt", Number, RecordLists -> True]

{{1, 4}, {2, 3}, {3, 2}, {4, 11}, {5, 29}, {6, 54},
 {7, 59}, {8, 35}, {9, 41}, {10, 46}, {11, 25},
 {12, 47}, {13, 60}, {14, 54}, {15, 46}, {16, 49},
 {17, 46}, {18, 41}, {19, 34}, {20, 22}, {21, 6}}
```

We can now use *Mathematica* to manipulate the data. For example, here we use ListPlot, which is discussed in detail in Chapter 4, to graph the ordered pairs in ruthruns in two different ways: in lp1, the points appear slightly larger than normal because the option PlotStyle->PointSize[0.02] is included in the ListPlot command; in lp2, consecutive points are connected with line segments because the option PlotJoined->True is included in the ListPlot command. Neither graphics object is displayed, because we include the option DisplayFunction->Identity in each command; the graphs are displayed side-by-side using Show together with GraphicsArray.

```
lp1 = ListPlot [ruthruns , PlotStyle -> PointSize [0.02 ],
  DisplayFunction -> Identity ];
lp2 = ListPlot [ruthruns ,
  PlotJoined -> True , DisplayFunction -> Identity ];
Show [GraphicsArray [{lp1 , lp2 }]]
```

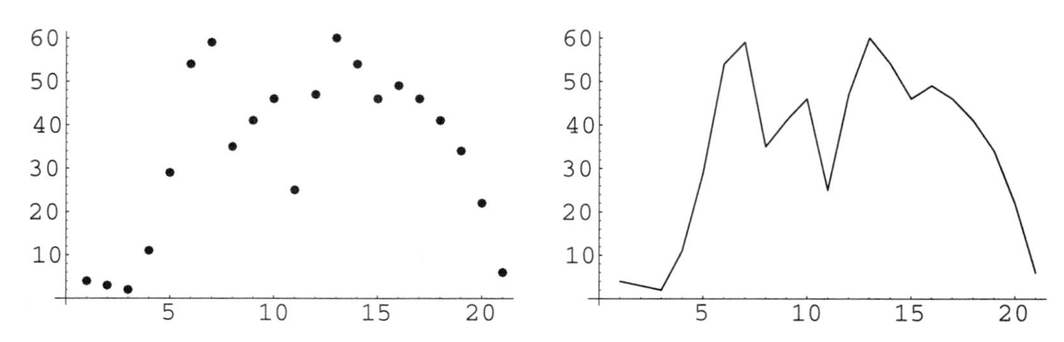

We can use statistical functions contained in the **Statistics** packages, such as **DataManipulation** and **DescriptiveStatistics**, to manipulate this data as well. First, we load these packages

```
<< Statistics'DataManipulation'
<< Statistics'DescriptiveStatistic
```

and then use `Column`, which is contained in the **DataManipulation** package, to extract the second column of `ruthruns`, naming the resulting list `runs`.

```
runs = Column [ruthruns , 2]

{4, 3, 2, 11, 29, 54, 59, 35, 41, 46, 25, 47, 60, 54,
 46, 49, 46, 41, 34, 22, 6}
```

We can now use a variety of *Mathematica* functions to obtain statistical information about the list of numbers, `runs`. First, we use `LocationReport`, which is contained in the **DescriptiveStatistics** package,

```
? LocationReport

LocationReport[list] gives the Mean, HarmonicMean,
  and Median location statistics for list.
```

to obtain the mean, harmonic mean, and median of the list. (These topics and the commands are discussed in detail in Chapter 2.)

```
LocationReport[runs] // N

{Mean→ 34., HarmonicMean→ 12.0332, Median→ 41.}
```

Similarly, we use `DispersionReport`, which is contained in the **Descriptive-Statistics** package as well,

```
? DispersionReport

DispersionReport[list] gives the Variance,
  StandardDeviation, SampleRange, MeanDeviation,
  MedianDeviation, and QuartileDeviation dispersion
  statistics for list.
```

to compute the variance, standard deviation, range, mean deviation, median deviation, and quartile deviation of `runs`.

```
DispersionReport  [runs ] // N

{Variance→ 373.7, StandardDeviation→ 19.3313,
 SampleRange→ 58., MeanDeviation→ 16.1905,
 MedianDeviation→ 13., QuartileDeviation→ 14.125}
```

Last, we use `ShapeReport`, which is also contained in the **DescriptiveStatistics** package,

```
? ShapeReport

ShapeReport[list] gives the Skewness, QuartileSkewness,
  and KurtosisExcess shape statistics for list.
```

to compute the skewness, quartile skewness, and kurtosis of `runs`.

```
ShapeReport  [runs ] // N

{Skewness→ -0.465361,
 QuartileSkewness→ -0.539823,
 KurtosisExcess→ -1.11344}
```

■

If the data in a data set contained in a text (ASCII) file consists of nonnumerical entries as well as numerical entries, we can still use `ReadList` to import the data into *Mathematica*, although the situation is slightly more difficult.

EXAMPLE 4: IMPORTING DATA. The following table shows the per capita consumption (in gallons) of selected beverages for several years from 1970 to 1993. The data is contained in the text (ASCII) file **Beverages.txt**. Import the data into *Mathematica*.

Commodity	**1970**	**1975**	**1980**	**1985**	**1989**	**1990**	**1991**	**1992**	**1993**
Milk (plain and flavored)	31.3	29.5	27.6	26.7	26.0	25.7	25.7	25.4	24.8
Tea	6.8	7.5	7.3	7.1	6.9	6.8	6.9	7.1	7.1
Coffee	33.4	31.4	26.7	27.4	26.3	27.0	27.1	26.9	26.0
Bottled water	(NA)	(NA)	2.4	4.5	7.4	8.0	8.0	8.2	9.2
Soft drinks	24.3	28.2	35.1	35.7	42.2	43.7	44.9	45.4	46.6
Selected juices (excludes vegetable juices)	(NA)	6.6	7.2	7.7	7.7	6.9	7.9	7.3	8.4
Beer (adult population)	30.6	33.9	36.6	34.6	33.9	34.9	33.2	32.6	32.4
Wine (adult population)	2.2	2.7	3.2	3.5	3.1	2.9	2.7	2.7	2.5
Distilled spirits (adult population)	3.0	3.1	3.0	2.6	2.2	2.2	2.0	2.0	1.9

Source: U.S. Dept. of Agriculture, *Economic Research Service, Food Consumption, Prices, and Expenditures*, annual; and unpublished data.

SOLUTION: The following screen shot shows how **Beverages.txt** looks in a typical text editor. Notice that records occur on successive lines and that entries are separated by tab marks.

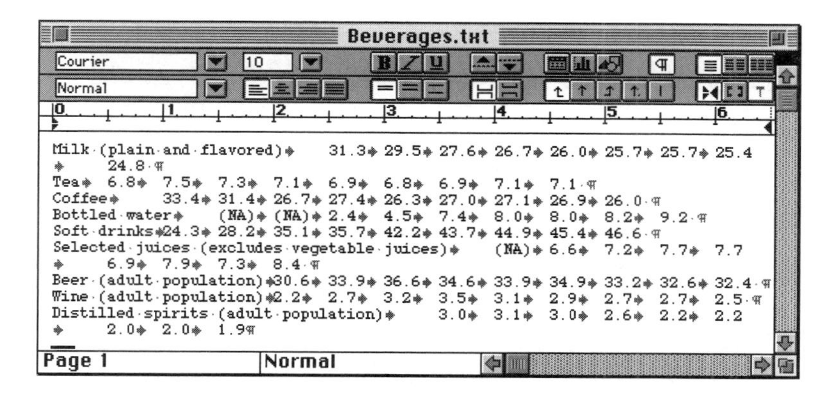

We view the file in *Mathematica* with `!!`.

```
!!"Beverages .txt"
```

```
Milk (plain and flavored)  31.3    29.5    27.6    26.7
    26.0    25.7    25.7    25.4    24.8
Tea 6.8 7.5 7.3 7.1 6.9 6.8 6.9 7.1 7.1
Coffee  33.4    31.4    26.7    27.4    26.3    27.0
27.1    26.9    26.0
Bottled water   (NA)    (NA)    2.4 4.5 7.4 8.0 8.0 8.2
9.2
Soft drinks 24.3    28.2    35.1    35.7    42.2    43.7
    44.9    45.4    46.6
Selected juices (excludes vegetable juices)(NA)    6.6
7.2 7.7 7.7 6.9 7.9 7.3 8.4
Beer (adult population) 30.6    33.9    36.6    34.6
33.9    34.9    33.2    32.6    32.4
Wine (adult population) 2.2 2.7 3.2 3.5 3.1 2.9 2.7 2.7
2.5
Distilled spirits (adult population)   3.0 3.1 3.0 2.6
2.2 2.2 2.0 2.0 1.9
```

As in Example 3, we will use `ReadList` to import the data into *Mathematica*.

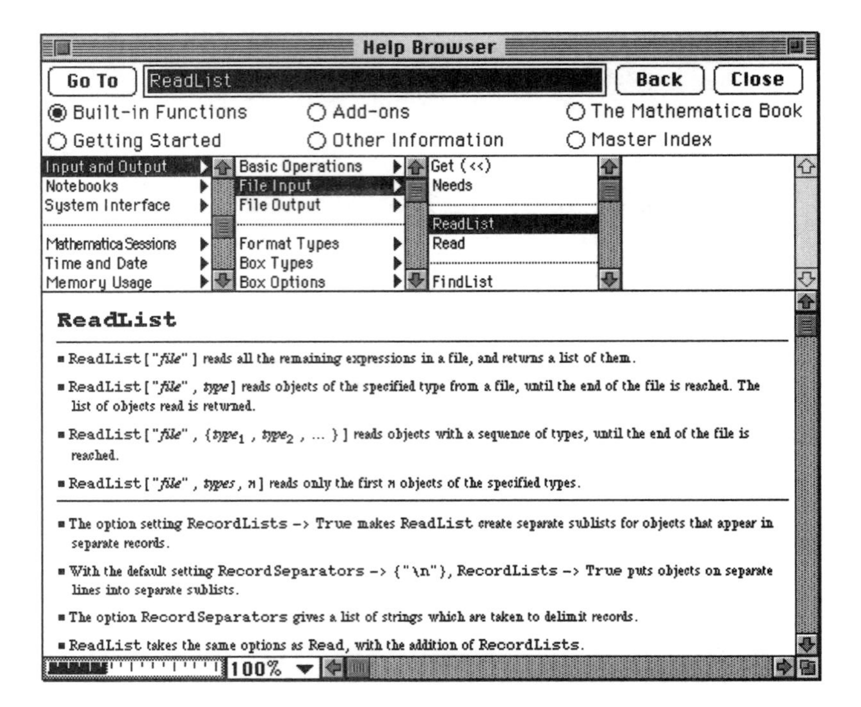

Because **Beverages.txt** contains both numerical and non-numerical data, we include `Word` in the `ReadList` command that follows. Also, because the entries are separated by tab marks, *not by spaces*, between characters, we include the option `WordSeparators->{\t"}`.

```
bevs = ReadList ["Beverages .txt ", Word ,
  WordSeparators  -> {"\t"}, RecordLists  -> True ]
```

```
{{Milk (plain and flavored), 31.3, 29.5, 27.6, 26.7,
  26.0, 25.7, 25.7, 25.4, 24.8 }, {Tea, 6.8, 7.5,
  7.3, 7.1, 6.9, 6.8, 6.9, 7.1, 7.1 }, {Coffee, 33.4,
  31.4, 26.7, 27.4, 26.3, 27.0, 27.1, 26.9, 26.0 },
 {Bottled water, (NA), (NA), 2.4, 4.5,
  7.4, 8.0, 8.0, 8.2, 9.2 }, {Soft drinks, 24.3,
  28.2, 35.1, 35.7, 42.2, 43.7, 44.9, 45.4, 46.6 },
 {Selected juices (excludes vegetable juices),
  (NA), 6.6, 7.2, 7.7, 7.7, 6.9, 7.9, 7.3, 8.4 },
 {Beer (adult population), 30.6, 33.9, 36.6, 34.6, 33.9,
  34.9, 33.2, 32.6, 32.4 }, {Wine (adult population),
  2.2, 2.7, 3.2, 3.5, 3.1, 2.9, 2.7, 2.7, 2.5 },
 {Distilled spirits (adult population), 3.0, 3.1,
  3.0, 2.6, 2.2, 2.2, 2.0, 2.0, 1.9}}
```

Although bevs is a valid *Mathematica* array, each entry is a string, including the numerical ones. Thus, we cannot perform arithmetic on the numerical entries at this point.

```
bevs = ReadList ["Beverages .txt ", Word ,
  WordSeparators  -> {"\t"}, RecordLists  -> True ]
```

```
{{"Milk (plain and flavored) ", "31.3", "29.5", "27.6",
  "26.7", "26.0", "25.7", "25.7", "25.4", "24.8 "},
 {"Tea", "6.8", "7.5", "7.3", "7.1", "6.9", "6.8",
  "6.9", "7.1", "7.1 "}, {"Coffee", "33.4", "31.4",
  "26.7", "27.4", "26.3", "27.0", "27.1", "26.9", "26.0 "},
 {"Bottled water", "(NA)", "(NA)",
  "2.4", "4.5", "7.4", "8.0", "8.0", "8.2", "9.2 "},
 {"Soft drinks", "24.3", "28.2", "35.1",
  "35.7", "42.2", "43.7", "44.9", "45.4", "46.6 "},
 {"Selected juices (excludes vegetable juices) ", "(NA)",
  "6.6", "7.2", "7.7", "7.7", "6.9", "7.9", "7.3", "8.4 "},
 {"Beer (adult population) ", "30.6", "33.9", "36.6",
  "34.6", "33.9", "34.9", "33.2", "32.6", "32.4 "},
 {"Wine (adult population) ", "2.2", "2.7",
  "3.2", "3.5", "3.1", "2.9", "2.7", "2.7", "2.5 "},
 {"Distilled spirits (adult population) ", "3.0",
  "3.1", "3.0", "2.6", "2.2", "2.2", "2.0", "2.0", "1.9"}}
```

To perform arithmetic on the numerical entries, we must convert them from strings to numbers. For example, to determine the mean amount of milk consumed for the years listed in the table, we first copy the data and then use ToExpression to convert the data from an array of strings to an array of numbers.

```
milk =
 ToExpression  [{"26.0 ", "25.7 ", "25.7 ", "25.4 ", "24.8  "}]
```

```
{26., 25.7, 25.7, 25.4, 24.8}
```

Then, we use Mean, which is contained in the **DescriptiveStatistics** package, to compute the mean of milk.

```
Mean[milk]
```

```
25.52
```

Similarly, if to compute the mean of each beverage consumed for the years 1989 to 1993, we first load the **DataManipulation** package and then use `ColumnTake` to extract the sixth through tenth columns of `bevs`, which corresponds to the data for these years.

```
consecyears = ColumnTake [bevs , {6, 10}]

{{26.0, 25.7, 25.7, 25.4, 24.8 },
 {6.9, 6.8, 6.9, 7.1, 7.1 },
 {26.3, 27.0, 27.1, 26.9, 26.0 },
 {7.4, 8.0, 8.0, 8.2, 9.2 },
 {42.2, 43.7, 44.9, 45.4, 46.6 },
 {7.7, 6.9, 7.9, 7.3, 8.4 },
 {33.9, 34.9, 33.2, 32.6, 32.4 },
 {3.1, 2.9, 2.7, 2.7, 2.5 },
 {2.2, 2.2, 2.0, 2.0, 1.9}}
```

We then use `ToExpression` to convert the data in `consecyears` from strings to numbers.

```
consecyears = ToExpression [consecyears ]

{{26., 25.7, 25.7, 25.4, 24.8},
 {6.9, 6.8, 6.9, 7.1, 7.1},
 {26.3, 27., 27.1, 26.9, 26.},
 {7.4, 8., 8., 8.2, 9.2},
 {42.2, 43.7, 44.9, 45.4, 46.6},
 {7.7, 6.9, 7.9, 7.3, 8.4},
 {33.9, 34.9, 33.2, 32.6, 32.4},
 {3.1, 2.9, 2.7, 2.7, 2.5},
 {2.2, 2.2, 2., 2., 1.9}}
```

Last, we use `Map` together with `Mean` to calculate the mean for each beverage.

```
Map[Mean, consecyears]

{25.52, 6.96, 26.66, 8.16, 44.56, 7.64, 33.4, 2.78, 2.06}
```

■

Output from Mathematica

Often you will wish to export results obtained with *Mathematica* to other programs. You can save *Mathematica* notebooks in a variety of formats by going to **File** under the *Mathematica* menu. In addition to selecting **Save** or **Save As**, you have the option of selecting **Save As Special...** and then selecting the format that describes how you wish to save your *Mathematica* notebook.

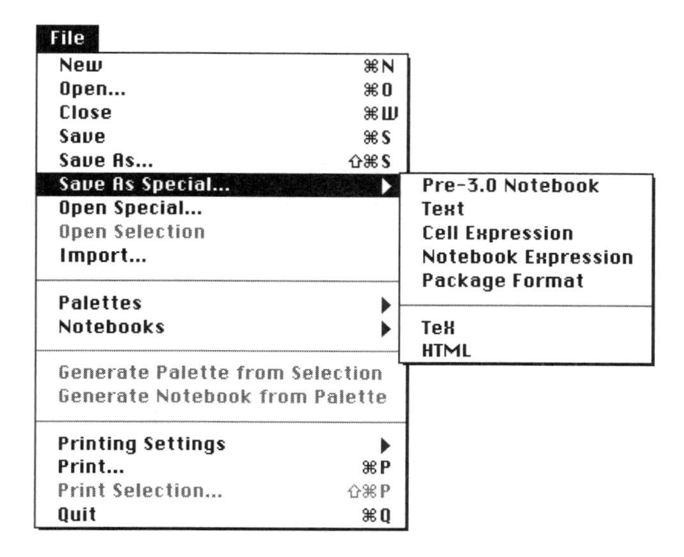

In general, from **File** individual and/or multiple cells contained within a *Mathematica* notebook, as well as the complete notebook, can be printed, saved, and/or copied to other programs in a variety of ways. It is also possible to output from *Mathamatica* by going to **Edit** under the *Mathematica* menu.

EXAMPLE 5: OUTPUTTING FROM EDIT. The following table shows the per capita income (in 1991 dollars), the unemployment rate (for September 1994), the poverty rate (for 1989), and the population (for 1990) for selected counties in Southeast Georgia. Construct a bar graph and copy it to a document on the desktop.

County	Per capita income	Unemployment	Poverty Rate	Population
Bryan	10826	4.8	17.2	15438
Bulloch	12767	4.0	28.3	43125
Candler	14214	6.2	33.1	7744
Chatham	18204	6.1	25.5	216935
Effingham	13199	4.4	16.3	25687

Continued

County	Per capita income	Unemployment	Poverty Rate	Population
Emanuel	12682	6.9	34.8	20546
Evans	15304	7.9	31.0	8724
Jenkins	11920	6.0	34.2	8247
Liberty	11470	9.5	23.0	52745
Long	8922	3.2	30.5	6202
Montgomery	13357	7.0	34.5	7163
Screven	12610	6.3	28.8	13842
Tattnall	14665	5.7	28.0	17722
Toombs	14963	6.6	32.3	24072
Treutlen	11684	8.1	34.8	5994
Wheeler	12144	8.7	38.8	4903

Source: Dan S. Rickman, "The Economic Impact of High School Noncompletion in Cedo Region 8," *Southern Economic Developer: Southeast Georgia Review* (Spring 1995), Volume 5, Number 2, pp. 3–13.

SOLUTION: First, we enter the data from the table into *Mathematica*. We name the resulting array `gadata`.

```
gadata = {{Bryan , 10826 , 4.8 , 17.2 , 15438 },
{Bulloch , 12767 , 4.0 , 28.3 , 43125 },
{Candler , 14214 , 6.2 , 33.1 , 7744 },
{Chatham , 18204 , 6.1 , 25.5 , 216935 },
{Effingham , 13199 , 4.4 , 16.3 , 25687 },
{Emanuel , 12682 , 6.9 , 34.8 , 20546 },
{Evans , 15304 , 7.9 , 31.0 , 8724 },
{Jenkins , 11920 , 6.0 , 34.2 , 8247 },
{Liberty , 11470 , 9.5 , 23.0 , 52745 },
{Long , 8922 , 3.2 , 30.5 , 6202 },
{Montgomery , 13357 , 7.0 , 34.5 , 7163 },
{Screven , 12610 , 6.3 , 28.8 , 13842 },
{Tattnall , 14665 , 5.7 , 28.0 , 17722 },
{Toombs , 14963 , 6.6 , 32.3 , 24072 },
{Treutlen , 11684 , 8.1 , 34.8 , 5994 },
{Wheeler , 12144 , 8.7 , 38.8 , 4903 }};
```

We view `gadata` in traditional row-and-column form with `TableForm`.

```
gadata // TableForm
```

Bryan	10826	4.8	17.2	15438
Bulloch	12767	4.	28.3	43125
Candler	14214	6.2	33.1	7744
Chatham	18204	6.1	25.5	216935
Effingham	13199	4.4	16.3	25687
Emanuel	12682	6.9	34.8	20546
Evans	15304	7.9	31.	8724
Jenkins	11920	6.	34.2	8247
Liberty	11470	9.5	23.	52745
Long	8922	3.2	30.5	6202
Montgomery	13357	7.	34.5	7163
Screven	12610	6.3	28.8	13842
Tattnall	14665	5.7	28.	17722
Toombs	14963	6.6	32.3	24072
Treutlen	11684	8.1	34.8	5994
Wheeler	12144	8.7	38.8	4903

To save this cell or a selection of cells in a file in a particular format or to copy a selected cell or two or more cells in a particular format, first go to **Edit** under the *Mathematica* menu and then select **Save Selection As...**, if you wish to save the selected cell(s) in a file in a particular format, or select **Copy As**, if you wish to copy the selected cell(s) using a particular format and then (immediately) paste the cell(s) into another application's file.

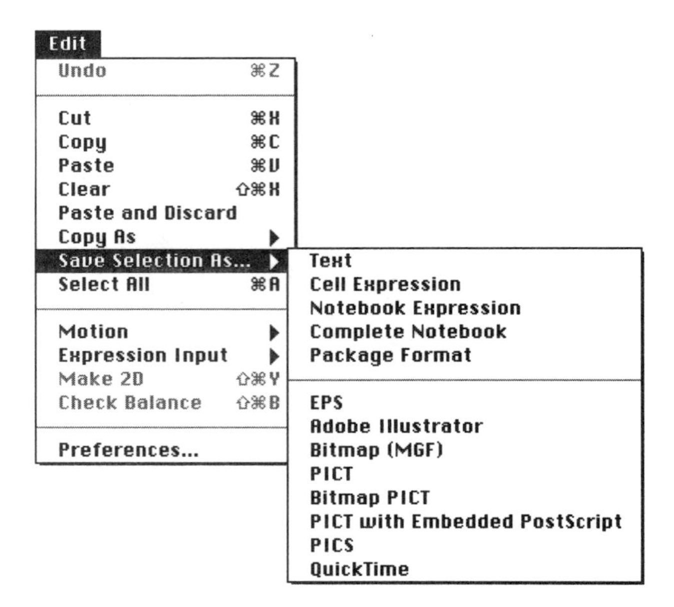

For example, **PICT with Embedded PostScript** is a popular option, especially when exporting graphics and typeset *Mathematica* input and output to other programs.

To illustrate, we use `Select` to extract the records from `gadata` for which the per capita income is greater than $23,000 and either the unemployment rate is greater than 7% or the poverty rate is greater than 25%. We name the resulting array `t1` and display it in traditional row-and-column form with `TableForm`.

```
t1 = Select [gadata ,
  And [#[[2]] > 13000 , Or [#[[3]] > 7, #[[4]] > 25]]&];
t1 // TableForm
```

```
Candler       14214    6.2    33.1    7744
Chatham       18204    6.1    25.5    216935
Evans         15304    7.9    31.     8724
Montgomery    13357    7.     34.5    7163
Tattnall      14665    5.7    28.     17722
Toombs        14963    6.6    32.3    24072
```

Next, we create a bar graph of the poverty rates for these counties using the `BarChart` function, which is contained in the **Graphics** package that is located in the **Graphics** folder (or directory). First, we load the **Graphics** package and use `?` to obtain basic information about the `BarChart` command.

```
<< Graphics'Graphics'
```

```
? BarChart
```

```
BarChart[list1, list2, ...] generates a bar chart of
  the data in the lists.
```

After loading the **DataManipulation** package, we use `Column` to extract the poverty rates for the counties listed in `t1`, naming the resulting array `t2`, and then use `BarChart` to graph the data in `t2`. Note how the bars are labeled.

```
<< Statistics'DataManipulation'
t2 = Column [t1, {4, 1}]
```

```
{{33.1, Candler}, {25.5, Chatham}, {31., Evans},
 {34.5, Montgomery}, {28., Tattnall}, {32.3, Toombs}}
```

```
BarChart [t2]
```

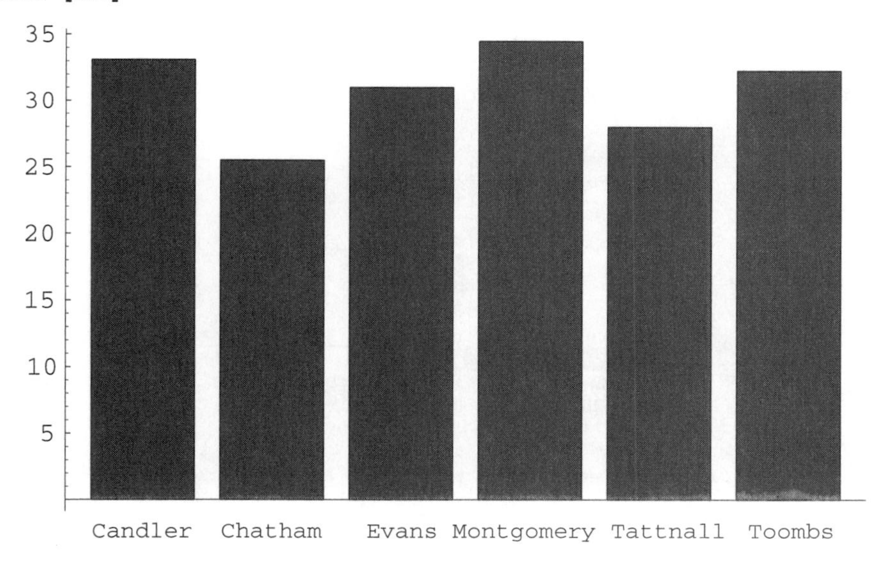

To copy this graphic to another file, first select the graphic

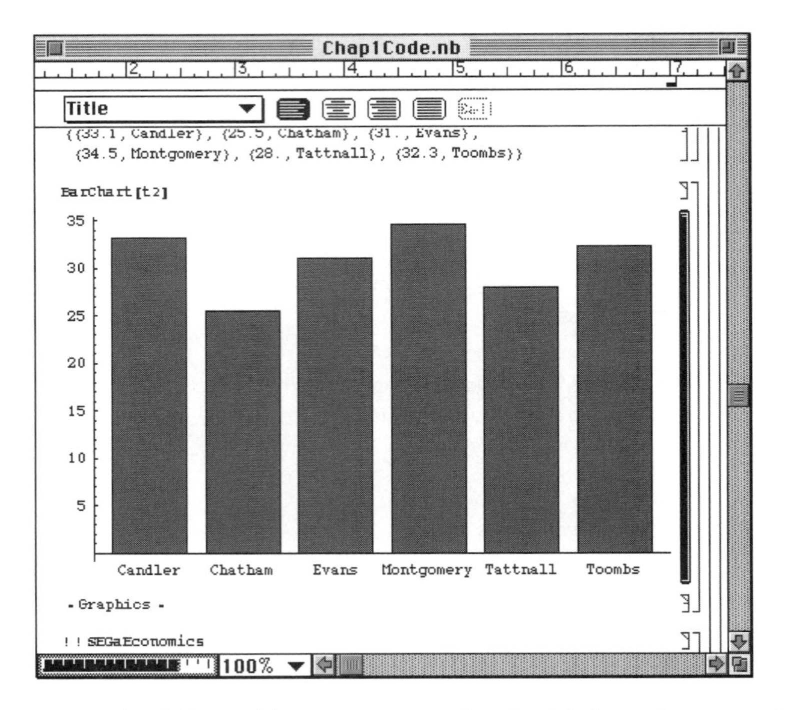

and then go to **Edit** followed by **Copy As** under the *Mathematica* menu. You can then choose the desired format. In many cases, **PICT with Embedded PostScript** produces excellent results on most printers.

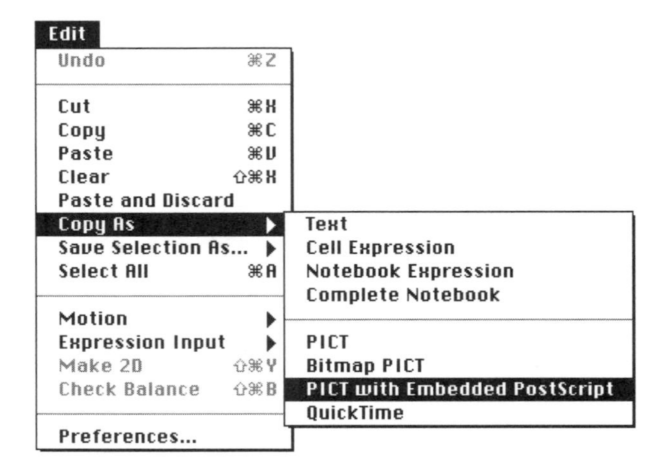

You can now paste the graph into another document. Note that if you had wished to save the graphic to a file, you would have followed the same procedure except that you would have selected **Save Selection As...** under **Edit** instead of **Copy As**.

■

CHAPTER 2

Univariate Methods for Describing Data

There are a large number of methods available to describe various properties of a data set. When the result is a number, it is called a **descriptive measure**. Most of these methods result in a measure of one of three characteristics: central tendency, variability, or shape. Together with a consideration of outliers, these characteristics are used to describe a data set. When we use the term **data set**, it will often refer to a sample; however, especially in this chapter, the data set may represent an entire population.

Most of the *Mathematica* functions presented in this chapter are contained in the standard *Mathematica* package **DescriptiveStatistics**, which is found in the **Statistics** folder (or directory). We use the **Help Browser** to obtain detailed information regarding the **DescriptiveStatistics** package.

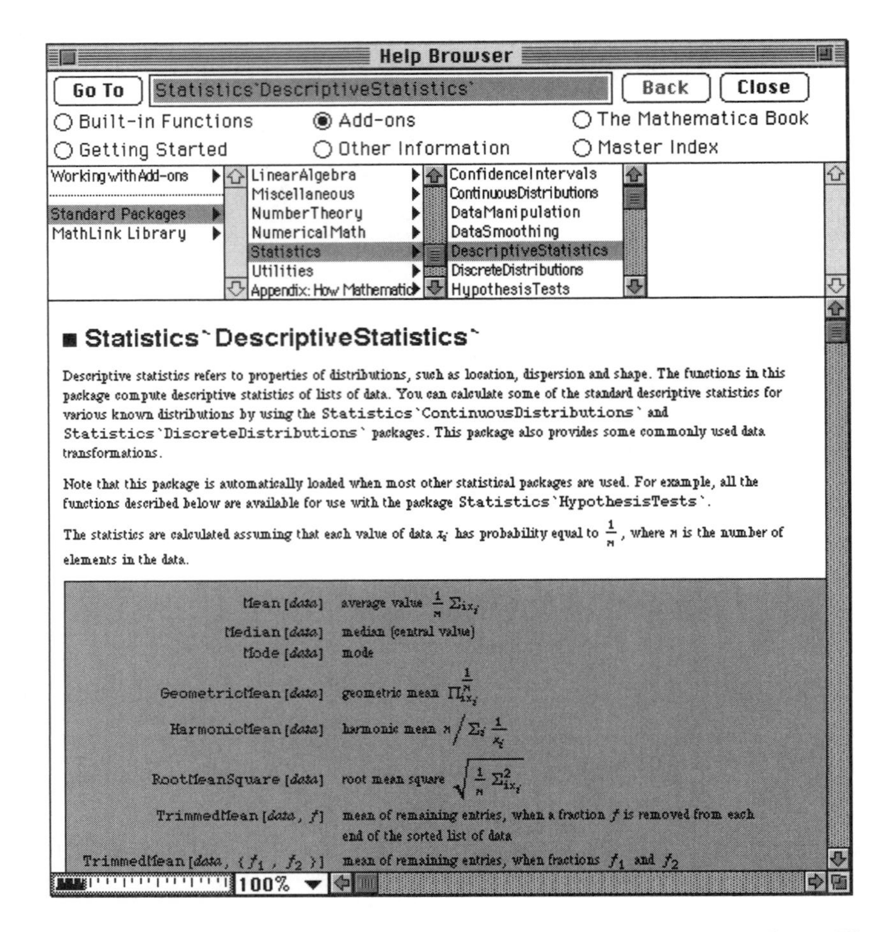

Remember that we must load packages before using commands in them. Thus, be sure to load the **DescriptiveStatistics** package by entering `<<Statistics`DescriptiveStatistics`` before using any commands contained in the package. Once you have loaded the package during your current *Mathematica* session, it is not necessary to reload it.

2.1 Measures of Location

In samples, as well as in populations, we often find a preponderance of values somewhere around the middle of the range of values. Whether or not this is the case, a description of the middle is called an average or, in statistical terms, a **measure of central tendency**. It indicates where, along the measurement scale, the sample or population is centered. As a group, **measures of location** include measures of center as well as measures corresponding to other parts of the data set. For example, the first quartile is a measure of location but not a measure of the center.

The various measures of location presented in this section are useful as parameters in that each describes a property of a population. The same *Mathematica* functions that would calculate the parameters can be used to calculate the sample statistics that are good estimates of them. Whether what we calculate is called a parameter or a statistic depends on whether the data set is a population or a sample.

The **mean** (or **arithmetic mean**) of a data set is defined as the sum of the data divided by the number of data values. For a population, the formula is

$$\mu = \frac{1}{n} \sum_{i=1}^{n} y_i,$$

where y_i is the ith value in the list of n values in the population.

For a sample, the formula is

$$\bar{y} = \frac{1}{n} \sum_{i=1}^{n} y_i,$$

where y_i is the ith value in the list of n sample values.

The **median** of a data set is the value that falls in the middle of an ordered list of the data. If the set contains an even number of values, then the median is the average of the two middle terms; if the set contains an odd number of values, then the median is the middle term.

The **mode** of a list of data is the value or values that occur most frequently. If no value occurs more than once, then the list of data has no mode. If the list has two modes, then the data set is bimodal; if it has three modes, it is trimodal. In general, a list with more than one mode is called multimodal.

The **harmonic mean** is the reciprocal of the mean of the set of values obtained by taking the reciprocal of each piece of data in the list. A general formula for the harmonic mean is

$$\frac{1}{\frac{1}{n} \sum_{i=1}^{n} \frac{1}{y_i}},$$

where y_i is the ith value in the list of n values. For example, the mean of the set $\{1/2, 1/3, 1/4\}$ is $13/36$, so the harmonic mean of the set $\{2, 3, 4\}$ is $36/13$. The harmonic mean is sometimes used when averaging rates. It is always less than the geometric mean unless all values are identical.

The **geometric mean** of a set of n values is the nth root of the product of the values. A general formula is given by

$$\sqrt[n]{\prod_{i=1}^{n} y_i},$$

where y_i is the ith value in the list. The geometric mean is the antilog of the mean of a set of values obtained by taking the log of each data value in the list. It is appropriate only when all values are positive. This measure is useful in averaging ratios, where it is desired to give each ratio equal weight, and in averaging percent changes. Unless all values are equal, the geometric mean is always less than the arithmetic mean.

A **trimmed mean** of a list is obtained by taking the mean of an ordered list after the high and low values are "trimmed" from the original list. For example, to compute the 10% trimmed mean of a data set, we first remove the top 10% and the bottom 10% of the ordered data and then we calculate the mean of the values that remain.

The **root mean square** or more appropriately the **quadratic mean** of a list of values is the square root of the mean of the squares of the values in the list. The formula is given by

$$\sqrt{\frac{1}{n}\sum_{i=1}^{n} y_i^2},$$

where y_i is the ith value in the list of n values.

We use the following commands, all of which are contained in the **DescriptiveStatistics** package, to compute means, medians, modes, harmonic means, geometric means, trimmed means, and quadratic means.

`Mean[list]` gives the mean of the entries in `list`.

`Median[list]` gives the median of the values in `list`.

`Mode[list]` yields the mode of the values in `list`. A list of modes is returned if the data is multimodal.

`HarmonicMean[list]` gives the harmonic mean of the values in `list`.

`GeometricMean[list]` gives the geometric mean of the values in `list`.

`TrimmedMean[list, f]` gives the mean of the (ordered) values from `list`, with a fraction f of values at each end dropped.

`TrimmedMean[list, {f1, f2}]` finds the mean after fractions `f1` and `f2` are trimmed off the bottom and top of the (ordered) values from `list`, respectively.

`RootMeanSquare[list]` gives the root mean square (quadratic mean) of values in `list`.

EXAMPLE 1: The salaries for the members of the Atlanta Braves major-league baseball team on opening day 1996 are listed in Table 2.1. Determine the following: (a) mean, (b) harmonic mean, (c) median, (d) mode, (e) geometric mean,

(f) trimmed mean with the highest three and lowest three salaries removed, (g) trimmed mean when the salaries over $4,000,000 and the minimum salaries of $109,000 are removed, and (h) quadratic mean of the salaries.

Player	Salary
Greg Maddux	$6,500,000
David Justice	$6,200,000
John Smoltz	$5,500,000
Tom Glavine	$5,250,000
Marquis Grissom	$4,800,000
Fred McGriff	$4,750,000
Steve Avery	$4,200,000
Jeff Blauser	$3,500,000
Mark Lemke	$1,500,000
Mark Wohlers	$1,400,000
Chipper Jones	$750,000
Rafael Belliard	$575,000
Jerome Walton	$500,000
Greg McMichael	$460,000
Dwight Smith	$350,000
Ryan Klesko	$315,000
Javy Lopez	$290,000
Pedro Borbon	$142,500
Brad Clontz	$137,500
Mike Mordecai	$135,000
Mike Bielecki	$125,000
Eddie Perez	$111,500
Jason Schmidt	$111,500
Ed Giovanola	$109,000
Tyler Houston	$109,000
Terrell Wade	$109,000

Table 2.1

SOLUTION: After loading the package, we enter the salaries in `braves` where each entry is given in millions. Next, we compute the mean, harmonic mean, and median of the data list, `braves`. (a) The mean is $1,843,460, (b) the harmonic mean is $273,986, and (c) the median is $480,000.

```
<<Statistics`DescriptiveStatistics`

braves={6.5,6.2,5.5,5.25,4.8,4.75,4.2,3.5,1.5,1.4,0.75,
0.575,0.5,0.46,0.35,0.315,0.29,0.1425,0.1375,0.135,
0.125,0.1115,0.1115,0.109,0.109,0.109};

Mean[braves]

1.84346

HarmonicMean[braves]

0.273986

Median[braves]

0.48
```

Notice that the list has 26 terms as we find with `Length`. Because 26 is an even number, the median is obtained by averaging the 13th and 14th terms of `braves`. These terms are extracted with bracket notation and averaged to see that the same result is obtained.

```
Length[braves]

26

braves[[13]]

0.5

braves[[14]]

0.46

(braves[[13]]+braves[[14]])/2

0.48
```

(d) The mode is $109,000 because three players make the minimum salary.

```
Mode[braves]

0.109
```

(e) The geometric mean is $650,286.

```
GeometricMean[braves]

0.650286
```

(f) If we remove the highest three and lowest three salaries, then the mean of the list that remains is $1,470,150.

```
TrimmedMean[braves,3/26]
1.47015
```

(g) Three players make the minimum salary, and seven players make over $4,000,000. We can remove the bottom 3 of the 26 salaries in `braves` and the top 7 to find a trimmed mean of $650,187.

```
TrimmedMean[braves,{3/26,7/26}]
0.650187
```

(h) The quadratic mean is $2,908,870.

```
RootMeanSquare[braves]
2.90887
```

■

We sometimes need to break a data set into a number of parts. Just as the median divides the ordered list into two halves, **quartiles** divide the ordered list of data into quarters (or four equal parts). In a similar manner, **deciles** divide the ordered list into tenths (10 equal parts), and **percentiles** divide the ordered list into hundredths (100 equal parts). Numbers that divide an ordered data set into equal parts are collectively called **quantiles**. The *Mathematica* function `Quantile` is used to find these **values of separation** of a data set. Often the value of a quantile would most appropriately be between two values, as is the case for the median when the number of values is even. The *Mathematica* function `InterpolatedQuantile` locates values of separation of a data set using linear interpolation. The method used may differ from the method(s) you have seen elsewhere.

`Quantile[list,q]` gives the `q`th quantile of the values in `list`. (The number Q given in the output indicates that the fraction `q` of the entries of `list` are less than or equal to Q.)

`InterpolatedQuantile[list,q]` yields the `q`th quantile of the probability distribution inferred by the values in `list`. The distribution is obtained through the linear interpolation of the values in `list`.

`Quartiles[list]` gives a list of the interpolated 0.25, 0.50, and 0.75 quantiles of the entries in `list`.

In addition to the **DescriptiveStatistics** package functions for calculating location statistics, *Mathematica* contains the built-in functions `Min` and `Max` for finding the minimum value and maximum value in a list of data.

`Min[list]` gives the numerically smallest value in `list`.

`Max[list]` gives the numerically largest value in `list`.

EXAMPLE 2: Using the salaries listed in Table 2.1, find (a) the interpolated quartiles, (b) the 6/26 quantile, (c) the 20th percentile (no interpolation), (d) the interpolated 20th percentile, and (e) the deciles of the list.

SOLUTION: (Note: If you are starting a new *Mathematica* session at this point, then you need to load the package **DescriptiveStatistics** and enter the data in `braves`. These steps are demonstrated in Example 1. If you are continuing your work from Example 1, then there is no need to carry out these steps again.)

(a) The interpolated quartiles are $135,000; $480,000; and $4,200,00.

```
Quartiles[braves]
{0.135, 0.48, 4.2}
```

(b) The 6/26-th quantile is $125,000. That is, 6 of the 26 salaries listed in Table 2.1 are less than or equal to $125,000.

```
Quantile[braves,6/26]
0.125
```

(c) The following calculation in *Mathematica* is equivalent to the result $\frac{1}{5}(26) = 5.2$.

```
(1/5) 26//N
5.2
```

To find the 20th percentile (1/5-th quantile) we round 5.2 up to the next integer, 6, and find that six elements of `braves` are less than or equal to the result, $125,000. The same result is obtained here that was found in (b).

```
Quantile[braves,0.2]
0.125
```

(d) The interpolated 20th percentile is $120,950.

```
InterpolatedQuantile[braves,0.2]
0.12095
```

(e) We can find the deciles by using the *Mathematica* function `Table` with `Quantile[braves,k/10]` for the values $k = 1, 2, \ldots, 9$. These are $109,000; $125,000; $137,500; $315,000; $460,000; $750,000; $3,500,000; $4,750,000; $5,500,000.

```
Table[Quantile[braves,k/10],{k,1,9}]
{0.109, 0.125, 0.1375, 0.315, 0.46, 0.75, 3.5, 4.75, 5.5}
```

Similarly, we can use InterpolatedQuantile. This yields $109,250; $120,950; $139,000; $312,500; $480,000; $815,000; $2,900,000; $4,765,000; $5,475,000.

```
Table[InterpolatedQuantile[braves,k/10],{k,1,9}]
{0.10925, 0.12095, 0.139, 0.3125, 0.48, 0.815, 2.9,
4.765, 5.475}
```

■

Reporting Location Statistics

The three location statistics, mean, harmonic mean, and median, are found simultaneously with LocationReport.

> LocationReport[list] gives the mean, harmonic mean, and median location statistics for the values in list.

However, we can define our own functions to display any combination of statistics we would like to find at once. The *Mathematica* code for LocationReport is:

```
LocationReport[list_] :=
   {
   Mean -> Mean[list] ,
   HarmonicMean -> HarmonicMean[list] ,
   Median -> Median[list]
   } /; VectorQ[list]
```

The built-in *Mathematica* function VectorQ[list] returns True if list is a vector and False otherwise. Including it in the definition of LocationReport causes the LocationReport[list] to be returned unevaluated if the argument inserted for list is not in the correct form of a vector, a $1 \times n$ matrix.

To compute a five-number summary of a data set (the minimum, first quartile, median, third quartile, and maximum), we can define a function such as fiveNumberSummary.fiveNumberSummary:

```
fiveNumberSummary[list_] :=
   {
   Min->Min[list],
   25thPercentile -> InterpolatedQuantile[list,0.25],
```

```
Median -> Median[list],
75thPercentile -> InterpolatedQuantile[list,0.75],
Max -> Max[list]
} /; VectorQ[list]
```

We can easily define functions in a similar way to specify that other location statistics be reported.

EXAMPLE 3: (a) Use `LocationReport` to find the mean, harmonic mean, and median of the data given in Table 2.1, and (b) find the five-number summary of this data.

SOLUTION: (a) After loading the **DescriptiveStatistics** package and the data in `braves`, all three descriptive measures can be obtained at once with `LocationReport`.

```
LocationReport[braves]
{Mean -> 1.84346, HarmonicMean -> 0.273986, Median -> 0.48}
```

(b) With `fiveNumberSummary`, we find that the minimum salary is \$109,000; the twenty-fifth percentile is \$135,000; the median is \$480,000; the seventy-fifth percentile is \$4,200,000; and the maximum salary is \$6,500,000.

```
fiveNumberSummary[list_] :=
   {
   Min->Min[list],
   25thPercentile-> InterpolatedQuantile[list,0.25],
   Median -> Median[list],
   75thPercentile-> InterpolatedQuantile[list,0.75],
   Max->Max[list]
   } /; VectorQ[list]

fiveNumberSummary[braves]

{Min -> 0.109, 25 thPercentile -> 0.135, Median -> 0.48,
  75 thPercentile -> 4.2, Max -> 6.5}
```

■

EXAMPLE 4: The text (ASCII) file Census.txt contains 738 records of persons 18 years of age and older taken from the 1990 Decennial Census Public Use Microdata 5% Samples for the State of Georgia. Five of the records from the file are listed below.

```
99001234024000778500100191100000011301 4
```

Each record is a a string of digits containing a value for each of the variables listed in Table 2.2. A value is found according to its position and length.

Variable Label	Position	Length
Age	6	2
Class of Worker 0=N/A 1=Emp. of a Private Company 2=Emp. of a Private Not for Profit Organization 3=Local Gov. Emp. 4=State Gov. Emp. 5=Fed. Gov. Emp. 6=Self Emp. in Own Not Incorp.d Business 7=Self Emp. in Own Incorp.d Business 8=Working Without Pay in Fam. Bus. or Farm 9=Unemp.	8	1
Number of Children Ever Born 00=N/A Less Than 15 Yrs./male 01=No Chld. 02=1 Child 03=2 Chld. 04=3 Chld. 05=4 Chld.	9	2

Continued

Variable Label	Position	Length
06=5 Chld.		
07=6 Chld.		
08=7 Chld.		
09=8 Chld.		
10=9 Chld.		
11=10 Chld.		
12=11 Chld.		
13=12 or More Chld.		
Usual Hours Worked Per Week in 1989	11	2
Wages or Salary Earned in 1989	13	6
Marital Status	19	1
0=Now Married, Except Separated		
1=Widowed		
2=Divorced		
3=Separated		
4=Never Married or Under 15 Yrs. Old		
Means of Transportation to Work	20	2
00=N/A		
01=Car, Truck or Van		
02=Bus or Trolley Bus		
03=Streetcar or Trolley Car		
04=Subway or Elevated		
05=Railroad		
06=Ferryboat		
07=Taxicab		
08=Motorcycle		
09=Bicycle		
10=Walked		
11=Worked at Home		
12=Other Method.		

Variable Label	Position	Length
Pers. Wgt *This code is used by the Bureau of the Census.*	22	4
Vehicle Occupancy 0=N/A 1=Drove Alone 2=2 People 3=3 People 4=4 People 5=5 People 6=6 People 7=7 to 9 People 8=10 or More People	26	1
School Enrollment 0=N/A Less Than 3. Yrs. Old 1=Not Attending School 2=Yes, Pub. School, Pub. Coll. 3=Yes, Private School, Private Coll.	27	1
Persons Serial Number *This code identifies the record but does not reveal the identity of the person.*	28	7
Sex 0=Male 1=Female	35	1
Weeks Worked in 1989	36	2
Educational Attainment 00=N/A Less Than 3 Yrs. Old 01=No School Completed 02=Nursery School 03=Kindergarten 04=1st, 2nd, 3rd, or 4th Grade	38	2

Continued

Variable Label	Position	Length
05=5th, 6th, 7th, or 8th Grade		
06=9th Grade		
07=10th Grade		
08=11th Grade		
09=12th Grade, No Diploma		
10=High School Graduate, Diploma, or Ged		
11=Some Coll., But No Degree		
12=Associate Degree in Coll., Occupational		
13=Associate Degree in Coll., Academic		
14=Bachelors Degree		
15=Masters Degree		
16=Professional Degree		
17=Doctorate Degree		

Table 2.2

For example, the record

990012340240007785001001911000000113014

indicates that this person is 23 years old,

99001**23**40240007785001001911000000113014

is a state government employee,

9900123**4**0240007785001001911000000113014

has one child,

99001234**02**40007785001001911000000113014

works 40 hours per week,

9900123402**40**007785001001911000000113014

earned $7,785 in 1989,

990012340240**007785**001001911000000113014

is married,

990012340240007785**0**01001911000000113014

drives a car, truck, or van to work

9900123402400077850**01**001911000000113014

and drives alone,

990012340240007785001001**9**1100000011301 4

is not attending school,

9900123402400077850010019**1**1000000113014

is a female,

99001234024000778500100191100000001**1**3014

who worked 30 weeks in 1989,

990012340240007785001001911000000011**30**14

and has a bachelor's degree.

9900123402400077850010019110000000113 0**14**

Determine: (a) the maximum, minimum, mean, harmonic mean, median, and mode of the ages given in the data set; (b) the quadratic mean of the salaries and the trimmed mean with the 0 salaries removed; and (c) the interpolated quartiles and 0.1 quantile for those who reported working more than 0 hours per week. In each case, find the five-number summary of the data.

SOLUTION: We use `ReadList` together with `Word` to read the records contained in **Census.txt** into *Mathematica* and name the resulting list `censusdata`. An abbreviated portion of `censusdata` is displayed with `Short`.

```
censusdata = ReadList["Census.txt", Word];
Short[censusdata, 3]
```

```
{990012340240007785001001911000000113014, <<736>>,
 990012310062016000001002521000044805210}
```

We learn that `censusdata` contains 738 records using `Length`.

```
Length[censusdata]
```

```
738
```

We extract the first element of `censusdata` using `Part` (`[[...]]`).

```
censusdata[[1]]
```

```
99001234024000778500100191100000011 3014
```

Because we included `Word` in the `ReadList` command, the elements of `censusdata` are strings. Therefore, before we perform arithmetic on variables extracted from records, we must be sure to first convert the variables from strings to numbers with `ToExpression`.

```
Head[censusdata[[1]]]
```

```
String
```

(a) The age variable is in positions 6 and 7 of each record. We use `Map` together with `StringTake` to extract the age from each record, convert the resulting list to a list of numbers with `ToExpression`, and name the resulting list `ages`. The first 10 entries of `ages` are viewed with `Take`.

```
ages =
 Map [StringTake [#, {6, 7}]&, censusdata] // ToExpression;
Take [ages , 10 ]
```

```
{23, 24, 41, 40, 39, 22, 21, 24, 22, 42}
```

The maximum and minimum ages are determined to be 89 and 18, respectively, with `Max` and `Min`.

```
Max[ages]
Min[ages]
```

```
89

18
```

For the remaining calculations, we use commands contained in the **DescriptiveStatistics** package. First we load the **DescriptiveStatistics** package. (Remember that you do not need to reload the **DescriptiveStatistics** package if you have already loaded it during your current *Mathematica* session. Similarly, if you have just completed Example 3, you do not need to redefine `fiveNumberSummary`.)

```
<< Statistics'DescriptiveStatistics'

fiveNumberSummary  [list_] :=
 {
 Min -> Min [list ],
 25 thPercentile  -> InterpolatedQuantile  [list , 0.25 ],
 Median  -> Median [list],
 75 thPercentile  -> InterpolatedQuantile  [list , 0.75 ],
 Max -> Max [list]
 }  /;  VectorQ [list]
```

Using `Mean`, `HarmonicMean`, `Median`, and `Mode`, we see that the mean age is approximately 40 years, the harmonic mean is approximately 34, the median is 36, and the mode is 25.

```
Mean[ages] // N

39.7995

HarmonicMean[ages] // N

33.6218

Median[ages]

36

Mode[ages]

25
```

The five-number summary is computed with `fiveNumberSummary`.

```
fiveNumberSummary [ages ]

{Min → 18, 25 thPercentile→ 26., Median → 36,
 75 thPercentile→ 51., Max → 89}
```

(b) The salaries variable is in positions 13 through 18 of each record. We use `Map` together with `StringTake` to extract the salary from each record, convert the resulting list to a list of numbers with `ToExpression`, and name the resulting list `salaries`. The last 5 entries of `salaries` are viewed with `Take`.

```
salaries =
 Map[StringTake[#, {13, 18}]&, censusdata] //
  ToExpression;
```

```
Take[salaries, -5]
```

```
{12000, 0, 35000, 9274, 16000}
```

The quadratic mean is determined to be $19633.40.

```
RootMeanSquare[salaries] // N
```

```
19633.4
```

We use `Count` to see that 215 records have 0 salary from employment.

```
Count[salaries, 0]
```

```
215
```

We then use `TrimmedMean` to calculate the trimmed mean of salaries when the lowest 215 salaries (the records with 0 salary) are removed.

```
TrimmedMean[salaries, {215/738, 0}] // N
```

```
18238.1
```

As in (a), we use `fiveNumberSummary` to find the five-number summary for salaries.

```
fiveNumberSummary [salaries]
```

```
{Min → 0, 25 thPercentile→ 0, Median → 10000,
 75 thPercentile→ 20000., Max → 192244}
```

(c) The usual number of hours worked each week variable is in positions 11 and 12 of each record. We use `Map` together with `StringTake` to extract the age from each record, convert the resulting list to a list of numbers with `ToExpression`, sort the result from lowest to highest with `Sort`, and name the resulting list `hours`.

```
hours = Map[StringTake[#, {11, 12}]&,
    censusdata] // ToExpression//
  Sort;
```

We use `Count` to see that 194 individuals reported working 0 hours.

```
Count[hours, 0]
```

```
194
```

The first 194 elements of hours (the 0's) are removed from `hours` with `Drop`.

```
hours = Drop[hours, 194];
```

The interpolated quartiles are then found with `Quartiles`.

```
Quartiles[hours]
```

```
{40, 40, 48}
```

while the 0.1 quantile is found with `Quantile`.

```
Quantile[hours, 1/10]
```

```
28
```

The five number summary is found with `fiveNumberSummary`.

```
fiveNumberSummary [hours]
```

```
{Min → 5, 25 thPercentile→ 40., Median → 40,
 75 thPercentile→ 48., Max → 99}
```

■

2.2 Measures of Dispersion

Measures of dispersion are also called **measures of variability** or **measures of spread**. They are used to describe the size of the differences among the values in a data set. Thus, the more spread out the values are, the larger the value for the measure of dispersion.

A measure of dispersion of a population is a parameter of the population, and the sample measure of dispersion that estimates the parameter is a statistic. The calculations for measures of dispersion for a population are often different from the calculations for the statistics. As a consequence, there are sometimes two *Mathematica* functions for the same measure of dispersion.

The **range** of a data set is the difference between the largest and smallest values in the data set. The *Mathematica* function `SampleRange`, which is contained in the **DescriptiveStatistics** package, will calculate the range of a sample or a population.

The **population standard deviation** equals the square root of the sum of the squares of the deviations from the mean of the values in the population divided by the number of values in the population, n. Standard notation for the population standard deviation is

$$\sigma = \sqrt{\frac{\sum_{i=1}^{n} (y_i - \mu)^2}{n}},$$

where y_i is the ith value in the list and μ the population mean.

The **population variance**, σ^2, is the square of the population standard deviation, σ.

The **sample standard deviation** equals the square root of the sum of the squares of the deviations from the mean of the values in the sample divided by $(n-1)$, where there are n values in the sample. Standard notation for the sample standard deviation is

$$s = \sqrt{\frac{\sum_{i=1}^{n} (y_i - \bar{y})^2}{n-1}},$$

where y_i is the ith value in the list, and $\bar{y}$ is the sample mean.

The **sample variance**, s^2, is the square of the sample standard deviation, s.

The **standard deviation of the sample mean**, $\sigma_{\bar{y}}$, measures the amount of sampling error to be expected when the mean μ of the population is estimated by the mean $\bar{y}$ of a sample from the population. The smaller the value of $\sigma_{\bar{y}}$, the smaller the sampling error tends to be. The value $\sigma_{\bar{y}}$ is usually called the **standard error of the mean**. It is calculated using the formula $\sigma_{\bar{y}} = \frac{\sigma}{\sqrt{n}}$, where n is the sample size and σ is the population standard deviation. The estimate of $\sigma_{\bar{y}}$ is $s_{\bar{y}} = \frac{s}{\sqrt{n}}$ where s is the sample standard deviation.

The **mean absolute deviation** of a sample of n values is given by $\frac{1}{n}\sum_{i=1}^{n} |y_i - \bar{y}|$, where y_i is the ith value of the set and $\bar{y}$ is the sample mean.

The **median absolute deviation** of a set of n values is the median of the values $|y_i - y_{\text{med}}|$, where y_i is the ith value of the set and y_{med} is the median.

The **interquartile range** of a set of data is the difference $Q_3 - Q_1$, where Q_3 is the upper quartile (75th percentile) and Q_1 is the lower quartile (25th percentile).

The **semi-interquartile range** also known as the **quartile deviation** is $\frac{Q_3 - Q_1}{2}$, which is the interquartile range divided by 2.

The **coefficient of variation** of a set of values is the ratio of the mean to the standard deviation. For a sample, the formula is $V = \frac{s}{\bar{y}}$. (V is frequently multiplied by 100 and expressed as a percent.) The mean and standard deviation have the same units, so V has no units. Consequently, it is useful to compare variability among sets of measurements having different units. The reciprocal of the coefficient of variation is called the **signal-to-noise ratio**, where the mean represents the "signal" and the standard deviation represents the "noise" (error) in the signal.

Use the following commands, all of which are contained in the **DescriptiveStatistics** package, to compute these values for a data set.

`SampleRange[list]` gives the range of the values in `list`.

`VarianceMLE[list]` gives the population variance of the entries in `list`. Division by n is used to calculate the population variance. If the values in `list` are a sample, the result is a maximum likelihood estimate of the population variance.

`StandardDeviationMLE[list]` gives the population standard deviation of the values in `list`. Division by n is used to calculate the population standard deviation. If the values in `list` are a sample, the result is a maximum likelihood estimate of the population standard deviation.

`Variance[list]` gives the sample variance of the values in `list`. Division by $(n-1)$ (instead of n) is used, giving an unbiased estimate of the population variance.

`StandardDeviation[list]` gives the sample standard deviation of the values in `list`. Division by $(n-1)$ is used. (Note: The documentation in *Mathematica* incorrectly labels s an unbiased estimate of σ, the population standard deviation. Actually, s consistently underestimates σ.)

`VarianceOfSampleMean[list]` gives an unbiased estimate of the variance of the sample mean, using the values in `list` as a sample from the population.

`StandardErrorOfSampleMean[list]` gives an estimate of the standard error (standard deviation) of the sample mean using the values in `list` as a sample from the population. (Note: The documentation in *Mathematica* incorrectly labels $s_{\bar{y}}$ an unbiased estimate of $\sigma_{\bar{y}}$.)

`MeanDeviation[list]` gives the mean absolute deviation about the mean of the values in `list`.

`MedianDeviation[list]` gives the median absolute deviation about the median of the values in `list`.

`InterquartileRange[list]` gives the difference between the upper and lower quartiles for the values in `list`. `This result is equal to that of Quantile[list,0.75]-Quantile[list,0.25]`.

`QuartileDeviation[list]` gives the quartile deviation, or semi-interquartile range, of the values in `list`.

`CoefficientOfVariation[list]` gives the coefficient of variation, defined as the ratio of the standard deviation to the mean of the values in `list`. (Note: If the values in `list` are not all positive, then `$Failed` is returned by this function. This is an unnecessary restriction.)

Several dispersion statistics are calculated simultaneously with the **DescriptiveStatistics** package command `DispersionReport`.

`DispersionReport[list]` gives the `Variance`, `StandardDeviation`, `SampleRange`, `MeanDeviation`, `MedianDeviation`, and `QuartileDeviation` dispersion statistics for `list`.

```
DispersionReport[list_] :=
    {
    Variance -> Variance[list] ,
    StandardDeviation -> StandardDeviation[list] ,
    SampleRange -> SampleRange[list] ,
    MeanDeviation -> MeanDeviation[list] ,
    MedianDeviation -> MedianDeviation[list] ,
    QuartileDeviation -> QuartileDeviation[list]
    } /; VectorQ[list]
```

As we did with `LocationReport`, we can define our own function to determine the particular statistics we would like to have found simultaneously by modifying the code of `DispersionReport`.

EXAMPLE 1: For the data in Table 2.1, calculate the (a) range, (b) standard deviation (both sample and population), (c) variance (both sample and population), (d) variance of the sample mean, (e) standard error of the sample mean, (f) mean absolute deviation, (g) median absolute deviation, (h) interquartile range, (i) semi-interquartile, and (j) coefficient of variation.

SOLUTION: (Be sure that you have defined `braves` as in Example 1, Section 2.1 before completing this example.) (a) After loading the **DescriptiveStatistics** package and the data in `braves` (if we are starting a new *Mathematica* session), we find that the range of the sample is \$6,391,000.

```
<<Statistics`DescriptiveStatistics`
SampleRange[braves]
```

```
6.391
```

(b) The sample standard deviation is \$2,294,710; the population standard deviation is \$2,250,150.

```
StandardDeviation[braves]
```

```
2.29471
```

```
StandardDeviationMLE[braves]
```

```
2.25015
```

(c) The sample variance is \$$5.26568 \times 10^{12}$; the population variance is \$$5.06315 \times 10^{12}$. Note that each salary was divided by 10^6. Because variance is based on squared salaries, to transform back requires multiplication by $10^{12} \times (10^6)^2$.

```
Variance[braves]
```

```
5.26568
```

```
VarianceMLE[braves]
```

```
5.06315
```

(d) An unbiased estimate of the variance of the sample mean is \$$2.02526 \times 10^{11}$. (e) An estimate of the standard error (or standard deviation) of the sample mean is \$450,029.

```
VarianceOfSampleMean[braves]
```

```
0.202526
```

```
StandardErrorOfSampleMean[braves]
```

```
0.450029
```

(f) The mean absolute deviation is \$1,996,330. (g) The median absolute deviation is \$369,750.

```
MeanDeviation[braves]
```

```
1.99633
```

```
MedianDeviation[braves]
```

```
0.36975
```

(h) The interquartile range is \$4,065,000. (i) The semi-interquartile range (quartile deviation) is \$2,032,500. and (j) The coefficient of variation is \$1,244,780.

```
InterquartileRange[braves]
```

```
4.065
```

```
QuartileDeviation[braves]
```

```
2.0325
```

```
CoefficientOfVariation[braves]
1.24478
```

We can compute several dispersion statistics with `DispersionReport`.

```
DispersionReport[braves]
{Variance -> 5.26568, StandardDeviation -> 2.29471,
 SampleRange -> 6.391, MeanDeviation -> 1.99633,
 MedianDeviation -> 0.36975, QuartileDeviation -> 2.0325}
```

■

EXAMPLE 2: The text (ASCII) file **GeorgiaData.txt** contains 599 records for municipalities in the state of Georgia. Several records from the file are listed below. Notice that the values are separated by commas. From left to right, the values represent the population size, the area in square kilometers, the area in square miles, the the population per square kilometer, the population per square mile, the name of the municipality, and its classification (city, town, village, CDP).

```
907,8.0,3.1,114.0,295.4,Abbeville,city
4519,12.0,4.6,377.0,976.4,Acworth,city
2131,14.4,5.6,147.9,383.0,Adairsville,city
5093,16.3,6.3,312.7,809.9,Adel,city
615,3.6,1.4,171.8,445.1,Adrian,city
```

(a) Compute the sample variance, sample standard deviation, range, mean absolute deviation, median absolute deviation, and quartile deviation for the population size of all municipalities.

(b) Compute the sample variance, population variance, sample standard deviation, population standard deviation, standard error of the sample mean, range, mean absolute deviation, median absolute deviation, quartile deviation, and coefficient of variation for the population size of just the cities and then for the population size of just the towns.

SOLUTION: We begin by using `ReadList` to read the data set into *Mathematica*, naming the resulting array `gadata`. The option `WordSeparators->{","}` instructs *Mathematica* that the variables are separated by commas. Note that because we have included `Word` in the `ReadList` command, every entry is read into *Mathematica* as a string. Therefore, to use *Mathematica* to perform arithmetic on the numeric entries, we must first convert them from strings to numbers.

```
gadata = ReadList ["GeorgiaData .txt", Word,
  WordSeparators -> {","}, RecordLists -> True ];
```

To convert the numerical entries in gadata from strings to numbers, we first load the **DataManipulation** package and then use ColumnTake together with ToExpression to convert the entries in the first five columns of gadata from strings to numbers, naming the resulting array numericdata. The entries in columns 6 and 7 are then extracted in stringdata.

```
<< Statistics'DataManipulation'

numericdata = ToExpression [ColumnTake [gadata, {1, 5}]];
stringdata = ColumnTake [gadata, {6, 7}];
```

We then use RowJoin to reassemble numericdata and stringdata into gadata. We use Take to view to the 100th through 109th entries in gadata.

```
gadata = RowJoin [numericdata , stringdata ];

Take [gadata , {100 , 109 }]

{{301, 8.5, 3.3, 35.4, 91.6, Centralhatchee, town},
 {7668, 8.1, 3.1, 943.2, 2442.8, Chamblee, city},
 {2865, 9.4, 3.6, 303.3, 785.6, Chatsworth, city},
 {4088, 19.5, 7.5, 209.4, 542.4, Chattanooga Valley, CDP},
 {312, 4.5, 1.7, 69.7, 180.5, Chauncey, town},
 {1072, 2.3, 0.9, 473.7, 1226.9, Chester, town},
 {2149, 4.2, 1.6, 516.1, 1336.7, Chickamauga, city},
 {1151, 4.8, 1.9, 238.5, 617.7, Clarkesville, city},
 {5385, 2.7, 1.1, 1975.4, 5116.3, Clarkston, city},
 {2464, 4., 1.6, 608.8, 1576.9, Claxton, city}}
```

To perform the analysis on the data, we load the **DescriptiveStatistics** package. (a) The populations of all municipalities are extracted with ColumnTake together with Flatten and the resulting list is named populations. Several elements of populations are viewed with Take.

```
<< Statistics'DescriptiveStatistics'

populations = ColumnTake[gadata, 1] // Flatten;

Take[populations, {200, 204}]

{8612, 279, 602, 1251, 2285}
```

DispersionReport is used to compute the sample variance, sample standard deviation, range, mean absolute deviation, median absolute deviation, and quartile deviation for the populations of the municipalities given in gadata. Note that DispersionReport returns exact values,

```
DispersionReport [populations]
```

$$\left\{\text{Variance}\rightarrow\frac{74687542987270}{179101},\ \text{StandardDeviation}\rightarrow\sqrt{\frac{74687542987270}{179101}},\ \text{SampleRange}\rightarrow 393995,\ \text{MeanDeviation}\rightarrow\frac{2368288124}{358801},\ \text{MedianDeviation}\rightarrow 975,\ \text{QuartileDeviation}\rightarrow\frac{14611}{8}\right\}$$

so we use N to obtain accurate approximations.

```
DispersionReport [populations] // N
```

```
{Variance→ 4.17014 × 10^8, StandardDeviation→ 20420.9,
 SampleRange→ 393995., MeanDeviation→ 6600.56,
 MedianDeviation→ 975., QuartileDeviation→ 1826.37}
```

To compute the sample variance, population variance, sample standard deviation, population standard deviation, standard error of the sample mean, range, mean absolute deviation, median absolute deviation, quartile deviation, and coefficient of variation for the population size of the cities and then the towns given in gadata, we modify the code of DispersionReport and define the function modifiedDispersionReport, which computes these values for the list list.

```
modifiedDispersionReport [list_] :=
 {
 Variance -> Variance[list],
 VarianceMLE -> VarianceMLE [list],
 StandardDeviation -> StandardDeviation [list],
 StandardDeviationMLE -> StandardDeviationMLE [list ],
 StandardErrorOfSampleMean ->
    StandardErrorOfSampleMean [list] ,
 SampleRange -> SampleRange [list],
 MeanDeviation -> MeanDeviation [list],
 MedianDeviation -> MedianDeviation [list],
```

```
QuartileDeviation   -> QuartileDeviation [list],
CoefficientOfVariation   ->  CoefficientOfVariation  [list]
}  /;  VectorQ [list]
```

The records for the cities in Georgia are then extracted from `gadata` with `Select` and the populations of the cities are extracted with `ColumnTake`.

```
gacities = Select[gadata, #[[7]] == "city"&];

citypops = ColumnTake[gacities, 1] // Flatten;
```

We then use `modifiedDispersionReport` to compute the sample variance, population variance, sample standard deviation, population standard deviation, standard error of the sample mean, range, mean absolute deviation, median absolute deviation, quartile deviation, and coefficient of variation for the populations of the cities in Georgia.

```
modifiedDispersionReport[citypops] // N

{Variance → 6.43041 〈 10^8,
 VarianceMLE → 6.41269 〈 10^8, StandardDeviation → 25358.2,
 StandardDeviationMLE → 25323.3,
 StandardErrorOfSampleMean → 1330.96,
 SampleRange → 393995., MeanDeviation → 8061.03,
 MedianDeviation → 1505., QuartileDeviation → 2051.63,
 CoefficientOfVariation → 3.68277}
```

We use the same steps to compute these values for the towns in Georgia.

```
gatowns = Select [gadata, #[[7]] == "town"&];

townpops = ColumnTake [gatowns, 1] // Flatten;

modifiedDispersionReport [townpops] // N

{Variance → 242846.,
 VarianceMLE → 241401., StandardDeviation → 492.794,
 StandardDeviationMLE → 491.326,
 StandardErrorOfSampleMean → 38.0199,
 SampleRange → 2758., MeanDeviation → 328.3,
 MedianDeviation → 160.5, QuartileDeviation → 200.75,
 CoefficientOfVariation → 1.01445}
```

■

2.3 Measures of Shape

Two frequently used measures of the shape of a distribution of values are measures of **skewness** (versus symmetry) and **kurtosis**. A distribution is **symmetric** if the mean and median are equal and the distribution of the values to the right of the mean is a mirror image of the distribution of the values to the left of the mean. A distribution with extreme values to the right of the mean is said to be **skewed right** or **positively skewed**. For this type of distribution, the mean is larger than the median. A distribution with extreme values to the left of the mean is said to be **skewed left** or **negatively skewed**. For this type of distribution, the mean is smaller than the median. Kurtosis is often characterized in terms of peakedness or flatness of a symmetric distribution. A normal distribution is said to be **mesokurtic**. A **leptokurtic** distribution has more values near the center and at the tail than does a normal distribution, whereas a **platykurtic** distribution has fewer values near the center and at the tails and more values around $\mu - \sigma$ and $\mu + \sigma$. A leptokurtic distribution would appear to be narrow and peaked. It can be pictured as the composite of two normal distributions with the same mean and different standard deviations. A platykurtic distribution would look wide and flat and can be pictured as the composite of two normal distributions with different means and the same standard deviation.

The ***r*th central moment**, or the rth moment about the mean of a population of n values, is found with

$$\frac{1}{n}\sum_{i=1}^{n}(y_i - \mu)^r,$$

where y_i is the ith value of the set and μ is the population mean. Note that the second moment about the mean is the variance. Skewness and kurtosis can be described using the third and fourth moments, respectively.

The **population skewness coefficient** of a list of n values is given by

$$\frac{1}{\sigma^3 n}\sum_{i=1}^{n}(y_i - \mu)^3,$$

where y_i is the ith value of the set, μ is the population mean, and σ the population standard deviation. This is equivalent to the third central moment divided by the standard deviation cubed. A positive value indicates positive skewness (skewed to the right). A negative value indicates negative skewness (skewed to the left). The coefficient is zero for a symmetric distribution.

We use the commands `CentralMoment` and `Skewness`, both of which are contained in the **DescriptiveStatistics** package, to compute population central moments and population skewness coefficients.

> `CentralMoment[list,r]` gives the `r`th central moment of the values in `list` with respect to their mean.
>
> `Skewness[list]` gives the coefficient of population skewness of the values in `list`.

The (unbiased) **sample skewness coefficient** of a list of n values equals $g_1 = \frac{k_3}{s^3}$, where

$$k_3 = \frac{n\sum_{i=1}^{n}(y_i - \bar{y})^3}{(n-1)(n-2)}$$

and s is the sample standard deviation. There is no package function to compute the sample skewness. However, we can define our own function, `sampleSkewness`, to do this.

```
sampleSkewness[list_]:=
   Length[list]/StandardDeviation[list]^3
      * Sum[(list[[i]]-Mean[list])^3,{i,1,Length[list]}]/
      ((Length[list]-1)(Length[list]-2))/; VectorQ[list]
```

Another approach involves using `Module` to define this function. We define `n`, `k3`, `numer`, and `denom` as local variables in the function so that the formula for the sample skewness is entered more easily and we are less likely to make a mistake in defining it.

```
sampleSkewness[list_]:=Module[{n,k3,numer,denom},
      n=Length[list];
      numer=n Sum[(list[[i]]-Mean[list])^3,{i,1,n}];
      denom=(n-1)(n-2);
      k3=numer/denom;
      k3/StandardDeviation[list]^3]/; VectorQ[list];
```

Pearson's first coefficient of skewness is given by $\frac{3}{s}(\bar{y} - y_{\text{mode}})$ where $\bar{y}$ is the mean, y_{mode} is the mode, and s is the standard deviation of the sample. A positive value for Pearson's first coefficient of skewness indicates that the mean is greater than the mode (positively skewed); a negative value for Pearson's first coefficient of skewness indicates that the mode is greater than the mean (negatively skewed).

Pearson's second coefficient of skewness is given by $\frac{3}{s}(\bar{y} - y_{\text{mode}})$ where $\bar{y}$ is the mean, y_{mode} is the median, and s is the standard deviation of the sample. A positive value for Pearson's second coefficient of skewness indicates that the mean is greater than the median (positively skewed); a negative value for Pearson's second coefficient of skewness indicates that the mean is less than the median (negatively skewed).

The **quartile coefficient of skewness,** $\frac{Q_3 + Q_1 - 2Q_2}{Q_3 - Q_1}$ (or Bowley coefficient of skewness), measures the symmetry between the first and third quartiles. The coefficient can take values between –1 and 1. A negative value indicates that the median is closer to the upper quartile than it is to the lower quartile and the distribution is skewed left. A value of zero indicates symmetry. A positive value indicates that the median is closer to the lower quartile than it is to the upper quartile and the distribution is skewed right.

The following commands, all of which are contained in the **DescriptiveStatistics** package, are used to compute these values:

`PearsonSkewness1[list]` gives Pearson's first coefficient of skewness of the values in `list`.

`PearsonSkewness2[list]` gives Pearson's second coefficient of skewness of the values in `list`.

`QuartileSkewness[list]` gives the quartile coefficient of skewness of the values in `list`.

The **population kurtosis coefficient** is

$$\frac{1}{\sigma^4 n}\sum_{i=1}^{n}(y_i-\mu)^4,$$

where y_i is the ith value of the set, μ is the population mean, and σ the population standard deviation. This value is equivalent to the fourth central moment divided by the variance squared. If the coefficient is less than 3, then the distribution is platykurtic. If it is greater than 3, then the distribution is leptokurtic. A normal (mesokurtic) distribution has a value of 3. Often 3 is subtracted from the kurtosis, and the result is called the kurtosis coefficient. The result is a measure which is negative for platykurtic distributions, zero for mesokurtic (normal) distributions, and positive for leptokurtic distributions.

The following commands, which are contained in the **DescriptiveStatistics** package, are used to compute kurtosis.

`Kurtosis[list]` gives the population kurtosis coefficient for the values in `list`.

`KurtosisExcess[list]` gives (`Kurtosis[list] - 3`) for the values in `list`.

The (unbiased) **sample kurtosis coefficient** is given by $g_2 = \frac{k_4}{s^4}$, where

$$k_4 = \frac{1}{(n-2)(n-3)}\left[\frac{n(n+1)}{n-1}\sum_{i=1}^{n}(y_i-\bar{y})^4 - 3\left(\sum_{i=1}^{n}(y_i-\bar{y})^2\right)^2\right]$$

and s is the sample standard deviation. This measure, which is negative for platykurtic distributions, zero for mesokurtic distributions, and positive for leptokurtic distributions, is like the kurtosis excess as defined by the *Mathematica* programmers for a population. There is no package function to calculate the sample kurtosis coefficient, so we define our own function `sampleKurtosis`.

```
sampleKurtosis[list_]:=Module[{n,one,two,k4},
        n=Length[list];
        one=n(n+1)/(n-1)Sum[(list[[i]]-
        Mean[list])^4,{i,1,n}];
        two=3Sum[(list[[i]]-Mean[list])^2,{i,1,n}]^2;
        k4=(one-two)/((n-2)(n-3));
        k4/StandardDeviation[list]^4]/; VectorQ[list]
```

A kurtosis measure based on octiles can be calculated using the formula $\frac{(O_7 - O_5) + (O_3 - O_1)}{O_6 - O_2}$ where O_k is the kth octile. This measure may range from zero, for extreme platykurtosis, to 1.233, for mesokurtosis, to infinity, for extreme leptokurtosis. *Mathematica* does not contain a function for finding this measure, so we must define one. Because O_1 is the same as the 12.5th percentile, O_2 the same as the 25th percentile, O_3 the same as the 37.5th percentile, etc., we can use `Quantile` in the definition of `quantileKurtosis`.

```
quantileKurtosis[list_]:=Module[{one,two,three},
          one=Quantile[list,0.875]-Quantile[list,0.625];
          two=Quantile[list,0.375]-Quantile[list,0.125];
          three=Quantile[list,0.75]-Quantile[list,0.25];
          (one+two)/three]/; VectorQ[list]
```

Several shape measures are reported simultaneously using `ShapeReport`, which is contained in the **DescriptiveStatistics** package.

> `ShapeReport[list]` gives the `Skewness`, `QuartileSkewness`, and `KurtosisExcess` shape statistics for the values in `list`.

In a manner similar to those in earlier sections, the code of `ShapeReport` can be modified to specify the statistics one would like to have calculated simultaneously.

```
ShapeReport[list_]:=
        {
        Skewness -> Skewness[list] ,
        QuartileSkewness -> QuartileSkewness[list] ,
        KurtosisExcess -> KurtosisExcess[list]
        }
```

EXAMPLE 1: Using the data in Table 2.1, compute the (a) second, third, and fourth, central moments, (b) coefficient of skewness, (c) Pearson's first coefficient of skewness, (d) Pearson's second coefficient of skewness, (e) quartile coefficient of skewness, (f) kurtosis coefficient, (g) kurtosis excess, (h) sample kurtosis coefficient, and (i) quantile coefficient of kurtosis.

SOLUTION: (a) If necessary, we load the DescriptiveStatistics package and enter the data in `braves`. Instead of finding the moments one at the time, we can use `Table` to find that they are 5.0615, 10.5736, and 55.2475.

```
Table[CentralMoment[braves, r],{r,2,4}]
  -16
{5.06315, 10.5736, 55.2475}
```

(b) The skewness coefficient is 0.928089. (c) Pearson's first coefficient of skewness is 2.26756. (d) Pearson's second coefficient of skewness is 1.78253, and (e) the quartile skewness is 0.830258. Each of these measures indicates positive skewness (skewed to the right).

```
Skewness[braves]
0.928089

PearsonSkewness1[braves]
2.26756

PearsonSkewness2[braves]
1.78253

QuartileSkewness[braves]
0.830258
```

(f) The kurtosis coefficient is 2.15512, (g) the kurtosis excess is –0.844884, (h) the sample kurtosis coefficient is –0.761407, and (i) the quantile coefficient of kurtosis is 0.991021. The coefficients suggest a platykurtic distribution.

```
Kurtosis[braves]
2.15512

KurtosisExcess[braves]
-0.844884
```

```
sampleKurtosis[list_]:=Module[{n,one,two,k4},
   n=Length[list];
   one=n(n+1)/(n-1)Sum[(list[[i]]-
   Mean[list])^4,{i,1,n}];
   two=3Sum[(list[[i]]-Mean[list])^2,{i,1,n}]^2;
   k4=(one-two)/((n-2)(n-3));
   k4/StandardDeviation[list]^4]/; VectorQ[list]
```

```
sampleKurtosis[braves]
-0.761407
```

```
quantileKurtosis[list_]:=Module[{one,two,three},
      one=Quantile[list,0.875]-Quantile[list,0.625];
      two=Quantile[list,0.375]-Quantile[list,0.125];
      three=Quantile[list,0.75]-Quantile[list,0.25];
      (one+two)/three]/; VectorQ[list]
```

```
quantileKurtosis[braves]
0.991021
```

Several shape statistics can be calculated at once with the command `ShapeReport`. Of course, we can define our own command to calculate any combination of statistics.

```
ShapeReport[braves]
{Skewness -> 0.928089, QuartileSkewness -> 0.830258,
 KurtosisExcess -> -0.844884}
```

■

2.4 Coding

Coding refers to the conversion of the original values by a simple arithmetic operation. Generally, coding employs a linear transformation such as adding (or subtracting) a constant or multiplying (or dividing) by a constant. When the data are coded by subtracting the mean and dividing by the standard deviation, the result is a list of **standardized values**. The standardized values have mean 0 and standard deviation (and variance) 1. **DescriptiveStatistics** contains functions for this type of coding.

`ZeroMean[list]` subtracts the mean from the values in `list` and returns values with zero mean.

`Standardize[list]` standardizes the values in `list` and returns values with zero mean and unit variance.

`Standardize[list, MLE -> True]` uses the population variance (maximum likelihood estimate of the variance). If `MLE -> False` (the default setting) is used, the sample variance (unbiased estimate) is used.

EXAMPLE 1: Using the data in Table 2.1, (a) subtract the mean from each value in the list, and (b) standardize the data. (Use both the unbiased estimate of the variance and the maximum likelihood estimate.)

SOLUTION: (a) Assuming that the package DescriptiveStatistics has been loaded and the data set entered in `braves`, the mean value 1.84346 is subtracted from each entry in `braves` with `ZeroMean`.

```
ZeroMean[braves]
{4.65654, 4.35654, 3.65654, 3.40654, 2.95654, 2.90654,
  2.35654, 1.65654, -0.343462, -0.443462, -1.09346,
 -1.26846, -1.34346, -1.38346, -1.49346, -1.52846,
 -1.55346, -1.70096, -1.70596, -1.70846, -1.71846,
 -1.73196, -1.73196, -1.73446, -1.73446, -1.73446}
```

(b) First, we standardize the data with an unbiased estimate of the variance. Notice that when we check that the mean of the new list `stan1` is zero, a small value close to zero is returned. The command `Chop`, which replaces values less than 10^{-10} with zero, returns the value of 0. Also, the standardized data set has variance 1.

```
stan1=Standardize[braves]
{2.02925, 1.89852, 1.59347, 1.48452, 1.28842, 1.26663,
 1.02695, 0.721895, -0.149676, -0.193254, -0.476515,
 -0.552777, -0.585461, -0.602892, -0.650829, -0.666081,
 -0.676976, -0.741254, -0.743433, -0.744523, -0.748881,
 -0.754764, -0.754764, -0.755853, -0.755853, -0.755853}

Mean[stan1]//Chop
0

Variance[stan1]
1.
```

Using the option `MLE->True`, we find again that the resulting values have mean approximately 0 and variance (using `VarianceMLE`) 1.

```
stan2=Standardize[braves,MLE->True]
{2.06944, 1.93611, 1.62502, 1.51392, 1.31393, 1.29171,
 1.04728, 0.736192, -0.15264, -0.197081, -0.485952,
 -0.563724, -0.597055, -0.614832, -0.663718, -0.679272,
 -0.690383, -0.755934, -0.758156, -0.759267, -0.763711,
 -0.769711, -0.769711, -0.770822, -0.770822, -0.770822}
```

```
Mean[stan2]//Chop
0
```

```
VarianceMLE[stan2]
1.
```

■

2.5 Expected Value Function

Given a list of n values $\{y_1, y_2, \ldots, y_n\}$, the arithmetic average is

$$\frac{1}{n}\sum_{i=1}^{n} y_i.$$

Given a function $f(y)$, with values $f(y_i)$, the mean or expected value of the function is defined by

$$\frac{1}{n}\sum_{i=1}^{n} f(y_i).$$

The **DescriptiveStatistics** package contains the command `ExpectedValue` that can be used to calculate the mean or expected value of a function.

`ExpectedValue[f, list]` computes the expected value of the pure function `f` with respect to the sample distribution of `list`.
`ExpectedValue[f, list, y]` yields the expected value of the function `f[y]` with respect to the sample distribution of `list`.

EXAMPLE 1: Given the data in Table 2.1, use the `ExpectedValue` command to compute (a) the arithmetic mean, (b) geometric mean, (c) harmonic mean, and (d) quadratic mean.

SOLUTION: (a) Assuming that the package **DescriptiveStatistics** has been loaded and the data set entered in `braves`, we let $f(y) = y$. Then the expected value of f with respect to the data set in `braves` is

$$\frac{1}{n}\sum_{i=1}^{n} f(y_i) = \frac{1}{n}\sum_{i=1}^{n} y_i,$$

the mean of the data.

```
f[y_]:=y
ExpectedValue[f[y],braves,y]
```

```
1.84346
```

The same result is obtained through the use of a pure function.

```
Clear[f]
f=Function[y,y];
ExpectedValue[f,braves]
```

```
1.84346
```

(b) With $f(y) = \ln y$, we use `ExpectedValue` to calculate `logm`, the mean of the $\ln y$ values. To find the geometric mean, we take the antilog of `logm` using the built-in function `Exp`, the exponential function. We verify our result with `GeometricMean`.

```
Clear[f]
f[y_]:=Log[y]
logm=ExpectedValue[f[y],braves,y]
```

```
-0.430342
```

```
Exp[logm]
```

```
0.650286
```

```
GeometricMean[braves]
```

```
0.650286
```

(c) In this case, we define $f(y) = \frac{1}{y}$ and calculate the mean of the reciprocals, `mrecip`. Evaluating the reciprocal of `mrecip` gives the harmonic mean, which we verify with `HarmonicMean`.

```
Clear[f]
f[y_]:=1/y
mrecip=ExpectedValue[f[y],braves,y]

3.64982

1/mrecip
0.273986

HarmonicMean[braves]
0.273986
```

(d) After defining $f(y) = y^2$, we find the mean of the squares, `msquare`. Using the built-in function `Sqrt`, we determine that the quadratic mean is 2.90887. We check our result with `RootMeanSquare`.

```
Clear[f]
f[y_]:=y^2
msquare=ExpectedValue[f[y],braves,y]

8.4615

Sqrt[msquare]
2.90887

RootMeanSquare[braves]
2.90887
```

■

CHAPTER 

Multivariate Methods for Describing Data

Throughout this chapter we will take advantage of the **MultiDescriptiveStatistics** package that is located in the **Statistics** folder (or directory). We use the **Help Browser** to obtain information about the commands contained in the package.

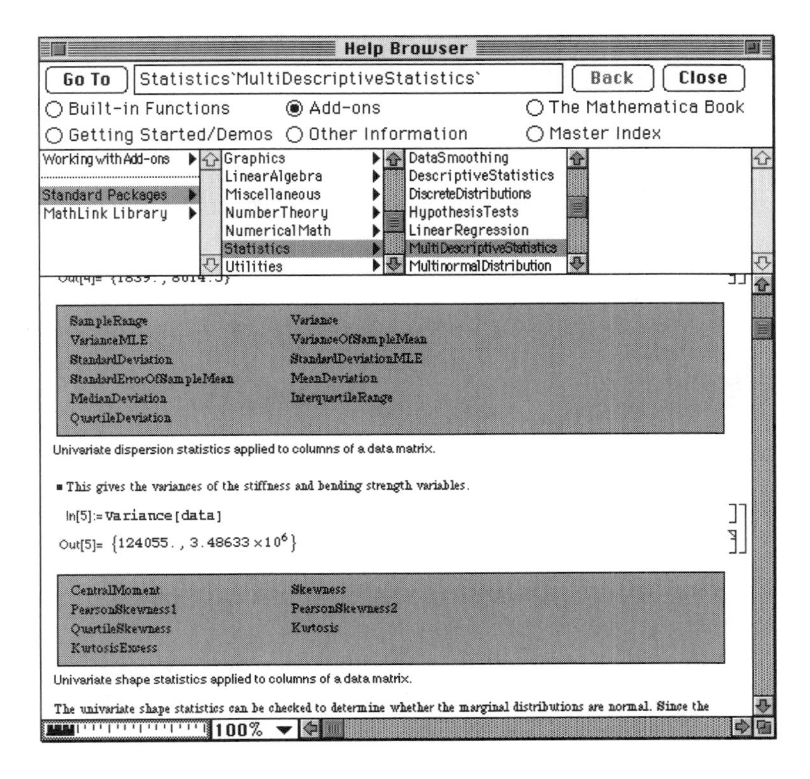

Often, several different variables are measured on the same experimental unit. For example, in the first data set in this chapter, for each team (experimental unit) in the Central Division of the National Basketball Association, values of the following four variables were recorded: winning percentage, points scored per game, opponents' points scored per game, and rebounding percentage. In Chapter 2, we discussed methods which were appropriate for describing one variable at a time. These are called **univariate methods**. Methods which include two variables simultaneously are called **bivariate methods**, and generally, when two or more variables are evaluated simultaneously, the methods are called **multivariate methods**. The *Mathematica* package **MultiDescriptiveStatistics** contains commands useful in analyzing multivariate data. In particular, it contains procedures for describing location, dispersion, and shape, as well as the interdependence between the variables of the multivariate data. We begin by discussing the commands available in **MultiDescriptiveStatistics** to extend the univariate methods that were presented in Chapter 2.

3.1 Extension of Univariate Methods

In the examples in Chapter 1 and Chapter 2, data are stored in arrays where a variable is represented by one or more columns and each experimental unit is represented by a row, called a record. With this in mind, it is straightforward to generalize to a data matrix. For an $n \times p$ matrix, the rows correspond to n experimental units, and the columns correspond to p variables. A row may be considered to be a vector containing a value of each variable, and, as such, each row may be called a **multivariate observation**. Thus, we may represent our data as the $n \times p$ matrix, $\mathbf{Y}$,

$$\mathbf{Y} = \begin{pmatrix} \mathbf{y}_1 \\ \mathbf{y}_2 \\ \vdots \\ \mathbf{y}_i \\ \vdots \\ \mathbf{y}_n \end{pmatrix} = \begin{pmatrix} y_{11} & y_{12} & \cdots & y_{1j} & \cdots & y_{1p} \\ y_{21} & y_{22} & \cdots & y_{2j} & \cdots & y_{2p} \\ \vdots & \vdots & & \vdots & & \vdots \\ y_{i1} & & \cdots & y_{ij} & \cdots & y_{ip} \\ \vdots & \vdots & & \vdots & & \vdots \\ y_{n1} & & \cdots & y_{nj} & \cdots & y_{np} \end{pmatrix},$$

where the n multivariate observations are denoted by the vectors $\mathbf{y}_1, \mathbf{y}_2, \ldots, \mathbf{y}_n$ with $\mathbf{y}i = (\mathbf{y}_{i1}\ \mathbf{y}_{i2} \cdots \mathbf{y}_{ip})$.

Note: In an effort to be consistent with the *Mathematica* convention of entering data in rows, we display vectors as rows, although it may be more conventional to display a vector as a column and its transpose as a row.

Measures of Location

In Chapter 2 we discussed the calculation of various measures of location for one variable at a time. The procedures discussed here calculate the measure of location for all p variables. We would obtain, for example, a vector of sample means, one for each of the p variables. The **sample mean vector** $\bar{\mathbf{y}}$ is given by

$$\bar{\mathbf{y}} = (\bar{y}_1 \bar{y}_2 \ldots \bar{y}_p)$$

where

$$\bar{y}_j = \frac{1}{n}\sum_{i=1}^{n} y_{ij}$$

is the mean of the jth column of $\mathbf{Y}$.

The **MultiDescriptiveStatistics** package contains the following commands to compute measures of location for the columns of a matrix. These are simple extensions of the univariate measures discussed in Chapter 2.

`Mean[matrix]` yields the mean vector by calculating the mean of each column in `matrix`.

`Median[matrix]` computes the median of each column in `matrix`.

`Mode[matrix]` gives the mode(s) of each column in `matrix`.

`GeometricMean[matrix]` gives the geometric mean of each column in `matrix`.

`HarmonicMean[matrix]` computes the harmonic mean of each column in `matrix`.

`RootMeanSquare[matrix]` calculates the root mean square quadratic mean of each column in `matrix`.

`TrimmedMean[matrix, f]` yields the mean of each column in `matrix`, with a fraction `f` of the entries at each end "trimmed." `TrimmedMean[matrix, {f1, f2}]` finds the mean with fractions `f1` and `f2` trimmed off the beginning and the end of each column, respectively.

`Quantile[matrix, q]` gives the `q`th quantile of each column in `matrix`.

`InterpolatedQuantile[matrix, q]` calculates the `q`th interpolated quantile of each column in `matrix`.

`Quartiles[matrix]` calculates the quartiles of each column in `matrix`.

EXAMPLE 1: The 1995-96 year end statistics for winning percentage, points scored per game, opponents' points scored per game, and rebounding percentage for the teams in the Central Division of the National Basketball Association are given in Table 3.1. Use this information to calculate location statistics of each column.

Team	Winning pct.	Points/game	Opponent points/game	Rebound pct.
Chicago	0.878	105.2	92.9	0.540
Indiana	0.634	99.3	96.1	0.517
Cleveland	0.573	91.1	88.5	0.490
Atlanta	0.561	98.3	97.1	0.499
Detroit	0.561	95.4	92.9	0.507
Charlotte	0.500	102.8	103.4	0.495
Milwaukee	0.305	95.5	100.9	0.487
Toronto	0.256	97.5	105.0	0.494

Table 3.1.

SOLUTION: We begin by entering the **MultiDescriptiveStatistics** package and defining the matrix `nbastats`. Notice that each row in the table represents a multivariate observation of four variables and is entered as a row vector into `nbastats`.

```
<<Statistics`MultiDescriptiveStatistics`
Clear[nbastats]

nbastats={{0.878, 105.2, 92.9, 0.54},{0.634, 99.3, 96.1,
  0.517},{0.573, 91.1, 88.5, 0.49},{0.561, 98.3, 97.1,
  0.499},{0.561, 95.4, 92.9, 0.507},{0.5, 102.8, 103.4,
  0.495},{0.305, 95.5, 100.9, 0.487},{0.256, 97.5, 105.,
  0.494}}

MatrixForm[nbastats]

0.878 105.2 92.9 0.54
0.634 99.3 96.1 0.517
0.573 91.1 88.5 0.49
0.561 98.3 97.1 0.499
0.561 95.4 92.9 0.507
0.5 102.8 103.4 0.495
0.305 95.5 100.9 0.487
0.256 97.5 105. 0.494
```

The mean vector is (0.5335, 98.1375, 97.1, 0.503625).

```
Mean[nbastats]
{0.5335, 98.1375, 97.1, 0.503625}
```

The median of each column is calculated with `Median`.

```
Median[nbastats]
{0.561, 97.9, 96.6, 0.497}
```

The geometric mean of the elements of each column of `nbastats` is given in the vector (0.500204, 98.0502, 96.9538, 0.503363). We also find the harmonic mean and root mean square vectors.

```
GeometricMean[nbastats]
{0.500204, 98.0502, 96.9538, 0.503363}
HarmonicMean[nbastats]
{0.464807, 97.963, 96.8075, 0.503107}
RootMeanSquare[nbastats]
{0.56338, 98.225, 97.2459, 0.503893}
```

■

Measures of Dispersion

The **MultiDescriptiveStatistics** package contains the following commands to compute measures of dispersion for the columns of a matrix. These are simple extensions of the univariate measures discussed in Chapter 2.

`SampleRange[matrix]` gives the range of each column in `matrix`.

`Variance[matrix]` computes an unbiased estimate of variance for each column in `matrix`.

`VarianceMLE[matrix]` calculates the maximum likelihood estimate of variance for each column in `matrix`.

`VarianceOfSampleMean[matrix]` returns an unbiased estimate of the variance of the sample mean of each column in `matrix`.

`StandardDeviation[matrix]` gives an estimate of the standard deviation of each column in `matrix`.

`StandardDeviationMLE[matrix]` returns the maximum likelihood estimate of the standard deviation of each column in `matrix`.

`StandardErrorOfSampleMean[matrix]` computes an estimate of the standard error of the sample mean for each column in `matrix`.

`MeanDeviation[matrix]` calculates the mean absolute deviation about the mean for each column in `matrix`.

`MedianDeviation[matrix]` yields the median absolute deviation about the median for each column in `matrix`.

`InterquartileRange[matrix]` returns the difference between the upper and lower quartiles for each column in `matrix`.

`QuartileDeviation[matrix]` gives the quartile deviation for each column in `matrix`.

EXAMPLE 2: Use the information from Table 3.1 to calculate measures of dispersion for each column.

SOLUTION: Loading the package **MultiDescriptiveStatistics** and defining the matrix `nbastats` are demonstrated in Example 1. The commands and corresponding vector output for various measures of dispersion are as follows.

```
SampleRange[nbastats]
{0.622, 14.1, 16.5, 0.053}

Variance[nbastats]
{0.0374563, 19.6255, 32.3971, 0.000309125}

VarianceMLE[nbastats]
{0.0327742, 17.1723, 28.3475, 0.000270484}

VarianceOfSampleMean[nbastats]
{0.00468204, 2.45319, 4.04964, 0.0000386406}

StandardDeviation[nbastats]
{0.193536, 4.43007, 5.69185, 0.017582}

StandardDeviationMLE[nbastats]
{0.181037, 4.14395, 5.32424, 0.0164464}

StandardErrorOfSampleMean[nbastats]
{0.0684254, 1.56627, 2.01237, 0.00621616}

MeanDeviation[nbastats]
{0.134875, 3.2625, 4.5, 0.0132813}

MedianDeviation[nbastats]
{0.067, 2.45, 4., 0.0085}
```

```
InterquartileRange[nbastats]
{0.201, 5.6, 9.25, 0.02}
QuartileDeviation[nbastats]
{0.1005, 2.8, 4.625, 0.01}
```

■

Measures of Shape

The **MultiDescriptiveStatistics** package contains the following commands to compute measures of shape for the columns of a matrix. These are simple extensions of the univariate measures discussed in Chapter 2.

`CentralMoment[matrix, r]` computes the `r`th population central moment for each column in `matrix`.

`Skewness[matrix]` gives the population coefficient of skewness for each column in `matrix`.

`PearsonSkewness1[matrix]` returns Pearson's first coefficient of skewness for each column in `matrix`.

`PearsonSkewness2[matrix]` calculates Pearson's second coefficient of skewness for each column in `matrix`.

`QuartileSkewness[matrix]` calculates the quartile coefficient of skewness for each column in `matrix`.

`Kurtosis[matrix]` calculates the population kurtosis coefficient for each column in `matrix`.

`KurtosisExcess[matrix]` calculates the population kurtosis coefficient, then subtracts 3 for each column in `matrix`.

EXAMPLE 3: Use the information from Table 3.1 to calculate measures of shape for each column.

SOLUTION: Loading the package **MultiDescriptiveStatistics** and defining the matrix `nbastats` are demonstrated in Example 1. The commands and corresponding vector output for various measures of shape are as follows. Note that with the command `PearsonSkewness1`, entries are returned as `{}` if all elements in the corresponding column are unique. In this case, all values for points scored per game and rebounding percentage are unique.

```
Skewness[nbastats]
{0.182577, 0.118638, 0.0105397, 1.17466}
PearsonSkewness1[nbastats]
{-0.426277, {}, 2.21369, {}}
PearsonSkewness2[nbastats]
{-0.426277, 0.160833, 0.263535, 1.13042}
QuartileSkewness[nbastats]
{-0.577114, 0.125, 0.2, 0.5}
Kurtosis[nbastats]
{2.65884, 2.33981, 1.83122, 3.26034}
```

■

3.2 Bivariate Measures of Association

Several techniques are available in *Mathematica* for evaluating relationships between two variables. If no relationship exists, they are said to be **independent**, whereas, if there is a relationship, the variables are said to be **associated**. **MultiDescriptiveStatistics** contains procedures for quantifying the degree or strength of the association between the values of a pair of variables. These procedures are used when both variables are quantitative. With categorical data, other measures of association can be employed as discussed in Chapter 12.

Covariance and Correlation

The **covariance** between two variables based on the set of paired values, $\{(x_1, y_1), (x_2, y_2), \ldots (x_n, y_n)\}$, where the values (x_i, y_i) are measured on the ith experimental unit, is calculated using

$$s_{xy} = \frac{1}{n-1}\sum_{i=1}^{n}(x_i - \bar{x})(y_i - \bar{y}) = \frac{1}{n-1}\left(\sum_{i=1}^{n} x_i y_i - \frac{\sum_{i=1}^{n} x_i \sum_{i=1}^{n} y_i}{n}\right).$$

When the variables are independent, we expect the covariance to be zero. A negative value for covariance reflects an inverse relationship where large values of one variable are associated with small values of the other. Positive covariance reflects a direct relationship where large values of one variable are associated with large values of the other variable. The magnitude of the covariance is a measure of the strength of the association.

A closely related measure of association is the **correlation coefficient**, also known as the **Pearson product-moment correlation coefficient**. It is denoted by r and may be calculated using the covariance and the respective standard deviations by $r = s_{xy}/s_x s_y$, or it may be calculated directly using

$$r = \frac{\sum_{i=1}^{n}(x_i - \bar{x})(y_i - \bar{y})}{\sqrt{\sum_{i=1}^{n}(x_i - \bar{x})^2 \sum_{i=1}^{n}(y_i - \bar{y})^2}} = \frac{\sum_{i=1}^{n} x_i y_i - \frac{\sum_{i=1}^{n} x_i \sum_{i=1}^{n} y_i}{n}}{\sqrt{\left(\sum_{i=1}^{n} x_i^2 - \frac{\left(\sum_{i=1}^{n} x_i\right)^2}{n}\right)\left(\sum_{i=1}^{n} y_i^2 - \frac{\left(\sum_{i=1}^{n} y_i\right)^2}{n}\right)}}.$$

The correlation coefficient measures the degree of linear association between the two variables. Note that $-1 \le r \le 1$, where $0 < r \le 1$ indicates positive correlation (large values of one variable are associated with large values of the other variable) and $-1 \le r < 0$ indicates negative correlation (large values of one variable are associated with a small values of the other variable). Values close to 1 or –1 indicate a high degree of linear relationship, and values close to zero indicate there is little linear association. When the variables have no linear relationship, we expect that $r = 0$.

The **MultiDescriptiveStatistics** package contains the following commands to compute covariance and correlation for bivariate data.

`Covariance[{x1, ...., xn}, {y1, ...., yn}]` gives the covariance between the `x` and `y` variables. Division by n–1 (rather than n) is used, giving an unbiased estimate of the population covariance. `Covariance[xlist, ylist,ScaleMethod -> method]` computes covariance using a measure of scale other than the default setting `StandardDeviation` (that is, `MeanDeviation`, `MedianDeviation`, `QuartileDeviation`); scale is standardized to give unity when the data is normal with zero mean and unit variance.

`CovarianceMLE[{x1, ...., xn},{y1, ...., yn}]` gives the population covariance between the `x` and `y` variables. Division by n (rather than $n-1$) is used. If the values are a sample, the result is a maximum likelihood estimate of the population covariance.

`Correlation[{x1, ...., xn}, {y1, ...., yn}]` gives the linear correlation coefficient between the `x` and `y` variables. `Correlation[xlist, ylist,ScaleMethod -> method]` computes correlation using a measure of scale other than `StandardDeviation` (that is, `MeanDeviation`, `MedianDeviation`, `QuartileDeviation`).

EXAMPLE 1: The batting average of opposing teams, the team batting average, the team earned run average (ERA), and the team winning percentage as of June 30, 1996, of the 14 teams in the National League are given in Table 3.2. (a) Compute the sample covariance between the winning percentage and the other three factors. (b) Compute the sample correlation between the winning percentage and the other three factors.

Team	Opponent batting average	Team batting average	Team earned run average	Winning percentage
Atlanta	0.240	0.276	3.33	0.625
Los Angeles	0.254	0.249	3.51	0.512
Florida	0.249	0.249	3.55	0.488
San Diego	0.245	0.260	3.65	0.524
Montreal	0.250	0.271	3.80	0.588
Chicago	0.252	0.241	4.20	0.475
St. Louis	0.254	0.269	4.22	0.513
New York	0.270	0.264	4.27	0.463
Houston	0.274	0.270	4.31	0.512
Philadelphia	0.264	0.240	4.48	0.405
Pittsburgh	0.280	0.259	4.53	0.450
Cincinnati	0.266	0.252	4.55	0.480
San Francisco	0.268	0.258	4.62	0.456
Colorado	0.286	0.293	5.86	0.506

Table 3.2

SOLUTION: After loading the **MultiDescriptiveStatistics**, we enter the four columns of data into the lists oppbavg (opponents' batting average), ownbavg (team batting average), teamera (team earned run average), and winpct (team winning percentage), respectively.

```
<<Statistics`MultiDescriptiveStatistics`

oppbavg={0.240,0.254,0.249,0.245,0.250,0.252,0.254,0.270,
0.274,0.264,0.280,0.266,0.268,0.286};

ownbavg={0.276,0.249,0.249,0.260,0.271,0.241,0.269,0.264,
0.270,0.240,0.259,0.252,0.258,0.293};

teamera={3.33,3.51,3.55,3.65,3.80,4.20,4.22,4.27,4.31,
4.48,4.53,4.55,4.62,5.86};

winpct={0.625,0.512,0.488,0.524,0.588,0.475,0.513,0.463,
0.512,0.405,0.450,0.480,0.456,0.506};
```

(a) We find that the covariance for the opponents' batting average and the team winning percentage is –0.000419495; the maximum likelihood estimate is –0.000389531.

```
Covariance[oppbavg,winpct]

-0.000419495

CovarianceMLE[oppbavg,winpct]

-0.000389531
```

The covariance for the team batting average and winning percentage is 0.000456181; the covariance obtained with the MeanDeviation scaling option is 0.000382815.

```
Covariance[ownbavg,winpct]

0.000456181

Covariance[ownbavg,winpct,ScaleMethod->MeanDeviation]

0.000382815
```

The covariance of the team ERA and winning percentage is –0.016711; with the QuartileDeviation scaling option it is –0.017053.

```
Covariance[teamera,winpct]

-0.016711

Covariance[teamera,winpct,ScaleMethod->
QuartileDeviation]

-0.017053
```

(b) The correlation for the opponent batting average and the winning percentage is –0.546816 and that for the team batting average and winning percentage is 0.564811.

```
Correlation[oppbavg,winpct]
-0.546816
Correlation[ownbavg,winpct]
0.564811
```

We see that the correlation for the team ERA and the winning percentage is –0.465908; using the `MedianDeviation` scaling option, we obtain –0.711771.

```
Correlation[teamera,winpct]
-0.465908
Correlation[teamera,winpct,ScaleMethod->MedianDeviation]
-0.711771
```

■

EXAMPLE 2: The text (ASCII) file **Census.txt** contains 738 records of person 18 years of age and older taken from the 1990 Decennial Census Public Use Microdata 5% Samples for the State of Georgia. (See Example 4 in Section 2.1 for a detailed description of this data set.) Calculate (a) the covariance between the ages and salaries; (b) the covariance between the salaries and usual hours worked; (c) the correlation between the ages and salaries; and (d) the correlation between the salaries and usual hours worked.

SOLUTION: As in Example 4 in Section 2.1, we first read the records contained in **Census.txt** into *Mathematica* and name the resulting list `censusdata`.

```
<<Statistics`MultiDescriptiveStatistics`
censusdata=ReadList["census.txt",Word];
```

The ages, salaries, and usual number of hours worked each week are then extracted from `censusdata` with `Map`, `StringTake`, and `ToExpression`.

```
ages=Map[StringTake[#,{6,7}]&,censusdata]//ToExpression;

salaries=Map[StringTake[#,{13,18}]&,censusdata]//
   ToExpression;

hours=Map[StringTake[#,{11,12}]&,censusdata]//
   ToExpression;
```

The covariance between the ages and salaries and the covariance between the salaries and usual hours worked are computed with `Covariance`; the correlation between the ages and salaries and the correlation between the salaries and usual hours worked are computed with `Correlation`.

```
Covariance[ages,salaries]//N
758683
Covariance[salaries,hours]//N
65527.7
Correlation[ages,salaries]//N
0.126394
Correlation[salaries,hours]//N
0.150373
```

■

Rank Correlation

The most common rank correlation coefficient of a set of data points is found by calculating the Pearson correlation coefficient on the ranks of those data points. It is called **Spearman's rank correlation coefficient** and denoted by r_s. Suppose that we have two variables x and y. Then the rank correlation coefficient is given by

$$r = \frac{\sum_{i=1}^{n}(u_i - \bar{u})(v_i - \bar{v})}{\sqrt{\sum_{i=1}^{n}(u_i - \bar{u})^2 \sum_{i=1}^{n}(v_i - \bar{v})^2}} = \frac{\sum_{i=1}^{n} u_i v_i - \frac{\sum_{i=1}^{n} u_i \sum_{i=1}^{n} v_i}{n}}{\sqrt{\left(\sum_{i=1}^{n} u_i^2 - \frac{\left(\sum_{i=1}^{n} u_i\right)^2}{n}\right)\left(\sum_{i=1}^{n} v_i^2 - \frac{\left(\sum_{i=1}^{n} v_i\right)^2}{n}\right)}},$$

where the values of u represent the ranks of the x values, and the values of v represent the ranks of the y values. The formula which appears in the *Mathematica* documentation is equivalent.

Another rank correlation coefficient is **Kendall's rank correlation coefficient**. It is also called Kendall's tau-b and is discussed along with other measures of association in Section 12.6

of this book. Except that the notation is different, the formula given in the *Mathematica* documentation is the same as that for Kendall's tau-b in Section 12.6.

The **MultiDescriptiveStatistics** package contains the following commands to compute rank correlation for two variables.

`SpearmanRankCorrelation[{x1, ...., xn}, {y1, ...., yn}]` calculates Spearman's rank correlation coefficient between the two lists of data in `{x1, ...., xn}` and `{y1, ...., yn}`.

`KendallRankCorrelation[{x1, ...., xn}, {y1, ...., yn}]` calculates Kendall's rank correlation coefficient between the two lists of data in `{x1, ...., xn}` and `{y1, ...., yn}`.

EXAMPLE 3: In Table 3.3, the test average (of three tests) and the final exam scores for 20 students in a calculus class are given. Determine Spearman's rank correlation coefficient and Kendall's rank correlation coefficient for these data.

Test average	Final exam score
55.667	47
84.667	94
91	85
75.667	89
66.667	72
79	82
74.667	82
65	77
80.333	80
91.667	95
71	72
49	61
41	52
88	93
79	84
79.667	91
61.333	73

Test average	Final exam score
56	35
61	46
72.667	88

Table 3.3

SOLUTION: After loading the **MultiDescriptiveStatistics** package, we enter the test averages in `testavg` and the final exam scores in `fscore`. We then check that these two lists contain the same number of entries. We use `Length` to see that each contains 20 scores. It is important that scores from the ith row of the table appear in the ith position of their respective lists. Because the two lists must be the same length, calculating the length provides a partial check for errors in data entry.

```
testavg={55.667,84.667,91,75.667,66.667,79,74.667,65,
80.333,91.667,71,49,41,88,79,79.667,61.333,56,61,72.667};

fscore={47,94,85,89,72,82,82,77,80,95,72,61,52,93,84,91,
73,35,46,88};

Length[testavg]

20

Length[fscore]

20
```

Spearman's rank correlation coefficient is 0.874294, whereas Kendall's rank correlation coefficient is found to be 0.684352.

```
SpearmanRankCorrelation[testavg,fscore]//N

0.874294

KendallRankCorrelation[testavg,fscore]//N

0.684352
```

■

EXAMPLE 4: Use the data given in Table 3.2 to find Spearman's rank correlation coefficient and Kendall's rank correlation coefficient for (a) the opponent's batting average and the team winning percentage; (b) the team batting average and the winning percentage; and (c) the team ERA and the winning percentage.

SOLUTION: Unless it has been done during the current session, we must first load the **MultiDescriptiveStatistics** package. After doing this, we enter the data in `oppbavg`, `ownbavg`, `teamera`, and `winpct`.

```
oppbavg={0.240,0.254,0.249,0.245,0.250,0.252,0.254,0.270,
0.274,0.264,0.280,0.266,0.268,0.286};

ownbavg={0.276,0.249,0.249,0.260,0.271,0.241,0.269,0.264,
0.270,0.240,0.259,0.252,0.258,0.293};

teamera={3.33,3.51,3.55,3.65,3.80,4.20,4.22,4.27,4.31,
4.48,4.53,4.55,4.62,5.86};

winpct={0.625,0.512,0.488,0.524,0.588,0.475,0.513,0.463,
0.512,0.405,0.450,0.480,0.456,0.506};
```

Next, we find Spearman's rank correlation coefficient for (a), (b), and (c).

```
SpearmanRankCorrelation[oppbavg,winpct]//N
-0.573788

SpearmanRankCorrelation[ownbavg,winpct]//N
0.615639

SpearmanRankCorrelation[teamera,winpct]//N
-0.620462
```

Finally, we find Kendall's rank correlation coefficient for (a), (b), and (c).

```
N[KendallRankCorrelation[oppbavg,winpct]]
-0.433333

KendallRankCorrelation[ownbavg,winpct]//N
0.433333

KendallRankCorrelation[teamera,winpct]//N
-0.441996
```

■

EXAMPLE 5: The text (ASCII) file **Hominoid.txt** is presented in Section 1.5 (Example 2), where the symbol `null` is used to represent missing data. Calculate the correlation between (i) the crown length and crown width; (ii) the crown width and enamel thickness; and (iii) the crown length and enamel thickness.

SOLUTION: We begin by loading the **MultiDescriptiveStatistics** and **DataManipulation** packages

```
<<Statistics`MultiDescriptiveStatistics`

<<Statistics`DataManipulation`
```

and then using `ReadList`to load the file **Hominoid.txt** into *Mathematica*, naming the resulting array `hominoiddata`. We include the option `WordSeparators->{"\t"}` to indicate that values are separated by tab marks. `ToExpression` is used to convert numerical entries from strings to numbers. The first five elements of `hominoiddata` are viewed with `Take`.

```
hominoiddata=ReadList["Hominoid.txt",Word,WordSeparators
->{"\t"},

   RecordLists->True]//ToExpression;

Take[hominoiddata,5]

{{Gg,1,0,26,14.6,16.1,0.9},{Gg,1,0,27,16.2,17.3,0.93},
 {Gg,1,0,18,17.3,16.4,null},{Gg,2,0,16,15.5,16.1,null},
 {Gg,3,0,17,15.6,16.8,0.83}}
```

The crown length, crown width, and average enamel thickness are contained in the fifth through seventh columns of `hominoiddata`, which are extraced from `hominoiddata` with `ColumnTake`. Records that contain nonnumeric entries (that is, `null`) are dropped with `DropNonNumeric`.

```
step1=ColumnTake[hominoiddata,{5,7}];

step2=DropNonNumeric[step1]

{{14.6,16.1,0.9},{16.2,17.3,0.93},{15.6,16.8,0.83},
 {16.2,13.6,0.89},{18.8,14.7,1.14},{18.1,14.8,1.05},
 {16.3,14.4,0.62},{18.2,15.5,1.01},{14.8,15.2,0.75},
 {15.1,14.9,0.83},{14.9,15.5,0.77},{14.9,15.8,0.81},
 {13.9,14.3,0.84},{16.3,14.6,0.96},{18.3,14.7,0.93},
 {16.5,14.6,0.73},{13.1,13.2,0.83},{12.2,13.9,1.13},
 {14.8,12.7,1.08},{15.6,13.5,0.98},{13.,12.3,0.94},
 {12.8,13.1,1.05},{11.7,13.5,1.19},{11.5,12.9,1.02},
 {11.2,13.1,1.07},{11.7,13.8,1.19},{13.5,12.3,0.93},
 {13.1,11.1,1.17},{14.,11.8,0.89},{14.8,12.9,1.36},
 {9.4,10.9,0.67},{8.9,11.1,1.01},{11.,10.9,0.78},
 {11.7,10.1,1.47},{9.8,9.6,1.63},{10.8,9.7,1.21},
 {10.7,11.8,1.45},{10.2,12.5,1.8},{9.9,12.6,1.15},
 {8.6,10.9,0.82},{8.7,11.,1.12},{10.3,9.6,0.92},
 {9.9,9.1,1.4},{11.4,9.7,1.3}}
```

To calculate the correlations, we extract the lengths, widths, and thicknesses from step2 with `Column`.

```
lengths=Column[step2,1];

widths=Column[step2,2];

thicknesses=Column[step2,3];
```

We then use `Correlation` to calculate the correlation between (i) the crown length and crown width; (ii) the crown width and enamel thickness; and (iii) the crown length and enamel thickness.

```
Correlation[lengths,widths]

0.758012

Correlation[lengths,thicknesses]

-0.373726

Correlation[widths,thicknesses]

-0.454025
```

■

3.3 Multivariate Methods with No Univariate Analogues

The multivariate extensions of the univariate descriptive procedures presented in Section 3.1 provide an easy-to-understand transition between univariate and multivariate methods. It is often the case, however, that more information is obtained for multivariate data using methods for which there are no one-dimensional analogues. Here, we discuss measures that require special definitions when working with multivariate data.

Multivariate Location Measures

The mean is probably the most widely used measure of location. As shown at the beginning of this chapter, the mean of a data matrix is a p-dimensional vector whose entries are the column means. When the columns of a data matrix are considered simultaneously, exactly the same mean vector is calculated. This equivalence does not hold for other measures of location such as the median or mode. The **MultiDescriptiveStatistics** package contains the following commands for these measures of location.

`SpatialMedian[matrix]` gives the multivariate median vector found by minimizing the sum of the Euclidean distances between the vector and rows of `matrix`.

`SimplexMedian[matrix]` gives the p-dimensional median as that vector which minimizes the sum of the volumes of the p-dimensional simplices formed with the vector and all possible combinations of p rows from `matrix`.

`ConvexHullMedian[matrix]` gives the p-dimensional median as the mean of the p-dimensional vectors lying on the innermost layer of the convex hull layers of the n p-dimensional points. (**Note:** This command fails if $p > 2$.)

`MultivariateTrimmedMean[matrix, f]` gives the mean of the p-dimensional vectors, with a fraction f of the most outlying vectors dropped. When `f` = 0, `MultivariateTrimmedMean` gives the mean. As `f` $\to$ 1.0, `MultivariateTrimmedMean` approaches the multivariate median given with `ConvexHullMedian`. (**Note:** This command fails if $p > 2$.)

`MultivariateMode[matrix]` gives the mode of the n p-dimensional vectors. A list of modes are returned if the data are bimodal, trimodal, or multimodal.

EXAMPLE 1: Consider the data in the first two columns in Table 3.1. (a) Determine the coordinate-wise mean vector, the coordinate-wise median vector, the median vector found by minimizing the Euclidean distances, the median vector found by minimizing the sum of the volumes of all possible simplices having the median vector as a vertex, and the median vector found by ordering the data based on the convex hull layer on which they lie, and taking the median to equal the mean of the data lying on the innermost layer. (b) Investigate the multivariate mode of this data. (c) Compute the 20%, 40%, 60%, and 80% trimmed means.

SOLUTION: (a) After loading the **MultiDescriptiveStatistics** package, we use `Drop` to eliminate the information contained in the last two columns of Table 3.1. We name this list `twocols`. Next, we find the spatial median, naming the result `sp`. We also find the simplex median, `sm`; the convex hull median, `cm`; the mean, `mn`; and the median, `med`.

```
<<Statistics`MultiDescriptiveStatistics`

twocols=Drop[Transpose[nbastats],-2]//Transpose

{{0.878,105.2},{0.634,99.3},{0.573,91.1},{0.561,98.3},
 {0.561,95.4},{0.5,102.8},{0.305,95.5},{0.256,97.5}}
```

```
sp=SpatialMedian[twocols]
```

```
{0.560978,98.3}
```

```
sm=SimplexMedian[twocols]
```

```
{0.54617,98.1378}
```

```
cm=ConvexHullMedian[twocols]
```

```
{0.585333,97.6667}
```

```
mn=Mean[twocols]
```

```
{0.5335,98.1375}
```

```
med=Median[twocols]
```

```
{0.561,97.9}
```

(b) There are no repeated ordered pairs in the list `twocols`, so the empty set is returned with `MultivariateMode`. To illustrate the use of this command, we enter another list, `newlist`, with repeated values and find the result of `MultivariateMode`.

```
MultivariateMode[twocols]
```

```
{}
```

```
newlist={{1,3},{1,2},{1,3},{1,3},{1,2},{1,1}};
MultivariateMode[newlist]
```

```
{1,3}
```

(c) We find that the 10% trimmed mean is {0.541793,98.0622}. We also find the range from no trimming (0%) to the convex hull median (100%) with a `Table` command using increments of 10%.

```
MultivariateTrimmedMean[twocols,0.1]
```

```
{0.541793,98.0622}
```

```
mult=Table[MultivariateTrimmedMean[twocols,0.1*i],
{i,0,10}]
```

```
{{0.5335,98.1375},{0.541793,98.0622},{0.550087,97.9868},
 {0.55838,97.9115},{0.566673,97.8362},{0.574967,
 97.7608},{0.58326,97.6855},{0.585333,97.6667},
 {0.585333,97.6667},{0.585333,97.6667},{0.585333,
 97.6667}}
```

■

The **MultiDescriptiveStatistics** package contains the following commands to calculate other multivariate location measures.

EllipsoidQuantile[{{y11, ..., y1p}, ..., {yn1, ..., ynp}},vq] gives the locus of the qth quantile of the p-variate data, where the data have been ordered using ellipsoids centered on Mean[{{y11, ...,vy1p}, ..., {yn1, ..., ynp}}]. The fraction of the data lying inside of this locus is q.

EllipsoidQuartiles[{{y11, ..., y1p}, ..., {yn1, ..., ynp}}] returns a list of the loci of the quartiles (q = 0.25, 0.50, 0.75) of the p-variate data, where the data have been ordered using ellipsoids centered on Mean[{{y11, ..., y1p}, ..., {yn1, ..., ynp}}].

PolytopeQuantile[{{y11, ..., y1p}, ..., {yn1, ..., ynp}}, q] yields the locus of the qth quantile of the p-variate data, where the data have been ordered using convex hulls centered on ConvexHullMedian[{{y11, ..., y1p}, ..., {yn1, ..., ynp}}]. The fraction of the data lying inside of this locus is q. (**Note:** This command may fail if $p > 2$.)

PolytopeQuartiles[{{y11, ..., y1p}, ..., {yn1, ..., ynp}}] returns a list of the loci of the quartiles (q = 0.25, 0.50, 0.75) of the p-variate data, where the data have been ordered using convex hulls centered on ConvexHullMedian[{{y11, ..., y1p}, ..., {yn1, ..., ynp}}]. (**Note:** This command may fail if $p > 2$.)

EXAMPLE 2: Consider the data in the first two columns of Table 3.1. (a) Graph the quartile contours of this data assuming that the distribution is elliptically contoured. (b) Graph the quartile contours using linear interpolation between convex layers of the data.

SOLUTION: Assuming that the **MultiDescriptiveStatistics** and **Graphics** packages and the data have been entered earlier, we begin by taking the transpose of twocols, naming the list {winpct,ppg}. We also find the minimum and maximum values in the lists winpct and ppg, where we name them {min1,max1} and {min2,max2}, respectively. Next, we use EllipsoidQuartiles to obtain a list, qrts, of the loci of the quartiles of our bivariate data. Finally, we use Show to plot the quartile contours.

```
{winpct,ppg}=Transpose[twocols]
```

```
{{0.878,0.634,0.573,0.561,0.561,0.5,0.305,0.256},
 {105.2,99.3,91.1,98.3,95.4,102.8,95.5,97.5}}
```

```
{{min1,max1},{min2,max2}}=Map[{Min[#],Max[#]}&,
 {winpct,ppg}]
```

```
{{0.256,0.878},{91.1,105.2}}
```

```
qrts=EllipsoidQuartiles[twocols]

{Ellipsoid[{0.5335,98.1375},{2.30273,0.0887251},
 {{-0.0205858,-0.999788},{-0.999788,0.0205858}}],
Ellipsoid[{0.5335,98.1375},{5.23531,0.201719},
 {{-0.0205858,-0.999788},{-0.999788,0.0205858}}],
Ellipsoid[{0.5335,98.1375},{6.89005,0.265477},
 {{-0.0205858,-0.999788},{-0.999788,0.0205858}}]}

Show[Graphics[qrts], Frame -> True, AspectRatio -> 1,
  PlotRange -> {{min1, max1}, {min2, max2}}]
```

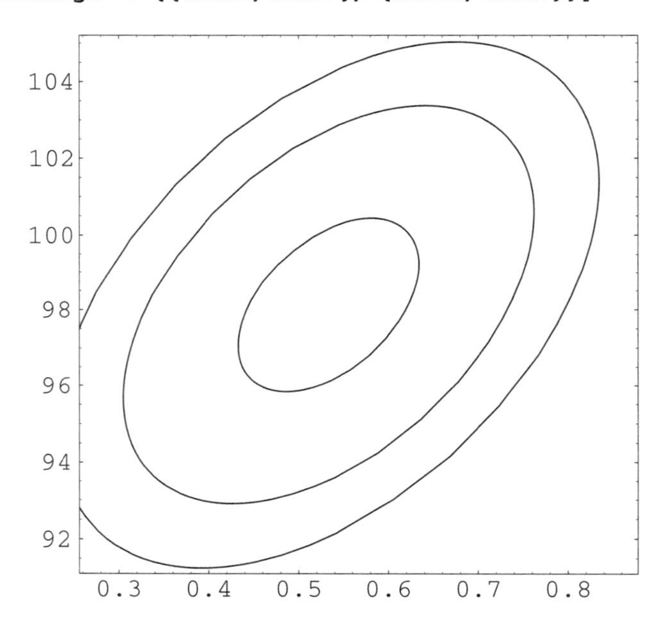

(b) By employing `PolytopeQuartiles`, we use linear interpolation between convex layers of the data in `twocols` to obtain a list of the loci of the quartiles. We name this list `qp` and then use `Show` to view the result.

```
qp=PolytopeQuartiles[twocols]

{Polytope[{{0.617778,98.7556},{0.569111,98.0889},
 {0.569111,96.1556}},-Connectivity-],
 Polytope[{{0.671801,99.8924},{0.665977,100.373},
 {0.557796,99.3232},{0.536601,98.935},{0.5,97.6235},
 {0.5098,97.0829},{0.553511,94.7024},
 {0.581252,95.4937}},-Connectivity],
```

```
Polytope[{{0.7749,102.546},{0.72993,102.52},
{0.528898,101.062},{0.487804,100.205},
{0.378,97.5617}, {0.4074,96.2914},{0.538532,93.3071},
{0.577126,93.2969}},-Connectivity-]}
```

```
Show[Graphics[qp], Frame ->True, AspectRatio ->1,
 PlotRange ->{{min1, max1}, {min2, max2}}]
```

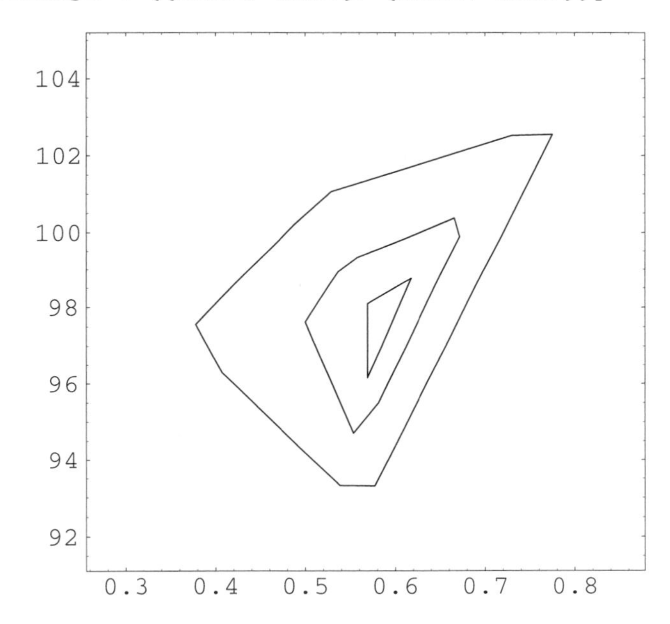

■

Multivariate Dispersion Measures

As we have seen, vector-valued dispersion measures may be obtained for multivariate data by a simple extension of the appropriate univariate measures. Measures of dispersion of p-variate data may also be in the form of a scalar (a single value) or a matrix. An important matrix measure of dispersion is the **covariance matrix**, also known as the **variance matrix** or the **variance–covariance matrix**. It is a $p \times p$ matrix $\mathbf{S}$ where the entries on the main diagonal are variances, and the off-diagonal entries are covariances.

$$\mathbf{S} = \begin{pmatrix} s_{11} & s_{12} & \cdots & s_{1p} \\ s_{21} & s_{22} & \cdots & s_{2p} \\ \vdots & \vdots & & \vdots \\ s_{p1} & s_{p2} & \cdots & s_{p} \end{pmatrix}.$$

For example, s_{22} is the variance for the variable represented by the second column of the data matrix, and s_{12} (as well as s_{21}) is the covariance for the variables represented by columns 1 and 2 of the data matrix.

Mathematica also provides the capability of calculating another form of the covariance matrix. The variables may be partitioned (i.e., separated) into two groups of size p and q, respectively, and two corresponding data matrices formed. One data matrix would be $n \times p$, and the second matrix would be $n \times q$. The covariance matrix for these two data matrices would be $n \times p$ of the form

$$\mathbf{S} = \begin{pmatrix} s_{11} & s_{12} & \cdots & s_{1q} \\ s_{21} & s_{22} & \cdots & s_{2q} \\ \vdots & \vdots & & \vdots \\ s_{p1} & s_{p2} & \cdots & s_{pq} \end{pmatrix}$$

and contain only covariances. For example, s_{11} is the covariance for variables represented by the first column of the $n \times p$ data matrix and the first column of the $n \times q$ data matrix, and s_{12} is the covariance for variables represented by the first column of the $n \times p$ data matrix and the second column of the $n \times q$ data matrix.

The **MultiDescriptiveStatistics** package contains the following commands that are used to find matrix-valued dispersion statistics.

`CovarianceMatrix[{{x11, ..., x1p}, {x21, ...,vx2p}, ..., {xn1, ..., xnp}}]` calculates the $p \times p$ covariance matrix for the data matrix formed from the n p-dimensional vectors. Division by $n - 1$ (rather than n) is used, giving unbiased estimates of the population variances and covariances. The option `ScaleMethod` has settings `StandardDeviation` (default setting), `MeanDeviation`, `MedianDeviation`, and `QuartileDeviation`, as discussed before with `Covariance` and `CovarianceMLE`. Commands are returned unevaluated if the setting `ScaleMethod->MedianDeviation` is included and does not yield a positive definite covariance matrix.

`CovarianceMatrix[{{x11, ..., x1p}, ..., {xn1, ..., xnp}},{{y11, ..., y1q}, ..., {yn1, ..., ynq}}]` yields the $p \times q$ covariance matrix between the n p-dimensional vectors and the n q-dimensional vectors giving unbiased estimates of the population covariances .

`CovarianceMatrixMLE[{{x11, ..., x1p}, {x21, ..., x2p}, ..., {xn1, ..., xnp}}]` calculates the $p \times p$ covariance matrix for the data matrix formed from the n p-dimensional vectors. Division by n (rather than $n-1$) is used, giving maximum likelihood estimates of the population variances and covariances.

`CovarianceMatrixMLE[{{x11, ..., x1p}, ..., {xn1, ..., xnp}},{{y11, ..., y1q}, ..., {yn1, ..., ynq}}]` yields the $p \times q$ covariance matrix between the n p-dimensional vectors and the n q-dimensional vectors giving maximum likelihood estimates of the population covariances.

`CovarianceMatrixOfSampleMean[{{x11, ..., x1p}, {x21, ... x2p}, ..., {xn1, ..., xnp}}]` gives the estimate of the covariance matrix of the sample mean vector $\bar{x}$. This is accomplished by dividing each entry in the covariance matrix output from `CovarianceMatrix` by n.

`DispersionMatrix[{{x11, ..., x1p}, {x21, ... x2p}, ..., {xn1, ..., xnp}}]` returns the estimate of the $p \times p$ dispersion matrix for data matrix formed from the n p-dimensional vectors using the p-variate points inside the convex hull of the n p-dimensional vectors. The result of this command is guaranteed to be a positive definite matrix. `DispersionMatrix` is a robust alternative to `CovarianceMatrix`. (**Note:** This command fails if $p > 2$.)

EXAMPLE 3: Consider the information in the first two columns of Table 3.1. Find the covariance matrix (unbiased estimate), the covariance matrix using the `MedianDeviation ScaleMethod` option setting, the maximum likelihood estimate of the covariance matrix, the dispersion matrix, and the covariance matrix of the sample mean.

SOLUTION: Assuming that the data in `twocols` are entered, we use `CovarianceMatrix` to find the unbiased estimate of the covariance matrix. Notice that the elements along the diagonal of the result equal the coordinate-wise variance of `twocols` (computed along the columns). Similarly, the off-diagonal elements equal the covariance of the winning percentage (`winpct`) and the points scored per game (`ppg`).

```
CovarianceMatrix[twocols]
```

```
{{0.0374563,0.403493},{0.403493,19.6255}}
```

```
Variance[twocols]
```

```
{0.0374563,19.6255}
```

```
{winpct,ppg}=Transpose[twocols]
```

```
{{0.878,0.634,0.573,0.561,0.561,0.5,0.305,0.256},
 {105.2,99.3,91.1,98.3,95.4,102.8,95.5,97.5}}
```

```
Covariance[winpct,ppg]
```

```
0.403493
```

Next, we find a covariance matrix using the `ScaleMethod->MedianDeviation` option setting. The result is a positive definite matrix. Otherwise, the command would be returned unevaluated. We also find the maximum likelihood estimate of the covariance matrix with `CovarianceMatrixMLE`, the dispersion matrix with `DispersionMatrix`, and the covariance matrix of the sample mean with `CovarianceMatrixOfSampleMean`.

```
CovarianceMatrix[twocols,ScaleMethod->MedianDeviation]
```

```
{{0.00986731,-0.00788876},{-0.00788876,13.1942}}
```

```
CovarianceMatrixMLE[twocols]
```

```
{{0.0327743,0.353056},{0.353056,17.1723}}
```

```
DispersionMatrix[twocols]
```

```
{{0.0146593,0.491989},{0.491989,33.8629}}
```

```
CovarianceMatrixOfSampleMean[twocols]
```

```
{{0.00468204,0.0504366},{0.0504366,2.45319}}
```

■

The **MultiDescriptiveStatistics** package also contains commands that are used to find scalar-valued measures of dispersion that consider all p variables simultaneously. Given the sample covariance matrix **S**, one way to measure this overall variance (or multivariate scatter) is through the **generalized variance**, found by taking the determinant of **S**.

Generalized Variance $= |\mathbf{S}|$

Another measure of the overall variability is called the **total variance**, which equals the trace of **S**, the sum of the diagonal elements of **S**.

Total Variance $= tr(\mathbf{S}) = s_{11} + s_{22} + \dots s_{pp}$.

The commands that compute these measures and several other scalar-valued multivariate dispersion statistics are as follows.

`GeneralizedVariance[{{x11, ..., x1p}, {x21, ... x2p}, ..., {xn1, ..., xnp}}]` gives the generalized variance of the n p-dimensional vectors. This is also equivalent to calculating the product of the variances of the principal components of the data.

`TotalVariation[{{x11, ..., x1p}, {x21, ... x2p}, ..., {xn1, ..., xnp}}]` gives the total variation of the n p-dimensional vectors. This is equivalent to calculating the sum of the variances of the principal components of the data.

`ConvexHullArea[data]` yields the area of the region enclosed by the convex hull of the bivariate data.

`MultivariateMeanDeviation[{{x11, ..., x1p}, {x21, ... x2p}, ..., {xn1, ..., xnp}}]` computes the scalar mean of the Euclidean distances between the p-variate mean and the n p-dimensional vectors.

`MultivariateMedianDeviation[{{x11, ..., x1p}, {x21, ... x2p}, ..., {xn1, ..., xnp}}]` calculates the scalar median of the Euclidean distances between the p-variate median and the n p-dimensional vectors. This command contains the option `MedianMethod` with settings `Median` (default setting), `SpatialMedian`, `SimplexMedian`, and `ConvexHullMedian`.

EXAMPLE 4: For the data in the first two columns in Table 3.1, calculate the generalized variance, the total variation, the multivariate mean deviation, and the multivariate median deviation.

SOLUTION: Assuming that the **MultiDescriptiveStatistics** package and the data in `twocols` have been loaded, we find that the generalized variance and the total variation are approximately 0.572293 and 19.663, respectively. To verify that the generalized variance equals the determinant of the covariance matrix, we find the covariance matrix, naming it `cmat`. We then compute the determinant of `cmat` with `Det`. We also show that the trace of `cmat`, given by `cmat[[1,1]] + cmat[[2,2]]`, equals the total variation.

```
twocols=Drop[Transpose[nbastats],-2]//Transpose

{{0.878,105.2},{0.634,99.3},{0.573,91.1},{0.561,98.3},
 {0.561,95.4},{0.5,102.8},{0.305,95.5},{0.256,97.5}}

GeneralizedVariance[twocols]

0.572293
```

```
TotalVariation[twocols]
19.663

cmat=CovarianceMatrix[twocols]
{{0.0374563,0.403493},{0.403493,19.6255}}

Det[cmat]
0.572293

cmat[[1,1]]+cmat[[2,2]]
19.663
```

We find that the area of the convex hull of the data in `twocols` is approximately 4.07995. The mean deviation is found to be approximately 3.27288. The median deviation is approximately 2.45681. With `MultivariateMedianDeviation`, we can use the option `MedianMethod` with settings `Median` (default setting), `SpatialMedian`, `SimplexMedian`, and `ConvexHullMedian`. We illustrate the use of the default setting `MedianMethod->Median` to find that the same result is obtained as was found with `MultivariateMedianDeviation` without a specified option setting. In addition, we find the value of `MultivariateMedianDeviation` with each of the four settings for `MedianMethod`.

```
ConvexHullArea[twocols]
4.07995

MultivariateMeanDeviation[twocols]
3.27288

MultivariateMedianDeviation[twocols]
2.45681

MultivariateMedianDeviation[twocols,
 MedianMethod->Median]
2.45681

Map[MultivariateMedianDeviation[twocols,
 MedianMethod->#]&,{Median,SpatialMedian,
 SimplexMedian,ConvexHullMedian}]
{2.45681,2.85579,2.69331,2.22576}
```

■

Multivariate Shape Measures

The **MultiDescriptiveStatistics** package contains several commands that compute multivariate shape measures which consider all variables of the data simultaneously.

`CentralMoment[mat,{r1,r2,...,rp}]` yields the rth central moment given by

$$\frac{1}{n}\sum_{i=1}^{n}(x_{i1}-\bar{x}_1)^{r_1}(x_{i2}-\bar{x}_2)^{r_2}\ldots(x_{ip}-\bar{x}_p)^{r_p},$$

where $r_1+r_2+\ldots+r_p=r$ and `mat` is an $n\times p$ matrix.

`MultivariatePearsonSkewness1[mat]` computes the multivariate Pearson's first coefficient of skewness, $9(\mu_{\text{mean}}-\mu_{\text{mode}})'\Sigma^{-1}(\mu_{\text{mean}}-\mu_{\text{mode}})$, where Σ is an unbiased estimate of the population covariance. (**Note:** This command should not be used if there are not many replications in `mat`.)

`MultivariatePearsonSkewness2[mat]` gives the multivariate Pearson's second coefficient of skewness, $9(\mu_{\text{mean}}-\mu_{\text{median}})'\Sigma^{-1}(\mu_{\text{mean}}-\mu_{\text{median}})$, where Σ is an unbiased estimate of the population covariance.

`MultivariateSkewness[mat]` calculates the multivariate skewness,

$$\frac{1}{n^2}\sum_i\sum_j((x_i-\bar{x})'\Sigma^{-1}(x_j-\bar{x}))^3,$$

where Σ is an maximum likelihood estimate of the population covariance.

`MultivariateKurtosis[mat]` yields the multivariate kurtosis,

$$\frac{1}{n}\sum_i((x_i-\bar{x})'\Sigma^{-1}(x_i-\bar{x}))^2,$$

where Σ is an maximum likelihood estimate of the population covariance.

`MultivariateKurtosisExcess[mat]` yields the multivariate kurtosis excess given by `MultivariateKurtosis[mat]` $-p(p+2)$.

EXAMPLE 5: Calculate the multivariate shape statistics for the data concerning teams in the National League of Major-League Baseball given in Table 3.2.

SOLUTION: Assuming that the **MultiDescriptiveStatistics** package is loaded, we enter the data, naming the data matrix `mat`. We then find several fourth moments.

```
oppbavg={0.240,0.254,0.249,0.245,0.250,0.252,0.254,
  0.270,0.274,0.264,0.280,0.266,0.268,0.286};
```

```
ownbavg={0.276,0.249,0.249,0.260,0.271,0.241,0.269,
  0.264,0.270,0.240,0.259,0.252,0.258,0.293};

teamera={3.33,3.51,3.55,3.65,3.80,4.20,4.22,4.27,4.31,
  4.48,4.53,4.55,4.62,5.86};

winpct={0.625,0.512,0.488,0.524,0.588,0.475,0.513,0.463,
  0.512,0.405,0.450,0.480,0.456,0.506};

mat={oppbavg,ownbavg,teamera,winpct}

CentralMoment[Transpose[mat],{1,1,1,1}]
```

3.59532×10^{-6}

```
CentralMoment[Transpose[mat],{0,0,0,4}]

0.000028551

CentralMoment[Transpose[mat],{0,2,2,0}]

0.00022901
```

When we elicit the use of `MultivariatePearsonSkewness1`, the command is returned unevaluated because of the lack of repeated values in `mat`. The command `MultivariatePearsonSkewness2` may be more appropriate, although we receive a message warning of a possible significant numerical error. We recieve a similar warning message when we use `MultivariateSkewness`, `MultivariateKurtosis`, and `MultivariateKurtosisExcess`.

```
MultivariatePearsonSkewness1[mat]

Inverse::luc:Result for Inverse of badly conditioned
matrix <<1>> may contain significant numerical errors.

MultivariatePearsonSkewness1[{{0.24,0.254,0.249,0.245,
0.25,0.252,0.254,0.27,0.274,0.264,0.28,0.266,0.268,
0.286},{0.276,0.249,0.249,0.26,0.271,0.241,0.269,0.264,
0.27,0.24,0.259,0.252,0.258,0.293},{3.33,3.51,3.55,
3.65,3.8,4.2,4.22,4.27,4.31,4.48,4.53,4.55,4.62,5.86},
{0.625,0.512,0.488,0.524,0.588,0.475,0.513,0.463,0.512,
0.405,0.45,0.48,0.456,0.506}}]

MultivariatePearsonSkewness2[mat]

Inverse::luc:Result for Inverse of badly conditioned
matrix <<1>> may contain significant numerical errors.
```

-2.76251×10^{12}

```
MultivariateSkewness[mat]

Inverse::luc:Result for Inverse of badly conditioned
matrix <<1>> may contain significant numerical errors.

-36.5576

MultivariateKurtosis[mat]

Inverse::luc:Result for Inverse of badly conditioned
matrix <<1>> may contain significant numerical errors.

21.4465

MultivariateKurtosisExcess[mat]

Inverse::luc:Result for Inverse of badly conditioned
matrix <<1>> may contain significant numerical errors.

-202.554
```

■

3.4 Multivariate Measures of Association

The **MultiDescriptiveStatistics** package includes several commands to calculate matrix-valued association measures. One such measure is the **correlation matrix**, a symmetric $p \times p$ matrix based on the Pearson correlation coefficient given by

$$\mathbf{R} = (r_{ij}) = \begin{pmatrix} 1 & r_{12} & \cdots & r_{1p} \\ r_{21} & 1 & \cdots & r_{2p} \\ \vdots & \vdots & & \vdots \\ r_{p1} & r_{p2} & \cdots & 1 \end{pmatrix}.$$

Note that the main diagonal contains ones, because the correlation coefficient for a column of the data matrix and itself is 1. Also, the matrix is symmetric, because the correlation between the ith and jth variables, is the same as the correlation between the jth and ith variables i. e. $r_{ij} = r_{ji}$. Finally, the matrix **R** can be obtained from the covariance matrix **S**. If we let

$$\mathbf{D}_s = \begin{pmatrix} s_1 & 0 & \dots & 0 \\ 0 & s_2 & \dots & 0 \\ \vdots & \vdots & & \vdots \\ 0 & 0 & \dots & s_p \end{pmatrix},$$

then

$$\mathbf{R} = \mathbf{D}_s^{-1}\mathbf{S}\mathbf{D}_s^{-1} \text{ and } \mathbf{S} = \mathbf{D}_s\mathbf{R}\mathbf{D}_s.$$

The commands that compute the correlation matrix and related multivariate measures of association are as follows.

`CorrelationMatrix[{{x11, ..., x1p}, {x21, ..., x2p}, ..., {xn1, ..., xnp}}]` gives the $p \times p$ correlation matrix of the n p-dimensional vectors. The option `ScaleMethod` has settings `StandardDeviation` (default setting), `MeanDeviation` `MedianDeviation`, and `QuartileDeviation`, as was the case with `CovarianceMatrix`.

`CorrelationMatrix[{{x11, ..., x1p}, ..., {xn1, ..., xnp}},{{y11, ..., y1q}, ..., {yn1, ..., ynq}}]` gives the $p \times q$ correlation matrix between the n p-dimensional vectors and the n q-dimensional vectors.

`AssociationMatrix[{{x11, ..., x1p}, {x21, ..., x2p}, ..., {xn1, ..., xnp}}]` gives the $p \times p$ association matrix of the n p-dimensional vectors inside the convex hull of the n p-dimensional vectors. `AssociationMatrix` is a robust alternative to `CorrelationMatrix`. (**Note:** This command fails if $p > 2$.)

EXAMPLE 1: Use the information given in Table 3.2 to find the correlation matrix for the opponents' batting average, the team batting average, the team ERA, and the team winning percentage. Also, compute the association matrix relating the team ERA and the team winning percentage.

SOLUTION: We assume that the **MultiDescriptiveStatistics** package is loaded, and using the four lists of data that we entered in Section 3.3, Example 5 (`oppbavg`, `ownbavg`, `teamera`, and `winpct`), we form a matrix `mat`.

```
mat={oppbavg,ownbavg,teamera,winpct};
```

However, we need for each observation to be represented by a column vector instead of a row vector. Therefore, we compute the correlation matrix for this data using the transpose of mat. Because there are four columns in the transpose of mat, these matrices are 4×4. Next, we view cor in the form of a matrix with MatrixForm.

```
cor=CorrelationMatrix[Transpose[mat]]

{{1.,0.277869,0.863743,-0.546816},{0.277869,1.,
 0.356205,0.564811},{0.863743,0.356205,1.,-0.465908},
 {-0.546816,0.564811,-0.465908,1.}}

cor // MatrixForm

MatrixForm =
```

$$\begin{pmatrix} 1. & 0.277869 & 0.863743 & -0.546816 \\ 0.277869 & 1. & 0.356205 & 0.564811 \\ 0.863743 & 0.356205 & 1. & -0.465908 \\ -0.546816 & 0.564811 & -0.465908 & 1. \end{pmatrix}$$

Because AssociationMatrix fails if $p > 2$, we compute the association matrix using only the columns teamera and winpct.

```
AssociationMatrix[Transpose[{teamera,winpct}]]

{{1.,-0.771478},{-0.771478,1.}}
```

■

3.5 Multivariate Data Transformations

The **MultiDescriptiveStatistics** package contains the following data transformation commands.

ZeroMean[matrix] for each column subtracts the mean of the column from the column entries, resulting in a matrix whose columns have zero means.

Standardize[matrix,Decorrelate->True] transforms the data in matrix so that the covariance matrix of the transformed matrix is the identity matrix.

PrincipalComponents[matrix] transforms the data in matrix so that the covariance matrix of the transformed matrix is a diagonal matrix with diagonal elements ordered from largest to smallest. (**Note:** This command fails if $p > 2$.)

EXAMPLE 1: Consider the data in Table 3.2 concerning teams in the National League of Major League Baseball. Calculate the covariance matrix for this data using (a) no option setting, (b) `ZeroMean`, and (c) `Standardize` with the `Decorrelate->True` option setting. Then, (d) consider only the information concerning team ERA and team winning percentage to find the covariance matrix using the transformation `PrincipalComponents`.

SOLUTION: Assume that the **MultiDescriptiveStatistics** package is loaded, and using the four lists of data that we entered in Section 3.3, Example 5 (`oppbavg`, `ownbavg`, `teamera`, and `winpct`), form a matrix `mat`. We then find the covariance matrix and notice that there is no difference in the result when the data are transformed with `ZeroMean`.

```
mat={oppbavg,ownbavg,teamera,winpct};

trans=Transpose[mat];

CovarianceMatrix[trans]

{{0.000189209,0.0000553516,0.00764088,-0.000419495},
 {0.0000553516,0.00020972,0.00331747,0.000456181},
 {0.00764088,0.00331747,0.413596,-0.016711},
 {-0.000419495,0.000456181,-0.016711,0.00311049}}

CovarianceMatrix[ZeroMean[trans]]

{{0.000189209,0.0000553516,0.00764088,-0.000419495},
 {0.0000553516,0.00020972,0.00331747,0.000456181},
 {0.00764088,0.00331747,0.413596,-0.016711},
 {-0.000419495,0.000456181,-0.016711,0.00311049}}
```

(c) When we use `Chop` with the following result named `cm`, we observe that the 4×4 identity matrix is obtained as expected.

```
cm=CovarianceMatrix[Standardize[trans,
   Decorrelate->True]]

{{1., -1.14972 × 10^-15, -1.62263 × 10^-16, -3.24527 × 10^-16},
 {-1.14972 × 10^-15, 1., 7.515336 × 10^-16, 9.7785 × 10^-16},
 {-1.62263 × 10^-16, 7.515336 × 10^-16, 1., -5.46571 × 10^-16},
 {-3.24527 × 10^-16, 9.7785 × 10^-16, -5.46571 × 10^-16, 1.}}

Chop[cm]

{{1.,0,0,0},{0,1.,0,0},{0,0,1.,0},{0,0,0,1.}}
```

When we consider only the information concerning team ERA and team winning percentage, we obtain a diagonal matrix (after we use `Chop`) with elements ordered from largest to smallest.

```
mat2={teamera,winpct}

cmtwo=CovarianceMatrix[PrincipalComponents
   [Transpose[mat2]]]
```

$\{\{0.414275,\ 4.27009 \times 10^{-18}\},$
$\{4.27009 \times 10^{-18},\ 0.0024313\}$

```
Chop[cmtwo]
```

```
{{0.414275,0},{0,0.0024313}}
```

■

CHAPTER 4

Tabular and Graphical Methods for Presenting Data

A variety of methods are available for organizing and presenting data. For example, a frequency distribution is a grouping of data into subsets showing the number of observations in each subset. It can be presented as a frequency table or graphically. With categorical data each subset or category may consist of a single attribute. For quantitative data, a subset or class may be a single value or a range of values. Some of the methods discussed are only used with quantitative data. Others can be used with both categorical and quantitative data. Useful *Mathematica* functions are contained in the standard *Mathematica* packages **DataManipulation**, **Graphics**, **Graphics3D**, and **Legend**. **DataManipulation** is found in the **Statistics** folder (or directory), and the other packages are in the **Graphics** folder (or directory).

4.1 Univariate Procedures

Bar Graphs

A bar graph may be used to summarize the values of a categorical variable or a discrete quantitative variable. Each category or class is represented by a bar whose height or area corresponds to frequency or relative frequency. The bar graph can be oriented so that the bars are either vertical or horizontal. The bars are usually separated from each other. (Histograms, which are bar graphs for continuous variables, are discussed in a later subsection.) We use the `BarChart` function, from the **Graphics** package located in the **Graphics** folder (or directory), to construct bar graphs. The **Help Browser** can be used to obtain detailed information about the `BarChart` function along with its options. This help feature also includes an example to demonstrate the syntax of the function.

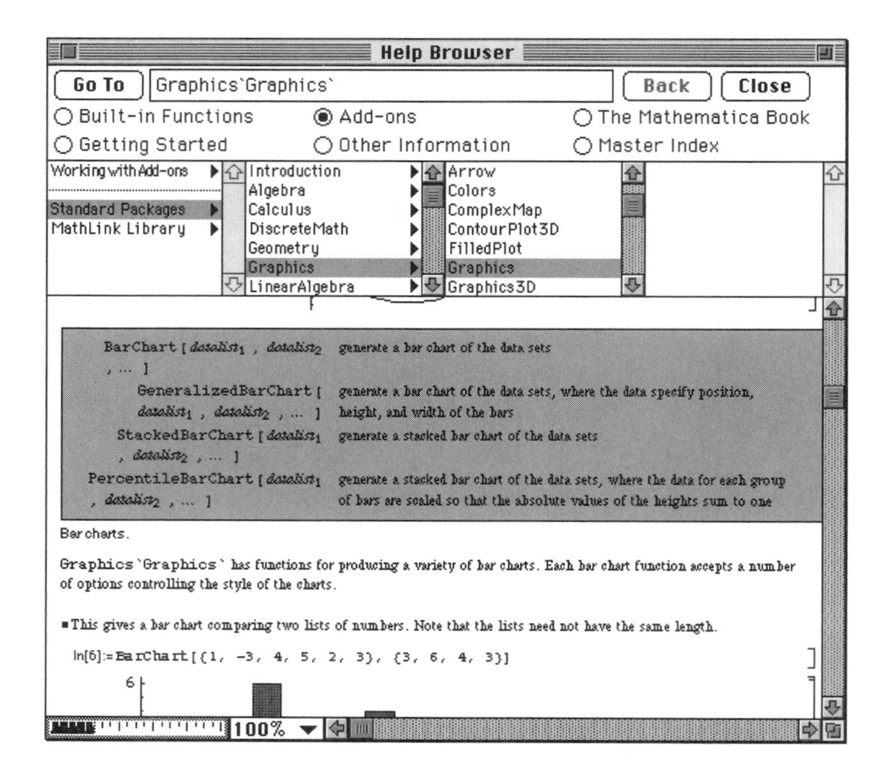

Graphics`Graphics` contains several other useful functions which will be used through this chapter.

EXAMPLE 1: Table 4.1 shows estimates of the number (i.e., frequency) of employees in various industries (i.e., categories) in southeast Georgia during the fourth quarters of 1993 and 1992. Make a bar chart to compare years for each category.

Average employment level	Fourth quarter 1993	Fourth quarter 1992
Agriculture and mining	6,242	5,828
Construction	18,865	18,531
Durable goods manufacturing	30,707	29,495
Nondurable goods manufacturing	49,846	51,353
Transportation and public utilities	21,595	22,028
Finance institutions and real estate	14,429	14,174
Trade	102,301	97,322
Services	87,368	84,352
Government	98,728	97,000

Source: Economic Indicators, *Southern Economic Developer: Southeast Georgia Economic Review*, Fall 1994, Volume 5, Number 1, pp. 12-15.

Table 4.1

SOLUTION: We first enter the data into *Mathematica*. (See Section 1.5, Example 1, for help to enter data.) The list of frequencies corresponding to the fourth quarter of 1992 is named `fourth92`; the list corresponding to the fourth quarter of 1993 is named `fourth93`.

```
fourth92={5828,18531,29495,51353,22028,14174,97322,
   84352,97000};

fourth93={6242,18865,30707,49846,21595,14429,102301,
   87368,98728};
```

After loading the **Graphics** package, we use `BarChart` to make bar charts representing the fourth quarter of 1992 and the fourth quarter of 1993. We then display both quarters on the same graph.

```
<< Graphics`Graphics`
```

```
BarChart[fourth92]
```

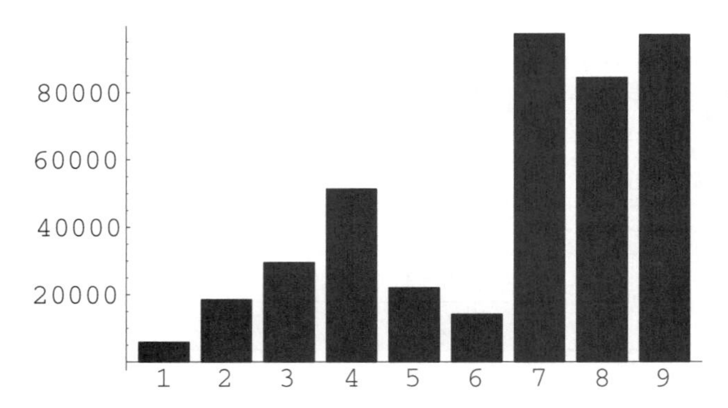


```
BarChart[fourth93]
```

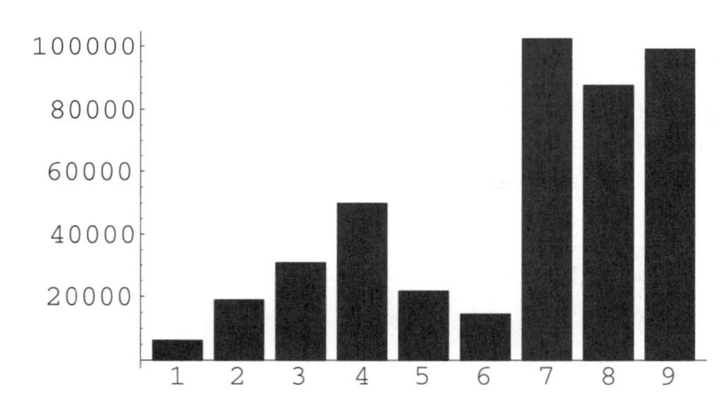


```
BarChart[fourth92, fourth93]
```

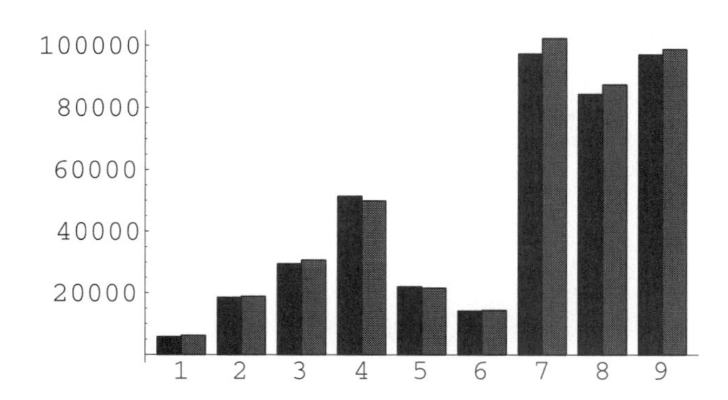

For the combined graph, on a color monitor, the bars corresponding to the data in `fourth92` appear in red, and the bars corresponding to the data in `fourth93` appear in blue.

If we include the option `BarOrientation->Horizontal`, we obtain a horizontal bar chart.

```
BarChart [fourth92, fourth93,
 BarOrientation -> Horizontal]
```

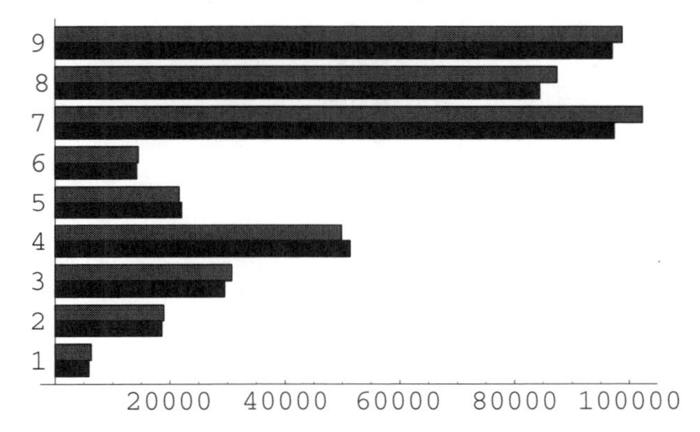

We illustrate the use of several other options in `b1`, `b2`, `b3`, and `b4`. All four graphs are displayed together using `Show` together with `GraphicsArray`. Notice that in `BarStyle`, `GrayLevel[0.3]` yields a darker shade of gray in shading the bars than does `GrayLevel[0.6]`; `BarLabels->list` labels the bars with the items given in `list`; and `BarSpacing->0.k` increases the spacing between the bars if `k` is positive whereas it decreases the spacing if `k` is negative. `BarGroupSpacing` indicates spacing between groups of bars. For both `BarSpacing` and `BarGroupSpacing` the value `0.k` specifies space as a fraction of the width of a bar.

```
b1=BarChart[fourth92,fourth93,BarStyle->{GrayLevel[0.3],
    GrayLevel[0.6]},
    DisplayFunction->Identity];
b2=BarChart[fourth92,fourth93,
    BarLabels->{"A","C","D","N","T","F","T","S","G"},
    DisplayFunction->Identity];
b3=BarChart[fourth92,fourth93,BarSpacing->0.2,
    BarEdges->None,
    DisplayFunction->Identity];
b4=BarChart[fourth92,fourth93,BarSpacing->
    -0.2,BarGroupSpacing->.5,
    DisplayFunction->Identity];
```

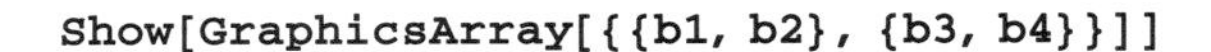

```
Show[GraphicsArray[{{b1, b2}, {b3, b4}}]]
```

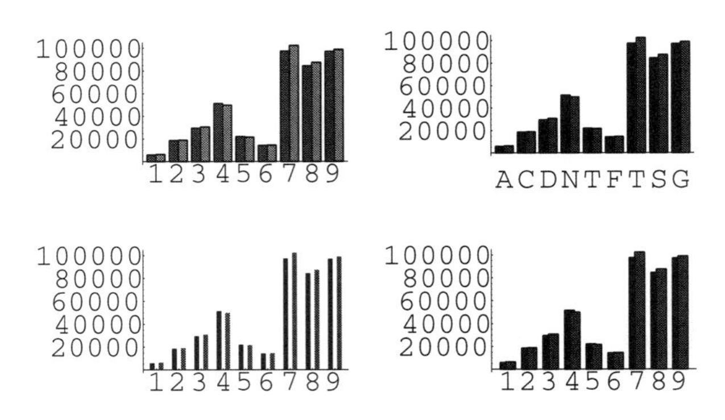

■

When looking at measurements repeated over time, as in Example 1, time would be called a grouping variable. In general, a set of measurements may be divided into two or more parts based on the values of the **grouping variable**. To include the grouping variable in the bar graph, in effect, a bar graph is generated for each value of the **grouping variable**, and these bar graphs are displayed together. Variations of this approach include stacked bars and percentile bars.

EXAMPLE 2: Table 4.2 contains estimates of the number (frequency) of employed persons 25 years of age and older for each of several occupation categories for the year 1994. The grouping variable is gender. Generate (a) a bar chart and (b) a stacked bar chart for these data.

Occupation	Women	Men
All occupational groups	47618	56523
Managerial and professional specialty	15110	16708
Executive, administrative, and managerial	6555	8883
Professional specialty occupations	8556	7825
Teachers, except college and university	2972	1002
Teachers, college and university	324	442
Technical, sales, and administrative support	19448	11161
Technicians and related support	1790	1661

Occupation	Women	Men
Sales occupations	5163	6346
Administrative support, including clerical	12495	3153
Service occupations	7594	4760
Precision production, craft, and repair	1128	10946
Operators, fabricators, and laborers	3724	10593
Farming, forestry, and fishing	614	2356

Source: 1995 Digest of Education Statistics

Table 4.2

SOLUTION: We first enter the data in the table into *Mathematica* and name the resulting lists `women` and `men`, respectively.

```
women={47618 ,15110 ,6555 ,8556 ,2972 ,324 ,19448 ,
 1790 ,5163 ,12495 ,7594 ,1128 ,3724 ,614 };

men={56523 ,16708 ,8883 ,7825 ,1002 ,442 ,11161,
 1661 ,6346,3153,4760,10946,10593,2356};
```

After loading the **Graphics** package, we use `?` to obtain basic information about the `StackedBarChart` function, which we use to generate stacked bar charts. Note that `StackedBarChart` has the same options as `BarChart`.

```
<<Graphics`Graphics`

?StackedBarChart

StackedBarChart[list1, list2, ...] generates a stacked
  bar chart of the data in the lists."
```

`BarChart` and `StackedBarChart` are used to generate a bar chart and stacked bar chart, with the results named `sb1` and `sb2`, respectively. We illustrate several of the options for the `StackedBarChart` command in `sb3` and `sb4`. No graphs are displayed because we include `DisplayFunction->Identity` in each command. Finally, all four graphics are displayed together as a graphics array using `Show` together with `GraphicsArray`. Note that `PlotRange->All` is needed to view the full range of values on the vertical axis in these bar graphs.

```
sb1=BarChart[women,men,DisplayFunction->Identity];

sb2=StackedBarChart[women,men,PlotRange->All,
  DisplayFunction->Identity];
```

```
sb3=StackedBarChart[men,women,PlotRange->All,
  DisplayFunction->Identity];
sb4=StackedBarChart[men,women,PlotRange->All,
  DisplayFunction->Identity,BarOrientation->Vertical];
Show[GraphicsArray[{{sb1,sb2},{sb3,sb4}}]]
```

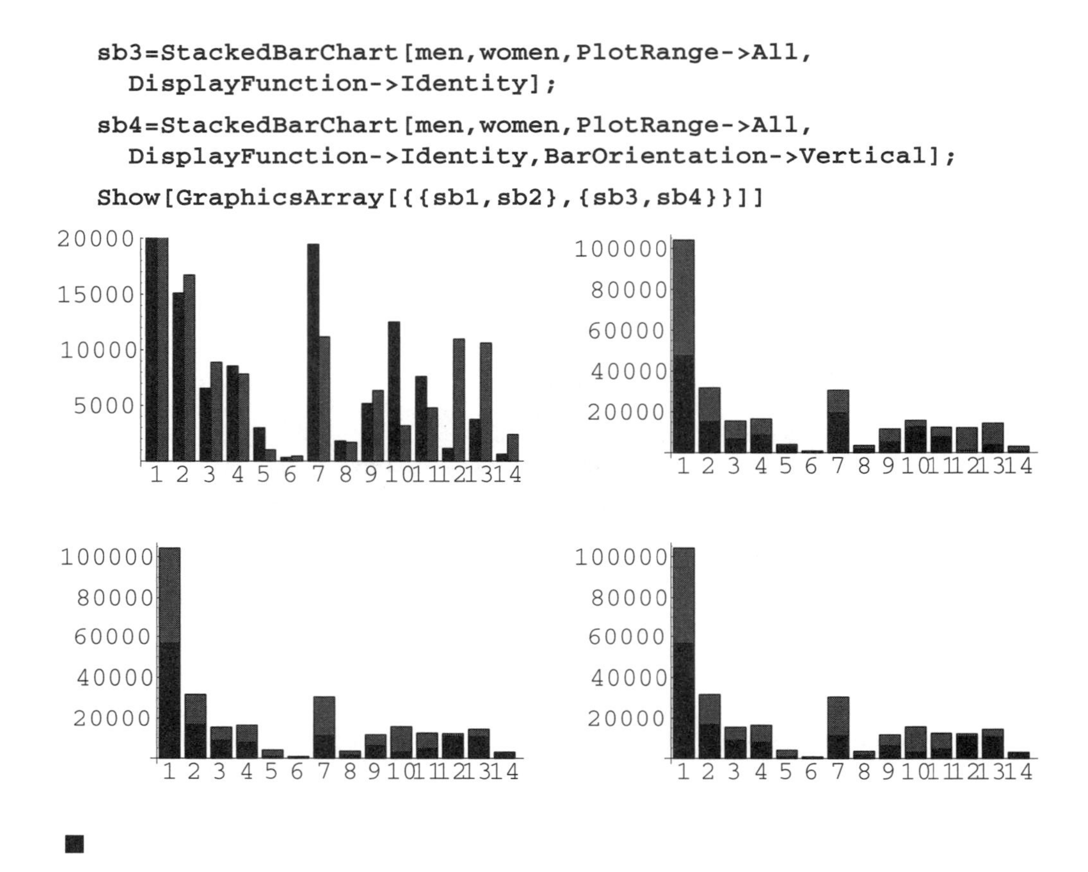

■

Bar Graphs for Descriptive Measures

Bar charts may be generated using values other than frequencies. For example, mean monthly earnings for each occupational category could be used to determine bar heights. Values of any suitable descriptive measure may be used in this way. See the Bar Graphs subsection above for examples of generating bar charts.

Pie Charts

Pie charts are useful to present relative frequencies or other ratios. A **pie chart** divides a circle (the pie) into slices, one corresponding to each category or classification. A pie chart may be created for any set of numbers where the area of a slice would represent one of the numbers as a proportion of the total of all of the numbers. We use the `PieChart`function, from the **Graphics** package that is located in the **Graphics** folder (or directory), to construct pie charts.

EXAMPLE 3: Table 4.3 summarizes the spending by the U.S. Federal government for the year 1995. Create a pie chart to represent this data.

Category	Outlay (in billions of dollars)
Discretionary (national defense)	274
Discretionary (international)	20
Discretionary (domestic)	252
Social Security	333
Medicare	157
Medicaid	89
Means-tested entitlements (except Medicaid)	92
Other programmatic	114
Net interest	232

Source: Office of Management and Budget

Table 4.3

SOLUTION: We first enter the data into *Mathematica*, naming the resulting list `spending`.

```
spending={274,20,252,333,157,89,92,114,232};
```

After checking that the **Graphics** package is loaded, we use `PieChart` to construct a pie chart representing the data. We use the option `PieLabels` to label each component of the pie chart, and the option `PieExploded` is used to emphasize the portion of the budget spent on payment of interest on the national debt. On a color monitor, the pie appears in various colors. Notice that in `PieExploded->{{9,0.1}}`, `9` indicates that the ninth element of the list `spending` should be emphasized, while 0.1 indicates the degree of separation of the exploded piece.

```
<< Graphics'Graphics'

PieChart [spending ,
 PieLabels  -> {"Defense ", "International  ",
   "National ", "SS", "Medicare ", "Medicaid ",
   "Entitlements  ", "Other ", "Net  Interest "},
 PieExploded  -> {{9, 0.1 }}]
```

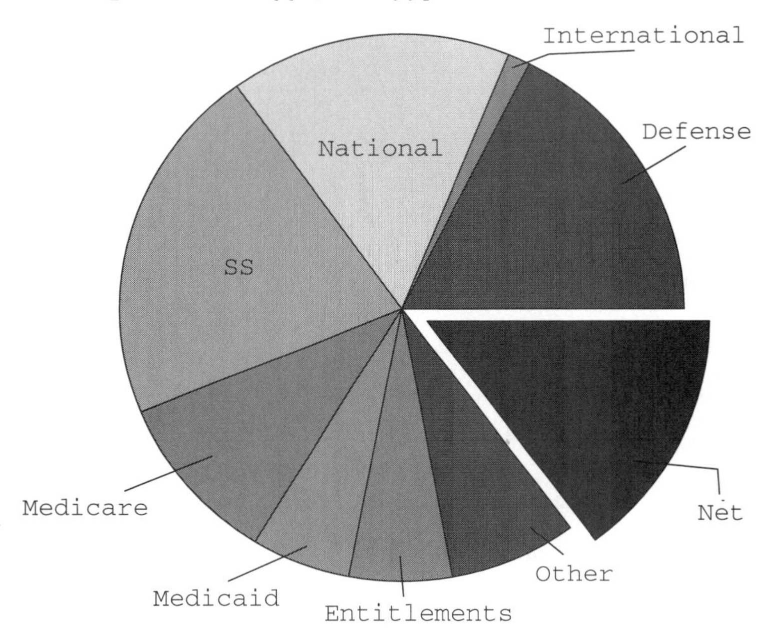

Next, we illustrate how to use `PieStyle` to shade the components of the pie chart in various shades of gray. `PieExploded` is used to emphasize the portion of the budget dedicated to discretionary spending (defense, international, and national). Notice that the piece of the pie corresponding to international spending (the second element in `spending`) is emphasized more than those corresponding to defense spending and national spending because of `{2,0.3}` in the `PieExploded` option.

```
grays = Table [GrayLevel [i], {i, 0.4 , 1, 0.6 / 8}];
PieChart [spending ,
 PieLabels  -> {"Defense ", "International  ",
   "National ", "SS", "Medicare ", "Medicaid ",
   "Entitlements  ", "Other ", "Net  Interest "},
 PieExploded  -> {{1, 0.1 }, {2, 0.3 }, {3, 0.1 }},
 PieStyle  -> grays ]
```

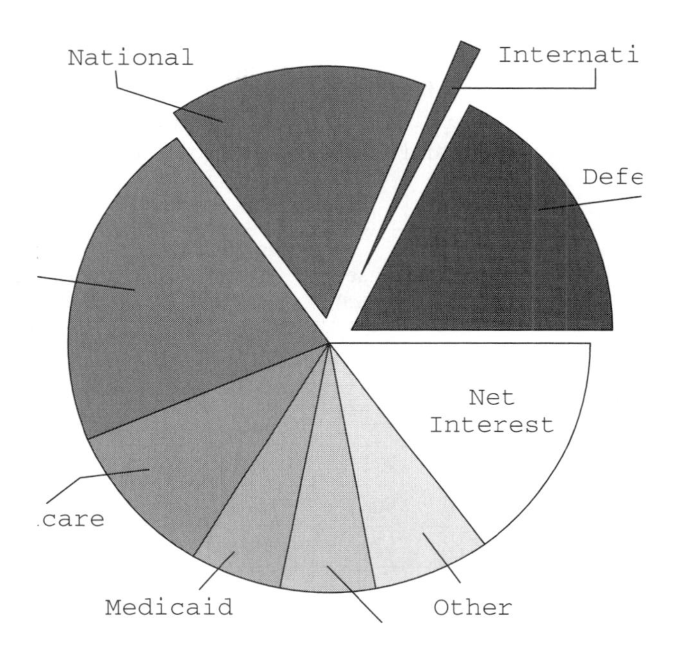

■

Frequency Tables and Histograms

In its simplest form, a **frequency table** lists the subsets or classes into which a set of values are grouped and the corresponding number or frequency of occurrences of values in each subset. The data can be categorical so that the class consists of a single attribute or a set of attributes. If the attributes have an inherent order, the classes should follow this order. The data can be quantitative so that the classes consist of a single number or a range of numbers. In the case of classes containing a range of values, the midpoint, sometimes called the **class mark**, may be used to represent the class. Each data value must belong to one, and only one, class. Whenever feasible, all classes should have the same width. For quantitative data, the **class width** is the difference between the midpoints of two consecutive classes. In addition to or in place of frequency, a table may list **relative frequency** or **cumulative frequency**. The relative frequency is calculated by dividing the frequency for a class by the total of all frequencies. Cumulative frequency only makes sense if the data can be ordered.

A **histogram** is a bar graph for a quantitative data set. Each class is represented by a bar where height or area corresponds to frequency, relative frequency, or cumulative frequency. The scale on the axis often corresponds to lower class limits or to class midpoints. For a histogram, the bars usually touch. The number of classes should be small enough to provide an effective summary but large enough to display the relevant characteristics of the data set.

EXAMPLE 4: Table 4.4 lists selected motor vehicle statistics for the 50 U.S. states. Generate (a) a frequency table, a bar graph, and a pie chart for the categorical variable Safety Belt Use Laws, and (b) a frequency table, a cumulative frequency table, a relative frequency table, and a histogram for the quantitative variable State Gas Tax.

State	State Gas Tax [a]	Safety Belt Use Law[b]	Licensed Drivers[c]	Registered Motor Vehicles[d]	Licensed Drivers[e]	Gallons of Fuel[f]	Miles[g]
Alabama	13	s	681	927	0.74	701	16.14
Alaska	8	s	571	868	0.66	670	12.45
Arizona	18	s	653	771	0.85	697	18.01
Arkansas	18.7	s	732	616	1.19	1098	13.22
California	16	s	667	737	0.91	669	17.66
Colorado	22	s	620	958	0.65	538	16.02
Connecticut	26	p	674	798	0.84	567	17.68
Delaware	19	s	728	790	0.92	733	16.97
Florida	11.6	s	714	846	0.84	610	16.48
Georgia	7.5	s	691	847	0.82	788	16.83
Hawaii	16	p	612	696	0.88	501	20.86
Idaho	21	s	699	1046	0.67	563	16.61
Illinois	19	s	638	689	0.93	786	13.47
Indiana	15	s	650	787	0.82	760	16.19
Iowa	20	p	674	948	0.71	635	13.75
Kansas	17	s	692	812	0.85	743	15.29
Kentucky	15.4	no	652	790	0.83	780	14.82
Louisiana	20	s	610	710	0.86	723	17.39
Maine	19	no	722	795	0.93	643	17.47
Maryland	18.5	s	703	754	0.93	643	17.47
Massachusetts	21	no	703	619	1.14	697	17.77
Michigan	15	s	693	776	0.89	649	17.33
Minnesota	20	s	578	802	0.72	629	17.65
Mississippi	18.2	p	732	729	1.00	813	16.00
Missouri	11.03	s	721	763	0.94	840	15.52
Montana	20	s	755	979	0.77	684	15.54

State	State Gas Tax [a]	Safety Belt Use Law[b]	Licensed Drivers[c]	Registered Motor Vehicles[d]	Licensed Drivers[e]	Gallons of Fuel[f]	Miles[g]
Nebraska	24.1	no	790	877	0.79	694	14.54
Nevada	21.5	s	704	710	0.99	890	13.46
New Hampshire	18.6	no	761	853	0.89	577	18.05
New Jersey	10.5	s	722	731	0.99	654	15.94
New Mexico	17	p	709	859	0.83	768	16.15
New York	22.89	p	570	567	1.01	655	16.01
North Carolina	22.3	p	686	779	0.88	665	15.56
North Dakota	17	no	665	986	0.67	640	14.66
Ohio	21	s	685	775	0.88	665	15.56
Oklahoma	17	s	724	842	0.86	767	16.28
Oregon	22	p	778	860	0.90	677	16.16
Pennsylvania	22.4	s	665	671	0.99	682	15.76
Rhode Island	26	s	669	670	1.00	598	17.49
South Carolina	16	s	680	723	0.94	869	15.70
South Dakota	18	no	707	1011	0.70	652	15.22
Tennessee	20	s	684	911	0.75	681	15.43
Texas	20	p	656	754	0.87	772	16.42
Utah	19	s	607	700	0.87	730	16.64
Vermont	16	no	732	820	0.89	695	18.18
Virginia	17.7	s	709	798	0.89	705	17.28
Washington	23	s	694	875	0.79	609	17.23
West Virginia	20.35	no	716	683	1.05	822	15.32
Wisconsin	22.2	s	680	751	0.91	668	18.05
Wyoming	9	s	735	1164	0.63	869	12.71

[a] (in cents per gallon) as of April 1, 1992
[b] as of August 1, 1991 (s= secondary; p= primary; no=None)
[c] per 1,000 Resident Population
[d] per 1,000 Resident Population
[e] per Registered Motor Vehicle
[f] per Vehicle
[g] per Gallon

Source: *The World Almanac and Book of Facts*, 1993

Table 4.4

SOLUTION: First, we must enter the data into *Mathematica*, naming the resulting array `motorvehicle`. (See Section 1.5, Examples 2 and 3, if you need help to import data.) Note that the datafile contains letter codes `no`, `p`, and `s`. If any of these symbols were defined previousely during your current *Mathematica* session, the definition(s) must be cleared using `Clear`. We display the first five records using `Take`.

```
Take[motorvehicle,5]
{{"Alabama",13,s,681,927,0.74,701,16.14},
 {"Alaska",8,s,571,868,0.66,670,12.45},
 {"Arizona",18,s,653,771,0.85,697,18.01},
 {"Arkansas",18.7,s,732,616,1.19,1098,13.22},
 {"California",16,s,667,737,0.91,669,17.66}}
```

After loading the **DataManipulation** package, we use `Column` to extract the third column of `motorvehicle`, corresponding to Safety Belt Use Laws, naming the resulting list `seatbelt`.

```
seatbelt=Column[motorvehicle,3]
{s,s,s,s,s,s,p,s,s,s,p,s,s,s,p,s,no,s,
 no,s,no,s,s,p,s,s,no,s,no,s,p,p,p,no,
 s,s,p,s,s,s,no,s,p,s,no,s,s,no,s,s}
```

The `Frequencies` function that is contained in the **DataManipulation** package can be used to generate frequency tables. This function is particularly useful when there are a relatively small number of distinct elements occurring in a list. It can be useful for categorical data and for quantative data where the classes are single-valued. Its output is a list in the correct format for use in various plotting functions.

```
?Frequencies
"Frequencies[list] gives a list of the distinct
   elements in list, together with the frequencies
   with which they occur."
```

In this case, we use `Frequencies` to see that, in 1991, 9 states did not have seat belt laws, 9 states had primary seat belt laws, and 32 states had secondary seat belt laws.

```
Frequencies[seatbelt]
{{9,no},{9,p},{32,s}}
```

To display this array in traditional row-and-column form, we have written the following procedure, called **`countTable`**, using `Column`, `Frequencies` and `TableForm`. The first column of the output identifies the categories, and the second column contains the frequencies. The input includes the name of the list followed (if you choose) by any options that can be used with `TableForm`. The procedure **countTable** can be found on the disk that comes with this book. After loading the disk, click on its icon. Then locate the procedure name, click on the right most cell bracket, press enter, and return to the *Mathematica* work sheet.

```
Clear[countTable]

countTable[datalist_,opts___]:=Module
[{column1,column2,gooddata},
  column1=Column[Frequencies[datalist],2];
  column2=Column[Frequencies[datalist],1];
  gooddata=Transpose[{column1,column2}];
  TableForm[gooddata,opts]]

countTable[seatbelt,TableHeadings->{{},
  {"Category","Frequency"}}]
```

Category	Frequency
no	9
p	9
s	32

After loading the **Graphics** package, we use BarChart to create a bar chart representing this data. The bars are labeled and the plot is given a title using the options BarLabels and PlotLabel, respectively. Notice that in this application the bar graph was generated using ordered pairs (see the output of Frequencies above), whereas in Example 1, single values were used.

```
<< Graphics'Graphics'

BarChart [Frequencies [seatbelt ],
 BarLabels -> {"None ", "Primary ", "Secondary "},
 PlotLabel -> "Seat Belt Use Laws (1991 )",
 PlotRange -> All ]
```

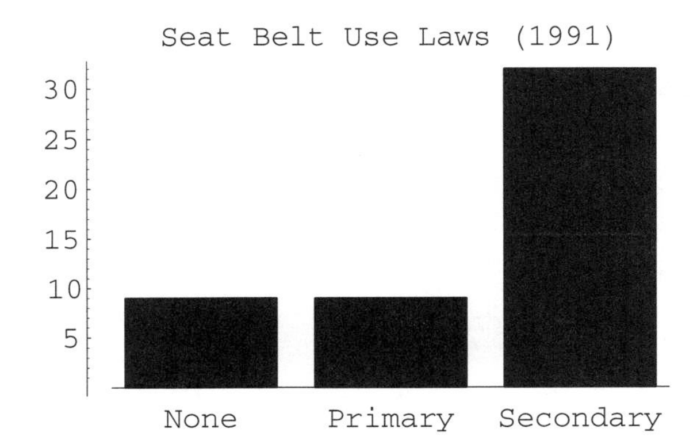

If we use the ordered pairs with PieChart, the slices of the pie are labeled using the second value of the ordered pair. In order to use PieLabels to specify labels for the slices, we use Column to place the frequencies in a list called belt and create the pie chart using belt.

```
belt = Column [Frequencies [seatbelt ], 1];

PieChart [belt ,
 PieLabels -> {"None ", "Primary ", "Secondary "},
 PlotLabel -> "Seat Belt Use Laws (1991 )"]
```

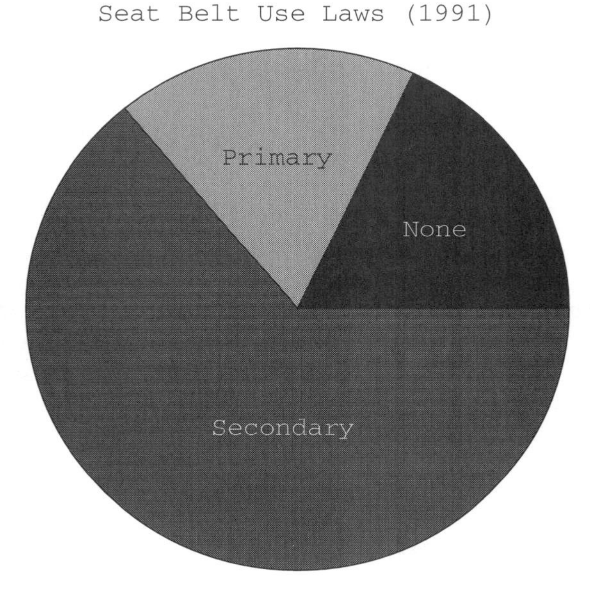

Next, we use Column to extract the second column of motorvehicle, corresponding to the gas taxes of the 50 states, naming the resulting list gastax.

```
gastax=Column[motorvehicle,2]

{13,8,18,18.7,16,22,26,19,11.6,7.5,16,21,19,15,20,17,
 15.4,20,19,18.5,21,15,20,18.2,11.03,20,24.1,21.5,
 18.6,10.5,17,22.89,22.3,17,21,17,22,22.4,26,16,18,
 20,20,19,16,17.7,23,20.35,22.2,9}
```

To construct a histogram for this data, we first determine the classes by finding the minimum and maximum values, deciding on the number of classes (we chose to have 10 classes); calculating the class width; and, using Table, determining the lower limits of the classes.

```
minimum=Min[gastax]
7.5

maximum=Max[gastax]
26

classes=10;
classwidth=(maximum-minimum)/(classes)
1.85

lowerlimits=Table[i,{i,minimum,maximum, classwidth}]
{7.5,9.35,11.2,13.05,14.9,16.75,18.6,20.45,22.3,
 24.15,26.}
```

Both BinCounts and RangeCounts in the **DataManipulation** package may be used to determine the number of data values in each class. We use ? to display basic information about RangeCounts, and then use RangeCounts to store the frequencies in the list named counts. From counts we learn that the number of elements of gastax less than 7.5 (the minimum) is zero; the number equal to or greater than 7.5 and less than 9.35 is 3; the number equal to or greater than 9.35 and less than 11.2 is 2; and so on. Finally, the number of elements of gastax equal to 26 (the maximum) is 2. There are no elements greater than the maximum. Because RangeCounts includes the lower limit, these classes are said to be closed on the left and open on the right. BinCounts, on the other hand, creates classes that are open on the left and closed on the right.

```
?RangeCounts
RangeCounts[{x1, x2, ...}, {c1, c2, ...}] gives a list
 of the number of elements in the data {x1, x2, ...}
 that lie between successive cutoffs. RangeCounts[{{x1,
 y1}, {x2, y2}, ...}, {xc1, xc2, ...}, {yc1, yc2, ...}]
 gives a 2-dimensional array of range counts for the
 bivariate data {{x1, y1}, {x2, y2}, ...}.  In general,
 RangeCounts gives a p-dimensional array of range counts
 for p-variate data.

counts=RangeCounts[gastax,lowerlimits]
{0,3,2,2,0,7,9,13,7,5,0,2}
```

RangeList is used to generate a list of the elements of gastax that fall in each of the classes. Notice that the first class is empty, because there are no values below the minimum, and the last class only contains the maximum value(s).

```
RangeLists[gastax,lowerlimits]
```

```
{{},{8,7.5,9},{11.03,10.5},{13,11.6},{},{16,16,15,15.4,
 15,16,16},{18,17,18.5,18.2,17,17,17,18,17.7},{18.7,19,
 19,20,20,19,20,20,18.6,20,20,19,20.35},{22,21,21,21.5,
 21,22,22.2},{24.1,22.89,22.3,22.4,23},{},{26,26}}
```

The following steps are used to modify `counts` before generating a frequency table and a histogram. We delete the leading zero corresponding to an empty class of values less than the minimum and combine the two occurrences of the maximum value, 26, into the class to the left. This class has no elements in the current example. In effect, all of our classes will be closed on the left and open on the right except the last class, which will be closed on both sides.

```
step1=Drop[counts,1]
```

```
{3,2,2,0,7,9,13,7,5,0,2}
```

```
step2=Take[step1,-2]
```

```
{0,2}
```

```
step3=Apply[Plus,step2]
```

```
2
```

```
step4=Drop[step1,-2]
```

```
{3,2,2,0,7,9,13,7,5}
```

```
frequencylist=Append[step4,step3]
```

```
{3,2,2,0,7,9,13,7,5,2}
```

The midpoints of the classes are computed using `Table`. Next, using `TableForm`, we first display a frequency table. Then we calculate cumulative frequencies using the function `CumulativeSums` from the **DataManipulation** package and display both frequencies and cumulative frequencies in a table.

```
midpoints=Table[i,{i,minimum+classwidth/2,maximum,
classwidth}]
```

```
{8.425,10.275,12.125,13.975,15.825,17.675,19.525,
 21.375,23.225,25.075}
```

```
Transpose[{midpoints,frequencylist}]//TableForm
```

```
  8.425      3
 10.275      2
 12.125      2
 13.975      0
```

```
  15.825        7
  17.675        9
  19.525       13
  21.375        7
  23.225        5
  25.075        2
```

```
cumulative=CumulativeSums[frequencylist]
{3,5,7,7,14,23,36,43,48,50}
```

```
Transpose[{midpoints,frequencylist,cumulative}]//
  TableForm
  8.425      3      3
 10.275      2      5
 12.125      2      7
 13.975      0      7
 15.825      7     14
 17.675      9     23
 19.525     13     36
 21.375      7     43
 23.225      5     48
 25.075      2     50
```

Relative frequencies are calculated by dividing each frequency by the number of elements in the data set. The built-in function `Length` is used to count the number of elements in `gastax`. The function `N` is used to display relative frequencies in numerical form rather than as ratios.

```
relativefrequency=frequencylist/Length[gastax]//N
{0.06,0.04,0.04,0,0.14,0.18,0.26,0.14,0.1,0.04}
```

```
Transpose[{midpoints,relativefrequency}]//TableForm
  8.425      0.06
 10.275      0.04
 12.125      0.04
 13.975      0
 15.825      0.14
 17.675      0.18
```

```
19.525        0.26
21.375        0.14
23.225        0.1
25.075        0.04
```

We will generate a histogram using `frequencylist` and label the bars with the values given in `midpoints`. To do so, we limit the values of the midpoints to two places in `midpoints1`, form the list `{frequencylist,midpoints1}`, and then use `Transpose` to interchange the rows and columns of this list, naming the result `tograph1`. Notice that `tograph1` is a list of ordered pairs. When we use `BarChart`, the first element of each ordered pair will be the height of the bar; the second element will be the label of the bar.

```
midpoints1=Map[N[#,2]&,midpoints]
{8.4,10.,12.,14.,16.,18.,20.,21.,23.,25.}
```

```
tograph1=Transpose[{frequencylist,midpoints1}]
{{3,8.4},{2,10.},{2,12.},{0,14.},{7,16.},{9,18.},
 {13,20.},{7,21.},{5,23.},{2,25.}}
```

The histogram is then generated with `BarChart`. Note how we use `BarChart` together with the `BarSpacing` option to construct the graph so that the bars touch, as is typical with a histogram. We also us options `PlotLabel` and `AxesLabel`.

```
BarChart [tograph1 , BarSpacing  -> -.2,
 PlotLabel  -> "State  Gas  Tax  (1992)",
 AxesLabel  -> {"Gas  Tax", "States "}]
```

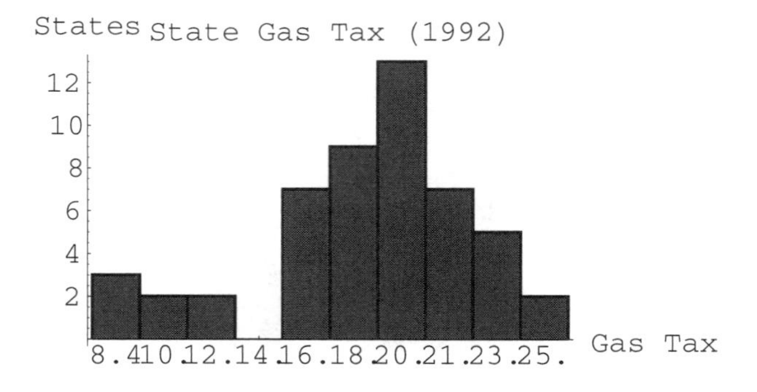

In addition to `BarChart`, the **Graphics`Graphics`** package contains `GeneralizedBarChart`, which we can learn about by using `?`. Each bar is specified according to position height and width. For the current example we store this information in `tograph` where bar height is determined by relative frequency. Options used in the `GeneralizedBarChart` command are `AxesOrigin` to keep the vertical axis to the left of the bars and `Ticks` to specify the values to be displayed on the horizontal axis. The default is `Automatic` for both axes.

```
? GeneralizedBarChart
```

```
GeneralizedBarChart[{{pos1, height1, width1},
  {pos2, height2, width2},...}] generates a bar
  chart with the bars at the given positions,
  heights, and widths.
```

```
tograph = Table [{midpoints [[i]],
    relativefrequency [[i]], classwidth },
   {i, 1, classes }];
```

```
GeneralizedBarChart [tograph , PlotRange -> All ,
 AxesOrigin -> {.98 minimum , 0}, Ticks ->
  {{7.5 , 11.2 , 14.9 , 18.6 , 22.3 , 26}, Automatic }]
```

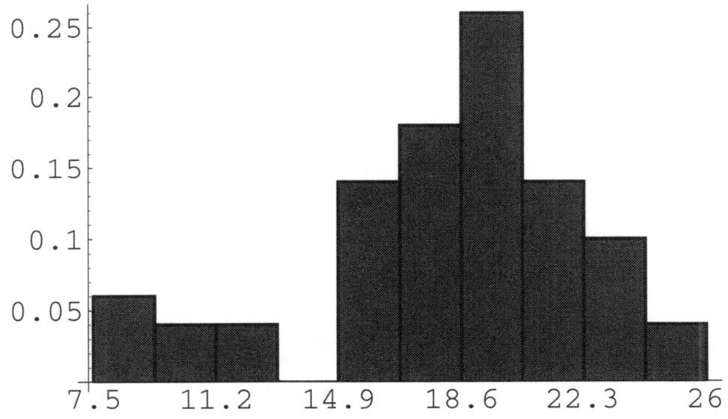

■

We have written several procedures to display frequency tables and histograms. The procedures **frequencyTable**, **cumulativeFrequencyTable**, and **relativeFrequencyTable** each accept a list of data followed by the desired number of classes as input. If the number of classes is not entered, 10 classes will be assumed. Any options that can be used with `TableForm` can then be entered. The procedures **frequencyHistogram**, **cumulativeFrequencyHistogram**, and **relativeFrequencyHistogram** each accept a data list, the number of classes (again the default is 10), and any options that can be used with `GeneralizedBarChart`. The procedure **relativeFrequencyHistogramAlternate** accepts a data list, the number of classes, a minimum, a maximum, and any options that can be used with `GeneralizedBarChart`. The minimum and/or maximum may be specified to exclude extreme values from the histogram.

To access any of the author-defined procedures presented in this book, go to the disk provided with the book, locate the procedure name, click on the rightmost cell bracket, press enter, and return to the *Mathematica* work sheet.

EXAMPLE 5: Use the data in Table 4.4 from Example 4 for Miles per Gallon to show (a) a frequency table, cumulative frequency table, and relative frequency table, and (b) a frequency histogram, cumulative frequency histogram, and relative frequency histogram.

SOLUTION: We assume that the array `motorvehicle` is open. After loading the **DataManipulation** package we use `Column` to extract the seventh column of `motorvehicle`, corresponding to Miles per Gallon, naming the resulting list `mpg`. We then find the minimum and maximum values to be 12.45 and 20.87, respectively.

```
<<Statistics`DataManipulation`

mpg=Column[motorvehicle,8]

{16.14,12.45,18.01,13.22,17.66,16.02,17.68,16.97,16.48,
 16.83,20.86,16.61,13.47,16.19,13.75,15.29,14.82,17.39,
 17.47,17.47,17.77,17.33,17.65,16.,15.52,15.54,14.54,
 13.46,18.05,15.94,16.15,16.01,15.56,14.66,15.56,16.28,
 16.16,15.76,17.49,15.7,15.22,15.43,16.42,16.64,18.18,
 17.28,17.23,15.32,18.05,12.71}

Min[mpg]

12.45

Max[mpg]

20.86
```

After loading them from the disk, each of **`frequencyTable`**, **`cumulativeFrequencyTable`**, and **`relativeFrequencyTable`** is run on `mpg`. The number of classes in each case is specified to be eight. Both cumulative and relative frequency tables include the frequencies. The output of the frequency column can be supressed by deleting `"frequencylist"` from the last line of the procedure code.

```
frequencyTable[mpg,8,TableHeadings->{{},
 {"MPG","Frequency"}}]
```

MPG	Frequency
12.9756	5
14.0269	2
15.0781	10
16.1294	15
17.1806	12
18.2319	5
19.2831	0
20.3344	1

```
cumulativeFrequencyTable[mpg,8,
  TableHeadings->{{},{"MPG","Frequency",
  "Cumulative Freq."}}]
```

MPG	Frequency	Cumulative Freq.
12.9756	5	5
14.0269	2	7
15.0781	10	17
16.1294	15	32
17.1806	12	44
18.2319	5	49
19.2831	0	49
20.3344	1	50

```
relativeFrequencyTable[mpg,8,
  TableHeadings->{{},{"MPG","Frequency",
  "Relative Freq."}}]
```

MPG	Frequency	Relative Freq.
12.9756	5	0.1
14.0269	2	0.04
15.0781	10	0.2
16.1294	15	0.3
17.1806	12	0.24
18.2319	5	0.1
19.2831	0	0
20.3344	1	0.02

After loading the procedures from the disk, each of **frequencyHistogram**, **cumulativeFrequencyHistogram**, and **relativeFrequencyHistogram** is run on mpg, again specifying eight classes. We also illustrate the use of **relativeFrequencyHistogramAlternate**. This procedure requires the user to specify the minimum and maximum. It can be useful to exclude extreme values from the display. Notice that no graphs are displayed, because we include DisplayFunction->Identity in each command. Finally, all four graphics are displayed together as a graphics array using Show together with GraphicsArray.

```
gr1=frequencyHistogram[mpg,8,PlotLabel->"MilesPerGallon
  by State",AxesLabel->{"MPG",""},
  DisplayFunction->Identity];
gr2=cumulativeFrequencyHistogram[mpg,8,
  DisplayFunction->Identity];
gr3=relativeFrequencyHistogram[mpg,8,
  Ticks->{{12.45,16.65,20.95},Automatic},
  DisplayFunction->Identity];
gr4=relativeFrequencyHistogramAlternate[mpg,8,13,18.5,
  AxesLabel->{"MPG",""},PlotLabel->"MilesPerGallon by
  State",DisplayFunction->Identity];
```

```
Show [GraphicsArray [{{gr1 , gr2 }, {gr3 , gr4 }}]]
```

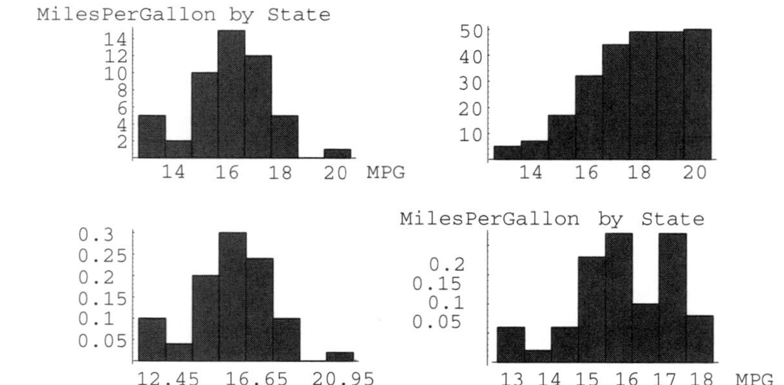

■

Polygons

Frequency polygons and histograms are similar in purpose. The frequency polygon has a distinct advantage over the histogram in comparing two or more frequency distributions. In addition, a cumulative frequency polygon, also called an **ogive**, is often preferred to a cumulative frequency histogram. To make a frequency polygon using quantitative data, a point is plotted at the height corresponding to the frequency of a class directly above the midpoint of the class. Neighboring points are connected by line segments. A polygon can be constructed in this way for any set of measurements. For example, for a set of means corresponding to the values of a grouping variable, a point is plotted for each mean, and neighboring points are connected by line segments.

To generate "simple" polygons, we use the built-in *Mathematica* command `ListPlot`. We use `??` to obtain detailed information about the `ListPlot` command. Note that `ListPlot` has many of the same options as other *Mathematica* graphics commands.

```
??ListPlot

ListPlot[{y1, y2, ... }] plots a list of values. The x
 coordinates for each point are taken to be 1, 2, ... .
 ListPlot[{{x1, y1}, {x2, y2}, ... }] plots a list of
 values with specified x and y coordinates.

Attributes[ListPlot] = {Protected}

Options[ListPlot] = {AspectRatio -> GoldenRatio^(-1),
 Axes -> Automatic, AxesLabel -> None, AxesOrigin ->
 Automatic, AxesStyle -> Automatic, Background ->
 Automatic, ColorOutput -> Automatic, DefaultColor ->
 Automatic, Epilog -> {}, Frame -> False, FrameLabel ->
 None, FrameStyle -> Automatic, FrameTicks -> Automatic,
 GridLines -> None, ImageSize -> Automatic, PlotJoined ->
 False, PlotLabel -> None, PlotRange -> Automatic,
 PlotRegion -> Automatic, PlotStyle -> Automatic,
 Prolog -> {}, RotateLabel -> True, Ticks -> Automatic,
 DefaultFont :> $DefaultFont, DisplayFunction :>
 $DisplayFunction, FormatType :> $FormatType,
 TextStyle :> $TextStyle}
```

EXAMPLE 6: Table 4.5 shows the mean verbal and mathematics SAT scores, by selected characteristics of the parents, of college-bound seniors for the years 1988, 1991, and 1995. (a) Plot the verbal scores for parents' highest education level for the three years. (b) Plot the verbal scores for parents' income for the three years. (c) Plot the math scores for parents' highest education level for the three years. (d) Compare verbal and math scores for each year.

	1988			1991			1995		
Characteristic	Percent	Verbal	Math	Percent	Verbal	Math	Percent	Verbal	Math
Parents' income									
Less than $10,000	5	363	418	5	353	415	5	354	415
$10,000–19,999	13	393	440	11	379	434	10	380	433
$20,000–29,999	17	416	460	15	404	452	13	405	454
$30,000–39,999	20	429	473	18	418	466	15	420	468
$40,000–49,999	14	441	487	13	430	480	12	431	482
$50,000–59,999	10	449	497	11	440	491	11	440	493
$60,000–69,999	6	456	505	7	449	500	9	448	502
$70,000 or more	15	469	523	19	469	528	25	471	533
Parents' highest education level									
No high school diploma	4	347	410	5	339	409	5	338	405
High school diploma	37	402	446	38	395	443	35	397	444
Associate's degree	7	414	457	8	407	454	8	411	458
Bachelor's degree	27	446	496	27	442	497	28	448	505
Graduate degree	24	476	524	23	476	528	25	481	540

Source: *1995 Digest of Educational Statistics*

Table 4.5

SOLUTION: First, we enter the scores on the verbal section of the SAT for the years 1988, 1991, and 1995 corresponding to parents' income in `verbalXincome88`, `verbalXincome91`, and `verbalXincome95`. (Read the name of the first list as "verbal by income 88.") We enter the verbal scores corresponding to parents' hightest education level in `verbalXeducation88`, `verbalXeducation91`, and `verbalXeducation95`. We first demonstrate `ListPlot` for a list of values using the latter three "education" lists. Notice in `lp2` the use of `PlotStyle->PointSize[r]` to increase the size of the plotted points, where r is the radius of the shaded circular region and represents a fraction of the total width of the graph. Larger values of r produce larger points. In lp3 we use AxesOrigin and PlotRange to locate the point at which the two axes cross and to specify the range of values on the two axes. Note that both options must be used to guarantee that the horizontal and vertical axes intersect. We have also used PlotJoined->True to cause the data points to be connected with line segments. All three plots are shown simultaneously as has been done several times in earlier examples. Notice that, when the input to ListPlot is a simple list of numbers, the first coordinate of each point plotted is the position in the list and the second coordinate is the number from the list.

```
verbalXincome88={363,393,416,429,441,449,456,469};
verbalXincome91={353,379,404,418,430,440,449,469};
verbalXincome95={354,380,405,420,431,440,448,471};
verbalXeducation88={347,402,414,446,476};
verbalXeducation91={339,395,407,442,476};
verbalXeducation95={338,397,411,448,481};
lp1=ListPlot[verbalXeducation88,
  DisplayFunction->Identity];
lp2=ListPlot[verbalXeducation91,
  PlotStyle->PointSize[0.02],
    DisplayFunction->Identity];
lp3=ListPlot[verbalXeducation95,AxesOrigin->{.5,330},
    PlotRange->{{.5,5},{330,480}},PlotJoined->True,
    DisplayFunction->Identity];

Show[GraphicsArray[{{lp1,lp2},{lp3}}]]
```

460
440
420
400
380
2 3 4 5

460
440
420
400
380
360
2 3 4 5

480
460
440
420
400
380
360
340
1 2 3 4 5

To plot the verbal scores corresponding to parents' income level, we first make a list called `midpts` with approximate midpoints of each of the income ranges in Table 4.5 expressed in thousands. Then, as in earlier examples, we use `Transpose` to create ordered pairs for input into `ListPlot`. Again the output is in the form of a graphics array. Notice the use of the option `Ticks` in `lp2` to cause the horizontal axis to be labeled with the midpoints. The labels in the other two plots result from the default, which is `Automatic`.

```
midpts={5,15,25,35,45,55,65,75};
vxi88=Transpose[{midpts,verbalXincome88}];
vxi91=Transpose[{midpts,verbalXincome91}];
vxi95=Transpose[{midpts,verbalXincome95}];
lp1=ListPlot[vxi88,DisplayFunction->Identity];
lp2=ListPlot[vxi91,PlotStyle->PointSize[0.02],
    Ticks->{{15,25,35,45,55,65,75},
      Automatic},DisplayFunction->Identity];
lp3=ListPlot[vxi95,AxesOrigin->{0,340},
    PlotRange->{{0,75},{340,480}},
    PlotJoined->True,DisplayFunction->Identity];
Show[GraphicsArray[{{lp1,lp2},{lp3}}]]
```

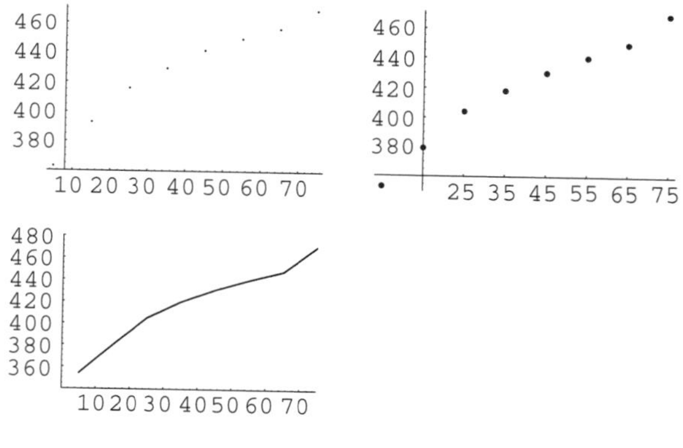

We enter the math SAT scores as the verbal scores were entered for the respective years. In this case, however, we show how these data are displayed together. In the graph named `lstp1a`, we plot the data in `mathXeducation88` using a light shade of gray with `GrayLevel[0.5]` to graph the data points. Similarly, in `lstp1b`, we use `GrayLevel[0.5]` to graph the line segments connecting the data points. Neither graph is displayed initially because we use the

`DisplayFunction->Identity` option. In the graph named `lstp2a`, we graph the data points using the same point size as before. However, the points are black. In `lstp2b`, we use `Dashing[{0.01}]` to produce dashed line segments between the data points. Then, in the graphs named `lstp3a` and `lstp3b`, we produce the graphics objects to display the data in `mathXeducation95` joined with solid line segments. We use `Show` with the `DisplayFunction->$DisplayFunction` to display the three graphs together.

```
mathXincome88={418,440,460,473,487,497,505,523};
mathXincome91={415,434,452,466,480,491,500,528};
mathXincome95={415,433,454,468,482,493,502,533};
mathXeducation88={410,446,457,496,524};
mathXeducation91={409,443,454,497,528};
mathXeducation95={405,444,458,505,540};

lstp1a=ListPlot[mathXeducation88,PlotStyle->
  {PointSize[0.02],GrayLevel[0.5]},
   DisplayFunction->Identity];
lstp1b=ListPlot[mathXeducation88,PlotStyle->
   GrayLevel[0.5],PlotJoined->True,
   DisplayFunction->Identity];

lstp2a=ListPlot[mathXeducation91,PlotStyle->
   PointSize[0.02],DisplayFunction->Identity];
lstp2b=ListPlot[mathXeducation91,PlotStyle->
   Dashing[{0.01}],PlotJoined->True,
   DisplayFunction->Identity];

lstp3a=ListPlot[mathXeducation95,PlotStyle->
   PointSize[0.02],DisplayFunction->Identity];
lstp3b=ListPlot[mathXeducation95,PlotJoined->
   True,DisplayFunction->Identity];

Show[lstp1a,lstp1b,lstp2a,lstp2b,lstp3a,lstp3b,
   DisplayFunction->$DisplayFunction,PlotRange->All
```

```
Show [lstp1a , lstp1b , lstp2a , lstp2b , lstp3a ,
 lstp3b , DisplayFunction -> $DisplayFunction ,
 PlotRange -> All ]
```

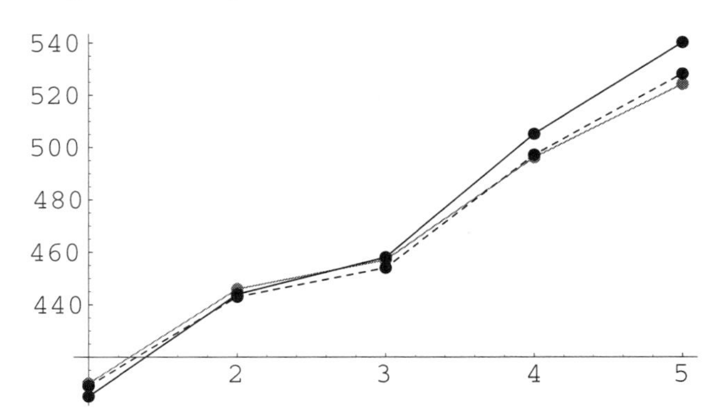

Instead of carrying out the numerous commands needed to create the previous graph showing the three sets of data together, we can take advantage of the **Graphics** package command `MultipleListPlot`. *Mathematica* plots the data contained the the lists simultaneously. We show how this command is used to plot the verbal SAT score data and include the `PlotJoined->True` option so that the data points are joined with line segments. Note that *Mathematica* automatically selects different styles for the line segments. We also use the `PlotLegend` option so that *Mathematica* will include a legend to describe the year from which the data was drawn.

```
<< Graphics'MultipleListPlot'

MultipleListPlot
 [verbalXeducation88 , verbalXeducation91 ,
 verbalXeducation95 , PlotJoined -> True ,
 PlotLegend -> {"1988 ", "1991 ", "1995 "}]
```

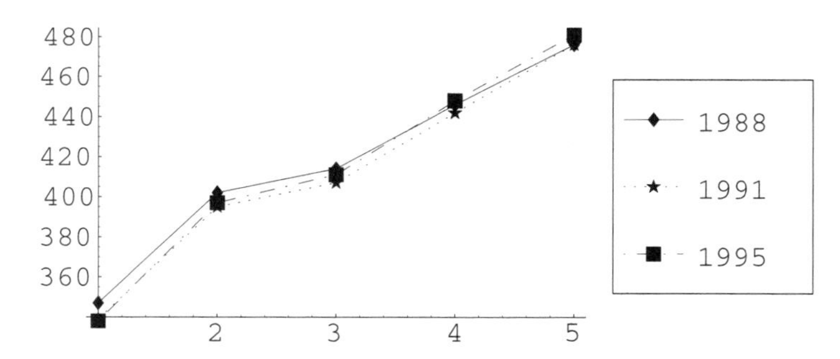

Finally, we use `MultipleListPlot` to compare verbal and math scores for the year 1995, and using a graphics array for the years 1988 and 1991. Note the use of `PlotLabel` to title each graph.

```
mpl1=MultipleListPlot[verbalXeducation88,
   mathXeducation88,PlotJoined->True,PlotLabel->
   "SAT Scores For 1988",DisplayFunction->Identity];
mpl2=MultipleListPlot[verbalXeducation91,
   mathXeducation91,PlotJoined->True,PlotLabel->
   "SAT Scores For 1991",DisplayFunction->Identity];
MultipleListPlot[verbalXeducation95,mathXeducation95,
   PlotJoined->True,PlotLabel->"SAT Scores For 1995",
   PlotLegend->{"Verbal","Math"}]
Show[GraphicsArray[{{mpl1,mpl2}}]]
```

SAT Scores For 1995

500
450
400
2 3 4 5

Verbal
Math

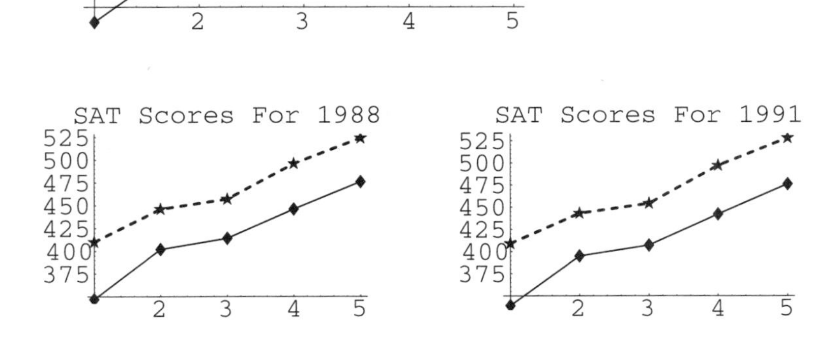

■

Box-and-Whiskers Plots

A **box-and-whiskers plot** or **box plot** consists of a rectangle (the box) with one end at the lower (or first) quartile Q_1 and the other end at the upper (or third) quartile Q_2. A line parallel to the ends cuts the box at the median. Lines (whiskers) extend in opposite directions from the ends of the box to the minimum and maximum. If there are outliers, the whiskers extend to the "adjacent" values, and values more extreme than the "adjacent" values are plotted. When referring to a box plot, the ends of the box located at the first and third quartiles are often called the **lower hinge** and the **upper hinge**, respectively.

The box-and-whiskers plot is a visual representation of the distribution of a data set. It provides information about the important characteristics of a distribution, namely center, variability, shape, and outliers. The median is a measure of the center. Its location with respect to the hinges and the difference in length of the whiskers indicate shape. The length of the box (the **interquartile range (IQR)** where IRQ = $Q_3 - Q_1$) and the length of the whiskers indicate variability. Finally, outliers are identified and plotted separately. A display containing two or more box plots is often used for visual comparison of the corresponding distributions.

Box-and-whiskers plots may be used to identify **outliers**, which are defined to be individual values located outside the overall pattern of the data. The **lower inner fence** is the lower quartile minus 1.5 (IQR); the **upper inner fence** is the upper quartile plus 1.5 (IQR); and the **adjacent values** are the two values in the data set that are closest to and inside the inner fences. Values in the data set that are more extreme than the inner fences are outliers. In practice, outliers may not belong with the data set because of data handling errors or other considerations If there is not a good reason to exclude them, they should be included with the data and given special consideration when appropriate. The **lower outer fence** is the lower quartile minus 3 (IQR), and the **upper outer fence** is the upper quartile plus 3 (IQR). Values between the inner and outer fences are called "outside" and are plotted separately with a special symbol. Values outside the outer fences are called "far out" and are plotted separately usually with a different special symbol.

We have written four procedures to display box-and-whiskers plots. They make use of the two *Mathematica* packages **Statistics`DataManipulation`** and **Statistics`DescriptiveStatistics`**. To use any of these procedures go to the disk provided with the book, locate the cell entitled **Box and Whiskers**, click on the rightmost cell bracket (as shown in the following figure, the cell marker will turn black), press **ENTER**, and return to the *Mathematica* work sheet.

■ **Box and Whiskers**

The procedures available in **Box and Whiskers** are as follows.

`boxplotLists[lists_,options___]`: Displays a box plot if the input, `lists`, is a single list of numbers. If the input is a list containing two or more lists, a graphic is displayed containing a box plot for each list. The box plots are horizontal.

`boxplotListsV[lists_,options___]`: Same as `boxplotLists` except the box plots are vertical.

`boxplotArray[dataset_,columns_,options___]`: Displays a graphic containing a box plot for each column of the array `dataset` which is indicated in the list `columns`. This procedure will ignore nonnumeric data. The box plots are horizontal.

`boxplotArrayV[dataset_,columns_,options___]`: Same as `boxplotArray`, except the box plots are vertical.

All of these procedures will accept any options that can be used with `Graphics` or `Show`. Information about options, for example, for `Graphics` is displayed using `??`.

```
??Graphics
```

```
Graphics[primitives, options] represents a two-
dimensional graphical image.

Attributes[Graphics] = {Protected, ReadProtected}

Options[Graphics] = {AspectRatio -> GoldenRatio^(-1),
Axes -> False, AxesLabel -> None, AxesOrigin ->
Automatic, AxesStyle -> Automatic, Background ->
Automatic, ColorOutput -> Automatic, DefaultColor ->
Automatic, Epilog -> {}, Frame -> False, FrameLabel ->
None, FrameStyle -> Automatic, FrameTicks -> Automatic,
GridLines -> None, ImageSize -> Automatic, PlotLabel ->
None, PlotRange -> Automatic, PlotRegion -> Automatic,
Prolog -> {}, RotateLabel -> True, Ticks -> Automatic,
DefaultFont :> $DefaultFont, DisplayFunction :>
$DisplayFunction, FormatType :> $FormatType,
TextStyle :> $TextStyle}
```

EXAMPLE 7: Table 4.6 shows the percent of undergraduate degrees awarded in the sciences for the years 1985, 1990, and 1991 for selected countries. (The symbol "null" is used to represent missing values in the table.) Construct a box plot for the percentages of all science degrees awarded in each of the years 1985, 1990, and 1991.

Country	All science degrees			Natural sciences			Math and computer science			Engineering		
	1985	1990	1991	1985	1990	1991	1985	1990	1991	1985	1990	1991
Australia	null	null	21.6	null	null	15.9	null	null	null	null	null	5.7
Austria	16.8	19.6	20.1	5.0	5.3	5.9	4.1	5.2	4.8	7.7	9.0	9.5
Belgium	14.7	null	32.2	4.6	null	4.3	1.7	null	1.7	8.4	null	26.3

Continued

Country	All science degrees			Natural sciences			Math and computer science			Engineering		
Canada	17.1	16.4	15.5	4.9	6.0	5.7	4.5	4.2	3.7	7.7	6.2	6.1
Denmark	22.5	26.1	27.6	6.3	4.4	6.1	null	null	0.1	16.2	21.7	21.4
Finland	39.3	33.5	34.5	7.7	4.1	4.2	6.3	5.9	6.6	25.3	23.4	23.7
Germany (former West)	23.8	31.3	31.5	5.0	7.2	7.3	2.3	3.5	3.9	16.5	20.5	20.2
Ireland	28.8	34.1	28.5	12.8	14.1	12.4	4.0	6.3	4.4	12.0	13.7	11.6
Italy	19.5	19.7	19.8	8.1	7.6	7.5	3.1	3.9	3.8	8.3	8.3	8.5
Japan	22.7	23.5	23.5	2.4	2.4	2.4	null	null	null	20.3	21.0	21.1
Nether-lands	21.8	21.1	21.4	8.5	7.1	6.5	1.2	1.6	1.6	12.1	12.4	13.3
New Zealand	20.5	19.5	16.3	11.7	8.2	7.1	5.5	5.5	4.0	3.3	5.8	5.2
Norway	6.1	12.9	12.3	2.5	2.1	1.8	1.8	0.6	0.6	1.8	10.2	9.9
Portugal	null	24.3	null	6.5	6.7	null	null	7.0	null	null	10.5	null
Spain	13.9	15.0	15.4	5.5	5.7	5.3	1.3	2.6	2.9	7.0	6.7	7.1
Sweden	15.4	24.0	24.3	2.6	4.1	4.2	1.6	4.7	4.9	11.3	15.2	15.2
Switzer-land	20.2	23.0	22.7	10.3	11.2	11.0	2.1	3.7	3.8	7.9	8.1	7.9
Turkey	23.0	20.6	21.3	3.6	4.6	4.9	1.6	2.1	2.3	17.8	13.8	14.1
United States	21.7	16.9	15.9	6.3	5.1	5.1	5.5	4.0	3.6	9.8	7.8	7.2

Source: *1995 Digest of Education Statistics*

Table 4.6

SOLUTION: After loading the functions in **Box and Whiskers** (found on the disk provided with this book), we enter the information contained in Table 4.6 and name it `sciencedegrees`. We display the last five records of `sciencedegrees` using `Take`.

```
Take[sciencedegrees,-5]
```

```
{{Spain,13.9,15.,15.4,5.5,5.7,5.3,1.3,2.6,2.9,7.,6.7,
 7.1},{Sweden,15.4,24.,24.3,2.6,4.1,4.2,1.6,4.7,4.9,
```

```
11.3,15.2,15.2},{Switzerland,20.2,23.,22.7,10.3,11.2,
11.,2.1,3.7,3.8,7.9,8.1,7.9},{Turkey,23.,20.6,21.3,
3.6,4.6,4.9,1.6,2.1,2.3,17.8,13.8,14.1}, {United
States,21.7,16.9,15.9,6.3,5.1,5.1,5.5,4.,3.6,
9.8,7.8,7.2}}
```

The information concerning all science degrees for the years 1985, 1990, and 1991 is in columns 2, 3, and 4 of `sciencedegrees`. To extract these columns we use `ColumnTake`, which is in the package **Statistics`DataManipulation`**. Because this package is automatically loaded with the box plot functions, we do not have to load it; however, for completeness, we show it as being loaded. First we take the percents for 1985 in column two and name the transpose `column2`. Using `Map` we then apply `DropNonNumeric` to remove the null entries. The list that results from this command is then assigned the name `column2`. We generate `column4` for 1991 in the same way. Using a graphics array, we display in the first row a box plot for each year using **`boxplotLists`** and in the second row a box plot for each year using **`boxplotListsV`**. Notice the use of the option `PlotLabel` for the vertical box plots.

```
<<Statistics`DataManipulation`
column2=ColumnTake[sciencedegrees,{2}]//Transpose;
column2=Map[DropNonNumeric,column2]
```

```
{{16.8,14.7,17.1,22.5,39.3,23.8,28.8,19.5,22.7,21.8,
  20.5,6.1,13.9,15.4,20.2,23.,21.7}}
```

```
column4=ColumnTake[sciencedegrees,{4}]//Transpose;
column4=Map[DropNonNumeric,column4]
```

```
{{21.6,20.1,32.2,15.5,27.6,34.5,31.5,28.5,19.8,23.5,21.
4,16.3,12.3,15.4,24.3,22.7,21.3,15.9}}
```

```
bp1=boxplotLists[column2,PlotRange->All,
  DisplayFunction->Identity];

bp2=boxplotLists[column4,PlotRange->All,
  DisplayFunction->Identity];

bp3=boxplotListsV[column2,PlotLabel->"Year 1985",
  PlotRange->All,DisplayFunction->Identity];

bp4=boxplotListsV[column4,PlotLabel->"Year 1991",
  PlotRange->All,DisplayFunction->Identity];

Show[GraphicsArray[{{bp1,bp2},{bp3,bp4}}]]
```

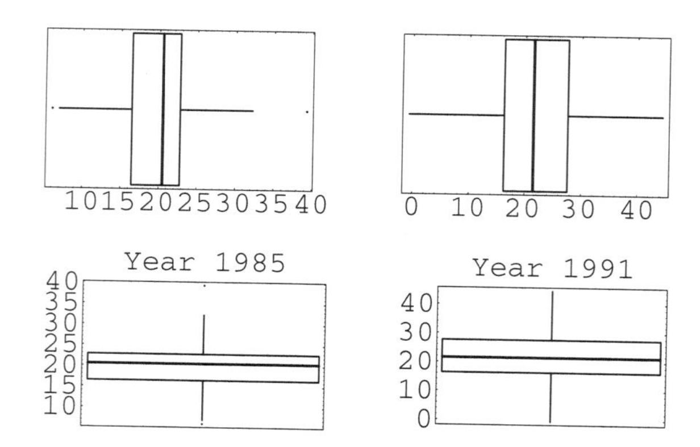

In order to display the boxplots for both years together, we join `column1` and `column2` into one list called `both` using `Join` and generate both horizontal and vertical displays in an array. Notice that the order of of display is bottom to top in the horizontal display and left to right in the vertical display.

```
both = Join [column2 , column4 ];

bothhbp = boxplotLists [both ,
  PlotRange -> All , DisplayFunction -> Identity ];
bothhbpv = boxplotListsV [both ,
  PlotRange -> All , DisplayFunction -> Identity ];
Show [GraphicsArray [{bothhbp , bothhbpv }]]
```

0 10 20 30 40

40 30 20 10 0

It is possible to take columns two, three, and four of `sciencedegrees` at once and, using `Transpose`, to generate a list called `percents` containing a list for each of the three years. We then display the horizontal and vertical versions of the combined boxplots.

```
percents=ColumnTake[sciencedegrees,{2,4}]//Transpose;

percents=Map[DropNonNumeric,percents]

{{16.8,14.7,17.1,22.5,39.3,23.8,28.8,19.5,22.7,21.8,
 20.5,6.1,13.9,15.4,20.2,23.,21.7},{19.6,16.4,26.1,
 33.5,31.3,34.1,19.7,23.5,21.1,19.5,12.9,24.3,15.,24.,
 23.,20.6,16.9},{21.6,20.1,32.2,15.5,27.6,34.5,31.5,
 28.5,19.8,23.5,21.4,16.3,12.3,15.4,24.3,22.7,21.3,15.9}}
```

```
percentsbp=boxplotLists[percents,PlotRange->All,
  DisplayFunction->Identity];
percentsbpv=boxplotListsV[percents,PlotRange->All,
  DisplayFunction->Identity];
Show[GraphicsArray[{percentsbp,percentsbpv}]]
```

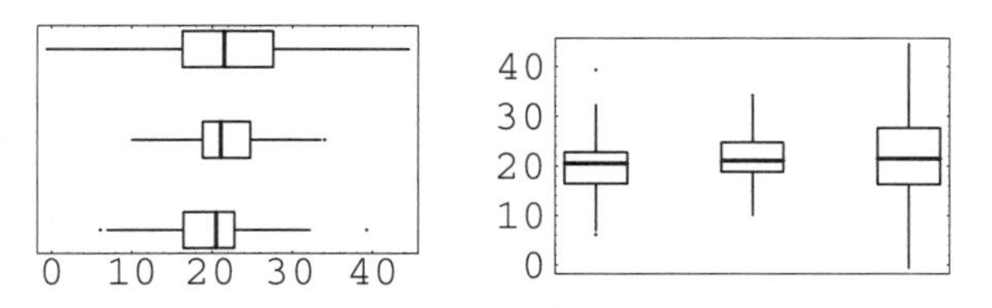

When the data are in the format of a rectangular array like `sciencedegrees`, the procedures **boxplotArray** and **boxplotArrayV** can be used directly by specifying a list of column numbers. The nonnumerical entries (null) are ignored.

```
databp = boxplotArray [sciencedegrees ,
  {2, 3, 4}, DisplayFunction -> Identity ];
databpv = boxplotArrayV [sciencedegrees ,
  {2, 3, 4}, DisplayFunction -> Identity ];
Show [GraphicsArray [{databp , databpv }]]
```

Finally, we illustrate the capabilities of the box plot procedures by using some of the available options. For example, the frame is removed by specifying `Frame->False`. Note that the boxplots are located above positions 0, 1, and 2. Thus, in order to place the dates below the horizontal axis, the origin was moved to the left and the ticks at locations 0, 1, and 2 were defined to be the corresponding years.

```
boxplotArrayV [sciencedegrees , {2, 3, 4},
 PlotLabel -> "Worldwide Science Degrees ",
 Frame -> False , Axes -> {True , True },
 AxesOrigin -> {-.5, Automatic },
 AxesLabel -> {"Years ", "Percent "}, Ticks ->
  {{{0, "1988 "}, {1, "1991 "}, {2, "1995 "}},
   Automatic }]
```

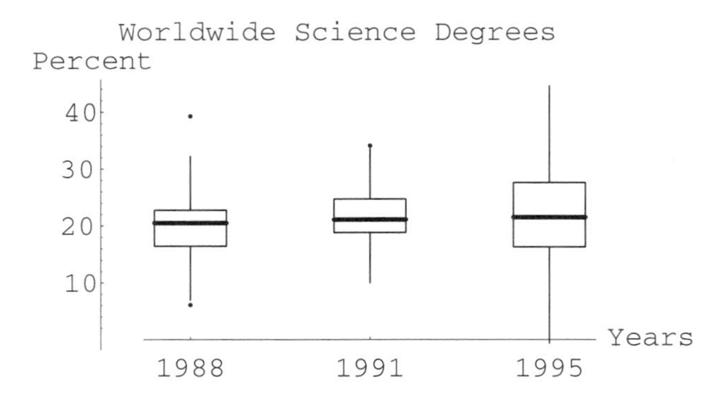

■

Stem-and-Leaf Charts

A **stem-and-leaf chart** is used to summarize and present the values of a quantitative variable. The stem corresponds to the leading digit or digits and the leaf to the trailing digit. A stem is equivalent to one class of a frequency table, and the leaves corresponding to that stem are listed in the chart adjacent to the stem. The frequency of the class represented by the stem can be found by counting the number of leaves. One advantage of a stem-and-leaf chart over a frequency table is that the original values may be preserved. That is, it may be possible to resurrect the data values from the chart. On the other hand, there is sometimes not much flexibility in choosing the stems, and the data values may have to be truncated before making the chart. A second advantage is the visual effect. The chart approximates a horizontal histogram of the data values. A disadvantage is that a stem-and-leaf chart is *not* suitable for large data sets. Finally, when the number of leaves per stem is large, the stems may be split into two or five parts. When splitting into two parts, each stem is listed twice. Leaves in the range 0 to 4 are placed after the first, and leaves in the range of 5 to 9 are placed after the second. When splitting into five parts, the leaf ranges are 0 to 1, 2 to 3, 4 to 5, 6 to 7, and 8 to 9.

We have written two procedures to construct and display a stem-and-leaf chart. One, called **`stemAndLeaf`**, requires an input list of numbers. The second, called **`stemAndLeafArray`**, accepts an array name and a column number. It will ignore nonnumeric entries in the column. Both procedures identify Outside Values and Far Outside Values in the data. Recall that data values between the inner and outer fences are called "outside" and data values outside the outer fences are called "far out." Such values are often called outliers. See the discussion in the subsection on box-and-whiskers plots. Both procedures will also output a five-number summary (minimum, first quartile, median, third quartile, and maximum) if the option `printStatistics->True` is specified. Note that the output may be double-spaced. If this happens because of each output line being in its own cell, the output can be compressed by selecting these cells and choosing Merge Cells from the Cell menu.

EXAMPLE 8: Table 4.4 (see Example 4) lists selected motor vehicle statistics for the 50 states. Generate stem-and-leaf charts for (a) State Gas Tax, (b) Registered Motor Vehicles per Thousand, (c) Licensed Drivers per Vehicle, and (d) Miles per Gallon.

SOLUTION: In Example 4, we named the array which resulted from entering the data from Table 4.4 into *Mathematica* `motorvehicle`. We assume it is open in *Mathematica*, and we also load the procedure **`stemAndLeaf`** into *Mathematica*. To do this, access the disk provided with the book, locate the procedure name, click on the rightmost cell bracket next to the name, press **ENTER**, and return to the *Mathematica* work sheet. Next we load the **DataManipulation** package and use `Column` to create four lists, one for each of columns 2, 5, 6, and 8 of `motorvehicle`. We then display a stem-and-leaf chart for each list.

```
<<Statistics`DataManipulation`

gastax=Column[motorvehicle,2];
registeredmv=Column[motorvehicle,5];
drivers=Column[motorvehicle,6];
mpg=Column[motorvehicle,8];
```

State Gas Tax: The gas tax rates range from 7.5 cents to 26.0 cents. Notice that the stems represent the units and tens part of the rates and the leaves represent the first decimal place. On the output, the value of the leaves is indicated by `Leaf Unit`. Thus, `Leaf Unit:0.1` indicates the first decimal place. Notice that some stems, for example 12 and 14, are missing. This may happen when there are no corresponding leaves. Be careful not to be misled about shape if too many stems are missing. The values 7.5 and 8.0 are indicated as being between the lower inner and outer fences by being listed as Outside Values.

```
stemAndLeaf[gastax]

Title: Stem-and-Leaf Plot
Leaf Unit: 0.1
   7      5
   8      0
   9      0
  10      5
  11      06
  13      0
  15      004
```

```
 16      0000
 17      00007
 18      002567
 19      0000
 20      0000003
 21      0005
 22      002348
 23      0
 24      1
 26      00
Outside Values: {7.5,8}
Far Outside Values: {}
```

Registered Motor Vehicles per Thousand: The number of registered motor vehicles per thousand ranges from 567 to 1164. In order to use a manageable number of stems, the units place is truncated and the leaves are taken from the tens place. This is indicated on the output by `Leaf Unit:10`. Notice that the stems have been split into two parts. Two equal stem values may or may not be output at the beginning or the end of the chart depending on the leaf values. In this output, only one stem value of 5 was needed, but both stem values of 11 were required. The maximum value, 1164, is indicated as falling between the upper inner and outer fences by being listed as an Outside Value.

```
stemAndLeaf[registeredmv,printStatistics->True]
```

```
Title: Stem-and-Leaf Plot
N: 50
Minimum: 567.
First Quartile: 729.
Median: 790.
Third Quartile: 860.
Maximum: 1164.
Leaf Unit: 10.
   5     6
   6     11
   6     77889
   7     0112233
```

```
   7      555677778999999
   8      012444
   8      556677
   9      124
   9      578
  10      14
  10
  11
  11      6
Outside Values: {1164}
Far Outside Values: {}
```

Licensed Drivers per Vehicle: The licensed drivers per registered vehicle range from 0.63 to 1.19. The leaves are the second decimal place as indicated by `Leaf Unit:0.01`. The stems have been split into two parts, and in this example both parts were needed both at the beginning and at the end. The maximum value, 1.19, is indicated as being "outside."

```
stemAndLeaf[drivers]
```

```
Title: Stem-and-Leaf Plot
Leaf Unit: 0.01
   6      3
   6      5677
   7      0124
   7      5799
   8      223344
   8      5566778889999
   9      011233344
   9      999
  10      001
  10      5
  11      4
  11      9
Outside Values: {1.19}
Far Outside Values: {}
```

Miles per Gallon: The miles per gallon values range from 12.45 to 20.86. Both of these values are identified as "outside." Notice that the second decimal place is truncated, and the leaves are from the first decimal place.

```
stemAndLeaf[mpg,printStatistics->True]
```

```
Title: Stem-and-Leaf Plot
N: 50
Minimum: 12.45
First Quartile: 15.43
Median: 16.155
Third Quartile: 17.39
Maximum: 20.86
Leaf Unit: 0.1
  12      47
  13      2447
  14      568
  15      22345555779
  16      00011112446689
  17      22334446667
  18      0001
  20      8
Outside Values: {12.45,20.86}
Far Outside Values: {}
```

Note that these stem-and-leaf charts could have been generated directly using **stemAndLeafArray**. For example, to make a chart for gas taxes, we indicate column 2 of `motorvehicle`. Note that the brackets, {}, are necessary.

```
stemAndLeafArray[motorvehicle,{2}]
```

```
Title: Stem-and-Leaf Plot
Leaf Unit: 0.1
   7      5
   8      0
   9      0
  10      5
  11      06
```

```
  13      0
  15      004
  16      0000
  17      00007
  18      002567
  19      0000
  20      0000003
  21      0005
  22      002348
  23      0
  24      1
  26      00
Outside Values: {7.5,8}
Far Outside Values: {}
```

■

EXAMPLE 9: Table 4.6 (see Example 7) shows the percent of undergraduate degrees awarded in the sciences for the years 1985, 1990, and 1991 for selected countries. (The symbol "null" is used to represent missing values in the table.) Make a stem-and-leaf chart for the percents of all science degrees awarded in each of the years 1985, 1990, and 1991.

SOLUTION: In Example 7, we named the array which resulted from entering the data from Table 4.6 into *Mathematica* `sciencedegrees`. We assume it is open in *Mathematica*, and we also load the procedure **`stemAndLeafArray`** into *Mathematica*. The percents of all science degrees for the years 1985, 1990, and 1991 are in columns 2, 3, and 4, respectively. Using **`stemAndLeafArray`** we display a chart for each year.

```
stemAndLeafArray[sciencedegrees,{2}]
```

```
Title: Stem-and-Leaf Plot
Leaf Unit: 1.
  0      6
  1      34
  1      5679
```

```
   2      00112233
   2      8
   3
   3      9
Outside Values: {6.1,39.3}
Far Outside Values: {}
```

stemAndLeafArray[sciencedegrees,{3}]

```
Title: Stem-and-Leaf Plot
Leaf Unit: 1.
   1      2
   1      566999
   2      013344
   2      6
   3      134
Outside Values: {33.5,34.1}
Far Outside Values: {}
```

stemAndLeafArray[sciencedegrees,{4}]

```
Title: Stem-and-Leaf Plot
Leaf Unit: 1.
   1      2
   1      55569
   2      0111234
   2      78
   3      124
Outside Values: {}
Far Outside Values: {}
```

Note that it is possible with **stemAndLeafArray**, *if it makes sense*, to combine the data from two or more columns and display a stem-and-leaf chart for the combined data. To illustrate this we generate a chart for the percents of all science degrees for the three years 1985, 1990, and 1991 combined.

stemAndLeafArray[sciencedegrees,{2,3,4}]

```
Title: Stem-and-Leaf Plot
Leaf Unit: 1.
```

```
    0     6
    1     2234
    1     555556666799999
    2     000011111122233333444
    2     6788
    3     112344
    3     9
  Outside Values: {39.3}
  Far Outside Values: {}
```

■

4.2 Bivariate Procedures

Scatter Plots

A **scatter plot** is formed by plotting an ordered pair of measurements as a point for each experimental unit. When the x-variable is categorical and the y-variable is quantitative, the scatter plot provides a visual comparison of the distribution of the y-values for each x-value. When the variables are both quantitative, the overall pattern of the scatter plot can be used to investigate whether there is a relationship between the variables, whether the relationship is linear or takes some other form, and, if there is a linear relationship, whether it is positive or negative. See also Section 3.2 for a discussion of the correlation coefficient, which is used to quantify the strength and direction of the linear relationship between two quantitative variables.

In *Mathematica*, the built-in function `ListPlot` is used to create a scatter plot. To obtain information on `ListPlot` type `??ListPlot` or refer to the subsection on polygons in Section 4.1.

EXAMPLE 1: Table 4.4 lists selected motor vehicle statistics for the 50 states. Generate a scatter plot to compareSeat Belt Use Laws (a categorical variable) for each of the quantitative variables, State Gas Tax, Registered Motor Vehicles per Thousand, Licensed Drivers per Vehicle, and Miles per Gallon.

SOLUTION: In Example 4 of Section 4.1 we named the array which resulted from entering the data from Table 4.4 into *Mathematica* `motorvehicle`. Assuming it is open, we load the **DataManipulation** package and use `Column` to create five lists, one for each of columns 3, 2, 5, 6, and 8.

```
<<Statistics`DataManipulation`

seatbelt=Column[motorvehicle,3];
gastax=Column[motorvehicle,2];
registeredmv=Column[motorvehicle,5];
drivers=Column[motorvehicle,6];
mpg=Column[motorvehicle,8];
```

After coding the values of Seat Belt Use Laws as 1 for none, 2 for secondary, and 3 for primary, we use `Transpose` to create four lists of ordered pairs to plot. Finally, we define the plots and display them together as a graphics array.

```
no=1;
s=2;
p=3;

sbxgt=Transpose[{seatbelt,gastax}];
sbxrmv=Transpose[{seatbelt,registeredmv}];
sbxd=Transpose[{seatbelt,drivers}];
sbxmpg=Transpose[{seatbelt,mpg}];

sp1=ListPlot[sbxgt,AxesOrigin->{0,7},
  PlotRange->{{0,3},{7,26}},Ticks->{{{1,"No"},
  {2,"Sec"},{3,"Pri"}},Automatic},
  PlotStyle->PointSize[0.02],DisplayFunction->Identity];

sp2=ListPlot[sbxrmv,AxesOrigin->{0,560},
  PlotRange->{{0,3},{560,1165}},Ticks->{{{1,"No"},
  {2,"Sec"},{3,"Pri"}},Automatic},
  PlotStyle->PointSize[0.02],DisplayFunction->Identity];

sp3=ListPlot[sbxd,AxesOrigin->{0,.6},
  PlotRange->{{0,3},{.6,1.2}},Ticks->{{{1,"No"},
  {2,"Sec"},{3,"Pri"}},Automatic},
PlotStyle->PointSize[0.02],DisplayFunction->Identity];

sp4=ListPlot[sbxmpg,AxesOrigin->{0,12.4},
  PlotRange->{{0,3},{12.4,20.9}},Ticks->{{{1,"No"},
  {2,"Sec"},{3,"Pri"}},Automatic},
  PlotStyle->PointSize[0.02],DisplayFunction->Identity];

Show[GraphicsArray[{{sp1,sp2},{sp3,sp4}}]]
```

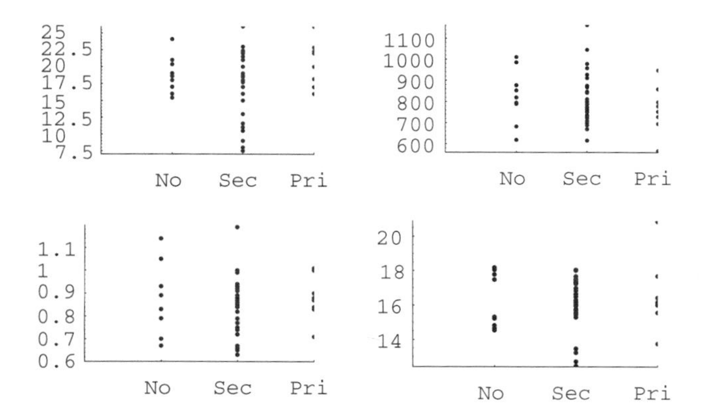

■

EXAMPLE 2: The ASCII data file **MaxMin.txt** contains the maximum and minimum temperature for various locations in the United States on December 1, 1995. Display a scatter plot of maximum versus minimum temperatures.

SOLUTION: The following screen shot shows how a portion of **MaxMin.txt** looks in a typical text editor. The first entry corresponds to the latitude, the second entry to the longitude, the third entry to the maximum, and the fourth entry to the minimum. Note that the entries are separated by tab marks. We first read the data into *Mathematica* using `ReadList` and name the list `tempdata`. The function `ToExpression` converts the data from string to numeric.

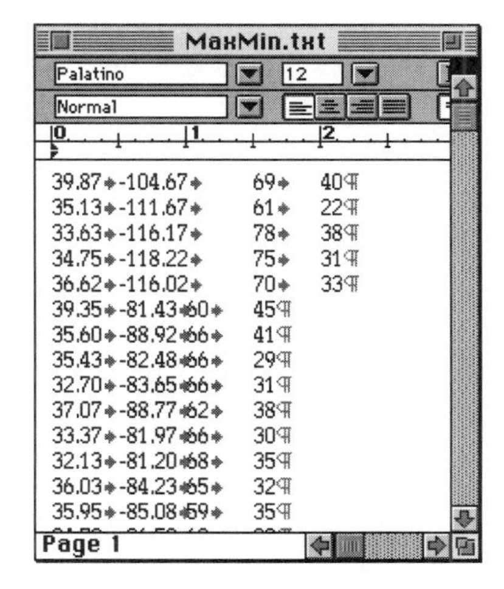
MaxMin.txt

Palatino 12

Normal

```
39.87♦-104.67♦	69♦	40¶
35.13♦-111.67♦	61♦	22¶
33.63♦-116.17♦	78♦	38¶
34.75♦-118.22♦	75♦	31¶
36.62♦-116.02♦	70♦	33¶
39.35♦-81.43♦60♦	45¶
35.60♦-88.92♦66♦	41¶
35.43♦-82.48♦66♦	29¶
32.70♦-83.65♦66♦	31¶
37.07♦-88.77♦62♦	38¶
33.37♦-81.97♦66♦	30¶
32.13♦-81.20♦68♦	35¶
36.03♦-84.23♦65♦	32¶
35.95♦-85.08♦59♦	35¶
```

Page 1

```
tempdata=ReadList["MaxMin.txt",Word,
  RecordLists->True,WordSeparators->{"\t"}];
tempdata=ToExpression[tempdata];
```

The length of the list `tempdata` is found to be 322 with `Length`. `First` and `Last` give the first and last elements in `tempdata`, respectively.

```
Length[tempdata]
```

```
322
```

```
First[tempdata]
Last[tempdata]
```

```
{39.87,-104.67,69,40}
{47.38,-92.85,24,2}
```

After loading the **DataManipulation** package, we extract the third and fourth columns of `tempdata` with `Column`, where these two columns correspond to the maximum and minimum temperatures, respectively. We calculate the minimum and maximum for each list using `Min` and `Max`.

```
<<Statistics`DataManipulation`
```

```
maxtemps=Column[tempdata,3];
mintemps=Column[tempdata,4];
```

```
Min[maxtemps]
```

```
19
```

```
Max[maxtemps]
```

```
85
```

```
Min[mintemps]
```

```
0
```

```
Max[mintemps]
```

```
68
```

To make a list of ordered pairs for use with `ListPlot` we use `ColumnTake` to extract the third and fourth columns together and name the result `toplot`. We then plot the points (x, y), where x corresponds to the maximum temperature and y corresponds to the minimum temperature. Mindful of the respective minimum and maximum values in `maxtemps` and `mintemps`, we use the `ListPlot` options `AxesOrigins` and `PlotRange` to indicate the location of the origin and the length of the axes. In addition, we use `PlotStyle->PointSize[0.01]` to instruct that the data points be graphed with a slightly larger point size than that produced with the default setting.

```
toplot = ColumnTake [tempdata , {3, 4}];

ListPlot [toplot , AxesOrigin -> {15, -5},
 PlotRange -> {{15, 90}, {-5, 70}},
 PlotStyle -> PointSize [0.01 ]]
```

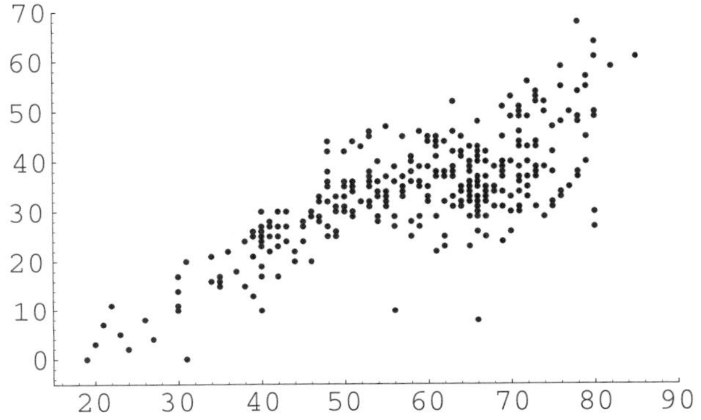

The scatter plot suggests a linear relationship between maximum and minimum temperatures, and we will emphasize this by including a straight line segment in the plot. Note that because there is no clear dependence relationship, a least-squares regression line is not appropriate. The equation of the line to use here is calculated in the following.

```
<<Statistics`DescriptiveStatistics`
meanforx=Mean[maxtemps];
meanfory=Mean[mintemps];
stdevforx=StandardDeviation[maxtemps];
stdevfory=StandardDeviation[mintemps];
slope=Divide[stdevfory,stdevforx];
intercept=meanfory-(slope)(meanforx);
f[x_]=intercept+slope(x)//N

-13.4967+0.809819 x
```

Next we create a plot of the line called `plotline`, being careful to use the same origin and ranges for the axes. The output is suppressed using `DisplayFunction->Identity`. We name the scatter plot `plotpoints` and similarly suppress output. Finally, we use `Show` and `DisplayFunction->$DisplayFunction` to display the graph. Notice that the relationship is positive. After loading the MultiDescriptiveStatistics package, we find that the correlation coefficient is 0.74.

```
plotline =
 Plot [f [x], {x, 15, 90}, AxesOrigin -> {15, -5},
  PlotRange -> {{15, 90}, {-5, 70}},
  DisplayFunction -> Identity ];
plotpoints =
 ListPlot [toplot , AxesOrigin -> {15, -5},
  PlotRange -> {{15, 90}, {-5, 70}},
  PlotStyle -> PointSize [0.01 ],
  DisplayFunction -> Identity ];
Show [plotline , plotpoints ,
 DisplayFunction -> $DisplayFunction,
 PlotRange -> All ]
```

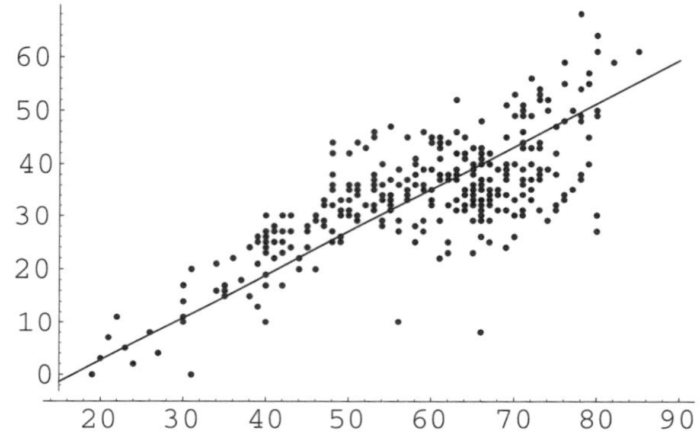

```
<< Statistics'MultiDescriptiveStatistics'
```

```
corr = Correlation [maxtemps , mintemps ] // N
```

```
0.735667
```

■

Contingency Tables (Cross Tabulations)

A two-dimensional frequency table, also called a **contingency table** or a **cross tabulation**, is a table of frequencies where each row represents the values for one variable, and each column represents the values for the second variable. The variables can be categorical or quantitative, although, for investigating relationships between variables, the contingency table is usually used when one or both of the variables are categorical. Relationships among the variables are described by calculating appropriate percentages from the frequencies. For quantative data the values are divided into classes, and each row (column) corresponds to a class.

We have written two procedures to construct and display a contingency table and a third procedure to display a three-dimensional bar chart corresponding to the contingency table. They make use of the *Mathematica* packages **Statistics`DataManipulation`** and **Graphics`Graphics3D`**. To use any of these procedures, go to the disk provided with the book, locate the cell entitled **Contingency Table**, click on the rightmost cell bracket, press **ENTER**, and return to the *Mathematica* work sheet.

The procedures available in **Contingency Table** are as follows.

`contingencyTableCategory[dataset_,{rows_,columns_},opts___]`: Displays a contingency table. Given a data set, `dataset`, and two columns in the data set each containing values of a categorical variable, the rows of the contingency table are defined by the first column number, `rows`, and the columns of the contingency table are defined by the second column number, `columns`. Any option available for `TableForm` can also be entered.

`contingencyTableClass[dataset_,{rows_,limits1_},{columns_,limits2_},opts___]`: Displays a contingency table. Given a data set, `dataset`, and two columns in the data set, the rows of the contingency table are defined by the first column number, `rows`, where `limits1` is a list of class limits. The columns of the contingency table are defined by the second column number, `columns`, where `limits2` is a list of class limits. Any option available for `TableForm` can also be entered.

`contingencyGraphCategory[dataset_,{rows_,columns_},opts___]`: Displays a three-dimensional bar chart corresponding to the contingency table generated by `contingencyTableCategory`. The input is the same as for `contingencyTable-Category`, except the options available correspond to those for `BarChart3D`.

EXAMPLE 3: A data set with information about students in an introductory statistics class is entered into *Mathematica* and named `dataset`. The first column contains values of the categorical variable gender where m means male and f means female. Age is stored in the second column. The third column contains values of the categorical variable hair color. (a) Create a cross tabulation using gender for rows and hair color for columns. (b) Create a corresponding three-dimensional bar chart.

SOLUTION: After entering the data set, we open **Contingency Table** on the disk which comes with the book, and use **`contingencyTableCategory`** to generate the table. Notice that the procedure automatically labels the rows and columns.

```
dataset={{f,20,brown},{f,20,brown},{f,21,black},{m,23,
 black},{f,21,brown},{f,22,red},{f,30,black},{f,24,
 brown},{f,22,brown},{m,23,blond},{f,22,red},{m,21,
 black},{f,26,blond},{m,25,brown},{m,32,brown},{f,23,
 blond},{f,23,red},{m,20,brown},{f,20,blond},{m,22,
 brown},{f,23,blond},{m,22,black},{f,20,blond},{m,21,
 brown},{f,24,red},{f,20,brown},{m,44,brown}}
```

```
contingencyTableCategory[dataset,{1,3}]
```

	black	blond	brown	red
f	2	5	6	4
m	3	1	6	0

The default labels on the rows and columns may bechanged or omitted by using the option `TableHeadings`.

```
contingencyTableCategory[dataset,{1,3},TableHeadings->
{{"Female","Male"},{"Black","Blond","Brown","Red"}}]
```

	Black	Blond	Brown	Red
Female	2	5	6	4
Male	3	1	6	0

We display the bar chart as follows.

```
contingencyGraphCategory[dataset, {1, 3}]
```

{black, blond, brown, red}

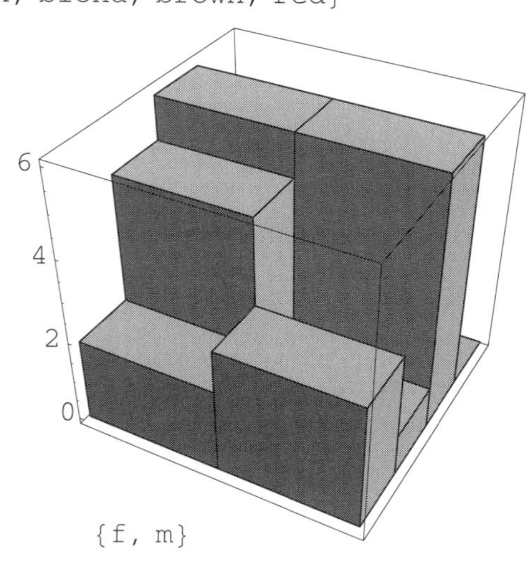

■

EXAMPLE 4: Table 4.4 lists selected motor vehicle statistics for the 50 states. (a) Generate a contingency table to compare quantitative variables State Gas Tax as rows with Miles per Gallon as columns. (b) Generate a contingency table to compare Seat Belt Use Laws (a categorical variable) as rows with Miles per Gallon as columns.

SOLUTION: In Example 4 of Section 4.1 we named the array which resulted from entering the data from Table 4.4 into *Mathematica* `motorvehicle`. Assuming it is open, we load **Contingency Table** from the disk which comes with the book. Note that even though we show the **DataManipulation** package being loaded, this is done automatically when **Contingency Table** is loaded. First, we illustrate the operation of **`contingencyTableClass`** then use it to generate the contingency tables. The illustration begins by using `Column` to creat two lists, one for column 2 called `gastax` and one for column 8 called `mpg`. Using `Table` we then create two lists of limits to be used to define the rows and columns of the contingency table.

```
<<Statistics`DataManipulation`
gastax=Column[motorvehicle,2]
```

```
{13,8,18,18.7,16,22,26,19,11.6,7.5,16,21,19,15,20,17,
 15.4,20,19,18.5,21,15,20,18.2,11.03,20,24.1,21.5,18.6,
 10.5,17,22.89,22.3,17,21,17,22,22.4,26,16,18,20,20,19,
 16,17.7,23,20.35,22.2,9}
```

```
mpg=Column[motorvehicle,8]
```

```
{16.14,12.45,18.01,13.22,17.66,16.02,17.68,16.97,16.48,
 16.83,20.86,16.61,13.47,16.19,13.75,15.29,14.82,17.39,
 17.47,17.47,17.77,17.33,17.65,16.,15.52,15.54,14.54,
 13.46,18.05,15.94,16.15,16.01,15.56,14.66,15.56,16.28,
 16.16,15.76,17.49,15.7,15.22,15.43,16.42,16.64,18.18,
 17.28,17.23,15.32,18.05,12.71}
```

```
limits1=Table[i,{i,Min[gastax],Max[gastax],
 (Max[gastax]-Min[gastax])/10}]
```

```
{7.5,9.35,11.2,13.05,14.9,16.75,18.6,20.45,22.3,
 24.15,26.}
```

```
limits2=Table[i,{i,Min[mpg],Max[mpg],
 (Max[mpg]-Min[mpg])/8}]
```

```
{12.45,13.5012,14.5525,15.6038,16.655,17.7062,18.7575,
 19.8087,20.86}
```

The procedure **contingencyTableClass** includes a function called arrangeclasses which takes a list of limits and assigns the data values to the classes defined by these limits. For example, the first class for gastax will contain the values equal to or greater than 7.5 and less than 9.35, the second class will contain the values equal to or greater than 9.35 and less than 11.2, and so forth. The last class will contain values equal to or greater than 24.15 and less than or equal to 26. Note that the classes are closed on the left and open on the right, except for the last class, which is closed on both sides. The 11 values in limits1 will define 10 classes corresponding, in this example, to 10 rows. Once the classes are constructed values in the first class are categorized as 1, values in the second class are categorized as 2, and so forth. The contingency table is then generated. The following output of arrangeclasses is transparent to the user of **contingencyTableClass**.

```
bins10=arrangeclasses[gastax,limits1]
```

```
{{8,7.5,9},{11.03,10.5},{13,11.6},{},{16,16,15,15.4,15,
  16,16},{18,17,18.5,18.2,17,17,17,18,17.7},{18.7,19,19,
  20,20,19,20,20,18.6,20,20,19,20.35},{22,21,21,21.5,21,
  22,22.2},{24.1,22.89,22.3,22.4,23},{26,26}}
```

```
bins20=arrangeclasses[mpg,limits2]
```

```
{{12.45,13.22,13.47,13.46,12.71},{13.75,14.54},{15.29,
  14.82,15.52,15.54,15.56,14.66,15.56,15.22,15.43,
  15.32},{16.14,16.02,16.48,16.61,16.19,16.,15.94,
  16.15,16.01,16.28,16.16,15.76,15.7,16.42,16.64},
 {17.66,17.68,16.97,16.83,17.39,17.47,17.47,17.33,
  17.65,17.49,17.28,17.23},{18.01,17.77,18.05,18.18,
  18.05},{},{20.86}}
```

We now show examples of generating the contingency table using different lists of limits. Notice that column numbers and lists of limits are all that is required to use **contingencyTableClass**. Also, there are no default row and column headings with this procedure. To generate the second contingency table we first load the package **DescriptiveStatistics** in order to use Median to define the limits in limits11.

```
contingencyTableClass
 [motorvehicle,{2,limits1},{8,limits2},
  TableHeadings->{{r1,r2,r3,r4,r5,r6,r7,r8,r9},
  {c1,c2,c3,c4,c5,c6,c7}}]
```

	c1	c2	c3	c4	c5	c6	c7
r1	2	0	0	0	1	0	0
r2	0	0	1	1	0	0	0
r3	0	0	0	2	0	0	0

```
r4   0  0  1  2  2  1  1
r5   0  0  3  3  2  1  0
r6   2  1  3  2  4  1  0
r7   1  0  1  3  0  2  0
r8   0  1  1  2  1  0  0
r9   0  0  0  0  2  0  0
```

```
<<Statistics`DescriptiveStatistics`

limits11={Min[gastax],Median[gastax],Max[gastax]};
limits22={Min[mpg],Median[mpg],Max[mpg]};

contingencyTableClass[motorvehicle,{2,limits11},
 {8,limits22},  TableHeadings->{{"Below Median",
  "Median & Up"},{"Below Median","Median & Up"}}]
```

	Below Median	Median & Up
Below Median	13	12
Median & Up	12	13

When using **contingencyTableClass** and one (or both) of the variables is categorical, there are two vital facts to remember. First, the categories have to be coded as 1, 2, 3, and so on, and these codes represent rows or columns of the contingency table. Notice that none, secondary, and primary form an order of increasing emphasis by the law on seat belt usage. We chose to code them as 1, 2, 3, respectively. Second, the list of limits must end with a value which is higher than the maximum value. Thus, even though our codes are 1, 2, and 3, the list of limits, `{1,2,3,4}`, extends to 4.

```
limits8=Table[i,{i,Min[mpg],Max[mpg],
 (Max[mpg]-Min[mpg])/3}];

no=1;
s=2;
p=3;

contingencyTableClass[motorvehicle,{3,{1,2,3,4}},
  {8,limits8},TableHeadings->{{"None","Secondary",
   "Primary"},{"Low","Medium","High"}}]
```

	Low	Medium	High
None	4	4	1
Secondary	5	27	0
Primary	1	7	1

■

4.3 A Multivariate Procedure

Scatter Plot Matrices

A **scatter plot matrix** (SPLOM) provides a convenient way to look at several variables in one display. It is a matrix where each cell off the main diagonal is a scatter plot, and each cell on the main diagonal is a histogram. Thus, the distribution of each variable is on the main diagonal, and a comparison between each pair of variables is contained in a cell off the main diagonal. An alternative to histograms on the main diagonal could be box plots which identify outliers.

The conventional way to to generate several plots and display them together is to generate each plot separately, including the option `DisplayFunction->Identity` in each `Plot` command, and then combine the plots with the function `Show`, including the option `DisplayFunction->$DisplayFunction` in the `Show` command. The *Mathematica* package **Graphics** contains two functions that automate this process. `DisplayTogether` can handle any plotting commands that will accept `DisplayFunc-tion ->Identity` as an option. The resulting plots must be able to be combined with `Show` for this function to be effective. `DisplayTogetherArray` accepts the same kind of inputs but displays them in an array. The input plot commands should be entered in an array whose structure matches that of the desired output array.

EXAMPLE 1: Table 4.6 shows the percent of undergraduate degrees awarded in the sciences for the years 1985, 1990, and 1991 for selected countries. (The symbol "null" is used to represent missing values in the table.) Construct a scatter plot matrix for the percents of all science degrees awarded in the years 1985, 1990, and 1991.

SOLUTION: In Example 7 of Section 4.1 we named the array which resulted from entering the data from Table 4.6 into *Mathematica* `sciencedegrees`. We assume it is open in *Mathematica*, and, after loading the **DataManipulation** package, we use `Column` to extract columns 2, 3, and 4 corresponding to years 1985, 1990, and 1991, respectively. We also remove all records which contain a null.

```
<<Statistics`DataManipulation`

alldegrees=Column[sciencedegrees,{2,3,4}]

{{16.8,19.6,20.1},{17.1,16.4,15.5},{22.5,26.1,27.6},
 {39.3,33.5,34.5},{23.8,31.3,31.5},{28.8,34.1,28.5},
 {19.5,19.7,19.8},{22.7,23.5,23.5},{21.8,21.1,21.4},
 {20.5,19.5,16.3},{6.1,12.9,12.3},{13.9,15.,15.4},
```

```
{15.4,24.,24.3},{20.2,23.,22.7},{23.,20.6,21.3},
{21.7,16.9,15.9}}
```

After loading **Box and Whiskers** from the disk which comes with this book, we define the box plots which will be on the main diagonal of the SPLOM. We suppress output using `DisplayFunction->Identity`.

```
array11=boxplotArrayV[sciencedegrees,{2},
  PlotLabel->"1985",Frame->False,Axes->{True,True},
  AxesOrigin->{-.3,Automatic},PlotRange->{Automatic,
  {0,40}},Ticks->{{},{10,20,30,40}},
  DisplayFunction->Identity];
array22=boxplotArrayV[sciencedegrees,{3},
  PlotLabel->"1990",Frame->False,Axes->{True,True},
  AxesOrigin->{-.3,Automatic},PlotRange->{Automatic,
  {0,40}},Ticks->{{},{10,20,30,40}},
  DisplayFunction->Identity];
array33=boxplotArrayV[sciencedegrees,{4},
  PlotLabel->"1991",Frame->False,Axes->{True,True},
  AxesOrigin->{-.3,Automatic},PlotRange->{Automatic,
  {0,40}},Ticks->{{},{10,20,30,40}},
  DisplayFunction->Identity];
```

Next, we define the scatter plots which will complete the rest of the SPLOM, suppressing output using `DisplayFunction->Identity`. Then, using `Show`, we display the SPLOM.

```
toplot12=Column[alldegrees,{1,2}];
toplot13=Column[alldegrees,{1,3}];
toplot21=Column[alldegrees,{2,1}];
toplot23=Column[alldegrees,{2,3}];
toplot31=Column[alldegrees,{3,1}];
toplot32=Column[alldegrees,{3,2}];
array12=ListPlot[toplot12,AxesOrigin->{0,0},
  PlotRange->{{0,40},{0,40}},PlotStyle->PointSize[0.01],
  Ticks->{{10,20,30,40},{10,20,30,40}},
  DisplayFunction->Identity];
array13=ListPlot[toplot13,AxesOrigin->{0,0},
  PlotRange->{{0,40},{0,40}},PlotStyle->PointSize[0.01],
  Ticks->{{10,20,30,40},{10,20,30,40}},
  DisplayFunction->Identity];
```

```
array21=ListPlot[toplot21,AxesOrigin->{0,0},
  PlotRange->{{0,40},{0,40}},PlotStyle->PointSize[0.01],
  Ticks->{{10,20,30,40},{10,20,30,40}},
  DisplayFunction->Identity];
array23=ListPlot[toplot23,AxesOrigin->{0,0},
  PlotRange->{{0,40},{0,40}},PlotStyle->PointSize[0.01],
  Ticks->{{10,20,30,40},{10,20,30,40}},
  DisplayFunction->Identity];
array31=ListPlot[toplot31,AxesOrigin->{0,0},
  PlotRange->{{0,40},{0,40}},PlotStyle->PointSize[0.01],
  Ticks->{{10,20,30,40},{10,20,30,40}},
  DisplayFunction->Identity];
array32=ListPlot[toplot32,AxesOrigin->{0,0},
  PlotRange->{{0,40},{0,40}},PlotStyle->PointSize[0.01],
  Ticks->{{10,20,30,40},{10,20,30,40}},
  DisplayFunction->Identity];
```

```
Show [GraphicsArray
 [{{array11 , array12 , array13 },
   {array21 , array22 , array23 },
   {array31 , array32 , array33 }}]]
```

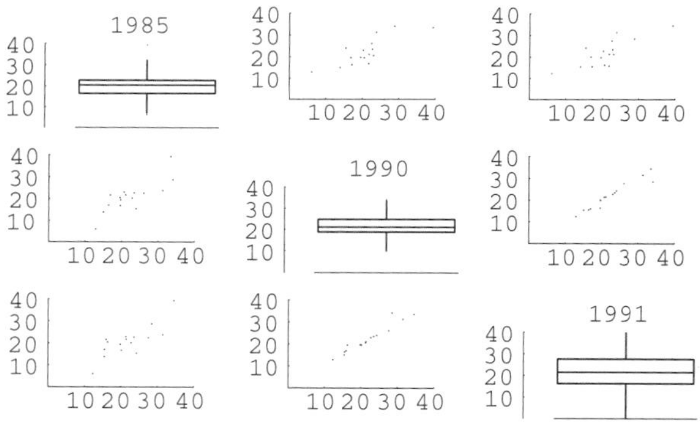

CHAPTER 5

Data Smoothing and Time Series: An Introduction

5.1 Time Series and Smoothing

A **univariate time series** is a sequence of observations taken on a single variable in time order. Let y_1 represent the first observation, y_2 represent the next observation taken after a fixed time increment, and so on to the nth observation, y_n. Then the time series may be represented by $y_1, y_2, \ldots, y_n$. These observations are often summarized graphically in a scatter plot where the values on the horizontal axis represent time.

EXAMPLE 1: The text file **Hours.dat** contains the average number of hours worked each week by workers in the United States for each month from January 1964 to February 1997. Note that records occur on consecutive lines. For each record, the first value is year; the second value is month, indicated by a number between 1 and 12; and the third value is average number of hours. These values are separated by tab marks. Plot the average number of hours worked for the 86 months beginning with January 1990 and ending with February 1997.

SOLUTION: We begin by using `ReadList` to load the data contained in the **Hours.dat** data file. Then using `Take` we display the first five records and the last five records. Notice that the average number of hours worked per week in January 1964 was 38.1, and the average for February 1997 was 34.5 hours per week.

```
hoursdatafile=ReadList["Hours.dat",Number,
 RecordLists->True];
```

```
Take[hoursdatafile,5]
```

```
{{1964,1,38.1},{1964,2,38.4},{1964,3,38.5},
 {1964,4,38.5},{1964,5,38.7}}
```

```
Take[hoursdatafile,-5]
```

```
{{1996,10,34.5},{1996,11,34.5},{1996,12,34.9},
 {1997,1,33.9},{1997,2,34.5}}
```

After loading **DataManipulation** located in the **Statistics** folder (or directory), we use `Column` to select the average hours per week from the third column of the data file and `Take` to select the last 86 values corresponding to the 86 months between January 1990 and February 1997.

```
<<Statistics`DataManipulation`
hoursdata=Column[hoursdatafile,3];
selecthours=Take[hoursdata,-86]
```

```
{34.1,34.2,34.3,34.3,34.4,34.8,34.9,34.8,34.7,34.3,
 34.3,34.7,33.7,33.9,33.9,34.,34.1,34.7,34.5,34.7,34.6,
 34.4,34.3,34.6,33.8,34.2,34.2,34.1,34.4,34.5,34.6,34.9,
 34.3,34.4,34.5,34.5,34.,34.1,34.,34.2,34.7,34.6,34.8,
 35.1,34.5,34.6,34.5,34.7,34.4,34.,34.4,34.5,34.8,34.8,
 34.9,35.,34.8,35.,34.6,34.8,34.4,34.2,34.2,34.3,34.2,
 34.6,34.8,34.8,34.6,34.7,34.4,34.5,33.4,34.1,34.2,34.1,
 34.3,34.9,34.6,34.8,34.9,34.5,34.5,34.9,33.9,34.5}
```

Finally, using `ListPlot` we define a plot of the points and a plot to join the points. Notice that the point size is increased to 0.02 from the default of 0.01, and the points and lines are lightened using `GreyLevel`. Output is initially suppressed, then the plot is displayed using `Show`.

```
p1=ListPlot[selecthours,PlotStyle->{GrayLevel[0.5],
   PointSize[0.02]},DisplayFunction->Identity];
```

```
p2=ListPlot[selecthours,PlotStyle->GrayLevel[0.5],
   PlotJoined->True,DisplayFunction->Identity];
p3=Show[p1,p2,DisplayFunction->$DisplayFunction]
```

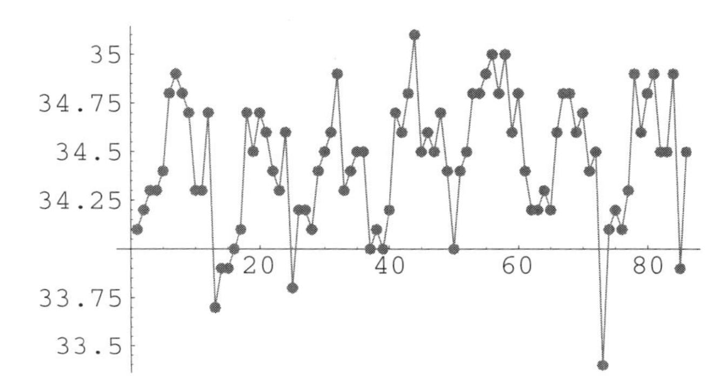

■

It is customary to classify the fluctuations of a time series into four basic types that can occur in any combination and account for the changes in the series over a period of time and give the series its irregular appearance. These four types or components of a time series are **secular trends**, cycles, seasonal variation, and irregular fluctuations. A secular trend is an upward or downward movement that characterizes a series over a period of time. It reflects the long-run growth or decline in the time series. **Cycle** refers to recurring up and down movements around trend levels. **Seasonal variations** are periodic patterns in a time series that complete themselves within a calendar year and are then repeated on a yearly basis. Finally, **irregular fluctuations** are erratic movements in a time series that follow no recognizable or regular pattern.

Because fluctuations of a time series due to seasonal variations and irregular variations can mask secular trends, a technique known as **smoothing** is often useful. In *Mathematica*, the package **DataSmoothing** located in the **Statistics** folder (or directory) contains four functions for smoothing time series data (or any other data that can be listed and plotted in a sequence corresponding to the regularly spaced values of a second variable). Information about the smoothing functions is available through the **Help Browser**.

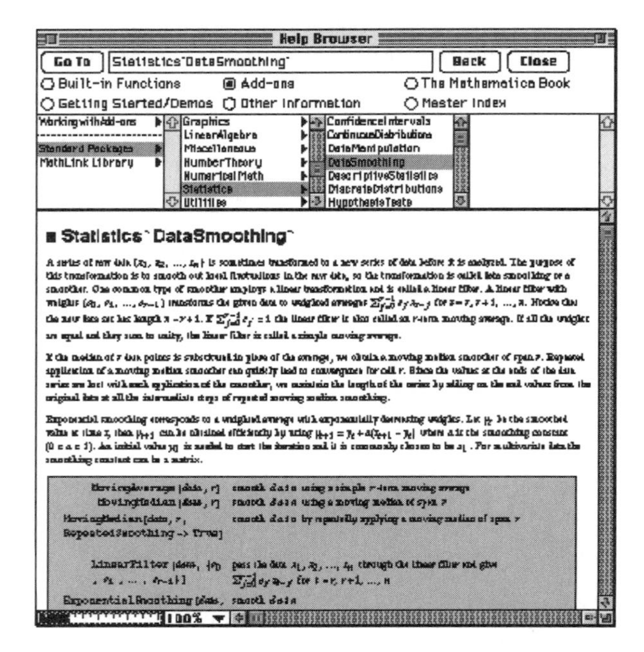

5.2 Univariate Smoothing Procedures

For problems in which the interest is mainly in the general behavior of a series, the time-honored approach is smoothing with a moving or running average. One way to think of smoothing is that it is the process of replacing each value in the series by the mean of itself and a specified number of adjacent values. Usually, an equal number of preceding and succeeding values are used in the calculation of the mean. For example, in a five-year moving average the value for a particular year is replaced by the mean of five values: itself, the values for the two preceding years, and those of the two succeeding years. Note that the number of values or terms in the mean, called the **period**, must be odd in order for this description to hold. If the period is even, then the moving average will correspond to a time midway between two of the original times. Centering these moving averages on the original times requires the calculation of two-period moving averages on the previously calculated moving averages. The results are called **centered moving averages**. Whenever possible, the period should be odd.

The general form of a moving average is

$$a_t = \frac{y_t + y_{t+1} + \ldots + y_{t-r+1}}{r},$$

where r is the period. The *Mathematica* documentation calls this an r-term moving average. The choice of r depends on the amount of smoothing desired. Increasing the value of r improves the smoothing. Notice that the choice $r = 1$ corresponds to no smoothing.

Moving Average Smoothing

The command `MovingAverage` is used to calculate what the *Mathematica* documentation calls simple moving averages. After loading the **DataSmoothing** package, we use `?` to display basic information about the command.

```
<<Statistics`DataSmoothing`
?MovingAverage
MovingAverage[datalist, r] smooths datalist using
  a simple r-term moving average. Each element in
  the result is the average of r elements from datalist.
```

For `datalist` = $\{y_1, y_2, \ldots, y_n\}$, the averages (means) are of the form

$$a_j = \frac{y_j + y_{j+1} + \ldots + y_{j-r+1}}{r}, \; j = 1, 2, \ldots, n - r + 1.$$

Notice that the smoothed series will have $r - 1$ fewer values than the original series.

EXAMPLE 1: The text file **NYSE.dat** contains New York Stock Exchange index closing values for 1995 for the composite, industrial, transportation, utility, and finance indexes. Note that records occur on consecutive lines, and the values are separated by tab marks. The first five entries in the file are displayed in the following table. (a) Use a moving average to smooth one month's values of the composite index. (b) Smooth the full year of closing values for the composite index using the moving average.

Date	Composite	Industrial	Transportation	Utility	Finance
950103	250.73	317.29	224.68	198.71	196.62
950104	251.46	317.88	225.82	199.09	198.24
950105	251.39	317.72	228.46	198.79	198.28
950106	251.59	317.80	229.78	198.97	198.80
950109	251.58	317.80	230.00	198.58	199.00

SOLUTION: Using `ReadList` and `Take` we load the data contained in the **NYSE.dat** and display the first five records. After being sure that **DataManipulation** is loaded, we use `Column` to select the composite index values from the second column of the data file and `Take` to select the last 20 values corresponding to December of 1995.

```
nysedatafile=ReadList["NYSE.dat",Number,
  RecordLists->True];
Take[nysedatafile,5]
{{950103,250.73,317.29,224.68,198.71,196.62},{950104,
  251.46,317.88,225.82,199.09,198.24},{950105,251.39,
  317.72,228.46,198.79,198.28},{950106,251.59,317.8,
  229.78,198.97,198.8},{950109,251.58,317.8,230.,
  198.58,199.}}
Length[nysedatafile]
252
<<Statistics`DataManipulation`
compositedata=Column[nysedatafile,2];
compositeDec95=Take[compositedata,-20]
{324.43,327.6,329.79,330.83,328.77,329.02,330.17,329.8,
  331.17,329.1,328.66,323.66,325.71,323.61,325.61,
  326.35,327.64,328.13,328.35,329.51}
```

(a) Using `MovingAverage` we define three series of moving averages having periods of size 3, 5 and 7 days, respectively. Then using `ListPlot` we define plots of the data points in `compositeDec95` and each of the smoothed series. They are displayed in a graphics array using `Show`.

```
dec95m3=MovingAverage[compositeDec95,3];
dec95m5=MovingAverage[compositeDec95,5];
dec95m7=MovingAverage[compositeDec95,7];

p1=ListPlot[compositeDec95,PlotStyle->PointSize[0.02],
   DisplayFunction->Identity];
p2=ListPlot[dec95m3,PlotStyle->PointSize[0.02],
   DisplayFunction->Identity];
p3=ListPlot[dec95m5,PlotStyle->PointSize[0.02],
   DisplayFunction->Identity];
p4=ListPlot[dec95m7,PlotStyle->PointSize[0.02],
   DisplayFunction->Identity];
```

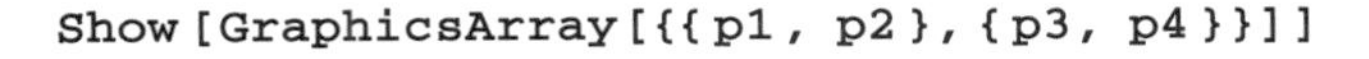

```
Show[GraphicsArray[{{p1, p2}, {p3, p4}}]]
```

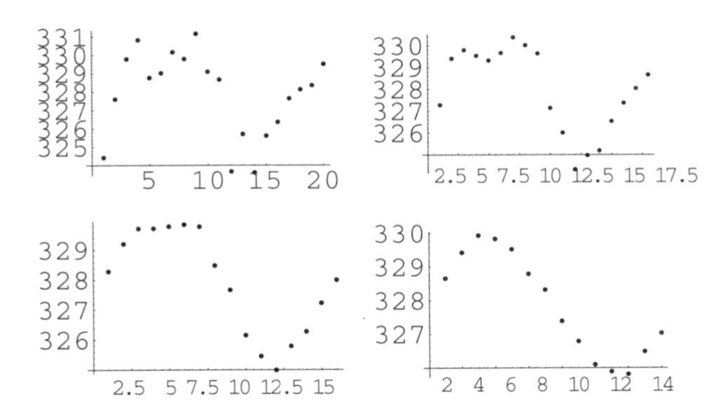

(b) Using the same steps for the entire year of composite values as were followed for the December values, we display the data points and three smoothed series having periods of 5, 15, and 31 days, respectively.

```
year95m5=MovingAverage[compositedata,5];
year95m15=MovingAverage[compositedata,15];
year95m31=MovingAverage[compositedata,31];

plot1=ListPlot[compositedata,DisplayFunction->Identity];
plot2=ListPlot[year95m5,DisplayFunction->Identity];
plot3=ListPlot[year95m15,DisplayFunction->Identity];
plot4=ListPlot[year95m31,DisplayFunction->Identity];

Show [GraphicsArray[
  {{plot1, plot2}, {plot3, plot4}}]]
```

■

Moving Median Smoothing

The command `MovingMedian` will calculate a smoothed time series using the same procedure as `MovingAverage`, except that the median of the r data points is calculated in place of the mean. The *Mathematica* documentation refers to `MovingMedian` as having span r. The option `RepeatedSmoothing->True` causes repeated smoothing until there is convergence, which means the final smoothed series would be unchanged if it was to be smoothed again. If this option is specified, only odd-numbered spans may be used. Otherwise, the repeated smoothing may not converge. Because the length of a smoothed series is reduced by r–1, in repeated smoothing the length of the successive smoothed series is maintained by adding the end values $((r-1)/2$ at each end) from the original data at all intermediate steps of repeated moving median smoothing.

EXAMPLE 2: The text file **NYSE2.dat** contains New York Stock Exchange index closes from January 2, 1990, to November 29, 1996, for the composite, industrial, transportation, utility, and finance indexes. Note that records occur on consecutive lines, and the values are separated by tab marks. (a) Smooth the closing values for the composite index using the moving median. (b) Smooth the closing values for the composite index using the moving median with option `RepeatedSmoothing->True`.

SOLUTION: Following the same procedure as for the earlier examples in this section, we use `ReadList` and `Take` to load the data contained in the **NYSE2.dat** and display the first five and last five records. After being sure that **DataManipulation** is loaded, we use `Column` to select the composite index values from the second column of the data file and `Take` to display the first and last five values in the resulting list, called `composite`.

```
nyse2datafile=ReadList["NYSE2.dat",Number,RecordLists-
>True];

Take[nyse2datafile,5]

{{900102,198.,236.68,182.25,102.92,158.17},{900103,
  197.8,236.52,181.5,102.41,158.71},{900104,196.29,
  235.13,181.,100.55,158.08},{900105,194.64,233.53,
  179.23,98.87,157.13},{900108,195.33,234.39,179.67,
  99.42,157.15}}

Take[nyse2datafile,-5]

{{961122,394.66,498.36,352.76,260.93,351.19},{961125,
  398.86,503.23,356.66,262.58,356.96},{961126,397.98,
  502.08,356.39,261.77,356.36},{961127,397.48,501.16,
  355.88,261.09,356.99},{961129,398.43,502.17,358.51,
  261.67,358.18}}
```

```
<<Statistics`DataManipulation`

composite=Column[nyse2datafile,2];

Take[composite,5]

{198.,197.8,196.29,194.64,195.33}

Take[composite,-5]

{394.66,398.86,397.98,397.48,398.43}
```

(a) After making sure that **DataSmoothing** is loaded, we use `MovingMedian` to define three series of moving medians having spans of size 31, 91, and 151 days, respectively. Then, using `ListPlot`, we define plots of the data points in `composite` and each of the smoothed series. They are displayed in a graphics array using `Show`.

```
<<Statistics`DataSmoothing`

mmed31=MovingMedian[composite,31];
mmed91=MovingMedian[composite,91];
mmed151=MovingMedian[composite,151];

plot1=ListPlot[composite,Ticks->{None,
  {200,250,300,350,400}},DisplayFunction->Identity];

plot2=ListPlot[mmed31,Ticks->{None,
  {200,250,300,350,400}},DisplayFunction->Identity];

plot3=ListPlot[mmed91,Ticks->{None,
  {200,250,300,350,400}},DisplayFunction->Identity];

plot4=ListPlot[mmed151,Ticks->{None,
  {200,250,300,350,400}},DisplayFunction->Identity];

Show[GraphicsArray[{{plot1, plot2}, {plot3, plot4}}]]
```

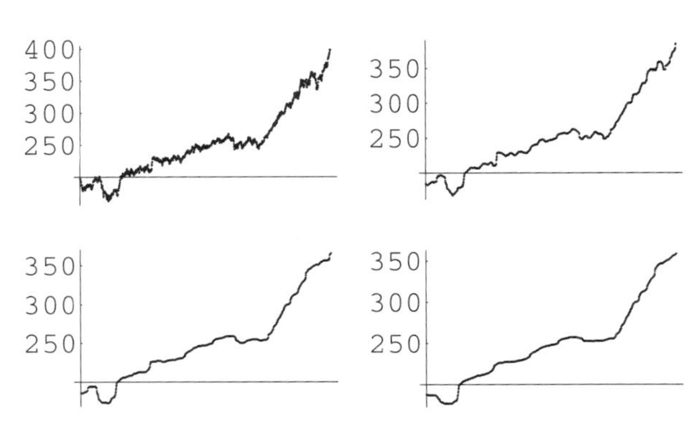

(b) The option `RepeatedSmoothing->True` is used with moving median smoothing having spans of 151 days. Note that the combination of a large data set and medians having large span will result in several minutes of computer processing to achieve convergence. The smoothed series will be plotted in red on a color monitor.

```
mmed151rs=MovingMedian[composite,151,
  RepeatedSmoothing->True];
```

```
plot5 =
 ListPlot[mmed151rs, PlotStyle -> RGBColor [1, 0, 0]]
```

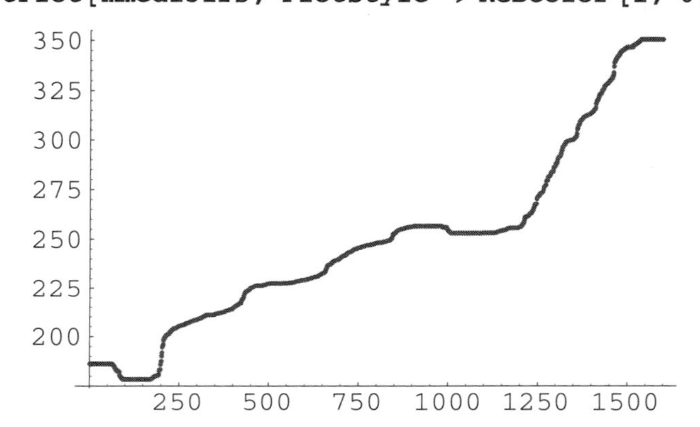


```
Show[plot1, plot5, DisplayFunction -> $DisplayFunction]
```

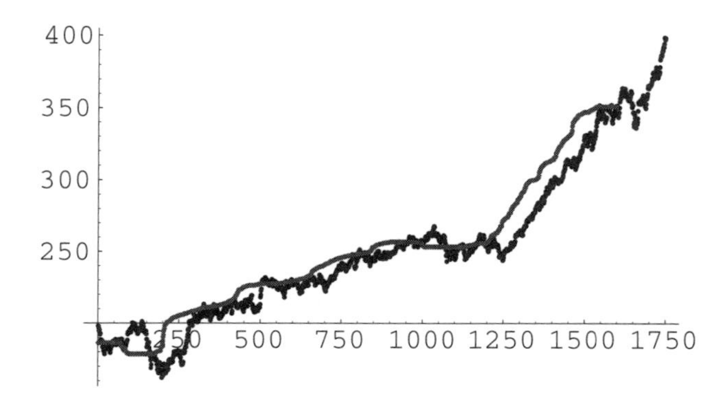

■

Linear Filter Smoothing

The procedure `LinearFilter` provides the capability to specify any linear combination to be used in calculating smoothed values. Basic information about the command is displayed using `?`.

```
<<Statistics`DataSmoothing`
?LinearFilter
LinearFilter[datalist, {c0, c1, c2, ...}] passes
  the data in datalist through the linear filter
  with weights {c0, c1, c2, ...}.  Each element in
  the resulting list is given by Sum[cj datalist[[t-j]],
  {j, 0, r-1}], for t = r, r+1, ... n, where
  n = Length[datalist] and r is the number of weights.
```

As an example, consider a time series represented by `datalist = ` $\{y_1, y_2, \ldots, y_n\}$ and suppose the number of terms used to determine each entry in the smoothed series is $r = 3$. Then a list of three weights must be specified. Suppose `weights = ` $\{c_0, c_1, c_2\}$. If the weights are {1/3, 1/3, 1/3}, then the output will be equivalent to that from `MovingAverage`. The *Mathematica* documentation refers to the terms of this smoothed series as **simple moving averages.** If the sum of the weights is one ($c_0 + c_1 + c_2 = 1$), then the terms of the smoothed series are weighted averages, and the *Mathematica* documentation refers to them as **moving averages.** In general, the terms of the smoothed series denoted by *sy* values would be calculated as follows:

$$\begin{cases} sy_1 = c_0 y_3 + c_1 y_2 + c_2 y_1 \\ sy_2 = c_1 y_4 + c_2 y_3 + c_3 y_2 \\ \vdots \\ sy_{t-2} = c_1 y_t + c_2 y_{t-1} + c_3 y_{t-2} \\ \vdots \\ sy_{n-2} = c_1 y_n + c_2 y_{n-1} + c_3 y_{n-2} \end{cases}$$

Notice the difference in ordering for the entry of the weights from what one might ordinarily expect.

EXAMPLE 3: The data file **Hours.dat** was introduced in Example 1 of Section 5.1. It contains the average number of hours worked each week by workers in the United States for each month from January 1964 to February 1997. In the earlier example we created a list called `selecthours` containing the average number of hours worked for the 86 months beginning with January 1990 and ending with February 1997. Use the linear filter to create weighted averages to smooth this series which (a) weight more recent values more heavily than earlier values; (b) weight all values equally; (c) weight earlier values more heavily than more recent values.

SOLUTION: We assume that both `selecthours` and **DataSmoothing** are loaded and use `LinearFilter` to create three smoothed series. The first, `lf1`, weights more recent values more heavily than earlier values; the second, `lf2`, weights all values equally; and the third, `lf3`, weights earlier values more heavily than more recent values. We also create the code to plot the points and to connect the points for each of the smoothed series.

```
lf1=LinearFilter[selecthours,{5/15,4/15,3/15,2/15,1/15}];
lf2=LinearFilter[selecthours,{3/15,3/15,3/15,3/15,3/15}];
lf3=LinearFilter[selecthours,{1/15,2/15,3/15,4/15,5/15}];
lfp1=ListPlot[lf1,PlotStyle->PointSize[0.015],
  DisplayFunction->Identity];
lfp11=ListPlot[lf1,PlotJoined->True,
  DisplayFunction->Identity];
lfp2=ListPlot[lf2,PlotStyle->PointSize[0.015],
  DisplayFunction->Identity];
lfp21=ListPlot[lf2,PlotJoined->True,
  DisplayFunction->Identity];
lfp3=ListPlot[lf3,PlotStyle->PointSize[0.015],
  DisplayFunction->Identity];
lfp31=ListPlot[lf3,PlotJoined->True,
  DisplayFunction->Identity];
```

(a) Using the code from Example 1 of Section 5.1 to plot the values in `selecthours` and the code just shown to plot the series `lf1`, we use `Show` to display series and the smoothed series together.

```
p1=ListPlot[selecthours,PlotStyle->{GrayLevel[0.5],
  PointSize[0.02]},DisplayFunction->Identity];
p2=ListPlot[selecthours,PlotStyle->GrayLevel[0.5],
  PlotJoined->True,DisplayFunction->Identity];
p3=Show[p1,p2,lfp1,lfp11,
  DisplayFunction->$DisplayFunction]
```

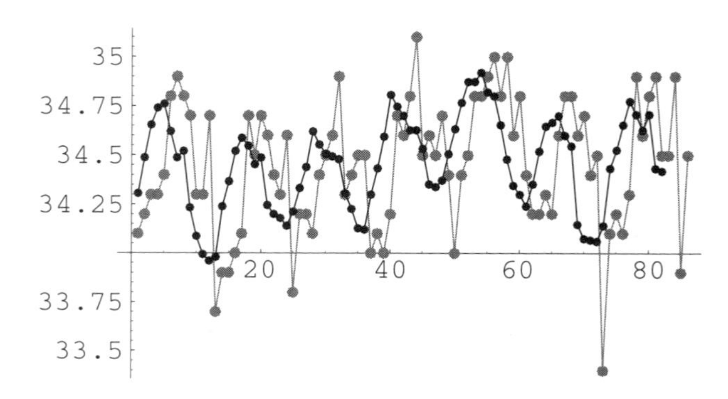

(b) In a similar manner, we display the smoothed series with equal weights, `lf2`.

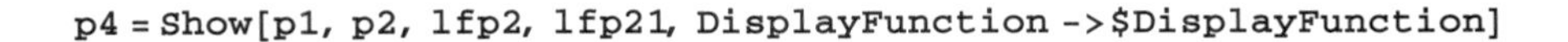

```
p4 = Show[p1, p2, lfp2, lfp21, DisplayFunction ->$DisplayFunction]
```

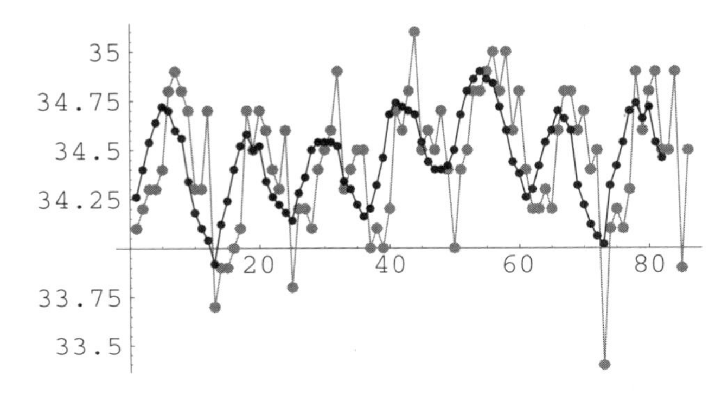

(c) Finally, we display the smoothed series `lf3`, which weights earlier values more heavily than more recent values.

```
p5 = Show[p1, p2, lfp3, lfp31, DisplayFunction -> $DisplayFunction]
```

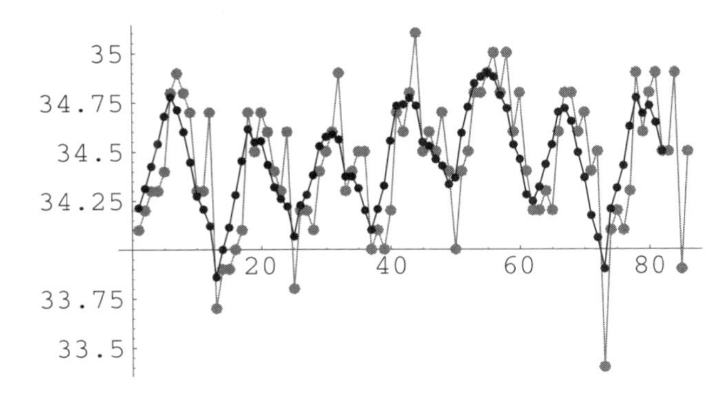

■

Exponential Smoothing

Exponential smoothing is used in forecasting (predicting future values of a time series). The technique weights the observed time series values, unequally with more recent observations being weighted more heavily than more remote observations. This unequal weighting is achieved through the use of a smoothing constant, w, which determines how much weight is given to each observation. To illustrate, we assume the smoothing constant, w, is between zero and one and consider the expression

$$\frac{y_t + wy_{t-1} + w^2y_{t-2} + w^3y_{t-3} + \ldots}{1/(1-w)}.$$

The coefficients 1, w, w^2, w^3, ... form a geometric series with sum $1/(1 - w)$. Thus, the weights, which are of the form $w^k (1 - w)$, $k = 1, 2, \ldots$, sum to 1, and, as noted in Subsection 5.2.3 on Linear Filter Smoothing, the result is a weighted average. Moreover, the most recent values receive heavier weights because with $0 < w < 1$, $1 > w > w^2 > \ldots$.

Observations are not available in the infinite past, so a truncated version of the earlier expression,

$$a_t = \frac{y_t + wy_{t-1} + \ldots + w^{t-1}y_1 + w^t y_0}{1/(1-w)},$$

is used to define the tth exponential smoothed value. The value y_0 is called the **starting value**. It is not part of the series and must be determined by the user. Its value will depend on the values in the series, the length of the series, and the reason for smoothing. Some of the possible choices include 0, y_1 (the first value in the series), the mean of the first several values in the series, and the mean of the entire series.

The *Mathematica* documentation contains a recursive formula for (using our notation) a_{t+1} in terms of a_t. In order to understand the origin of that formula and to explain its implications, we will begin by using the formula for a_t to derive the recursive formula. First, multiply both sides by $1/(1 - w)$ to get

$$\left(\frac{1}{1-w}\right)a_t = y_t + wy_{t-1} + \ldots + w^{t-1}y_1 + w^t y_0.$$

Next, multiply both sides by w, add a_{t+1} to both sides, and simplify:

$$\begin{aligned} w\left(\frac{1}{1-w}\right) + y_{t+1} &= y_{t+1} + w(y_t + wy_{t-1} + \ldots + w^{t-1}y_1 + w^t y_0) \\ &= y_{t+1} + wy_t + w^2y_{t-1} + \ldots + w^t y_1 + w^{t+1}y_0 \\ &= \left(\frac{1}{1-w}\right)a_{t+1}. \end{aligned}$$

Finally, multiply both sides by $(1 - w)$ and rearrange the terms to result in a recursion formula:

$$a_{t+1} = wa_t + (1-w)y_{t+1}.$$

The formula in the *Mathematica* documentation uses a to represent the smoothing constant, where $0 < a < 1$. Setting $w = 1 - a$ and substituting in the preceding formula, we determine that

$$a_{t+1} = ay_{t+1} + (1-a)a_t.$$

Except that the notation is slightly different, this formula is a simplified version of the formula in the *Mathematica* documentation. Our formula can provide insights into the role of a as the smoothing constant. For example, on one hand, as a_{t+1} gets close to 1, y_{t+1} gets close to y_{t+1}. This indicates that little smoothing has taken place. On the other hand, as a gets close to zero, there is little effect of y_{t+1} on a_{t+1}. This indicates that a substantial amount of smoothing has taken place because the value a_{t+1} is almost entirely composed of past observations.

Although it is beyond the scope of our presentation to discuss forecasting, we note that the basic equation for forecasts l lead time units into the future is

$$\hat{y}_{t+l} = a_t$$

Basic information about the procedure, ExponentialSmoothing, is displayed in the following output of the ? command.

```
?ExponentialSmoothing
ExponentialSmoothing[datalist, a] smooths datalist using
 an exponentially weighted average with smoothing
 constant a, taking the first entry of datalist as the
 starting value. ExponentialSmoothing[datalist, a, y0]
 smooths datalist taking the starting value to be y0. If
 yt is the smoothed result at time t, then the result at
 time t+1 is given by yt + a*(datalist[[t+1]] - yt,
 where a is the smoothing constant (0 < a < 1).
```

Notice that the smoothing constant a must be specified, and optionally the starting value, y_0, may also be specified.

EXAMPLE 4: The data file **Hours.dat** was introduced in Example 1 of Section 5.1 and used in Example 3 of this section. It contains the average number of hours worked each week by workers in the United States for each month from January 1964 to February 1997. In the earlier examples we used a list called `selecthours` containing the average number of hours worked for the 86 months beginning with January 1990 and ending with February 1997. Use exponential smoothing to smooth this series (a) for a smoothing constant close to 1 without and with an initial value and (b) for a smoothing constant close to zero without and with an initial value.

SOLUTION: We assume that both `selecthours` and **DataSmoothing** are loaded and define a starting value to be the mean of the values in `selecthours`. Note that it is necessary to load **DescriptiveStatistics** before calculating the mean. Next, we use `ExponentialSmoothing` to create four smoothed series. The first, em1, with smoothing constant 0.8, weights the more recent values most heavily; the second, em2, has the same smoothing constant and includes the starting value; the third, em3, with smoothing constant 0.2, weights earlier values more heavily than more recent values; and the fourth, em4, has the same smoothing constant and includes the starting value. We also create the code to plot the points and to connect the points for each of the smoothed series.

```
<<Statistics`DescriptiveStatistics`

startingvalue=Mean[selecthours]

34.45

em1=ExponentialSmoothing[selecthours,0.8];

plotem1=ListPlot[em1,PlotStyle->Thickness[0.015],
  DisplayFunction->Identity];

plotem11=ListPlot[em1,PlotJoined->True,
  DisplayFunction->Identity];

em2=ExponentialSmoothing[selecthours,0.8,startingvalue];

plotem2=ListPlot[em2,PlotStyle->Thickness[0.01],
  DisplayFunction->Identity];

plotem21=ListPlot[em2,PlotJoined->True,
  DisplayFunction->Identity];

em3=ExponentialSmoothing[selecthours,0.2];

plotem3=ListPlot[em3,PlotStyle->Thickness[0.015],
  DisplayFunction->Identity];

plotem31=ListPlot[em3,PlotJoined->True,
  DisplayFunction->Identity];

em4=ExponentialSmoothing[selecthours,0.2,startingvalue];

plotem4=ListPlot[em4,PlotStyle->Thickness[0.015],
  DisplayFunction->Identity];

plotem41=ListPlot[em4,PlotJoined->True,
  DisplayFunction->Identity];
```

Using the code from Example 1 of Section 5.1 to plot the values in selecthours and the code just listed to plot each of the smoothed series, we use Show to develop and display a graphics array with each of the smoothed series plotted on top of the original series.

```
p1=ListPlot[selecthours,PlotStyle->{GrayLevel[0.5],
  PointSize[0.02]},DisplayFunction->Identity];

p2=ListPlot[selecthours,PlotStyle->GrayLevel[0.5],
  PlotJoined->True, DisplayFunction->Identity];
```

```
p3=Show[p1,p2,plotem1,plotem11,
  DisplayFunction->Identity];
p4=Show[p1,p2,plotem2,plotem21,
  DisplayFunction->Identity];
p5=Show[p1,p2,plotem3,plotem31,
  DisplayFunction->Identity];
p6=Show[p1,p2,plotem4,plotem41,
  DisplayFunction->Identity];
```

```
Show[GraphicsArray[{{p3, p4}, {p5, p6}}]]
```

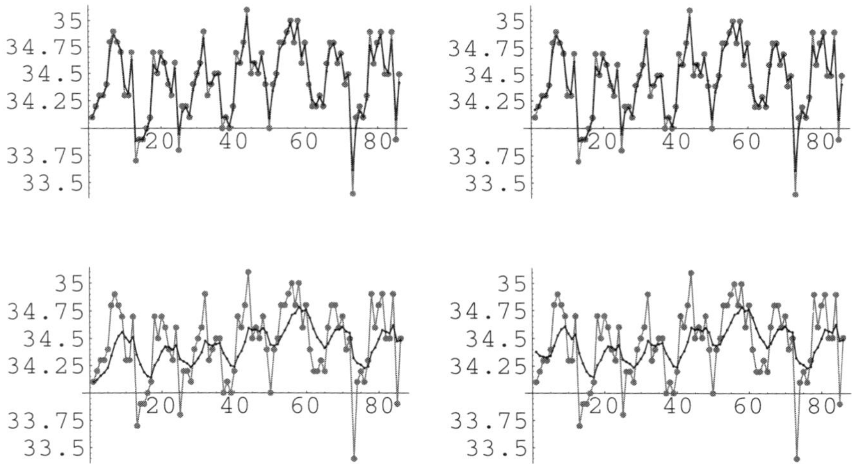

■

5.3 Multivariate Extension

A multivariate time series in *Mathematica* is a sequence of lists containing observations on m variables ($m \geq 2$) in time order. Let $\{y_{11}, y_{12}, \ldots, y_{1m}\}$ represent the list of first observations, $\{y_{21}, y_{22}, \ldots, y_{2m}\}$ represent the list of next observations taken after a fixed time increment, and so on to the nth list, $\{y_{n1}, y_{n2}, \ldots, y_{nm}\}$. Each of the functions in the **DataSmoothing** package will accept multivariate input. The smoothing is done for each dimension (column) of the data separately. For `LinearFilter`, different weights can be specified for each dimension. For exponential smoothing, the smoothing constant and starting value can be specified for each dimension.

CHAPTER 

Probability and Probability Distributions

6.1 Introduction

The setting for the study of probability is a **sample space**, which is defined to be the set of all possible outcomes of an experiment. Often, in introductory probability, the experiments discussed include tossing coins, rolling dice, and drawing cards from a deck of playing cards; however, any activity with a variable result that cannot be predicted ahead of time may qualify as an experiment. An element of a sample space is called a **sample point** or a **simple event**. A subset of the sample space is called an **event**.

Once a sample space has been defined, the next step is to assign probabilities to the events. The **probability** of an event is a measure of the likelihood that the event will occur. The sample space and assignment of probabilities form a mathematical model of the experiment, and together they are called a **probability space**. More formally stated, a probability is a numerically valued function that assigns a number $P(E)$ to every event, E, in the sample space, S, so that the following axioms hold:

1. $P(E) \geq 0$.
2. $P(S) = 1$.
3. For any sequence $E_1, E_2 \ldots$ of **mutually exclusive** events (that is for all $i \neq j$, E_i and E_j have no common outcomes),

$$P\left(\bigcup_{i=1}^{\infty} E_i\right) = \sum_{i=1}^{\infty} P(E_i).$$

All of the properties of probability follow from these three axioms. For example, three elementary properties that follow directly are as follows.

For the event E', the complement of the event E, $P(E') = 1 - P(E)$.
For any two events E and F, $P(E \cup F) = P(E) + P(F) - P(E \cap F)$.
For $F \subset E$, $P(F) \leq P(E)$.

A **random variable** is a function that assigns a real number to every element in the sample space S. Random variables are usually denoted by capital letters at the end of the alphabet, such as X, Y, Z. The numbers that a random variable assigns to the elements of S are called values of the random variable and denoted by lowercase letters such as x, y, z. A random variable X is said to be a **discrete random variable** if it can take on only a finite, or countably infinite, number of possible values x. In this case,

1. $P(X = x) = p(x) \geq 0$,
2. $\sum_x p(x) = 1$,

where the sum is over all possible values x. The function $p(x)$ is called the **probability function** of X, or it may be called the **probability mass function**. The probability distribution of a discrete random variable consists of the possible values of the random variable and their corresponding probabilities. These values and probabilities may be represented in a table, in graphical form, or by means of a formula.

A random variable X is said to be a **continuous random variable** if it can take on the infinite number of possible values associated with intervals of real numbers, and there is a function $f(x)$ called the **probability density function** (pdf) such that

1. $f(x) \geq 0$ for all x.
2. $\int_{-\infty}^{\infty} f(x)dx = 1$.
3. $P(a \leq X \leq b) = \int_a^b f(x)dx$.

Note that for a continuous random variable X, $P(X = a) = 0$ for any specific value a.

The *Mathematica* packages discussed in this chapter contain various functions to work with the discrete and continuous random variables and their probability distributions. The following discussions may be helpful in understanding the purpose and output of these functions.

PDF

For a discrete probability distribution in *Mathematica*, PDF represents the probability function $p(x)$. For a continuous distribution PDF represents the probability density function $f(x)$.

CDF

The **cumulative distribution function** (cdf), F, for the probability distribution of the random variable X is defined to be $F(x) = P(X \leq x)$. If X is discrete with probability function

$$p(x),\ F(x) = \sum_i p(x_i),$$

where the sum is taken over indices i that satisfy $x_i \leq x$. If X is continuous with probability density function $f(x)$, then $F(x) = \int_{-\infty}^{x} f(t)dt$. Note that in the continuous case the derivative of the cdf is the pdf, $F'(x) = f(x)$.

Quantile

In general, a **quantile,** x, of the random variable X is a possible value of X. In *Mathematica*, the quantile corresponding to a specified cumulative probability can be displayed using the `Quantile` function. For example, the $.50^{\text{th}}$ quantile is the median, m, of the distribution. Notice that $F(m) = .5$. In effect the `Quantile` function is the inverse of the cdf.

Domain

As used in the *Mathematica* packages discussed in this chapter, **domain** refers to the possible values of the random variable. That is, it refers to the domain of the probability function or the pdf and the cdf.

The next seven topics correspond to *Mathematica* functions that describe location, dispersion, or shape of a probability function or a pdf of the random variable X. See Chapter 2 for more information on these descriptive methods.

Mean

Discrete $\mu = \sum_x xp(x)$

Continuous $\mu = \int_{-\infty}^{\infty} xf(x)dx$

Variance

Discrete $\sigma^2 = \sum_x (x-\mu)^2 p(x)$

Continuous $\sigma^2 = \int_{-\infty}^{\infty} (x-\mu)^2 f(x)dx$

Standard Deviation

The standard deviation σ is found by taking the positive square root of the variance. That is, $\sigma = \sqrt{\sigma^2}$.

Skewness

Discrete $\eta^3 = \frac{1}{\sigma^3} \sum_x (x-\mu)^3 p(x)$

Continuous $\eta^3 = \frac{1}{\sigma^3} \int_{-\infty}^{\infty} (x-\mu)^3 f(x)d(x)$

Kurtosis

Discrete $\eta_4 = \frac{1}{\sigma^4} \sum_x (x-\mu)^4 p(x)$

Continuous $\eta_4 = \frac{1}{\sigma^4} \int_{-\infty}^{\infty} (x-\mu)^4 f(x)d(x)$

Kurtosis Excess

The kurtosis excess is found by subtracting 3 from the kurtosis. That is, calculate $\eta_4 - 3$.

Expected Value

Let $g(x)$ be some function of x. For example, $g(x) = x^2$. Then the expected value of the new random variable $g(X)$ is calculated as follows:

Discrete $E(g(X)) = \sum_x g(x)p(x)$

Continuous $E(g(X)) = \int_{-\infty}^{\infty} g(x)\mathrm{f}(x)dx$.

Note that if $g(x) = x$, then $E(g(X)) = E(X) = \mu$.

Characteristic Function

The expected value notation just used is useful for a discussion of the moment generating function and the more general characteristic function. For a random variable X, the expected values of powers of X are called **moments**. The first moment is $E(X) = \mu$, the mean of X. The second moment is $E(X^2)$, the third moment is $E(X^3)$, and so forth. Note that the expected values of powers of $(X - \mu)$ are called **central moments** or **moments about the mean**. The first central moment is $E(X - \mu) = 0$, the second central moment is $E((X-\mu)^2) = \sigma^2$ the variance of X, the third central moment is used to calculate skewness, and the fourth central moment is used to calculate kurtosis. Moments and central moments are, of course, related. For example, $\sigma^2 = E((X-\mu)^2) = E(X^2) - (E(X))^2$. That is, the second central moment or variance is calculated by subtracting the square of the first moment from the second moment.

The **moment generating function** is a function of the auxiliary variable t denoted by M, where $M(t) = E(e^{tX})$. Thus, if X is discrete with probability function $p(x)$, then $M(t) = \sum_x e^{tx} p(x)$, and if X is continuous with pdf $f(x)$, then $M(t) = \int_{-\infty}^{\infty} e^{tx} f(x) dx$. Notice that $M(0) = E(e^{0X}) = E(1) = 1$. That is, $M(t)$ always exists for $t = 0$. When $M(t)$ is also defined in an open neighborhood of zero, its derivatives evaluated at $t = 0$ are the moments of the random variable X. The first moment $E(X) = M'(0)$, the second moment $E(X^2) = M''(0)$, and so forth. A drawback to the use of the moment generating function is that it may not be defined in an open neighborhood of zero for a given random variable.

The **characteristic function** is a function of the auxiliary variable t and the imaginary quantity i ($i^2 = -1$). It is denoted by ϕ in the *Mathematica* documentation, where $\phi(t) = E(e^{itx})$. If X is discrete with probability function $p(x)$, $\phi(t) = \sum_x e^{itx} p(x)$, and if X is continuous with pdf $f(x)$, $\phi(t) = \int_{-\infty}^{\infty} e^{itx} f(x) dx$. The characteristic function is well defined for every random variable, and its derivatives evaluated at $t = 0$ are used to find the moments of the random variable.

In the packages discussed in this chapter, distributions are represented in **symbolic form**. That is, for the functions in these packages, the output is a formula if symbols are input. If numbers are input, then a number is output. This is illustrated in the examples that follow.

6.2 Discrete Random Variables and Distributions

Mathematica's **Statistics** package **DiscreteDistributions** contains eight discrete distributions: Bernoulli, binomial, discrete uniform, geometric, hypergeometric, logarithmic series, negative binomial (or Pascal), and Poisson. The **Help Browser** can be used to obtain information about these distributions as well as the functions which operate on them. This help feature may also include examples to demonstrate the syntax of the functions.

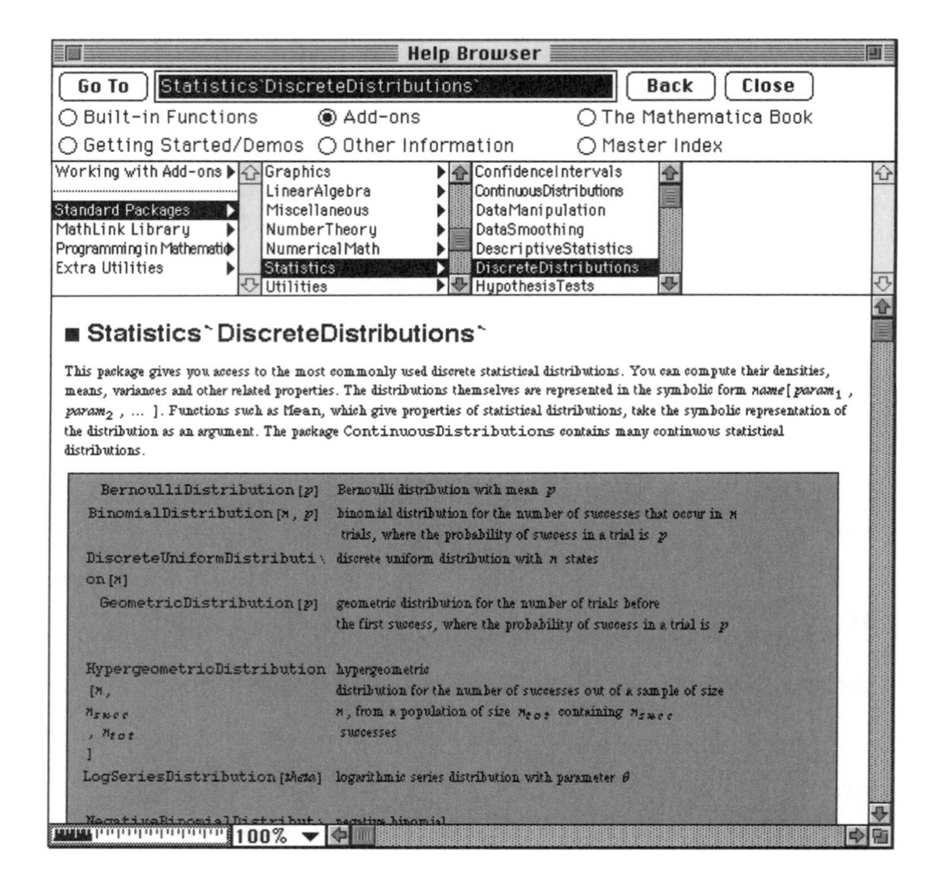

The functions in **DiscreteDistributions** include:

`Domain[dist]` DiscreteDistributions:Domain, which gives the domain of the distribution `dist`,

`PDF[dist,x]` DiscreteDistributions:PDF, which gives the probability function of the distribution `dist` evaluated at `x`;

`CDF[dist,x]` DiscreteDistributions:CDF, which gives the cumulative distribution function of the distribution `dist` evaluated at `x`; and

`CharacteristicFunction[dist,t]` DiscreteDistributions:CharacteristicFunction which gives the characteristic function (moment generating function) of the distribution `dist` as a function of the variable `t`.

Other commands (from the **DescriptiveStatistics** package) are redefined in this package to give location, dispersion, and shape measures of the distributions. This means that the **DescriptiveStatistics** package need not be loaded to use the following commands.

`Mean[dist]` DiscreteDistributions:Mean, yields the mean of the distribution `dist`.

`Variance[dist]` DiscreteDistributions:Variance, gives the variance of the distribution `dist`.

`StandardDeviation[dist]` DiscreteDistributions:StandardDeviation yields the standard deviation of the distribution `dist`.

`Skewness[dist]` DiscreteDistributions:Skewness, supplies the coefficient of skewness of the distribution `dist`.

`Kurtosis[dist]` DiscreteDistributions:Kurtosis, gives the coefficient of kurtosis of the distribution `dist`.

`KurtosisExcess[dist]` DiscreteDistributions:KurtosisExcess, gives the kurtosis excess (Kurtosis - 3) of the distribution `dist`.

`Quantile[dist,q]` DiscreteDistributions:Quantile, yields the qth quantile of the distribution `dist`.

`ExpectedValue[f,dist,x]` DiscreteDistributions:ExpectedValue, gives the expected value of the function f of x with respect to the distribution `dist`.

Bernoulli Distribution

A Bernoulli experiment or Bernoulli trial is an experiment that has exactly two possible outcomes, success ($X = 1$) or failure ($X = 0$). The probability of success is $P(X = 1) = p$, $0 < p < 1$. The probability of failure is $P(X = 0) = 1 - p$. Often the substitution $q = 1 - p$ is made, and the probability of failure is denoted by q. An example of a Bernoulli trial is tossing a coin which lands showing head (success) or tail (failure). A sequence of independent Bernoulli trials gives rise to other distributions including the binomial, negative binomial, and geometric distributions. For the Bernoulli distribution:

Domain	$\{0,1\}$
Probability function	$f(0) = 1 - p; f(1) = p$
Distribution function	$F(0) = 1 - p;\ F(1) = 1$
Characteristic function	$1 + p(e^{it} - 1)$
r^{th} moment	p

Mean	p
Variance	$p(1-p)$

`BernoulliDistribution[p]` represents the Bernoulli distribution with probability of success `p`.

EXAMPLE 1: Find the domain and the formulas for the probability function, distribution function, characteristic function, moments, mean, variance, and standard deviation of the Bernoulli distribution with probability of success p.

SOLUTION: After loading the **DiscreteDistributions** package, we find the domain to be the set containing zero and one. We then illustrate the symbolic output of *Mathematica*. Notice that the symbolic output may not be simplified. For example, the output of `ExpectedValue[x^r,BernoulliDistribution[p],x]` simplifies to p.

```
<<Statistics`DiscreteDistributions`

<< Statistics'DiscreteDistributions'

Domain[BernoulliDistribution[p]]

{0, 1}

PDF[BernoulliDistribution[p], x]

If[x == 0, 1 - p, p]

CDF[BernoulliDistribution[p], x]

1 - p

CharacteristicFunction[
 BernoulliDistribution[p], t]

1 - p + E^(I t) p

ExpectedValue[x ^ r,
 BernoulliDistribution[p], x]

0^r (1 - p) + p

Mean[BernoulliDistribution[p]]

p

Variance[BernoulliDistribution[p]]

(1 - p) p
```

```
StandardDeviation[
 BernoulliDistribution[p]]
```

$\sqrt{(1-p)\,p}$

■

Binomial Distribution

A binomial experiment consists of a finite number of independent, Bernoulli trials. Recall that each trial has two possible outcomes: success or failure; the probability of success on each trial is p. The random variable X is the number of successes. The distribution of the count X of successes in n independent Bernoulli trials with probability p of success is called the binomial distribution with parameters n and p, $B(n, p)$. An example of a binomial random variable is the number of heads when a fair coin is tossed 10 times. For the binomial distribution:

Domain	$\{0, 1, \ldots, n\}$
Probability function	$\binom{n}{x} p^x (1-p)^{n-x}$
Distribution function	$\sum_{i=0}^{x} \binom{n}{i} p^i (1-p)^{n-i}$
Moment generating function	$(1-p+pe^t)^n$
Moments	
Mean	np
Second	$np((n-1)p+1)$
Third	$np((n-1)(n-2)p^2+3p(n-1)+1)$
Central moments	
Variance	$np(1-p)$
Third	$np(1-p)(1-2p)$
Fourth	$np(1+3p(1-p)(n-2))$
Skewness	$\frac{(1-2p)}{\sqrt{np(1-p)}}$
Kurtosis	$3-\frac{6}{n}+\frac{1}{np(1-p)}$

`BinomialDistribution[n,p]` represents the binomial distribution with `n` independent trials and probability of success `p`.

EXAMPLE 2: During his career, Michael Jordan has been successful on approximately 84% of his free throw attempts. During Game 2 of the National Basketball Association Finals on June 7, 1996, Jordan missed 6 of his 16 attempts. (a) Calculate the probability that he would miss 6 attempts out of 16 in a single game. (b) Calculate the probability that he would miss 6 or more attempts out of 16 in a single game. (c) How unusual is it for him to shoot this poorly?

SOLUTION: Based on his career statistics, the probability that Michael Jordan will miss a free throw is 0.16. Let X represent the number of free throw shots missed out of 16 attempts. (a) We can use `PDF` to determine that the probability he would miss six attempts is 0.0234983.

```
PDF[BinomialDistribution[16,0.16], 6]
0.0234983
```

(b) We use `CDF` to determine the probability that he would miss 6 or more attempts out of 16. First we use CDF to compute $P(X \leq 5) = P(X = 0) + P(X = 1) + P(X = 2) + P(X = 3) + P(X = 4) + P(X = 5)$.

```
p5orless=CDF[BinomialDistribution[16,0.16], 5]
0.968471
```

Then we calculate $P(X \geq 6) = 1 - P(X \leq 5)$.

```
1-p5orless
0.031529
```

(c) This result indicates that Michael Jordan's missing 6 or more free throws out of 16 attempts is rare, because it happens only 3.15% of the time or roughly once every 32 games ($1/32 = 0.03125$).

■

Discrete Uniform Distribution

Consider a random experiment with n equally likely outcomes. Let X be the random variable obtained by assigning the numbers 1 through n to the n outcomes. This random variable is called uniform, because each outcome has the same chance of occurring. As an example consider rolling a fair die, and let X correspond to the number of dots on the upper face of the die. Then X can take on the values 1 through 6, and each of the six possible values has probability $\frac{1}{6}$. Note that a random variable Y with probability distribution $P(Y = k) = \frac{1}{b}$, $a \le k \le a + b - 1$ has a discrete uniform distribution. The mean of Y is $E(Y) = a + \frac{b-1}{2}$ while the variance is $\text{Var}(Y) = a + \frac{b^2-1}{12}$.

For the distribution of X:

Domain	$\{1, \ldots, n\}$
Probability function	$\frac{1}{n}$
Distribution function	$\frac{[x]}{n}$ where [ ] means "greatest integer in". In *Mathematica*, use `Floor[x]`.
Characteristic function	$\dfrac{e^{it}(1 - e^{\text{int}})}{(1 - e^{it})n}$
Moments	
Mean	$\dfrac{n+1}{2}$
Second	$\dfrac{(n+1)(2n+1)}{6}$
Third	$\dfrac{n(n+1)^2}{4}$
Central moments	
Variance	$\dfrac{n^2-1}{12}$
Third	0
Fourth	$\dfrac{(n^2-1)(3n^2-7)}{240}$
Skewness	0
Kurtosis	$\frac{3}{5}\left(3 - \dfrac{4}{n^2-1}\right)$

`DiscreteUniformDistribution[n]` represents the discrete uniform distribution with `n` possible outcomes.

EXAMPLE 3: Consider an experiment with 10 possible outcomes that are distributed uniformly, such as randomly selecting an integer on the interval $[-4, 5]$. Let the random variable Y represent the outcome of the experiment(a) Determine the domain of this distribution. (b) Find $P(Y = k), -4 \le k \le 5$. (c) Calculate $P(Y \le 4)$. (d) Determine the quartiles of this distribution.

SOLUTION: (a) In *Mathematica,* the domain of a discrete uniform distribution with 10 possible outcomes is $\{1, 2, \ldots, 10\}$. For this example, the 10 possible integers on the interval $[-4, 5]$ form the domain, $\{-4, -3, -2, -1, 0, 1, 2, 3, 4, 5\}$. The integer -4 corresponds to Outcome 1; the integer -3 corresponds to Outcome 2; and so forth.

```
Domain[DiscreteUniformDistribution[10]]
```

```
{1,2,3,4,5,6,7,8,9,10}
```

(b) To illustrate the necessary command, we use `PDF` to find that $P(X = 9) = 1/10$. Then, we use `Table` to find $P(X = k), 1 \le k \le 10$. Of course, $P(X = k) = 1/10$ for each value of k, $1 \le k \le 10$. Because of the correspondence between the values of X and those of Y, the output of the command `Table` represents $P(Y = k), -4 \le k \le 5$.

```
PDF [DiscreteUniformDistribution [10], 9]
```

$$\frac{1}{10}$$

```
Table [PDF [DiscreteUniformDistribution [10], j],
  {j, 1, 10}]
```

$$\left\{\frac{1}{10}, \frac{1}{10}, \frac{1}{10}, \frac{1}{10}, \frac{1}{10}, \frac{1}{10}, \frac{1}{10}, \frac{1}{10}, \frac{1}{10}, \frac{1}{10}\right\}$$

(c) We use `CDF` to calculate $P(X \le 9)$. This command evaluates $\sum_{k=1}^{9} P(X = k)$. As expected, $P(X \le 9) = 9/10$. Because 4 is the ninth value in the domain of Y, $P(Y \le 4) = 9/10$.

```
CDF [DiscreteUniformDistribution [10], 9]
```

$$\frac{9}{10}$$

(d) Unlike `Quantile`, the **DescriptiveStatistics** command `Quartile` is not automatically loaded with the **DiscreteDistributions** package. Therefore, we use `Quantile` to compute the quartiles for X. The first quartile (1/4-th quantile or 25th percentile) is 3. `Table` is then used to find all quartiles which are 3, 5, and 8. Using the correspondence between the values of X and those of Y, the quartiles for Y are –2, 0, and 3.

```
Quantile[DiscreteUniformDistribution[10],0.25]
```

```
3
```

```
Table[Quantile[DiscreteUniformDistribution[10],0.25k],
  {k,1,3}]
```

```
{3,5,8}
```

■

Geometric Distribution

Consider a sequence of independent Bernoulli trials. Recall that each trial has two possible outcomes; success or failure, and the probability of success on each trial is p. The geometric random variable may be defined in two ways. Some sources, including the *Mathematica* documentation, define it to be the number of trials (failures) before the first success. We denote this random variable by X. Note that X is also the number of trials before the first success after any given trial, and it is the number of trials between any success and the next success. The random variable X is sometimes described as the **waiting time** for a success for Bernoulli trials. Many introductory textbooks define the geometric random variable to be the number of the trial on which the first success occurs. We denote this random variable by Y, where $Y = X + 1$. The distribution of X (or of Y) is called the geometric distribution with parameter p. As an example of a geometric random variable, let X be the number of tails before a head occurs in repeated tossing of a fair coin. Equivalently, Y is the number of the trial on which the first head occurs. The geometric random variable can take on a countably infinite number of values.

For the random variable X:

Domain	$\{0, 1, 2, \ldots\}$
Probability function	$(1-p)^n p$
Distribution function	$1-(1-p)^{n+1}$
Moment generating function	$\dfrac{p}{1-(1-p)e^t}$

Moments

Mean $\dfrac{1-p}{p}$

Second $\dfrac{2+p(p-3)}{p^2}$

Third $\dfrac{6-p(p-3)(p-4)}{p^3}$

Central moments

Variance $\dfrac{1-p}{p^2}$

Third $\dfrac{(1-p)(2-p)}{p^3}$

Fourth $\dfrac{9(1-p)^2}{p^4}+\dfrac{1-p}{p^2}$

Skewness $\dfrac{(2-p)}{\sqrt{(1-p)}}$

Kurtosis $9+\dfrac{p^2}{1-p}$

For the random variable $Y = X + 1$:

Domain $\{1, 2, \ldots\}$

Probability function $(1-p)^{n-1}p$

Distribution function $1-(1-p)^n$

Moment generating function $\dfrac{pe^t}{1-(1-p)e^t}$

Moments

Mean $\dfrac{1}{p}$

Second $\dfrac{2-p}{p^2}$

Third $\dfrac{6+p(p-6)}{p^3}$

Central moments

Variance $\dfrac{1-p}{p^2}$

Third $\dfrac{(1-p)(2-p)}{p^3}$

Fourth	$\frac{9(1-p)^2}{p^4}+\frac{1-p}{p^2}$
Skewness	$\frac{(2-p)}{\sqrt{(1-p)}}$
Kurtosis	$9+\frac{p^2}{1-p}$

The *Mathematica* command `GeometricDistribution[p]` represents the geometric distribution with probability of success `p`.

EXAMPLE 4: Three sons are asked by a parent to perform a chore. Because no son is specified, each son tosses a coin to determine the odd person who will then complete the chore. In case all three get heads or all three get tails, they continue tossing until they reach a decision. Let p be the probability of a head and $q = 1 - p$ the probability of a tail for the coins used. (a) Find the probability that they reach a decision in exactly n tosses. (b) Find the probability that they reach a decision in less than n tosses. (c) If $p = \frac{1}{2}$, what is the minimum number of tosses required to reach a decision with probability 0.95?

SOLUTION: (a) The sons will reach a decision if one head and two tails or two heads and one tail result from tossing the coins. Using `BinomialDistribution`, we find the probability of one head and the probability of two heads in three tosses. We add them together using the built-in function `Sum`. After simplifying and rearranging the *Mathematica* output, we see that the probability of reaching a decision on any one trial is $3pq$, where $q = 1 - p$. Thus, the probability that no decision is reached is $1 - 3pq$.

```
pdecision=Sum[PDF[BinomialDistribution[3,p],k],{k,1,2}]
3(1 - p)^2 p + 3(1 - p)p^2

Simplify[pdecision]
-3(-1+p)p
```

The probability that n tosses are needed to make a decision is the same as the probability that the number of failures (or the waiting time) is $n - 1$ before a decision is made. We calculate this using `GeometricDistribution`. The answer is $(1 - 3pq)^{n-1}$. If $p = \frac{1}{2}$ and $n = 5$, the actual probability is .003.

```
PDF[GeometricDistribution[3pq],n-1]
3(1 - 3\pq)^(-1+n) pq
```

```
PDF[GeometricDistribution[3/4],5-1]//N
0.00292969
```

(b) The probability that fewer than n tosses are required to reach a decision is the same as the probability that the number of failures (or waiting time) is less than $n-1$ before a decision is made. To calculate $P(X < n-1) = P(X \leq n-2) = P(X=0) + P(X=1) + \ldots + P(X = n-2)$ we can again make use of `Sum` or, better yet, use `CDF`. The answer is $1-(1-3pq)^{n-1}$. If $p = \frac{1}{2}$ and $n = 5$, the actual probability is .996.

```
Sum[(3 pq) (1 - 3 pq) ^ (k), {k, 0, n - 2}]
-1 + (1 - 3 pq)^n + 3 pq
------------------------
       -1 + 3 pq

CDF[GeometricDistribution[3 pq], n - 2]
1 - (1 - 3 pq)^Floor[-1+n]

CDF[GeometricDistribution[3 / 4], 5 - 2] // N
0.996094
```

(c) In `geotable`, we compute $P(X \leq k)$, $k = 0, 1, 2, \ldots, 4$. Notice that the probability 0.984375 corresponding to $k = 2$ is over 0.95. Thus, $n - 1 = 2$, so the minimum number of tosses required to reach a decision with probability 0.95 is 3.

```
geotable=Table[CDF[GeometricDistribution[3/4],k]//N,
  {k,0,4}]
{0.75,0.9375,0.984375,0.996094,0.999023}
```

■

Hypergeometric Distribution

Suppose that in a set of N objects there are g objects of one type (called good or successes) and $N-g$ of another type (called defective or failures). Further suppose that n of the objects are sampled randomly with none of the sampled objects being replaced. If X counts the number of good objects in the sample, then X has a hypergeometric distribution. As an example of a hypergeometric random variable, let X be the number of face cards in a hand of five cards drawn from a deck of 52 playing cards. In this example, $N = 52$, $g = 12$, $N - g = 40$, and $n = 5$.

For the hypergeometric random variable X:

Domain	$\max[0, n - N + g] \le k \le \min[g, n]$
Probability function	$\dfrac{\binom{g}{k}\binom{N-g}{n-k}}{\binom{N}{k}}.$
Mean	$\dfrac{ng}{N}$
Variance	$n\left(\dfrac{g}{N}\right)\left(1 - \dfrac{g}{N}\right)\left(\dfrac{N-n}{N-1}\right)$

`HypergeometricDistribution[n, nsucc, ntot]` represents the hypergeometric distribution for samples of size `n` drawn without replacement from a population with `nsucc` successes (good) and total size `ntot`. (In the notation of the previous discussion, the total number of objects N is given by `ntot`; the number of good objects g is given by `nsucc`; and the number of objects selected is given by `n`.)

EXAMPLE 5: A certain product is packaged in lots of 50. Before accepting a lot when it arrives, a company has an inspector select five units from the lot at random. The lot is accepted if none of these units are defective. However, if one or more are found to be defective, then each product in the lot is inspected. Suppose that there are three defective products in the lot. What is the probability that the entire lot will have to be inspected?

SOLUTION: In this case, `n` = 5, `nsucc` = 47, and `ntot` = 50. The probability that the entire lot will have to be inspected is given by $P(X < 5) = 1 - P(X = 5)$. In `p5`, we find that $P(X < 5) = 1419/1960$. Then, we see that $P(X < 5) = 0.27602$. Notice that the answer is found directly using `CDF`.

```
p5 = PDF [HypergeometricDistribution [5, 47, 50], 5]
```

$$\frac{1419}{1960}$$

```
1 - p5 // N
```

```
0.27602
```

```
CDF [HypergeometricDistribution [5, 47, 50], 4] // N
```

```
0.27602
```

■

Logarithmic Series Distribution

According to the *Mathematica* documentation, the logarithmic series distribution is sometimes used to model the number of items of a product purchased by a buyer in a specified interval. Note that the parameter theta is limited to values between zero and one, $0 < \theta < 1$. For the random variable X with this distribution, the following apply.

Domain $\{1, 2, \ldots\}$

Probability function $\dfrac{\theta^k}{-k\ln(1-\theta)}$

Moment generating function $\dfrac{\ln(1-\theta e^t)}{\ln(1-\theta)}$

Moments

Mean $-\dfrac{\theta}{(1-\theta)\ln(1-\theta)}$

Second $-\dfrac{\theta}{(1-\theta)^2\ln(1-\theta)}$

Third $-\dfrac{\theta(1+\theta)}{(1-\theta)^3\ln(1-\theta)}$

Central moments

Variance $-\dfrac{\theta\left(1+\dfrac{\theta}{\ln(1-\theta)}\right)}{(1-\theta)^2\ln(1-\theta)}$

Third $-\dfrac{\theta\left(1+\theta+\dfrac{3\theta}{\ln(1-\theta)}+\dfrac{2\theta^2}{(\ln(1-\theta))^2}\right)}{(1-\theta)^3\ln(1-\theta)}$

Fourth $-\dfrac{\theta\left(1+4\theta+\theta^2+\dfrac{4\theta(1+\theta)}{\ln(1-\theta)}+\dfrac{6\theta^2}{(\ln(1-\theta))^2}+\dfrac{3\theta^3}{(\ln(1-\theta))^3}\right)}{(1-\theta)^4\ln(1-\theta)}$

Skewness $\dfrac{(1+\theta)+\dfrac{3\theta}{\ln(1-\theta)}+\dfrac{2\theta^2}{(\ln(1-\theta))^2}}{\sqrt{-\dfrac{\theta}{\ln(1-\theta)}}\sqrt{\left(1+\dfrac{\theta}{\ln(1-\theta)}\right)^3}}$

Kurtosis $\dfrac{1+4\theta+\theta^2+\dfrac{4\theta(1+\theta)}{\ln(1-\theta)}+\dfrac{6\theta^2}{(\ln(1-\theta))^2}+\dfrac{3\theta^3}{(\ln(1-\theta))^3}}{\dfrac{\theta}{\dfrac{\theta}{\ln(1-\theta)}\left(1+\dfrac{\theta}{\ln(1-\theta)}\right)^2}}$

`LogSeriesDistribution[theta]` represents the logarithmic series distribution with parameter `theta`.

EXAMPLE 6: Find and graph $P(X = k)$ for $k = 1, 2, 3, 4$ given a logarithmic series distribution with parameter values $\theta = 0.1, 0.35, 0.75$.

SOLUTION: In logtab, we find $P(X = k)$ for $k = 1, 2, 3, 4$ for the logarithmic series distribution with $\theta = 0.1$. (Note that we display the results in the form $\{k, P(X = k)\}$.) We will plot these points. In dots1, we generate a graphics object representing these points using a small PointSize. Then in p1, we create a ListPlot of the points in logtab. We use the option PlotJoined->True so that the points are joined by line segments making them easier to follow later in the example. Finally, in pd1, we show dots1 and the graph generated with ListPlot. Output is suppressed by DisplayFunction->Identity.

```
Clear[logtab]
logtab=Table[{k,PDF[LogSeriesDistribution[0.1],k]},
 {k,1,4}]
{{1,0.949122},{2,0.0474561},{3,0.00316374},
 {4,0.000237281}}
dots1=Graphics[{PointSize[0.015],Map[Point,logtab]}];
p1=ListPlot[logtab,PlotJoined->True,
DisplayFunction->Identity,Axes->True];
pd1=Show[dots1,p1,DisplayFunction->Identity,
Axes->Automatic];
```

We follow this same procedure to generate the same types of graphs for the other two cases in pd2 and pd3. We use the option DisplayFunction->$DisplayFunction so that the suppressed graphs in pd1, pd2, and pd3 can be viewed with Show.

```
Clear[logtab2]
logtab2=Table[{k,PDF[LogSeriesDistribution[0.35],k]},
 {k,1,4}]
{{1,0.812474},{2,0.142183},{3,0.033176},{4,0.00870871}}
dots2=Graphics[{PointSize[0.03],Map[Point,logtab2]}];
p2=ListPlot[logtab2,PlotJoined->True,
DisplayFunction->Identity,Axes->True];
pd2=Show[dots2,p2,DisplayFunction->Identity,
Axes->Automatic];
```

```
Clear[logtab3]
logtab3=Table[{k,PDF[LogSeriesDistribution[0.75],k]},
  {k,1,4}]
```

```
{{1,0.541011},{2,0.202879},{3,0.101439},{4,0.0570597}}
```

```
dots3=Graphics[{PointSize[0.04],Map[Point,logtab3]}];
p3=ListPlot[logtab3,PlotJoined->True,
DisplayFunction->Identity,Axes->True];
pd3=Show[dots3,p3,DisplayFunction->Identity,
Axes->Automatic];
```

```
Show[pd1, pd2, pd3, DisplayFunction -> $DisplayFunction]
```

0.8
0.6
0.4
0.2
1.5 2 2.5 3 3.5 4

■

Negative Binomial Distribution (Pascal Distribution)

Consider a sequence of independent Bernoulli trials. Recall that each Bernoulli trial has two possible outcomes, success or failure, and the probability of success on each trial is p. Like the geometric random variable, which is a special case, the negative binomial random variable may be defined in two ways. Some sources, including the *Mathematica* documentation, define it to be the number of failures before the nth success. We denote this random variable by X. It is sometimes described as the waiting time for the nth success for Bernoulli trials. Many introductory textbooks define the negative binomial random variable to be the number of the trial on which the nth success occurs. We denote this random variable by Y where $Y = X + n$. The distribution of X (or of Y) is called the negative binomial distribution with parameters n and p. This distribution is also referred to as the Pascal distribution. As an example of a negative binomial random variable, let X be the number of tails before the third head occurs in repeated tossing of a fair coin. Equivalently, Y is the number of the trial on which the third head occurs. In the case of $n = 1$, the negative binomial random variable is equivalent to the geometric random variable.

For the random variable X:

Domain	$\{0, 1, 2, \dots\}$
Probability function	$\binom{n+x-1}{n-1} p^n (1-p)^x$
Distribution function	$\sum_{i=1}^{x} \binom{n+i-1}{n-1} p^n (1-p)^i$
Moment generating function	$\left(\frac{p}{1-(1-p)e^t}\right)^n$
Mean	$\frac{n(1-p)}{p}$
Central moments	
Variance	$\frac{n(1-p)}{p^2}$
Third	$\frac{n(1-p)(2-p)}{p^3}$
Fourth	$\frac{3n(n+2)(1-p)^2}{p^4} + \frac{n(1-p)}{p^2}$
Skewness	$\frac{(2-p)}{\sqrt{n(1-p)}}$
Kurtosis	$3 + \frac{6}{n} + \frac{p^2}{n(1-p)}$

For the random variable $Y = X + n$:

Domain	$\{n, n+1, n+2, \dots\}$
Probability function	$\binom{x-1}{n-1} p^n (1-p)^{x-n}$
Distribution function	$\sum_{i=n}^{x} \binom{i-1}{n-1} p^n (1-p)^{i-n}$
Moment generating function	$\left(\frac{pe^t}{1-(1-p)e^t}\right)^n$

Moments	
Mean	$\frac{n}{p}$
Second	$\frac{n(n+1-p)}{p^2}$
Central moments	
Variance	$\frac{n(1-p)}{p^2}$
Third	$\frac{n(1-p)(2-p)}{p^3}$
Fourth	$\frac{3n(n+2)(1-p)^2}{p^4}+\frac{n(1-p)}{p^2}$
Skewness	$\frac{(2-p)}{\sqrt{n(1-p)}}$
Kurtosis	$3+\frac{6}{n}+\frac{p^2}{n(1-p)}$

`NegativeBinomialDistribution[n,p]` represents the negative binomial distribution for the specified number of successes `n` and probability `p`.

EXAMPLE 7: In a game, contestants throw darts at a target. They are allowed to throw darts until they miss the bullseye three times, and when they hit the bullseye, they win a prize. Based on the total area of the target, the probability of hitting the bullseye is 0.4; so that the probability of missing the bullseye is 0.6. (a) Find the probability that a contestant will get to toss a dart exactly five times. (b) Find the probability that a contestant will get to toss a dart at most five times.

SOLUTION: We will define miss the bullseye to be success in this problem, and consider the negative binomial distribution with probability $p = 0.6$ and $n = 3$. (a) The contestant will toss the dart exactly five times if the third miss occurs on the fifth trial. Thus, we are interested in finding $P(Y = 5)$. To use *Mathematica*, we translate to the random variable X the number of failures (bullseyes) in the five trials. In this instance there must be two bullseyes in the five trials. Thus, we calculate $P(X = 2)$ using `PDF`.

```
PDF[NegativeBinomialDistribution[3,0.6],2]
0.20736
```

(b) "At most five" means five or less. Following the same reasoning we calculate $P(Y \leq 5) = P(X \leq 2)$ using `CDF`.

```
CDF[NegativeBinomialDistribution[3,0.6],2]
0.68256
```

■

Poisson Distribution

The Poisson distribution was originally developed by the French mathematician and physicist Simeon Poisson as an approximation to the binomial distribution when $n \to \infty$ and $p \to 0$. It is often used to model the frequency with which a particular event occurs during a specified period of time or in a specific area or volume. The random variable X is the number of occurrences. The distribution of the count X is called the Poisson distribution with parameter λ. Examples of Poisson random variables include the number of flaws in a square yard of textile, the number of bacteria in a cubic centimeter of water, the number of arrivals at a gas station during a 10 minute time period, and the number of chocolate chips in a cookie selected from a batch of chocolate chip cookies.

For the Poisson distribution:

Domain	$\{0, 1, 2, \ldots\}$
Probability function	$\frac{\lambda^x e^{-\lambda}}{x!}$
Distribution function	$\sum_{i=0}^{x} \frac{\lambda^i e^{-\lambda}}{i!}$
Moment generating function	$e^{\lambda(e^t - 1)}$
Moments	
Mean	λ
Second	$\lambda + \lambda^2$
Third	$\lambda((\lambda + 1)^2 + \lambda)$
Central moments	
Variance	λ
Third	λ
Fourth	$\lambda(1 + 3\lambda)$
Skewness	$\frac{1}{\sqrt{\lambda}}$
Kurtosis	$3 + \frac{1}{\lambda}$

`PoissonDistribution[mu]` represents the Poisson distribution with parameter μ. We used the notation λ instead of μ in the preceding discussion. Notice that λ is the mean.

EXAMPLE 8: Suppose the average number of power failures per year is λ. (a) Find the probability of power failure during a specified week. (b) Find the probability of no power failure during the next month.

SOLUTION: (a) If the average number of power failures per year is λ, then the average number per week is $\frac{\lambda}{52}$. Using `PDF` we see that the probability of no power failure in the week is $e^{-\lambda/52}$, so the probability of one or more failures is $1 - e^{-\lambda/52}$.

```
PDF[PoissonDistribution[lambda / 52], 0]

E^(-lambda/52)
```

(b) If the average number of power failures per year is λ, then the average number per month is $\frac{\lambda}{12}$. Using `PDF` we see that the probability of no power failure in the month is $e^{-\lambda/12}$,

```
PDF[PoissonDistribution[lambda / 12], 0]

E^(-lambda/12)
```

■

EXAMPLE 9: For each day's operation the number of repairs, X, that machine A requires is a Poisson random variable with mean $0.10t$, where t denotes the number of hours of daily operation. The number of daily repairs, Y, for machine B is Poisson with mean $0.12t$. The daily cost of operating machine A is $10t + 30X^2$, and the daily cost for machine B is $8t + 30Y^2$. Which machine minimizes expected daily cost if a day consists of (a) 10 hours? (b) 20 hours?

SOLUTION: Using `ExpectedValue` and simplifying the *Mathematica* output, we learn that the expected cost for machine A is $13t + 0.3t^2$, and the expected cost for machine B is $11.6t + 0.432t^2$. When these expressions are evaluated at $t = 10$

machine B has the lower expected cost. For $t = 20$, machine A has the lower expected cost. Machine B is more economical for short time periods because of its smaller hourly operating cost. Machine A is more economical for longer time periods, because it tends to be repaired less frequently.

```
costforA  = ExpectedValue [10 t + 30 x ^ 2,
  PoissonDistribution[.10 t], x]
```

$$\frac{1}{10} E^{0.t} (130\,t + 3\,t^2)$$

```
costforB  = ExpectedValue [8 t + 30 y ^ 2,
  PoissonDistribution [.12 t], y]
```

$$\frac{2}{125} E^{0.t}\, t\, (725 + 27\,t)$$

```
Table [costforA, {t, 10, 20, 10}]

{160., 380.}

Table [costforB, {t, 10, 20, 10}]

{159.2, 404.8}
```

■

6.3 Continuous Random Variables and Distributions

Mathematica's **Statistics** package **ContinuousDistributions** contains commands for working with the following continuous distributions: the continuous beta distribution, the Cauchy distribution, the chi distribution, the chi-square distribution, the noncentral chi-square distribution, the exponential distribution, the extreme-value distribution, the F distribution, the noncentral F distribution, the gamma distribution, the normal (Gaussian) distribution, the half-normal distribution, the Laplace (double exponential) distribution, the log-normal distribution, the logistic distribution, the Pareto distribution, the Rayleigh distribution, the Student's t-distribution, the noncentral Student's t-distribution, the uniform distribution, and the Weibull distribution.

A subset of these distributions (`ChiSquareDistribution`, `FRatioDistribution`, `NormalDistribution`, and `StudentTDistribution`) are in the **NormalDistribution** package, so if a user only wants to work with one of these continuous distributions, then the

NormalDistribution package can be loaded instead of **ContinuousDistributions**. The **Help Browser** can be used to obtain information about the continuous distributions as well as the functions which operate on them. This help feature may also include examples to demonstrate the syntax of the functions.

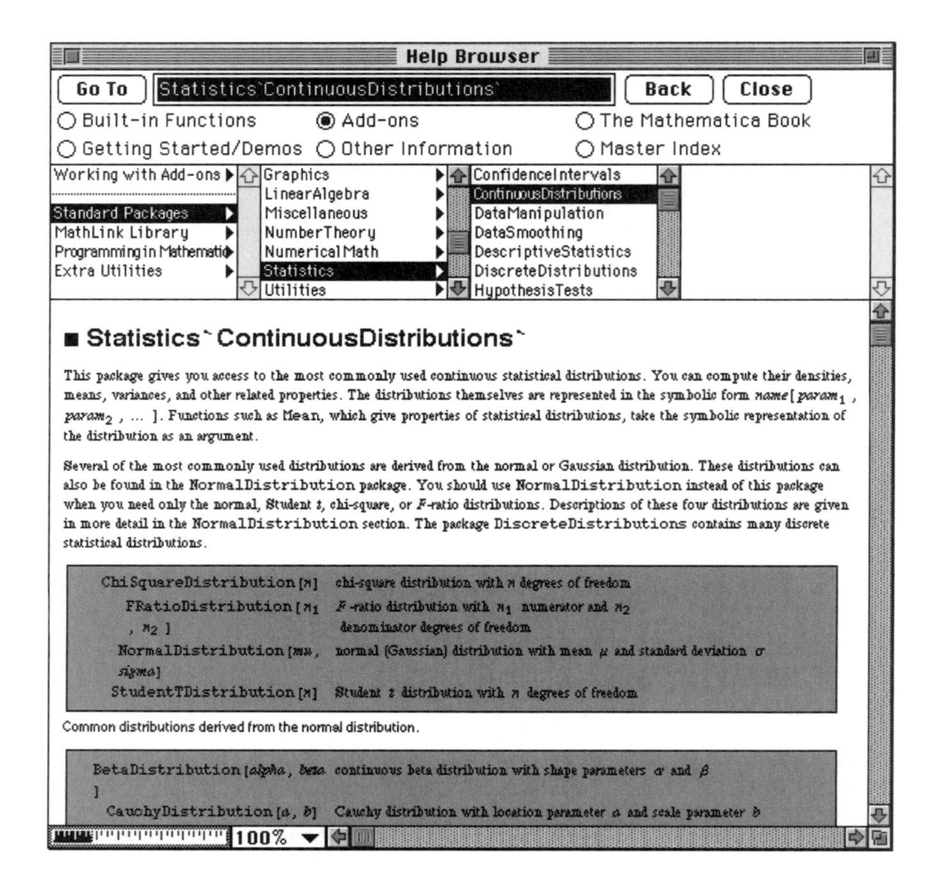

The functions that operate on the distributions in **ContinuousDistributions** and **NormalDistribution** include:

`Domain[dist]` ContinuousDistributions:Domain, which gives the domain of the distribution `dist`,

`PDF[dist,x]` ContinuousDistributions:PDF, which gives the probability density function of the distribution `dist` evaluated at `x`;

`CDF[dist,x]` ContinuousDistributions:CDF, which gives the cumulative distribution function of the distribution `dist` evaluated at `x`; and

`CharacteristicFunction[dist,t]` ContinuousDistributions:CharacteristicFunction, which gives the characteristic function (moment generating function) of the distribution `dist` as a function of the variable `t`.

Other commands (from the **DescriptiveStatistics** package) are redefined in these packages to give location, dispersion, and shape statistics of the distributions. This means that the **DescriptiveStatistics** package need not be loaded to use the following commands.

`Mean[dist]` ContinuousDistributions:Mean, yields the mean of the distribution `dist`.

`Variance[dist]` ContinuousDistributions:Variance, gives the variance of the distribution `dist`.

`StandardDeviation[dist]` ContinuousDistributions:StandardDeviation, yields the standard deviation of the distribution `dist`.

`Skewness[dist]` ContinuousDistributions:Skewness, supplies the coefficient of skewness of the distribution `dist`.

`Kurtosis[dist]` ContinuousDistributions:Kurtosis, gives the coefficient of kurtosis of the distribution `dist`.

`KurtosisExcess[dist]` ContinuousDistributions:KurtosisExcess, gives the kurtosis excess (Kurtosis - 3) of the distribution `dist`.

`Quantile[dist,q]` ContinuousDistributions:Quantile, yields the qth quantile of the distribution `dist`.

`ExpectedValue[f,dist,x]` ContinuousDistributions:ExpectedValue, gives the expected value of the function f of x with respect to the distribution `dist`.

Beta Distribution

The probability density function for the **standard form of the beta distribution** is

$$f(x) = x^{\alpha-1}(1-x)^{\beta-1}/B(\alpha, \beta)$$

with parameters $\alpha > 0$ and $\beta > 0$, where $B(\alpha, \beta)$ is called the **beta function** and is given by

$$B(\alpha, \beta) = \int_0^1 u^{\alpha-1}(1-u)^{\beta-1}du.$$

Standard form refers to the fact that the domain is $0 \le x \le 1$. There is a more general form of the density function which can be defined on any finite interval. The integral of the standard form pdf up to x (the cdf at x) is called the **incomplete beta function ratio**. If $B(\alpha, \beta)$ is removed from the incomplete beta function ratio, the result is called the **incomplete beta function**. Its formula is

$$B(\alpha, \beta) = \int_0^1 u^{\alpha-1}(1-u)^{\beta-1}du.$$

When $\alpha = \beta = 1/2$, the beta distribution may be called the **arcsine distribution**. This distribution is useful in the theory of random walks. Generally, the beta distribution is useful for modeling the behavior of certain random variables, such as proportions, which have values in a finite interval. The beta distribution arises in many other situations. For example, it is the distribution of $Y^2 = \dfrac{X_1^2}{(X_1^2 + X_2^2)}$, where X_1^2 and X_2^2 are independent and distributed according to the chi-square distribution with degrees of freedom v_1 and v_2, respectively. In this case $\alpha = v_1/2$ and $\beta = v_2/2$.

For the standard form of the beta distribution:

Mean $\dfrac{\alpha}{\alpha+\beta}$

Variance $\dfrac{\alpha\beta}{(\alpha+\beta)^2(\alpha+\beta+1)}$

`BetaDistribution[alpha, beta]` represents the continuous (standard form) Beta distribution with parameters `alpha` and `beta`.

EXAMPLE 1: Graph the pdf for the beta distribution with (a) $\alpha = 2$ and $\beta = 3$; (b) $\alpha = 3$ and $\beta = 2$ (c) $\alpha = 2$ and $\beta = 2$, (d) $\alpha = \frac{1}{2}$ and $\beta = 2$, (e) $\alpha = \frac{1}{2}$ and $\beta = \frac{1}{2}$.

SOLUTION: First, we load the **ContinuousDistributions** package. Then in `p1` though `p5` we define the graphs of the pdfs specified in (a) through (e), respectively. Output is suppressed by use of `DisplayFunction->Identity`. In `p12` we define a plot to display the graphs for (a) and (b) together. Finally, the graphs are displayed as a graphics array.

```
<<Statistics`ContinuousDistributions`

p1=Plot[PDF[BetaDistribution[2,3],x],{x,0,1},
  PlotRange->{{0,1},{0,2}},DisplayFunction->Identity];
p2=Plot[PDF[BetaDistribution[3,2],x],{x,0,1},
  PlotRange->{{0,1},{0,2}},DisplayFunction->Identity];
```

```
p3=Plot[PDF[BetaDistribution[2,2],x],{x,0,1},
  PlotRange->{{0,1},{0,2}},DisplayFunction->Identity];
p4=Plot[PDF[BetaDistribution[1/2,2],x],{x,0,1},
  PlotRange->{{0,1},{0,2}},DisplayFunction->Identity];
p5=Plot[PDF[BetaDistribution[1/2,1/2],x],{x,0,1},
  PlotRange->{{0,1},{0,2}},DisplayFunction->Identity];
p12=Show[p1,p2,DisplayFunction->Identity];

Show [GraphicsArray[{{p12 , p3}, {p4 , p5}}],
 DisplayFunction -> $DisplayFunction]
```

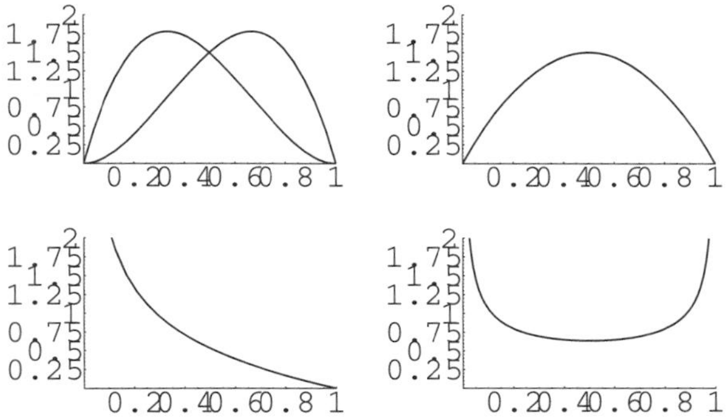

■

Cauchy Distribution

The probability density function for the Cauchy distribution with location parameter (median) a and scale parameter $b > 0$ is

$$f(x) = \frac{1}{\pi b\{1 + [(x-a)/b]^2\}}.$$

The domain is $-\infty < x < +\infty$. The pdf is bell-shaped. It is most notably different from the normal pdf in having higher and flatter tails. The distribution does not have finite moments and so does not possess a finite mean or variance. Consequently, there is no standardized form (having mean zero and standard deviation one) of the distribution. A standard form of the Cauchy distribution is obtained by setting $a = 0$ and $b = 1$. In this form the Cauchy is equivalent to a Student's t-distribution with one degree of freedom. The Cauchy distribution may be used to describe the distribution of points of impact of particles from a point source with a fixed straight line.

For the Cauchy distribution:

Distribution function	$\frac{1}{2} + \frac{\tan^{-1}((x-a)/b)}{\pi}$
Lower quartile	$a - b$
Upper quartile	$a + b$

`CauchyDistribution[a, b]` represents the Cauchy distribution with parameters `a` and `b`.

EXAMPLE 2: Graph the pdf of the standard form, ($a = 0$ and $b = 1$), of the Cauchy distribution together with the pdf of the normal distribution having the same quartiles.

SOLUTION: Assuming that the **ContinuousDistributions** package has been loaded, we first determine the mean and standard deviation of the normal distribution to be plotted. From the discussion and subsequent list just given, we calculate the quartiles of the standard form of the Cauchy distribution to be –1, 0, and 1, where its median is zero. Because the normal distribution is symmetric around its mean, its mean is equal to its median. Thus, we will need a normal distribution with mean zero. To determine the standard deviation, we first determine the third quartile (75th percentile) of the standard normal distribution (the normal distribution having mean zero and standard deviation one) using `Quantile`.

```
z=Quantile[NormalDistribution[0,1],.75]
0.67449
```

The following equation describes the relationship between a random variable X and its standardized form Z. Notice that a random variable is standardized by subtracting its mean and dividing by its standard deviation. The resulting standardized form has mean zero and standard deviation one.

$$Z = \frac{X - \mu_x}{\sigma_x}$$

For the current problem we are looking for a standard deviation of X. We know the value 1 is the third quartile of X and that X has mean zero. The corresponding third quartile for Z is 0.67449. Thus, we have

$$0.67449 = \frac{1-0}{\sigma_x} \text{ or } \sigma_x = \frac{1}{0.67449} = 1.48260.$$

Therefore, the normal distribution having the same quartiles as the standard form of the Cauchy distribution has mean zero and standard deviation 1.4826. Using `Plot` and `Show` we display the two pdfs using a solid line for the Cauchy and a dotted line for the normal.

```
p1= Plot[PDF[CauchyDistribution[0,1],x],{x,-5,5},
  PlotStyle->{GrayLevel[0]},DisplayFunction->Identity];
p2= Plot[PDF[NormalDistribution[0,1/.67449],x],{x,-5,5},
  PlotStyle->{Dashing[{0.02}]},DisplayFunction->Identity];

Show [p1, p2, DisplayFunction -> $DisplayFunction]
```

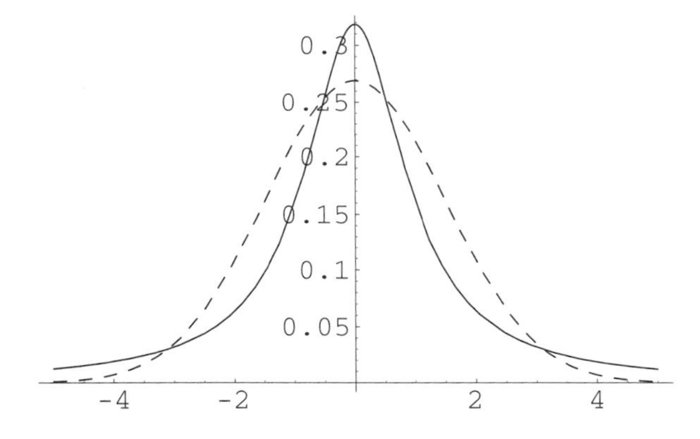

■

Chi Distribution

The probability density function for the chi distribution with $\nu > 0$ degrees of freedom is

$$f(x) = \frac{x^{\nu - 1} e^{-x^2/2}}{2^{(\nu/2) - 1}\Gamma(\nu/2)}.$$

The domain is $0 \le x < +\infty$. This is the distribution of the square root of a chi-square random variable with ν degrees of freedom. For example, because the sample variance for samples from a normal population has a chi-square distribution, the sample standard deviation for samples from a normal population has a chi distribution.

For the chi distribution:

Mean $$\frac{\sqrt{2}\,\Gamma\left(\frac{v+1}{2}\right)}{\Gamma\left(\frac{v}{2}\right)}$$

Variance $$2\left[\frac{\Gamma\left(\frac{v+2}{2}\right)}{\Gamma\left(\frac{v}{2}\right)}-\left(\frac{\Gamma\left(\frac{v+1}{2}\right)}{\Gamma\left(\frac{v}{2}\right)}\right)^2\right]$$

`ChiDistribution[n]` represents the Chi distribution with `n` degrees of freedom.

EXAMPLE 3: Graph the pdfs of the chi distribution with degrees of freedom 1, 2, and 5.

SOLUTION: Assuming that the **ContinuousDistributions** package has been loaded, we use `Plot` to create the three pdf plots but suppress the output using `DisplayFunction->Identity`. Using `Show` and `DisplayFunction->$DisplayFunction`, we display the three pdfs on the same graph.

```
chi1= Plot[PDF[ChiDistribution[1],x],{x,0,5},
  DisplayFunction->Identity];
chi2= Plot[PDF[ChiDistribution[2],x],{x,0,5},
  DisplayFunction->Identity];
chi3= Plot[PDF[ChiDistribution[5],x],{x,0,5},
  DisplayFunction->Identity];
```

```
Show [chi1 , chi2 , chi3 ,
 DisplayFunction -> $DisplayFunction]
```

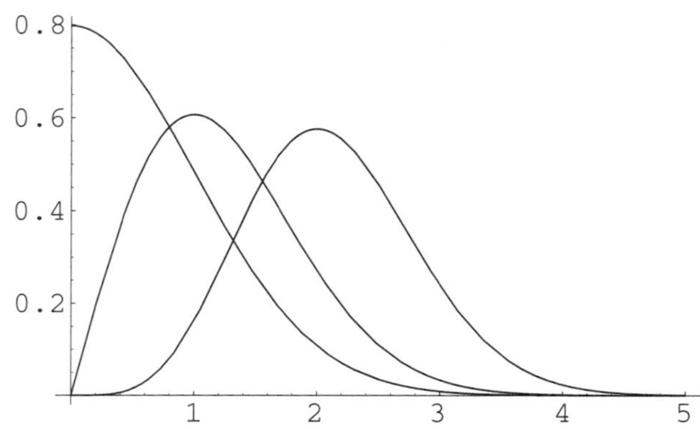

■

Chi-Square Distribution

The probability density function for the chi-square distribution with $v > 0$ degrees of freedom is

$$f(x) = \frac{x^{(v/2)-1}e^{-x/2}}{2^{v/2}\Gamma(v/2)}.$$

The domain is $0 \le x < +\infty$. The chi-square distribution is a particular form of the **gamma distribution** and represents its most important application in statistics. Given a set of independent random variables $Z_1, Z_2, \ldots Z_v$ each of which has a standard normal distribution (that is a normal distribution with mean zero and standard deviation one), then the distribution of $\sum_{i=1}^{v} Z_i^2$ is the chi-square distribution with v degrees of freedom. Chi-square distributions are encountered in statistical inference procedures based on sums of squares of independent normal random variables. For example, the sample variance for samples from a normal population has a chi-square distribution.

For the chi-square distribution:

Moment generating function	$(1-2t)^{-v/2}$ for $t < \frac{1}{2}$
Mean	v
Variance	$2v$

`ChiSquareDistribution[n]` represents the Chi-square distribution with `n` degrees of freedom.

EXAMPLE 4: Graph the pdfs of the chi-square distribution with degrees of freedom 2, 3, and 5.

SOLUTION: Assuming that the **ContinuousDistributions** (or **NormalDistribution**) package has been loaded, we use `Plot` to create the three pdf plots, but suppress the output using `DisplayFunction->Identity`. Then using `Show` and `DisplayFunction->$DisplayFunction`, we display the three pdfs on the same graph.

```
chisq1= Plot[PDF[ChiSquareDistribution[2],x],{x,0,12},
  DisplayFunction->Identity];
```

```
chisq2= Plot[PDF[ChiSquareDistribution[3],x],{x,0,12},
  DisplayFunction->Identity];
chisq3= Plot[PDF[ChiSquareDistribution[5],x],{x,0,12},
  DisplayFunction->Identity];

Show[chisq1, chisq2, chisq3,
 DisplayFunction -> $DisplayFunction]
```

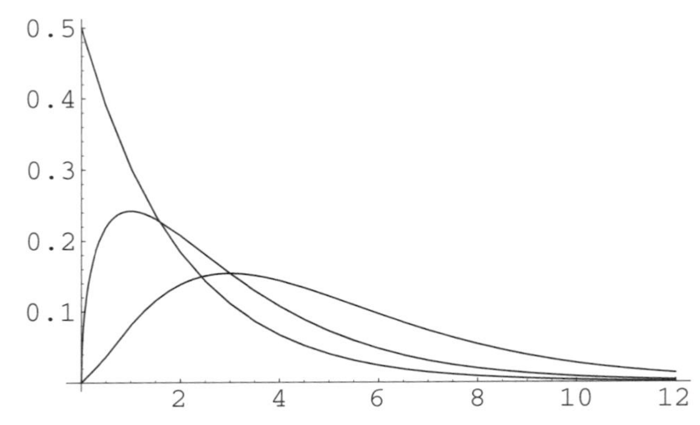

■

Noncentral Chi Square Distribution

The probability density function for the noncentral chi-square distribution with $v > 0$ degrees of freedom and noncentrality parameter $\lambda \geq 0$ is represented in different ways. One, for example, is as a mixture (sum) of (central) chi-square pdfs. The range is $0 \leq x < +\infty$. Given a set of independent normally distributed random variables $X_1, X_2, \ldots, X_v$ where X_i has mean μ_i and standard deviation one, $i = 1, 2, \ldots, v$, respectively, the distribution of $\sum_{i=1}^{v} X_i^2$ is noncentral chi-square with v degrees of freedom and noncentrality parameter $\lambda = \sum_{i=1}^{v} \mu_i^2$. That is, the sum of the squares of v normally distributed random variables with standard deviations one and nonzero means has the noncentral chi-square distribution.

When the means are all zero, $\lambda = 0$, and the noncentral chi-square is equivalent to the (central) chi-square.

For the noncentral chi-square distribution:

Moment generating function	$\dfrac{e^{\lambda t/(1-2t)}}{(1-2t)^{\nu/2}}$ for $t < \dfrac{1}{2}$
Mean	$\nu + \lambda$
Variance	$2(\nu + 2\lambda)$

`NoncentralChiSquareDistribution[n, lambda]` represents the non-central chi-square distribution with `n` degrees of freedom and noncentrality parameter `lambda`.

EXAMPLE 5: (a) Graph the pdfs of the noncentral chi-square distribution with degrees of freedom 2, 4, 6, and 8 and noncentrality parameter 5. (b) Graph the pdfs of the noncentral chi-square distribution with noncentrality parameter 1, 3, 5, and 7 and degrees of freedom 3.

SOLUTION: We assuming that the **ContinuousDistributions** package has been loaded. Then for both (a) and (b) the individual graphs are created using `Plot`, where the output is suppressed by using `DisplayFunction->Identity`. Using `Show` and `DisplayFunction->$DisplayFunction`, we display the four pdfs on the same graph.

(a)

```
nc1= Plot[PDF[NoncentralChiSquareDistribution[2,5],x],
   {x,0,30},DisplayFunction->Identity];
nc2= Plot[PDF[NoncentralChiSquareDistribution[4,5],x],
   {x,0,30},DisplayFunction->Identity];
nc3= Plot[PDF[NoncentralChiSquareDistribution[6,5],x],
   {x,0,30},DisplayFunction->Identity];
nc4= Plot[PDF[NoncentralChiSquareDistribution[8,5],x],
   {x,0,30},DisplayFunction->Identity];
Show[nc1,nc2,nc3,nc4,DisplayFunction->$DisplayFunction]
```

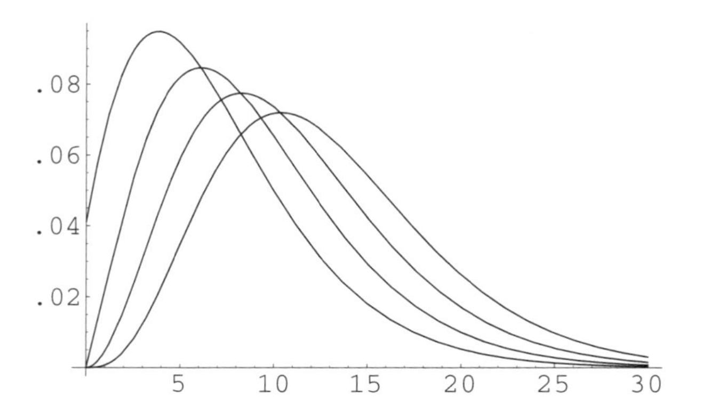

(b)

```
nc11= Plot[PDF[NoncentralChiSquareDistribution[3,1],x],
  {x,0,30},DisplayFunction->Identity];
nc21= Plot[PDF[NoncentralChiSquareDistribution[3,3],x],
  {x,0,30},DisplayFunction->Identity];
nc31= Plot[PDF[NoncentralChiSquareDistribution[3,5],x],
  {x,0,30},DisplayFunction->Identity];
nc41= Plot[PDF[NoncentralChiSquareDistribution[3,7],x],
  {x,0,30},DisplayFunction->Identity];
Show[nc11,nc21,nc31,nc41,
  DisplayFunction->$DisplayFunction]
```

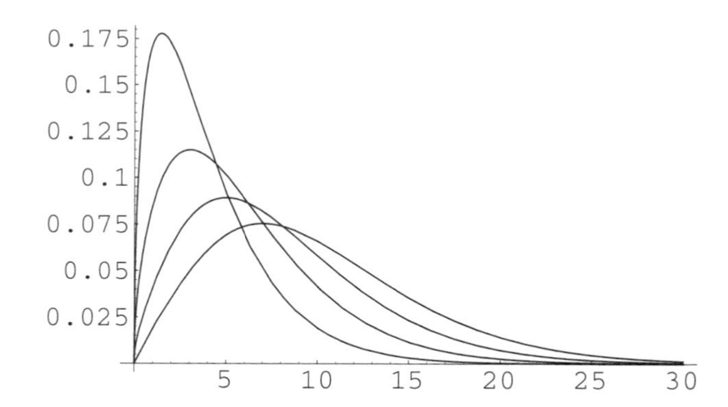

■

Exponential Distribution

The probability density function for the exponential distribution with parameter $\lambda > 0$ is

$$f(x) = \lambda e^{-\lambda x}.$$

The domain is $0 \le x < +\infty$. Note that the density function can be generalized for any domain of the form $\theta \le x < \infty$ where $\theta \ge 0$. When $\theta = 0$, as is the case with the pdf just shown, the distribution is sometimes called the **one-parameter exponential distribution**. When, in addition, $\lambda = 1$, it is called the **standard exponential distribution**. The exponential distribution is used to describe the distribution of the waiting time between an initial point in time and the occurrence of some phenomenon or the **waiting time** between two occurrences of a random phenomenon. It has a property that is often described as lack of memory. That is, the conditional probability of waiting t time units given no occurrence during the preceding a time units is the same as the unconditional probability of waiting t time units from the beginning. It is as if we start afresh, forgetting the fact that we have waited a time units already.

The exponential distribution is a special case of the gamma distribution.

For the exponential distribution:

Distribution function	$1 - e^{-\lambda x}$
Moment generating function	$\dfrac{\lambda}{\lambda - t}$ for $t < \lambda$
Mean	$\dfrac{1}{\lambda}$
Variance	$\dfrac{1}{\lambda^2}$

`ExponentialDistribution[lambda]` represents the (one-parameter) exponential distribution with scale parameter `lambda`.

EXAMPLE 6: (a) The magnitude of earthquakes recorded in a region of North America can be modeled by an exponential distribution with mean 2.4, as measured on the Richter scale. Find the probability that the next earthquake to strike this region will (a) exceed 3.0 on the Richter scale; (b) fall between 2.0 and 3.0 on the Richter scale.

SOLUTION: Because the mean is 2.4, the parameter of the exponential distribution is $\lambda = 1/2.4$. Assuming that the **ContinuousDistributions** package has been loaded, we use `Plot` to display the probability density function.

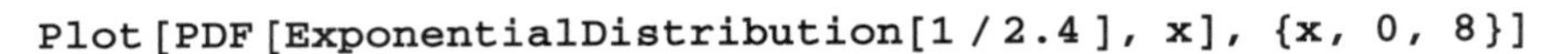

```
Plot[PDF[ExponentialDistribution[1 / 2.4 ], x], {x, 0, 8}]
```

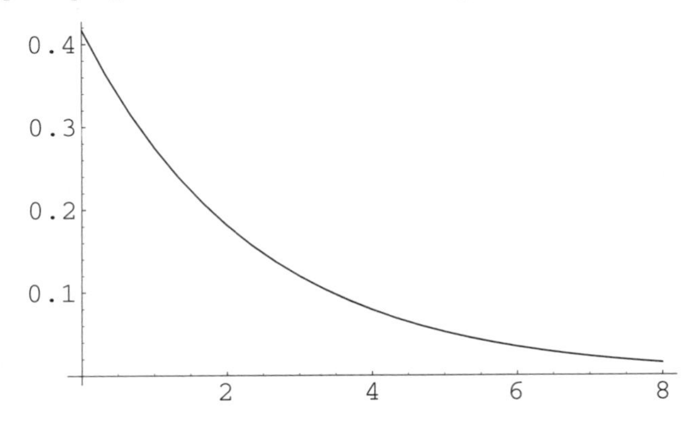

(a) The area under the pdf from 3 to infinity is the desired probability. Using `CDF` we find that $P(X \leq 3.0) = 0.713495$. Thus, $P(X > 3.0) = 1 - 0.713495 = 0.286505$.

```
prob3=CDF[ExponentialDistribution[1/2.4],3.0]
```

```
0.713495
```

```
1-prob3
```

```
0.286505
```

(b) To find $P(2.0 \leq X \leq 3.0)$, we subtract $P(X < 2)$ from $P(X \leq 3)$. Notice that the same result is calculated using the built-in function `Integrate`.

```
prob2=CDF[ExponentialDistribution[1/2.4],2.0]
```

```
0.565402
```

```
prob3-prob2
```

```
0.148093
```

```
Integrate[PDF[ExponentialDistribution[1/2.4],x],
  {x,2.0,3.0}]
```

```
0.148093
```

■

Extreme Value Distribution

The probability density function for the extreme value distribution with location parameter $\alpha(-\infty < \alpha < +\infty)$ and scale parameter $\beta > 0$ is

$$f(x) = 1/\beta e^{-\frac{x-\alpha}{\beta}} \exp\left[-\exp\left(-\frac{x-\alpha}{\beta}\right)\right].$$

The domain is $-\infty < x < +\infty$. The name "extreme value" is attached to this distribution because it can be obtained as a limiting distribution as $n \to \infty$ of the greatest value among n independent random variables, each having the same continuous distribution. By replacing X with $-X$, a limiting distribution of the least value is obtained. The *Mathematica* documentation refers to the extreme value distribution as the **Fisher–Tippett distribution**; other sources (with better justification) refer to it as the **Gumbel distribution**. The extreme value distribution is also sometimes called the **log-Weibull distribution**, because if X has an extreme value distribution, then e^X has a Weibull distribution. Note also that $e^{X/\beta}$ has an exponential distribution. The standard form of the extreme value distribution results when $\alpha = 0$ and $\beta = 1$. The corresponding pdf is then

$$e^{-x}\exp[-e^{-x}].$$

The extreme value distribution has been used in fields of application that include rainfall, flood flow, earth quakes, aircraft load, and microorganism survival time. The density function has the same shape for all values of α and β.

For the extreme value distribution:

Distribution function	$\exp\left[-\exp\left(-\frac{x-\alpha}{\beta}\right)\right]$
Moment generating function	$e^{t\alpha}\Gamma(1-\beta t)$ for $\lvert t\rvert < \frac{1}{\beta}$
Mean	$\alpha + 0.57721\beta$
Variance	$\frac{1}{6}\pi^2\beta^2$

`ExtremeValueDistribution[alpha, beta]` represents the extreme value distribution with parameters `alpha` and `beta`.

EXAMPLE 7: (a) Plot the pdf of the standard form of the extreme value distribution. (b) Determine the quartiles, coefficients of skewness and kurtosis, and the mode of the extreme value distribution with location parameter α and scale parameter β.

SOLUTION: (a) The standard form of the extreme value distribution has $\alpha = 0$ and $\beta = 1$. Assuming that the **ContinuousDistributions** package has been loaded, we use `Plot` to display the pdf.

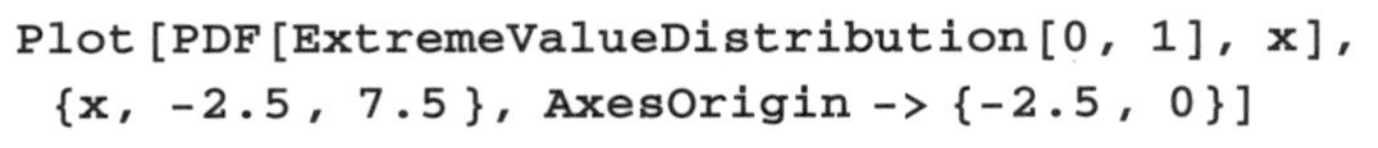

```
Plot[PDF[ExtremeValueDistribution[0, 1], x],
  {x, -2.5, 7.5}, AxesOrigin -> {-2.5, 0}]
```

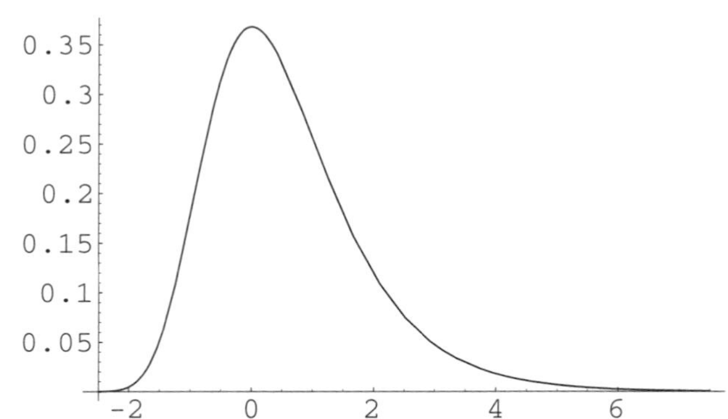

(b) Recall that `Quantile` is loaded with the **ContinuousDistributions** package. We use this command with 0.25, 0.50, and 0.75 to find the first, second, and third quartiles, respectively. The calculations are done using a `Table` command to generate all three quartiles. Notice that the quartiles corresponding to $\alpha = 0$ and $\beta = 1$ are calculated in the second `Table` command.

```
Table[Quantile[ExtremeValueDistribution[α, β], 0.25 k],
  {k, 1, 3}]
```

```
{α - 0.326634 β, α + 0.366513 β, α + 1.2459 β}
```

```
Table[Quantile[ExtremeValueDistribution[0, 1], 0.25 k],
  {k, 1, 3}]
```

```
{-0.326634, 0.366513, 1.2459}
```

Values of skewness and kurtosis coefficients are calculated in general. They are constants, because the shape of the pdf is the same for all values of α and β. The mode of the extreme value distribution is α. It is the value on the horizontal axis under the highest point of the pdf.

```
Skewness[ExtremeValueDistribution[α, β]] // N
1.13955
Kurtosis[ExtremeValueDistribution[α, β]] // N
5.4
```

■

F (Variance Ratio) Distribution

The probability density function for the F (variance ratio) distribution is

$$f(x) = \frac{v_1^{v_1/2} v_2^{v_2/2}}{B(v_1/2, v_2/2)} \frac{x^{(v_1/2)-1}}{(v_2 + v_1 x)^{(v_2+v_1)/2}},$$

where $B(\alpha,\beta) = \int_0^1 u^{\alpha-1}(1-u)^{\beta-1}du$ is the beta function, and the parameters v_1 and v_2 are positive integers called the degrees of freedom for the numerator and the denominator, respectively. The domain is $0 \le x < +\infty$. Often the pdf is expressed in terms of the gamma function, $\Gamma(\alpha) = \int_0^\infty u^{\alpha-1}e^{-u}du$, where the beta and gamma functions are related by

$$B(\alpha,\beta) = \frac{\Gamma(\alpha)\Gamma(\beta)}{\Gamma(\alpha+\beta)}$$

Given random variables X_1 and X_2 which are independent and have chi-square distributions with v_1 and v_2 degrees of freedom, respectively, then the ratio

$$F_{v_1 v_2} = \frac{X_1/v_1}{X_2/v_2}$$

has an F-distribution with numerator degrees of freedom v_1 and denominator degrees of freedom v_2. Note that the order of degrees of freedom is important. That is, the F-distribution with numerator degrees of freedom v_2 and denominator degrees of freedom v_1 is different from the F-distribution with numerator degrees of freedom v_1 and denominator degrees of freedom v_2. They are, however, related in the following way:

$$Quantile[FRatioDistribution[v_1, v_2], q] = \frac{1}{Quantile[FRatioDistribution[v_2, v_1], 1-q]}.$$

The importance of the F-distribution in statistical methodology derives mainly from its applicability to the distribution of ratios of independent estimates of variance such as the ratio of the variances from two independent samples and ratios of mean squares in the analysis of variance.

For the F-distribution:

Mean $\frac{\nu_2}{\nu_2 - 2}$ for $\nu_2 > 2$

Variance $\frac{2\nu_2^2(\nu_1 + \nu_2 - 2)}{\nu_1(\nu_2 - \nu_1)^2(\nu_2 - 4)}$ for $\nu_2 > 4$

`FRatioDistribution[n1, n2]` represents the F-distribution with `n1` numerator degrees of freedom and `n2` denominator degrees of freedom.

EXAMPLE 8: Graph the following pdfs of the F (variance ratio) distribution on the same plot. (a) F-distribution densities with degrees of freedom $\nu_1 = 10$, $\nu_2 = 1$, $\nu_1 = 10$, $\nu_2 = 10$, and $\nu_1 = 10$, $\nu_2 = 1000$; (b) F-distribution densities with degrees of freedom $\nu_1 = 1$, $\nu_2 = 20$, $\nu_1 = 10$, $\nu_2 = 20$, and $\nu_1 = 1000$, $\nu_2 = 20$.

SOLUTION: Assuming that the **ContinuousDistributions** (or the **NormalDistribution**) package has been loaded, we use `Plot` to display the pdfs.

(a)

```
Plot[{PDF[FRatioDistribution[10, 1], x],
   PDF[FRatioDistribution[10, 10], x],
   PDF[FRatioDistribution[10, 1000], x]},
   {x, 0, 4}]
```

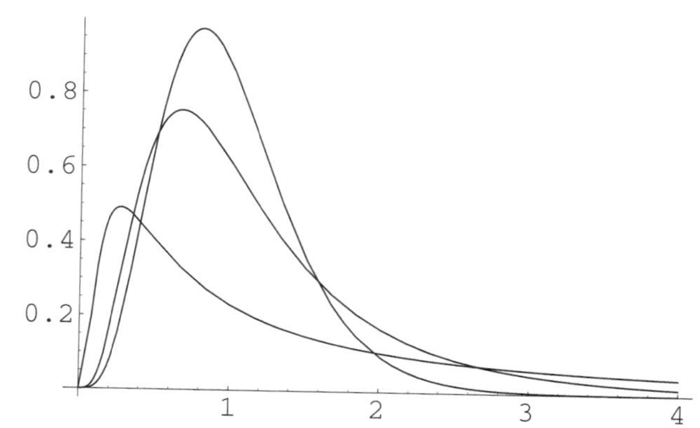

(b)

```
Plot [{PDF [FRatioDistribution [1, 20], x],
  PDF [FRatioDistribution [10, 20], x],
  PDF [FRatioDistribution [1000 , 20], x]},
  {x, 0, 4}]
```

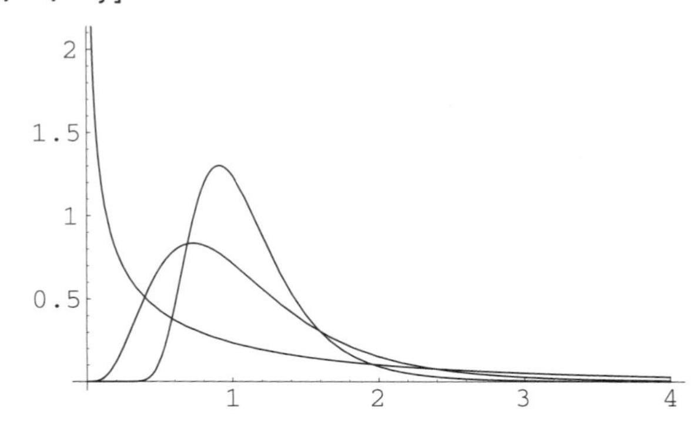

■

Noncentral F *(Variance Ratio) Distribution*

The probability density function for the noncentral F (variance ratio) distribution with numerator degrees of freedom v_1, denominator degrees of freedom v_2 , and (numerator) noncentrality parameter $\lambda \geq 0$ is represented in different ways. One, for example, is as a mixture (sum) of (central) F-distribution pdfs. The range is $0 \geq x < +\infty$. Note that v_1 and v_2 are positive integers. Suppose that X_1 and X_2 are independent random variables where X_1 has a noncentral chi-square distributions with v_1 degrees of freedom and noncentrality parameter $\lambda \geq 0$, and X_2 has a (central) chi-square distributions with v_2 degrees of freedom. Then the ratio

$$F_{v_1 v_2} = \frac{X_1/v_1}{X_2/v_2}$$

has a noncentral F-distribution with numerator degrees of freedom v_1, denominator degrees of freedom v_2, and (numerator) noncentrality parameter $\lambda \geq 0$. If both X_1 and X_2 have noncentral chi-square distributions with v_1 and v_2 degrees of freedom, respectively, and noncentrality parameters $\lambda_1 \geq 0$ and $\lambda_2 \geq 0$, respectively, then the preceding ratio has a **doubly noncentral *F*-distribution** with numerator degrees of freedom v_1, denominator degrees of freedom v_2, and noncentrality parameters $\lambda_1 \geq 0$ and $\lambda_2 \geq 0$. The noncentral F-distribution is used in the calculation of power for hypothesis tests used in the analysis of variance.

For the noncentral F-distribution:

Mean $\quad \dfrac{v_2(v_1 - \lambda)}{v_1(v_2 - 2)}$ for $v_2 > 2$

Variance $\quad 2\left(\frac{v_2}{v_1}\right)^2 \frac{(v_1-\lambda)^2+(v_1+2\lambda)(v_2-2)}{(v_2-2)^2(v_2-4)}$ for $v_2 > 4$

`NoncentralFRatioDistribution[n1, n2, lambda]` represents the noncentral F distribution with `n1` numerator degrees of freedom, `n2` denominator degrees of freedom, and numerator noncentrality parameter `lambda`.

EXAMPLE 9: Graph the following pdfs of the noncentral F (variance ratio) distribution on the same plot: noncentral F-distribution densities with degrees of freedom $v_1 = 40$, $v_2 = 4$ and noncentrality parameters $\lambda = 0.5$, $\lambda = 2$, and $\lambda = 5$.

SOLUTION: Assuming that the **ContinuousDistributions** package has been loaded, we use `Plot` to display the pdfs.

```
Plot[ {PDF [NoncentralFRatioDistribution
   [4, 40, .5], x],
   PDF [NoncentralFRatioDistribution [4, 40, 2], x],
   PDF [NoncentralFRatioDistribution [4, 40, 5], x]},
   {x,0, 6}]
```

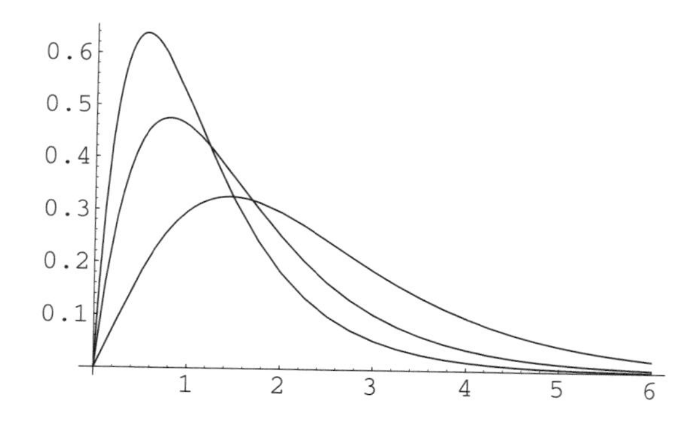

■

Gamma Distribution

The probability density function for the gamma distribution with shape parameter $\alpha > 0$ and scale parameter $\lambda > 0$ is

$$f(x) = \frac{1}{\Gamma(\alpha)\lambda^{\alpha}} x^{\alpha-1} e^{-x/\lambda},$$

where $\Gamma(\alpha) = \int_0^{\infty} u^{\alpha-1} e^{-u} du$ is called the **gamma function**. The domain is $0 \leq x +\infty$. Note that the density function can be generalized for any domain of the form $\lambda \leq x < \infty$ where $\lambda \geq 0$. If $\alpha = 1$, the above gamma pdf reduces to the pdf of an **exponential distribution**. When α is a positive integer, it is the pdf of an **Erlang distribution**. When $\alpha = v/2$ and $\lambda = 2$, the gamma distribution is called a **chi-square distribution** with v degrees of freedom. The **standard form of the gamma distribution** results when $\lambda = 1$, giving

$$f(x) = \frac{1}{\Gamma(\alpha)} x^{\alpha-1} e^{-x}.$$

The integral of this pdf up to x (the cdf of the standard form at x) is an incomplete gamma function ratio and is sometimes called the incomplete gamma function.

One of the most important properties of the gamma distribution is the reproductive property. This is illustrated by the following. Suppose $X_1, X_2, \ldots X_n$ represent independent gamma random variables with parameters a_i and λ, $i = 1, 2, \ldots, n,$, respectively. Then $\sum_{i=1}^{n} X_i$ also has a gamma distribution with parameters $\sum_{i=1}^{n} \alpha_i$ and λ. A second property of interest is that if X_1 and X_2 are independent gamma random variables with parameters α_1 and λ, $i = 1, 2$, respectively, then $X_1/(X_1 + X_2)$ has a beta distribution with parameters α_1 and α_2.

The gamma distribution is used to represent the distribution of the range in random samples from a normal population. It has been used to model physical situations such as waiting times in life testing. Finally, gamma distributions share with log-normal distributions the ability to closely approximate a normal distribution while representing an essentially positive random variable. The gamma distribution approaches normality as $\alpha \to \infty$.

For the gamma distribution:

Moment generating function	$(1 - \lambda t)^{-\alpha}$, $t < 1/\lambda$
Mean	$\alpha\lambda$
Variance	$\alpha\lambda^2$

`GammaDistribution[alpha, lambda]` represents the gamma distribution with shape parameter `alpha` and scale parameter `lambda`.

EXAMPLE 10: Based on previous experience, the length of time, X, to conduct periodic maintenance on a piece of machinery is known to follow a gamma distribution with $\alpha = 3$ and $\lambda = 2$ minutes. Suppose that a new repair person requires 20 minutes to check the machine. Does this time to perform a maintenance check seem to disagree with previous experience?

SOLUTION: Assuming that the **ContinuousDistributions** package has been loaded, we use `Plot` to display the pdf and cdf for the gamma density with $\alpha = 3$ and $\lambda = 2$. Notice that the area under the pdf to the left of 14 is quite large and the area to the right is correspondingly small. The area to the left of 14 under the pdf is the value of the cdf at 14. That the value is quite close to 1 is apparent from the plot of the cdf. The actual value for the area to the left, which is the probability that X is less than 14, is calculated using `CDF` to be 0.946713. Therefore, the probability of 14 or more minutes is 0.0532869. Thus, it is quite unusual for the maintenance check to take 14 minutes or longer (something that would ordinarily occur only about 5% of the time). In view of the fact that this is a new maintenance person, we would suspect that he or she is working more slowly than previous maintenance personnel.

```
Plot [PDF [GammaDistribution [3, 2], x], {x, 0, 20}]
```

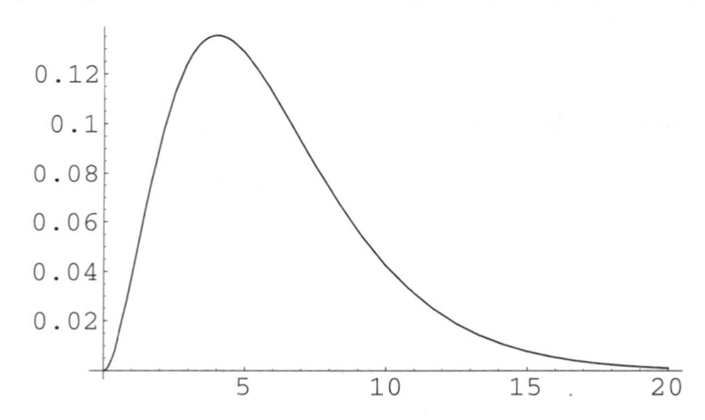


```
Plot [CDF [GammaDistribution [3, 2], x], {x, 0, 20}]
```

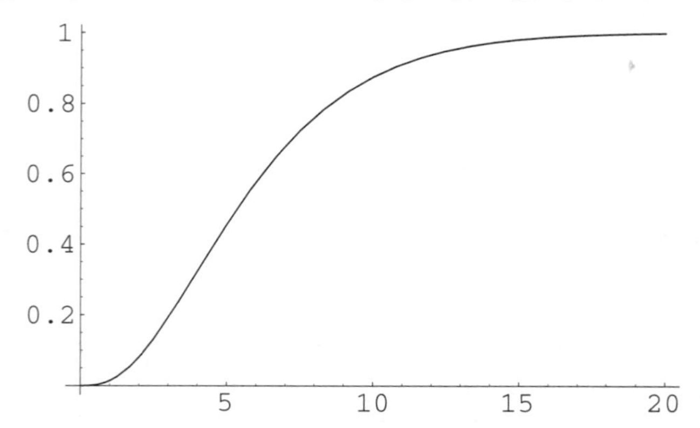


```
prob=CDF[GammaDistribution[2,3],14]//N
0.946713
```

```
1-prob
0.0532869
```

■

Normal (Gaussian) Distribution

The probability density function for the normal distribution with mean μ ($-\infty < \mu < \infty$) and standard deviation $\sigma > 0$ is

$$f(x) = \frac{1}{\sigma\sqrt{2\pi}}\exp\left[\frac{-(x-\mu)^2}{2\sigma^2}\right].$$

The domain is $-\infty < x < +\infty$). The standardized form, called **standard normal** or **unit normal**, has mean zero and standard deviation 1. Its pdf is ,

$$f(z) = \frac{1}{\sqrt{2\pi}}\exp\left[-\frac{1}{2}z^2\right],$$

where the random variable is often denoted by Z.

Of all the continuous distributions, the normal distribution is the most widely studied and frequently used. It is known by several different names, including Gaussian, Laplace, Laplace–Gauss, and deMoivre. It was originally derived as an approximation to a binomial distribution. One reason for the importance of the normal distribution is based on the **central limit theorem,** which states conditions under which the distributions of sums (or means) of random variables tend to the normal distribution as the number of variables in the sum (or mean) increases. More specifically, one form of the central limit theorem says that if a random variable Y can be regarded as a sum of a large number of independent random variables, that is, $Y = X_1 + X_2 + \ldots + X_n$, then no matter what the distribution of the X's values, the distribution of Y is approximately normal if certain conditions on the variances of the X's values are satisfied. The normal distribution is used to approximate many other distributions, and in turn other distributions are used to approximate the normal. For the latter case, consider the log-normal, gamma, logistic, Weibull, and Cauchy distributions. The normal distribution appears very often in probability models of real-world phenomena. Many measurements on living organisms and manufactured items, for example, are shown to have distributions for which normal pdfs are good approximations.

For the normal distribution:

Moment generating function $\quad \exp\left(\mu t + \frac{1}{2}\sigma^2 t^2\right)$

Mean	μ
Variance	σ^2

`NormalDistribution[mu, sigma]` represents the Normal (Gaussian) distribution with mean `mu` and standard deviation `sigma`.

EXAMPLE 11: The vehicles that cross a particular bridge are known to have an average weight of 4675 pounds, with a standard deviation of 345 pounds. Suppose there are 40 vehicles on the bridge at once. (a) Find the probability that the entire weight exceeds 190,000 pounds. (b) Find the number w such that the probability is .99 that the combined weight will not exceed w.

SOLUTION: The weights of the 40 vehicles can be thought of as values of 40 independent random variables, each having mean 4675 pounds and standard deviation 345 pounds. The sum of these 40 random variables, call it Y, is a random variable whose value is the total weight. The mean of the sum of random variables is the sum of the means, so the mean of Y is $40 \times 4675 = 187{,}000$. The variance of the sum of independent random variables is the sum of the variances, so the variance of Y is $40 \times 345^2 = 4{,}761{,}000$. The standard deviation is 2181.9716. By the central limit theorem, the distribution of Y is approximately normal. We could load the **ContinuousDistributions** to work with the normal distribution, but we load instead the **NormalDistribution** package.

```
<<Statistics`NormalDistribution`
```

(a) The function `CDF` is used to calculate $P(Y \leq 190{,}000)$. The probability that the entire weight exceeds 190,000 pounds is found by subtracting the value of the `CDF` from 1.

```
(1-CDF[NormalDistribution[187000,2181.9716],190000])//N
0.0845807
```

(b) We use the function `Quantile` to find the value of w.

```
Quantile[NormalDistribution[187000,2181.9716],.99]
192076.
```

■

Half-Normal Distribution

The probability density function for the half-normal distribution with variance parameter θ as defined by the *Mathematica* programmers is

$$f(x) = \frac{2\theta}{\pi}\exp\left[\frac{-\theta^2 x^2}{\pi}\right].$$

The domain is $0 \le x \le +\infty$. The half-normal distribution is a special case of a truncated normal distribution which can be defined on any interval $a \le x \le b$. The general form of the pdf is

$$\frac{1}{\sqrt{2\pi}\sigma}\exp\left(-\frac{1}{2\sigma^2}(x-\mu)^2\right)\left[\frac{1}{\sqrt{2\pi}\sigma}\int_a^b \exp\left(-\frac{1}{2\sigma^2}(x-\mu)^2\right)dt\right]^{-1}.$$

The half-normal distribution is generally defined on an interval $\mu \le x \le \infty$. In *Mathematica* the mean μ is set to zero.

`HalfNormalDistribution[theta]` represents the half-normal distribution with variance parameter `theta`.

EXAMPLE 12: (a) Use *Mathematica* to determine formulas for the pdf, the mean, and the variance of a half-normal distribution with variance parameter θ. (b) Graph the pdf for θ = 0.1, 0.2, 0.3, 0.4.

SOLUTION: (a) Assuming that the **ContinuousDistributions** package has been loaded, we find the pdf for the half-normal distribution and define this function as `f[θ,x]`. We also find that the mean is $\frac{1}{\theta}$ and the variance is $\frac{\pi - 2}{2\theta^2}$.

```
f[θ_, x_] = PDF[HalfNormalDistribution[θ], x]
```

$$\frac{2\,E^{-\frac{x^2\,\theta^2}{\pi}}\,\theta}{\pi}$$

```
Mean[HalfNormalDistribution[θ]]
```

$$\frac{1}{\theta}$$

```
Variance[HalfNormalDistribution[θ]]
```

$$\frac{-2+\pi}{2\,\theta^2}$$

(b) We define the function `pdfplot` that, when given a pdf, graphs the function over the interval $0 \le x \le 20$. In `graphs`, we graph the pdf for θ = 0.1, 0.2, 0.3, 0.4

where the output is suppressed in each case. Finally, we view the graphs simultaneously with Show.

```
pdfplot[θ_] :=
  Plot[f[θ, x], {x, 0, 20}, DisplayFunction -> Identity]
graphs = Table[pdfplot[θ], {θ, 0.1, 0.4, 0.1}];
```

```
Show[Evaluate[graphs],
  DisplayFunction -> $DisplayFunction]
```

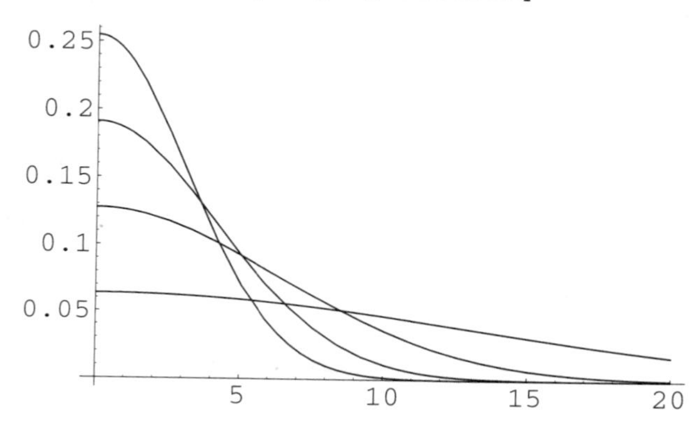

■

Laplace Distribution

The probability density function for the Laplace distribution with mean μ ($-\infty < x < +\infty$) and scale parameter $\beta > 0$ is

$$f(x) = \frac{1}{2\beta}\exp\{-|x-\mu|/\beta\}.$$

The domain is $-\infty < x < +\infty$. This distribution is called the **first law of Laplace** (the normal distribution is called the **second law of Laplace**). Because the Laplace distribution is the distribution of the difference of two independent random variables with identical exponential distributions, it is also known as the **double-exponential** distribution. Note that the distribution is symmetric about its mean, μ.

A **standard form of the Laplace distribution** is obtained by setting $\mu = 0$ and $\beta = 1$. The pdf then becomes

$$\frac{1}{2}e^{-|x|}.$$

This form is sometimes called **Poisson's first law of error**.

For the Laplace distribution:

Distribution function	$\begin{cases} \frac{1}{2}\left[-\exp\left(-\frac{\mu - x}{\beta}\right)\right] & x < \mu \\ 1 - \frac{1}{2}\left[-\exp\left(-\frac{\mu - x}{\beta}\right)\right] & x \geq \mu \end{cases}$
Moment generating function	$\frac{e^{\mu t}}{1 - \beta^2 t^2}$ for $|t| < \frac{1}{\beta}$
Mean	μ
Variance	$2\beta^2$

`LaplaceDistribution[mu, beta]` represents the Laplace (double exponential) distribution with mean `mu` and variance parameter `beta`.

EXAMPLE 13: Graph the pdfs for the Laplace distribution with (a) $\mu = 0$ and $\beta = 0.5, 1.0, 1.5$; (b) $\mu = 1, 3, 5$ and $\beta = 1$.

SOLUTION: Assuming that the **ContinuousDistributions** package has been loaded, we define the pdf for the Laplace distribution to be the function $g[\mu, \beta, x]$. We graph the probability density functions for (a) and (b) is a similar manner. In each case, the functions are graphed simultaneously using the `GrayLevel` option.

```
g[μ_, β_, x_] = PDF [LaplaceDistribution [μ, β], x]
```

(a) Notice that the width of the Laplace pdf increases with increasing values of β.

```
Plot [{g[0, .5, x], g[0, 1, x], g[0, 1.5, x]},
  {x, -10, 10}, PlotStyle ->
   {GrayLevel [0], GrayLevel [0.1], GrayLevel [0.5]}]
```

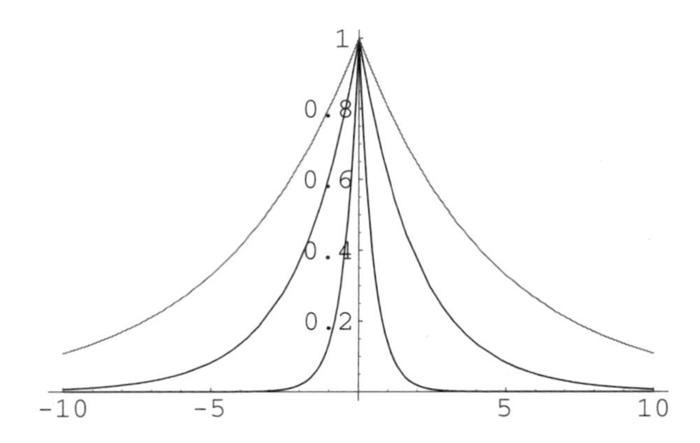

(b) Notice that the location is shifted to the right with increasing values of μ.

```
Plot [{g[1, 1, x], g[3, 1, x], g[5, 1, x]},
  {x, -7, 13}, PlotStyle ->
   {GrayLevel [0], GrayLevel [0.1], GrayLevel [0.5]}]
```

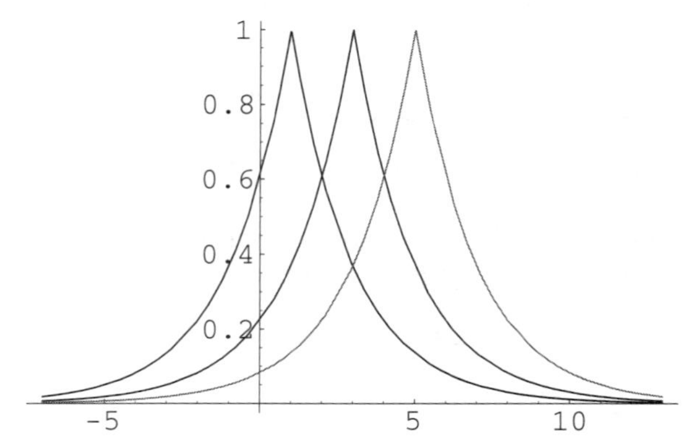

■

Log-Normal Distributions

The probability density function for the log-normal distribution with scale parameter μ and shape parameter $\sigma > 0$ is

$$f(x) = \frac{1}{x\sigma\sqrt{2\pi}}\exp\left\{\frac{-(\ln x - \mu)^2}{2\sigma^2}\right\}.$$

The domain is $0 \le x \le +\infty$. The random variable X has a log-normal distribution if $Y = \ln X$ has a normal distribution. Another way to think of this relationship is that if Y has a normal distribution, then $X = e^y$ has a log-normal distribution. In this sense a more descriptive name would be antilog-normal. The parameter μ is the mean of Y (the normal distribution), and the parameter σ is the standard deviation of Y. Whereas the normal distribution can represent the distribution of the sum of independent random variables (central limit theorem), the log-normal distribution represents the distribution of the product of independent random variables.

Other names have been used for the log-normal distribution, such as the **Galton–McAlister distribution** after some of the pioneers in its development. In economics the log-normal may be applied to production data and in this context may be called the **Cobb–Douglas distribution**. The log-normal distribution is a more realistic representation than the normal distribution of characteristics such as weight, height, and density. These quantities cannot take negative values, but the normal distribution ascribes positive probabilities to such events while the log-normal does not. Furthermore, by taking σ small enough, it is possible to construct a log-normal distribution to closely resemble any normal distribution.

Phenomena such as city population size, stock price fluctuations, and personal incomes have distributions with very long right tails. The log-normal and Pareto distributions have been used together to model such phenomena. The Pareto is used to fit the the tail of the distribution, and the log-normal is used to fit the rest of the distribution.

For the log-normal distribution:

Mean	$e^{\mu + 1/2\sigma^2}$
Variance	$e^{2\mu + \sigma^2}\left(e^{\sigma^2} - 1\right)$

`LogNormalDistribution[mu, sigma]` represents the log-normal distribution with mean parameter `mu` and variance (or standard deviation) parameter `sigma`.

EXAMPLE 14: Graph the pdfs for the log-normal distribution with (a) $\mu = 0$ and $\sigma = 0.3, 0.5, 1.0$; (b) $\mu = 2.0, 2.5, 3.0$ and $\sigma = 1$.

SOLUTION: Assuming that the **ContinuousDistributions** package has been loaded, we use the pdf for the log-normal distribution to define the function *LNrml*[μ, σ, x]. We graph the probability density functions for (a) and (b) is a similar manner. In each case, we graph the functions simultaneously using the `GrayLevel` option.

```
LNrml[μ_, σ_, x_] = PDF[LogNormalDistribution[μ, σ], x]
```

(a) Notice that the general shape of the log-normal distribution is skewed to the right, but it is more symmetric for small values of σ. (The darkest and highest curve corresponds to $\sigma = 0.3$.)

```
Plot[{LNrml[0, .3, x], LNrml[0, .5, x], LNrml[0, 1, x]},
  {x, 0, 5}, PlotStyle ->
  {GrayLevel[0], GrayLevel[0.1], GrayLevel[0.5]}]
```

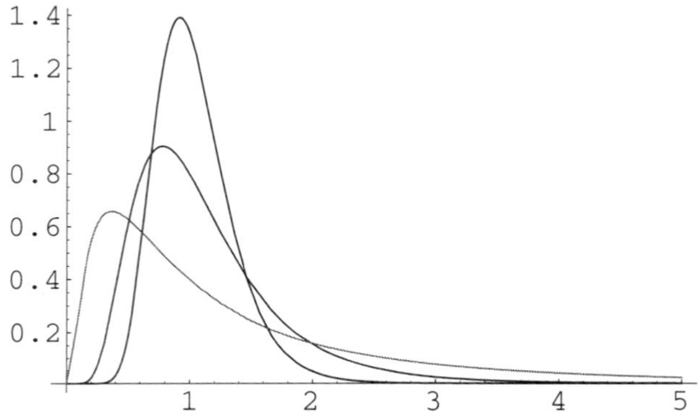

(b) Notice the dramatic changes in the curves corresponding to small changes in μ and σ.

```
Plot[{LNrml[2, 1, x], LNrml[2.5, 1, x], LNrml[3, 1, x]},
  {x, 0, 50}, PlotStyle ->
  {GrayLevel[0], GrayLevel[0.1], GrayLevel[0.5]}]
```

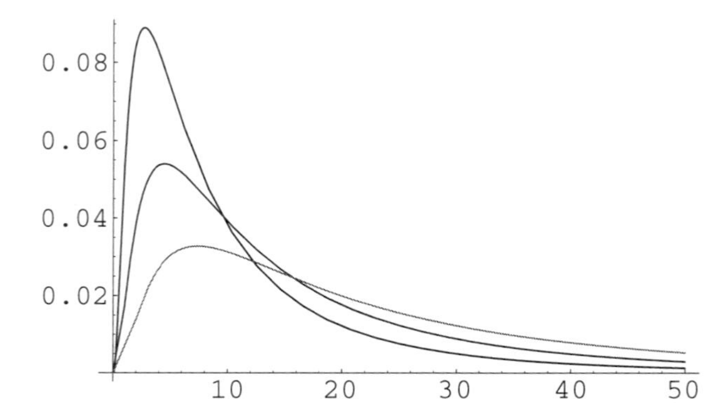

■

Logistic Distribution

The probability density function for the logistic distribution with mean parameter μ ($-\infty \le \mu \le \infty$) and scale parameter β is

$$f(x) = \frac{\exp[-(x-\mu)/\beta]}{\beta\{1+\exp[-(x-\mu)/\beta]\}^2}.$$

The domain is $-\infty \le x \le +\infty$. The logistic cumulative distribution function has been use extensively as a growth function with x representing time. The logistic distribution arises in statistics as the limiting distribution as $n \to \infty$ of the midrange (the average of the largest value and the smallest value) of a random sample of size n. A standard form of the distribution occurs when $\mu = 0$ and $\beta = 1$. The pdf then becomes

$$\frac{e^x}{(1+e^x)^2}.$$

The logistic pdf has a similar shape to that of the normal pdf and may qualify to replace the normal in suitable applications. The difference between the two pdfs is largely due to the higher tails of the logistic distribution.

For the logistic distribution:

Distribution function	$\frac{1}{1+\exp[-(x-\mu)/\beta]}$
Moment generating function	$\pi\beta t\frac{e^{\mu t}}{\sin(\pi\beta t)}$
Mean	μ
Variance	$\frac{\pi^2\beta^2}{3}$

`LogisticDistribution[mu, beta]` represents the logistic distribution with mean `mu` and scale parameter `beta`.

EXAMPLE 15: Graph the pdf of the standard form ($\mu = 0$ and $\beta = 1$) of the logistic distribution together with the pdf of the normal distribution having the same mean and standard deviation. Show the difference in the tails of the two distributions

---

SOLUTION: First, we must determine the mean and standard deviation of the normal distribution that will be plotted. Because the mean of the standard form of the logistic distribution is zero, the mean of the normal distribution will be the same. Assuming that the **ContinuousDistributions** package has been loaded, we calculate the standard deviation of the standard form of the logistic distribution to use as the standard deviation of the normal distribution.

```
sigma=StandardDeviation[LogisticDistribution[0,1]]//N
```

```
1.8138
```

Next we use `Plot` to create the graphs of the two pdfs. Notice that the normal pdf will be displayed using dashed lines. Output is suppressed using `DisplayFunction->Identity`. The pdfs are displayed together using `Show`.

```
p1= Plot[PDF[LogisticDistribution[0,1],x],{x,-5,5},
  DisplayFunction->Identity];
p2= Plot[PDF[NormalDistribution[0,sigma],x],{x,-5,5},
  PlotStyle->{Dashing[{0.02}]},
  DisplayFunction->Identity];
```

```
Show [p1, p2, DisplayFunction -> $DisplayFunction ]
```

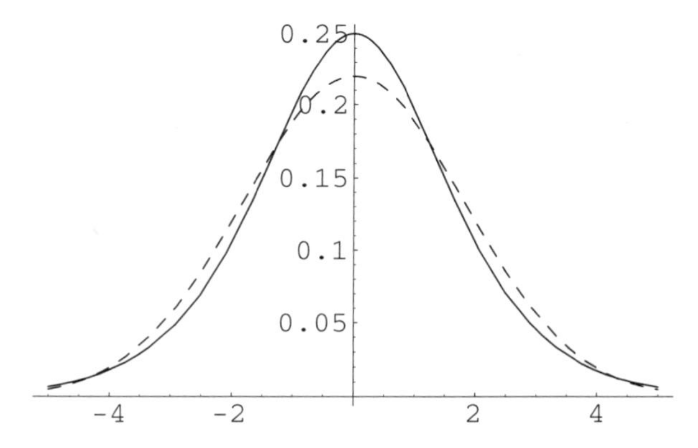

To display the tail behavior of the two pdfs, we change the range for the horizontal axis in the `Plot` commands to focus on the tails on the positive side. Because both pdfs are symmetric about zero, the tails on the negative side are similar.

```
p1= Plot[PDF[LogisticDistribution[0,1],x],{x,3,6},
  PlotStyle->{GrayLevel[0]},DisplayFunction->Identity];
p2= Plot[PDF[NormalDistribution[0,sigma],x],{x,3,6},
  PlotStyle->{Dashing[{0.02}]},
  DisplayFunction->Identity];
```

```
Show[p1, p2, DisplayFunction -> $DisplayFunction]
```

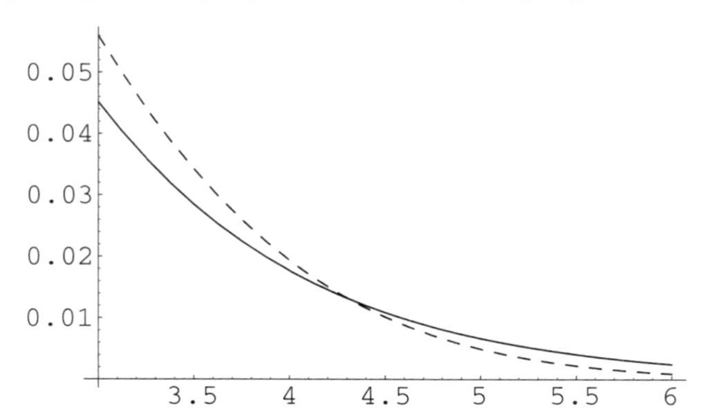

■

Pareto Distribution

The probability density function for the Pareto distribution with location parameter $k > 0$ and shape parameter $\alpha > 0$ is

$$f(x) = \frac{\alpha k^{\alpha}}{x^{\alpha+1}}.$$

The domain is $k \leq x < \infty$. The Pareto distribution was originally developed to model the distribution of income in a population. In the form $1 - CDF(x) = \left(\frac{k}{x}\right)^{\alpha}$, it gave the probability that income is greater than or equal to x, where k represents some minimum income. The shape parameter α was called **Pareto's constant**. The form of the distribution in *Mathematica* is the "Pareto distribution of the first kind." Two other forms were proposed. The Pareto distribution has been used in quality control to model loss. It is the basis for **Pareto charts** in quality control.

Phenomena such as city population size, stock price fluctuations, and personal incomes have distributions with very long right tails. The Pareto and log-normal distributions have been used together to model such phenomena. The Pareto is used to fit the tail of the distribution, and the log-normal is used to fit the rest of the distribution.

There is a class of distribution that follow the form of the Pareto distribution with $0 < \alpha < 2$. Such distributions are referred to as **stable Pareto distributions**.

For the Pareto distribution:

Distribution function	$1 - \left(\frac{k}{x}\right)^{\alpha}$
Mean	$k/(\alpha - 1)$ for $\alpha > 1$
Variance	$\alpha k^2/[(\alpha - 1^2)(\alpha - 2)]$ for $\alpha > 2$

`ParetoDistribution[k,alpha]` represents the Pareto distribution with location parameter `k` and shape parameter `alpha`.

EXAMPLE 16: Graph the pdfs for the Pareto distribution with (a) $k = 1$ and $\alpha = 0.5$, 1.0 and (b) $k = 10$ and $\alpha = 5, 10$.

SOLUTION: Assuming that the **ContinuousDistributions** package has been loaded, we define the pdf for the Pareto distribution to be the function $Ppdf[k, \alpha, x]$.

```
Ppdf [k_, α_, x_] = PDF [ParetoDistribution[k, α], x]
```

$k^{\alpha} x^{-1-\alpha} \alpha$

(a) We graph the probability density functions simultaneously using `Plot`.

```
Plot [{Ppdf [1, .5, x], Ppdf [1, 1, x]}, {x, 1, 5}]
```

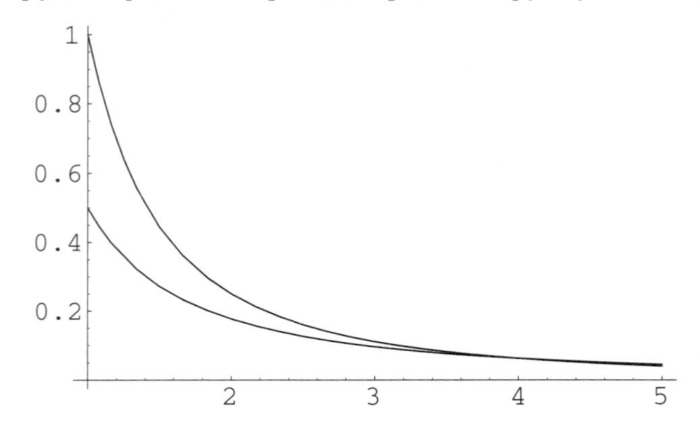

(b) We obtain a similar display to that in (a), again using `Plot`.

```
Plot[{Ppdf[10, 5, x], Ppdf[10, 10, x]}, {x, 10, 15}]
```

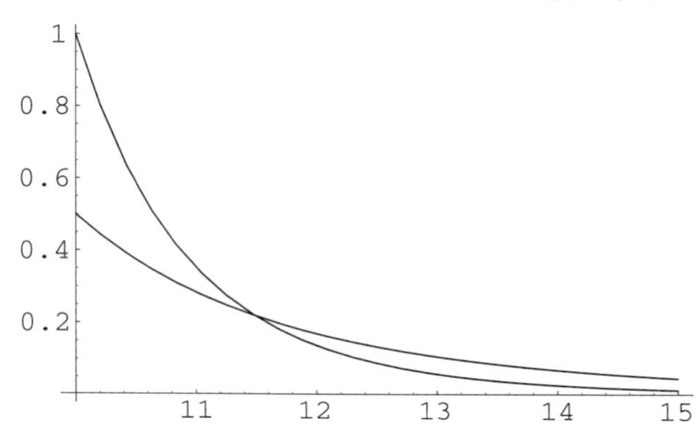

■

Rayleigh Distribution

The probability density function for the Rayleigh distribution with scale parameter $\sigma > 0$ is

$$f(x) = \frac{x}{\sigma^2}\exp\left[-\frac{x^2}{2\sigma^2}\right].$$

The domain is $0 \le x < \infty$.

For the Rayleigh distribution:

Distribution function	$1 - \exp[-x^2/2\sigma^2]$
Mean	$\sigma\left(\frac{\pi}{2}\right)^{\frac{1}{2}}$
Variance	$\left(2 - \frac{\pi}{2}\right)\sigma^2$

`RayleighDistribution[sigma]` represents the Rayleigh distribution with scale parameter `sigma`.

EXAMPLE 17: It has been suggested that the Rayleigh distribution with scale parameter σ corresponds to a Weibull distribution with shape parameter 2 and scale parameter σ. (a) Compare the pdf for the Rayleigh distribution with that of the Weibull distribution with shape parameter 2. (b) Graph the pdf for the Rayleigh distributions with scale parameters 1.0 and 2.0, respectively, together with the corresponding Weibull distributions with shape parameter 2.

SOLUTION: Assuming that the ContinuousDistributions package has been loaded, we define the functions `Rpdf[σ,x]` and `Wpdf[α,σ,x]` to represent the pdfs for the Rayleigh and Weibull distributions, respectively.

```
Rpdf [σ_, x_] = PDF [RayleighDistribution [σ], x]
```

$$\frac{E^{-\frac{x^2}{2\sigma^2}}\,x}{\sigma^2}$$

```
Wpdf [α_, σ_, x_] = PDF [WeibullDistribution [α, σ], x]
```

$$E^{-\left(\frac{x}{\sigma}\right)^{\alpha}}\,x^{-1+\alpha}\,\alpha\,\sigma^{-\alpha}$$

(a) We next display the pdf for the Weibull distribution with shape parameter 2 and scale parameter σ (`Wpdf[2,σ,x]`) so it can be compared to the pdf of the Rayleigh distribution with scale parameter σ.

```
Wpdf [2, σ, x]
```

$$\frac{2\,E^{-\frac{x^2}{\sigma^2}}\,x}{\sigma^2}$$

(b) Using `Plot` we define the plots of the two pdfs. Notice that the Weibull pdf is displayed using dashed lines. The function `Show` is used to display the two plots as a graphics array.

```
p1=Plot[{Rpdf[1,x],Wpdf[2,1,x]},{x,0,6},PlotRange->
  {{0,6},{0,1}},PlotStyle->{GrayLevel[0],
   Dashing[{0.02}]},DisplayFunction->Identity];
p2=Plot[{Rpdf[2,x],Wpdf[2,2,x]},{x,0,6},PlotRange->
  {{0,6},{0,1}},PlotStyle->{GrayLevel[0],
   Dashing[{0.02}]},DisplayFunction->Identity];
```

```
Show[GraphicsArray[{p1, p2}],
 DisplayFunction -> $DisplayFunction]
```

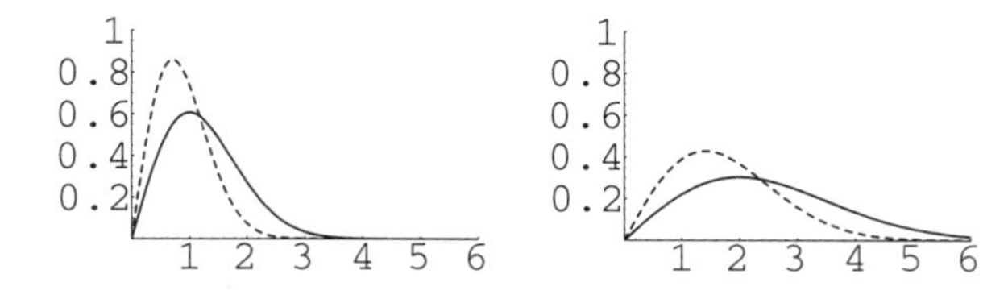

■

Student's t-Distribution

The probability density function for the Student's t-distribution with shape parameter v, a positive integer, is

$$f(x) = \frac{1}{\sqrt{v}\beta(v/2, 1/2)}\left(\frac{v}{v+x^2}\right)^{\frac{v+1}{2}},$$

where B is the beta function defined in subsection 6.3.1 and v is called degrees of freedom. The domain is $-\infty < x < \infty$. This pdf is bell-shaped and symmetric about zero. As $v \to \infty$ the t-distributions approach standard normal (normal with mean zero and standard deviation 1).

Let $Y_1, Y_2, \ldots, Y_n$ be independent random variables each having the same normal distribution with mean μ and standard deviation σ. Let

$$\bar{Y} = \frac{1}{n}\sum_{i=1}^{n} Y_i \text{ and let } S = \sqrt{\frac{1}{n-1}\sum_{i=1}^{n}(Y_i - \bar{Y})^2}.$$

In statistical inference the Y_i values represent a random sample from a normal population; $\bar{Y}$ represents the sample mean, and S represents the sample standard deviation. The ratio $T = \frac{(\bar{Y}-\mu)}{(S/\sqrt{n})}$ is said to have Student's t-distribution with $v = n - 1$ degrees of freedom. It can be reexpressed as a ratio of the form

$$\frac{Z}{\sqrt{\chi^2/v}}.$$

where Z and χ^2 are independent, Z is standard normal (normal with mean zero and standard deviation 1), and χ^2 is chi-square with degrees of freedom v, which leads to the density function listed at the beginning of this subsection.

The major applications of the t-distribution are in the construction of hypothesis tests and confidence intervals for means of one or two normal populations. In practice, even when the populations are not normal, good results are obtained as long as one can assume that the distributions of the sample means are approximately normal.

For the t-distribution:

Mean	0 for $\nu > 1$
Variance	$\frac{\nu}{\nu - 2}$ for $\nu > 2$

StudentTDistribution[n] represents Student's t-distribution with n degrees of freedom.

EXAMPLE 18: Graph the pdfs of the t-distributions with degrees of freedom 1, 5, and 10 together with the pdf of the standard normal distribution.

SOLUTION: Assuming that the **ContinuousDistributions** (or the **NormalDistribution**) package has been loaded, we define the functions studentpdf[υ,x] and normalpdf[μ,σ,x] to represent the pdfs for the t and normal distributions, respectively. We use Plot and Show to display the three t-distribution pdfs and the standard normal. The t-distribution pdfs are displayed using solid lines with progressive change from black to gray as degrees of freedom increase, and the standard normal is displayed as a dotted line.

```
studentpdf[ν_, x_] = PDF[StudentTDistribution[ν], x];

normalpdf[μ_, σ_, x_] = PDF[NormalDistribution[μ, σ], x];

p1 = Plot[studentpdf[1, x], {x, -3, 3}, PlotStyle ->
    {GrayLevel[0]}, DisplayFunction -> Identity];
p2 = Plot[studentpdf[5, x], {x, -3, 3}, PlotStyle ->
    {GrayLevel[0.1]}, DisplayFunction -> Identity];
p3 = Plot[studentpdf[10, x],
   {x, -3, 3}, PlotStyle -> {GrayLevel[0.5]},
   DisplayFunction -> Identity];
p4 = Plot[normalpdf[0, 1, x],
   {x, -3, 3}, PlotStyle -> {Dashing[{0.02}]},
   DisplayFunction -> Identity];
```

```
Show[p1, p2, p3, p4,
 DisplayFunction -> $DisplayFunction]
```

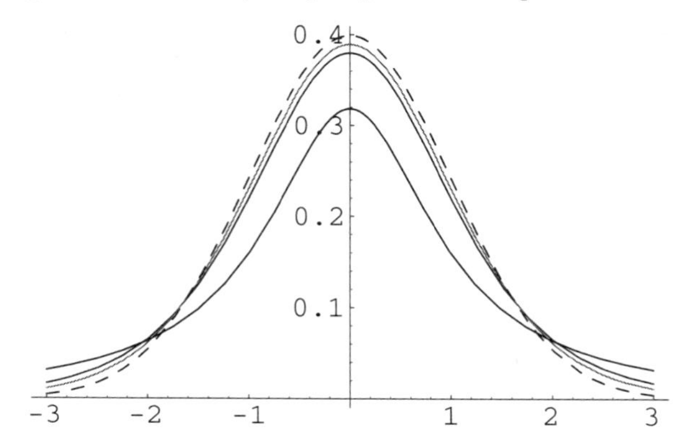

■

Noncentral Student's t-Distribution

The probability density function for the noncentral Student's *t*-distribution with shape parameter (degrees of freedom) ν, a positive integer, and noncentrality parameter λ, $-\infty < \lambda < \infty$, is represented in different ways, all of which are long and involved. The noncentral *t*-distribution arises as the distribution of the ratio

$$\frac{X}{\sqrt{\chi^2/\nu}},$$

where X and χ^2 are independent random variables distributed as normal with mean λ and standard deviation 1 and chi-square with degrees of freedom v, respectively. If $\lambda = 0$, the distribution is that of the (central) t with ν degrees of freedom. The noncentral *t*-distribution is used to calculate the power of hypothesis tests of means from one or two normal populations.

For the noncentral *t*-distribution:

Mean	$\dfrac{\lambda\sqrt{(\nu/2)}\,\Gamma((\nu-1)/2)}{\Gamma(\nu/2)}$ for $\nu > 1$
Variance	$\dfrac{\nu}{\nu-2}(1+\lambda^2) - \dfrac{\nu}{2}\lambda^2\left[\dfrac{\Gamma((\nu-1)/2)}{\Gamma(\nu/2)}\right]^2$ for $\nu > 2$

`NoncentralStudentTDistribution[n, lambda]` represents the noncentral Student's *t*-distribution with `n` degrees of freedom and noncentrality parameter `lambda`.

EXAMPLE 19: Graph the pdfs for the noncentral Student's *t*-distribution with degrees of freedom 5 and noncentrality parameter values of 0.5, 2.0, and 5.0.

SOLUTION: Assuming that the **ContinuousDistributions** package has been loaded, we graph the probability density functions simultaneously using `Plot`.

```
Plot [{PDF [NoncentralStudentTDistribution [5, .5], x],
  PDF [NoncentralStudentTDistribution [5, 2], x],
  PDF [NoncentralStudentTDistribution [5, 5], x]},
  {x, -3, 12}]
```

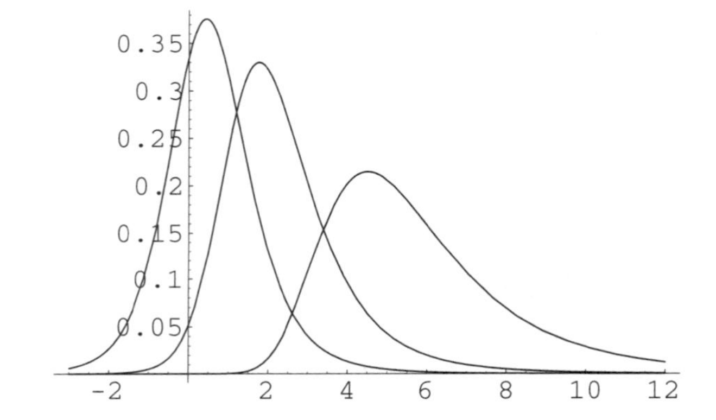

■

Uniform Distribution

The probability density function for the uniform distribution with location parameter *a*, the lower limit, and parameter *b*, the upper limit is

$$f(x) = \begin{cases} \dfrac{1}{b-a}, & a \le x \le b \\ 0, & \text{otherwise} \end{cases}.$$

Thus, the domain is $\alpha \le x \le b$. The uniform distribution is also known as the **rectangular distribution**. The name is based on the appearance of the graph of the pdf. A standard form called the **unit uniform** is obtained when $a = 0$ and $b = 1$. This is a special case of the standard form of the beta distribution (called simply the beta distribution in *Mathematica*) obtained by setting exponents α and β equal to 1.

Most computer random number generators give numbers that behave as though they were independent observations from the unit uniform distribution. Further operations are then needed to generate other kinds of random numbers. Note that by using its cdf any random variable can be expressed as a function of the unit uniform distribution. The uniform distribution is used widely in constructing mathematical models of physical, biological, and social phenomena.

For the uniform distribution:

Distribution function	$\frac{x-a}{b-a}$
Moment generating function	$\frac{e^{bt}-e^{at}}{t(b-a)}$
Mean	$\frac{a+b}{2}$
Variance	$\frac{(b-a)^2}{12}$

`UniformDistribution[min, max]` represents the uniform distribution on the interval `{min, max}`.

EXAMPLE 20: Starting at 5:00 A.M., every half-hour there is a flight from airport N to airport W. Suppose that for every flight there is always space for an additional passenger. A person who wants to fly to W arrives at N at a random time between 8:45 A.M. and 9:45 A.M. Find the probability that she waits (a) at most 10 minutes; (b) at least 15 minutes.

SOLUTION: Assume the passenger arrives at the airport X minutes after 8:45. Then X is a uniform random variable over the interval $[0, 60]$. In order to calculate the probability that the time between arrival at the airport and flight departure is between c and d minutes, we assume that the **ContinuousDistributions** package has been loaded and define the function `prob[c,d]` using *Mathematica's* `CDF` function.

```
prob[c_,d_]=CDF[UniformDistribution[0,60],d]
  -CDF[UniformDistribution[0,60],c];
```

(a) In order for the passenger to wait at most 10 minutes, she must arrive within 10 minutes of a departing flight. Thus, she must arrive between 8:50 and 9:00 or between 9:20 and 9:30. These translate to $5 \le X \le 15$ and $35 \le X \le 45$. Adding the probability of these two events results in the answer of $\frac{1}{3}$.

```
prob[5, 15] + prob[35, 45]
```

$\frac{1}{3}$

(b) The passenger will wait at least 15 minutes if she arrives between 9:00 and 9:15 or between 9:30 and 9:45. That is , if $15 \leq X \leq 30$ or $45 \leq X \leq 60$. Adding the probability of these two events results in the answer of $\frac{1}{2}$.

```
prob[15, 30] + prob[45, 60]
```

$\frac{1}{2}$

■

Weibull Distribution

The probability density function for the Weibull distribution with shape parameter $\alpha > 0$ and scale parameter $\beta > 0$ is

$$f(x) = (\alpha x^{\alpha-1}/\beta^{\alpha})\exp[-(x/\beta)^{\alpha}].$$

The domain is $0 \leq x < +\infty$. In its full generality, the Weibull distribution has a third parameter which corresponds to the smallest value of the random variable, call it v. Then the domain is $0 \leq v \leq x < \infty$ and in the formulas for the pdf and cdf x is replaced by $x - v$. In Mathematica Graphic! is always zero. The Weibull distribution is a generalization of the exponential distribution in the sense that if X has a Weibull distribution with parameters α, β, and v, then

$$Y = \left(\frac{X - v}{\beta}\right)^{\alpha}$$

has the exponential distribution with pdf e^{-y}. The standard form of the Weibull distribution is obtained by setting $\beta = 1$ (and $v = 0$).

The Weibull distribution is widely used in engineering applications. It was originally proposed for representing the distribution of the breaking strength of materials. Its use now extends to the distribution of the lifetime of objects consisting of many parts where failure occurs when any of the parts fails. Note that the value zero for Graphic! is by far the most frequently used, especially in representing the distribution of lifetimes.

For the Weibull distribution:

Distribution function $\quad 1 - \exp[-(x/\beta)^{\alpha}]$

Mean	$\beta\Gamma\left(\frac{\alpha+1}{\alpha}\right)$
Variance	$\beta^2\left(\Gamma\left(\frac{\alpha+2}{\alpha}\right)-\left\{\Gamma\left(\frac{\alpha+2}{\alpha}\right)\right\}^2\right)$

`WeibullDistribution[alpha, beta]` represents the Weibull distribution with shape parameter `alpha` and scale parameter `beta`.

EXAMPLE 21: Graph the pdf for the standard form (that is, $\beta = 1$) of the Weibull distribution with shape parameter $\alpha = 1, 2, 3, 4$.

SOLUTION: Assuming that the **ContinuousDistributions** package has been loaded, we define the function `plotWeibullpdf` to graph the pdf for given values of α and β. We then generate the graphs in `wgraphs` and display them simultaneously. Notice that the graphs become more symmetric as the value of α increases.

```
plotWeibullpdf[α_, β_] :=
 Plot[PDF[WeibullDistribution[α, β], x],
  {x, 0, 4}, DisplayFunction -> Identity];
wgraphs = Table[plotWeibullpdf[a, 1], {a, 1, 4}];
```

```
Show[Evaluate[wgraphs],
 DisplayFunction -> $DisplayFunction, PlotRange -> All]
```

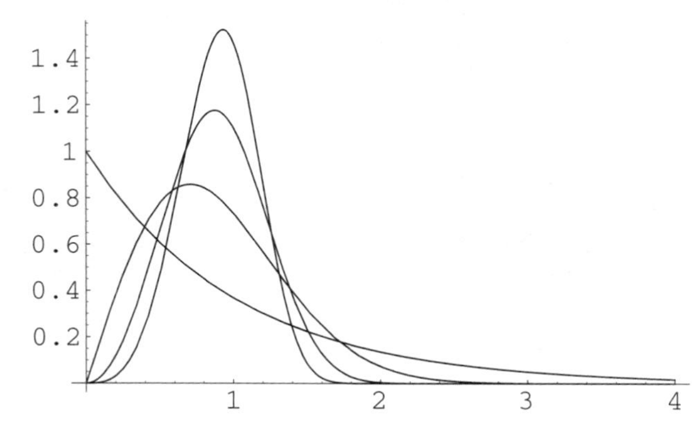

■

6.4 The Multivariate Normal Distribution and Related Distributions

Mathematica's **Statistics** package **MultinormalDistributions** contains commands for working with the following distributions: the multivariate normal distribution, the multivariate Student t-distribution, the Wishart distribution, the Hotelling T^2 distribution, and the distribution of the quadratic form of a multivariate normal. The **Help Browser** can be used to obtain information about these distributions as well as the functions which operate on them. This help feature may also include examples to demonstrate the syntax of the functions.

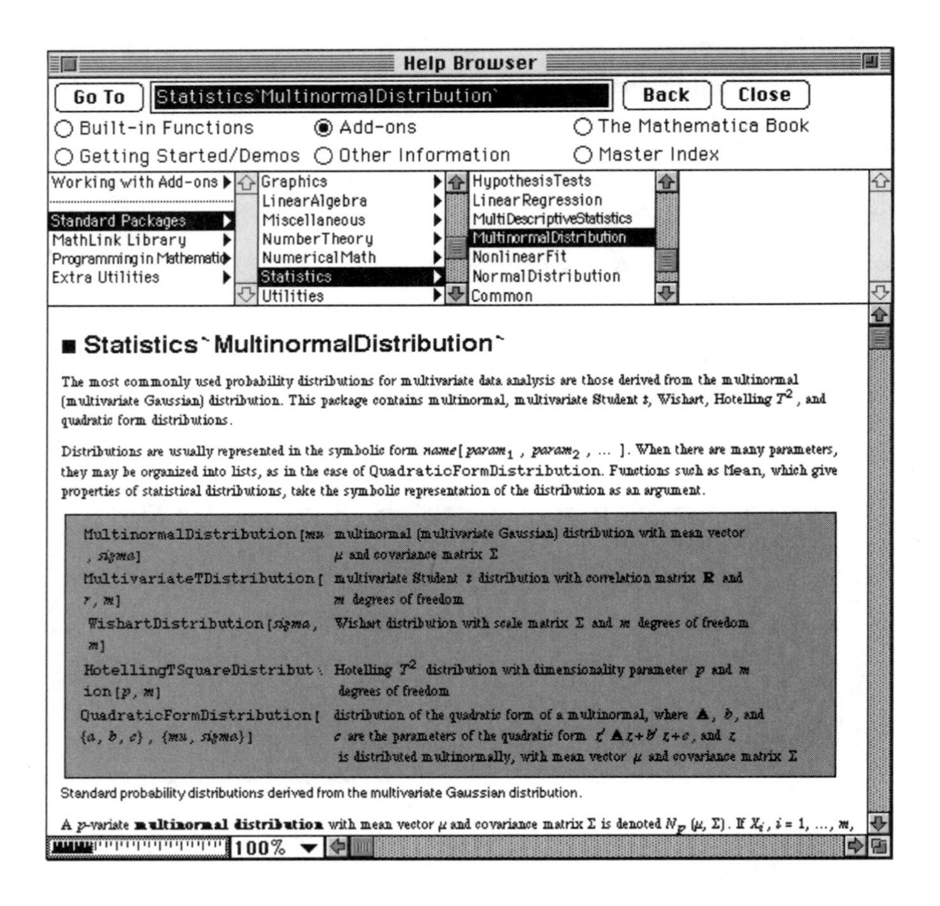

The functions for univariate distributions which are also available for use in **MultinormalDistributions** are:

`Domain[dist]` MultinormalDistributions:Domain

`PDF[dist,x]` MultinormalDistributions:PDF

`CDF[dist,x]` MultinormalDistributions:CDF

`CharacteristicFunction[dist,t]` MultinormalDistributions:CharacteristicFunction

`Mean[dist]` MultinormalDistributions:Mean

`Variance[dist]` MultinormalDistributions:Variance

`StandardDeviation[dist]` MultinormalDistributions:StandardDeviation

`Skewness[dist]` MultinormalDistributions:Skewness

`Kurtosis[dist]` MultinormalDistributions:Kurtosis

`KurtosisExcess[dist]` MultinormalDistributions:KurtosisExcess

`ExpectedValue[f,dist,x]` MultinormalDistributions:ExpectedValue.

In addition, the function `Quantile[dist,q]` MultinormalDistributions:Quantile is available only for use with the univariate distributions, `HotellingTSquareDistribution` and `QuadraticFormDistribution`.

The functions of vector-valued multivariate distributions include `EllipsoidQuantile[dist,q]` MultinormalDistributions:EllipsoidQuantile and its inverse `RegionProbability[dist,domain]` MultinormalDistributions:RegionProbability, to be used with the distributions `MultinormalDistribution` and `MultivariateTDistribution`. The other functions are:

`CovarianceMatrix[dist]` MultinormalDistributions:CovarianceMatrix

`CorrelationMatrix[dist]` MultinormalDistributions:CorrelationMatrix

`MultivariateSkewness[dist]` MultinormalDistributions:MultivariateSkewness

`MultivariateKurtosis[dist]` MultinormalDistributions:MultivariateKursosis

`MultivariateKurtosisExcess[dist]` MultinormalDistributions:MultivariateKursosisExcess

Information about any of these functions is easily available using the command `?`. For example, to find out about `CovarianceMatrix`, type

```
?CovarianceMatrix
```

```
CovarianceMatrix[{{x11, ..., x1p}, ..., {xn1, ...,
 xnp}}] gives the p x p covariance matrix of the n p-
 dimensional vectors.  Division by n-1 (rather than n)
```

```
is used, giving an unbiased estimate of the population
covariance (use CovarianceMatrixMLE for a maximum
likelihood estimate). CovarianceMatrix[{{x11, ...,
x1p}, ..., {xn1, ..., xnp}}, {{y11, ..., y1q}, ...,
{yn1, ..., ynq}}] gives the p x q covariance matrix
between the n p-dimensional vectors and the n q-
dimensional vectors. CovarianceMatrix[distribution]
gives the covariance matrix of the specified
multivariate statistical distribution.
```

Multivariate Normal (Multinormal) Distribution

If x has a multivariate normal (also often called **multinormal**) distribution with mean vector μ and covariance (also often called variance–covariance) matrix Σ, the probability density function is given by

$$f(\boldsymbol{x}) = \frac{1}{(\sqrt{2\pi})^p |\Sigma|^{1/2}} e^{-(\boldsymbol{x}-\mu)\Sigma^{-1}(\boldsymbol{x}-\mu)/2},$$

where p is the number of variables. (Note that $(\boldsymbol{x} - \mu)'$ denotes the transpose of the vector $(x - \mu)$.) Assuming that Σ is positive definite, the term $(\boldsymbol{x} - \mu)'\Sigma^{-1}(x - \mu)$ measures the squared generalized distance (also called the Mahalanobis distance) from **x** to μ. When **x** has density defined earlier as $f(\boldsymbol{x})$, then the multinormal distribution is denoted by $N_p(\mu,\Sigma)$.

Let $X_1, X_2, \dots X_p$ be random variables. That they have a multinormal distribution means, in a simple sense, that the individual variables are each normal and that their combinations are also normal. A combination is G $\sum_{i=1}^{p} a_i X_i$, where $a_1, a_2, \dots a_p$ are real numbers and not all zero. Thus multivariate normality implies that each variable is (univariate) normal. However, the converse is not necessarily true. That is, two or more univariate normal variables are not necessarily multinormal.

Nearly all of the commonly used multivariate inferential procedures are based on the multinormal distribution. Although real data may not often be exactly multinormal, the multinormal distribution will frequently serve as a useful approximation to the true distribution.

A **standard multinormal distribution** has $\mu_i = 0$ and $\sigma_i = 0$ for all i = 1, 2, ... p. The covariance matrix in this case is the correlation matrix, where the entry in the ith row and j^{th} column is 1 when $i = j$ and the simple correlation coefficient for X_i and X_j when $i \neq j$.

`MultinormalDistribution[mu,sigma]` MultinormalDistributions:MultinormalDistribution represents the multivariate normal (Gaussian) distribution with mean vector mu and covariance matrix sigma. For a p-variate random vector to be distributed MultinormalDistribution[mu, sigma], mu must be a p-variate vector and sigma must be a $p \times p$ symmetric positive definite matrix.

EXAMPLE 1: Assuming the covariance matrix for a multinormal distribution is diagonal, that is, the variables are not correlated, if the variances are all equal, the distribution is sometimes called the **circular normal**. If the variances are not all equal, the distribution is sometimes called **ellipsoidal normal**. (a) Plot the bivariate standard multinormal pdf where the variables are uncorrelated. (b) Plot the bivariate multinormal pdf where the variables are uncorrelated, with zero means, but where the variances are 1 and 2, respectively.

SOLUTION: (a) After loading the **MultinormalDistribution** package, we create the mean vector of zeros and call it mn and the covariance matrix, which is the identity matrix in this case, and call it sig. Next we define the function f[x,y] to represent the multinormal pdf with mean vector mn and covariance matrix sig. Finally, using Plot3D we display the pdf. The PlotPoints option is used to produce a smoother graph. Notice the circular pattern.

```
mn={0,0};
sig={{1,0},{0,1}};
f[x_,y_]:=PDF[MultinormalDistribution[mn, sig],{x,y}]

Plot3D[f[x, y], {x, -3, 3}, {y, -3, 3}, PlotPoints -> 25]
```

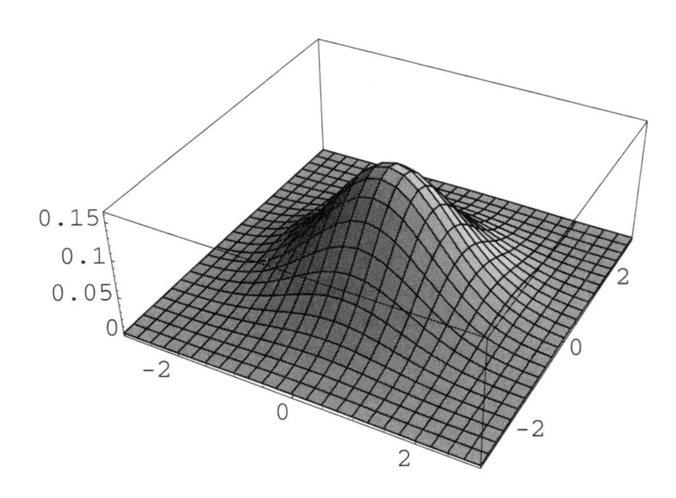

(b) After redefining the covariance matrix so that the variables have unequal variance, we display the pdf using Plot3D. Notice the elliptical pattern.

```
Clear [sig]
sig = {{1, 0}, {0, 2}};
Plot3D [f[x, y], {x, -3, 3}, {y, -6, 6}, PlotPoints -> 25,
  PlotRange -> All]
```

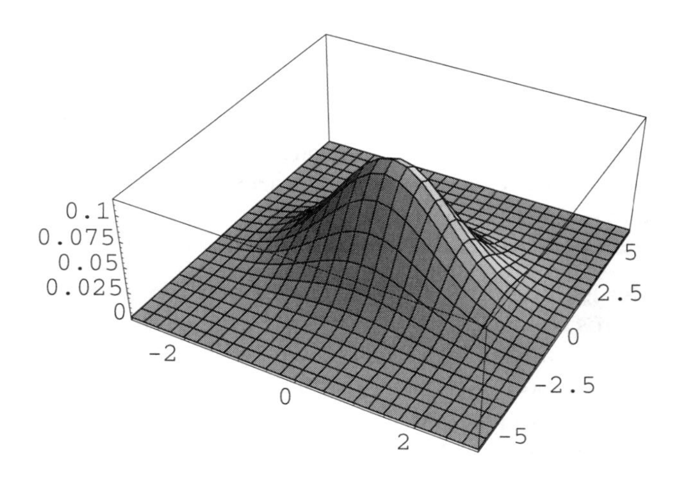

■

EXAMPLE 2: In Table 6.1, the test average (of three tests) and the final examination scores for 20 students in a calculus class are given. Determine the covariance matrix Σ and mean vector μ for this data. Also, graph the pdf and contour plot of the multinormal distribution having this mean vector and covariance matrix.

Test Average	Final Exam Score
55.667	47
84.667	94
91	85
75.667	89
66.667	72
79	82
74.667	82
65	77

Test Average	Final Exam Score
80.333	80
91.667	95
71	72
49	61
41	52
88	93
79	84
79.667	91
61.333	73
56	35
61	46
72.667	88

Table 6.1

SOLUTION: Assuming that the **MultinormalDistributions** package has been loaded, we enter the test averages in `testavg` and the final exam scores in `fscore`.

```
testavg={55.667,84.667,91,75.667,66.667,79,74.667,65,
  80.333,91.667,71,49,41,88,79,79.667,61.333,56,61,
  72.667};
```

```
fscore={47,94,85,89,72,82,82,77,80,95,72,61,52,93,84,
  91,73,35,46,88};
```

We find the covariance matrix and name the result `sig`; similarly, we find the mean vector and name this result `mn`. (Notice that the eigenvalues of the symmetric matrix `sig` are positive, so `sig` is positive definite, as required.)

```
Clear[sig,mn]
  sig=CovarianceMatrix[Transpose[{testavg,fscore}]]//N
```

```
{{195.948,207.526},{207.526,315.042}}
```

```
mn=Mean[Transpose[{testavg,fscore}]]//N
```

```
{71.1501,74.9}
```

```
Eigenvalues[sig]
```

```
{471.395,39.5955}
```

We define the pdf of the multinormal distribution with covariance matrix `sig` and mean vector `mn` in `f` and graph this function with `Plot3D` over the rectangular region [40, 100] × [30, 100].

```
Clear[f]
  f[x_,y_]:=PDF[MultinormalDistribution[mn, sig],{x,y}]
```

```
Plot3D [f[x, y], {x, 40, 100}, {y, 30, 100},
 PlotPoints -> 25]
```

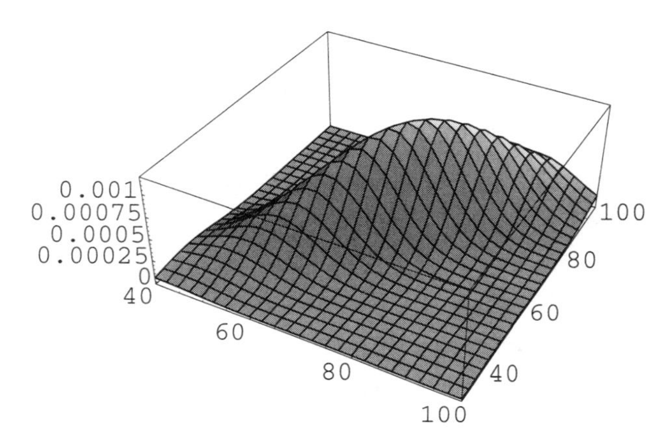

We can use the **3D ViewPoint Selector** (under **Input** on the *Mathematica* menu) to view the graph of the pdf from another perspective.

```
Plot3D [f[x, y], {x, 40, 100}, {y, 30, 100},
 PlotPoints -> 25, ViewPoint -> {2.376 , -0.028 , 2.409 }]
```

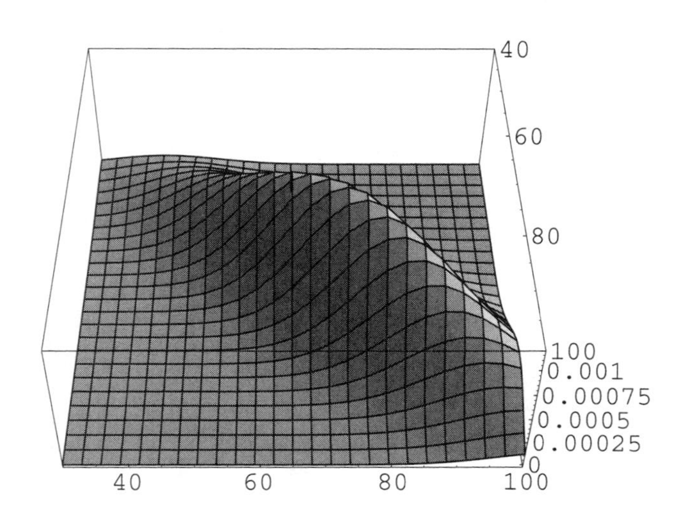

We use `ContourPlot` to graph the contours (or level curves) of the pdf. (The contours are obtained by setting the density function equal to a constant. In this case, when the bivariate normal density surface graphed with `Plot3D` is sliced at a constant height, the traces are ellipses.) `PlotPoints` produces a smoother graph while `ContourShading->False` indicates that no shading will be included in the graph.

```
ContourPlot[f[x, y], {x, 40, 100}, {y, 30, 100},
  PlotPoints -> 30, ContourShading -> False]
```

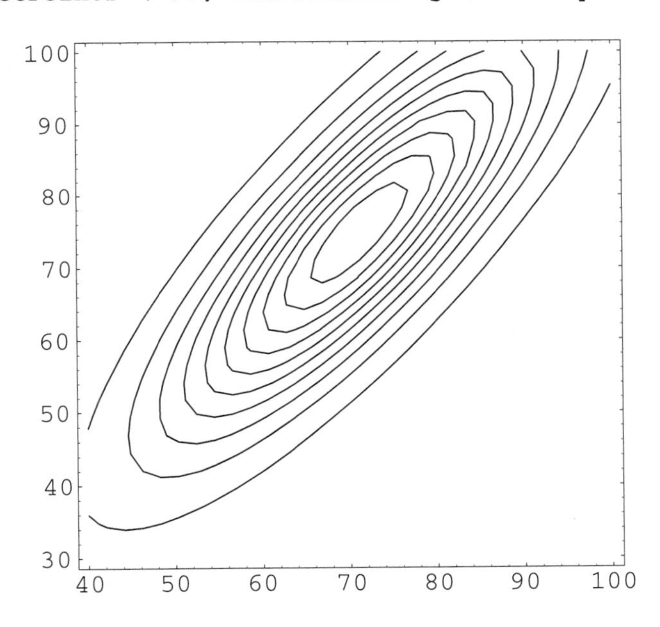

■

Multivariate t-Distribution

The probability density function of the general multivariate t-distribution is

$$p(x) = \frac{\Gamma((v+p)/2)}{(\pi v)^{p/2}\Gamma(v/2)|\mathbf{R}|^{1/2}}(1 + v^{-1}\mathbf{x}'\mathbf{R}^{-1}\mathbf{x})^{-(v+p)/2}$$

This form of the multivariate t-distribution arises as the joint distribution of random variables $X_1, X_2, \ldots X_p$, each having the form

$$X_i = \frac{Y_i}{\sqrt{S^2/v}},$$

where the random variables $Y_1, Y_2, \ldots Y_p$ have a standard multinormal distribution with covariance matrix **R** (which is the correlation matrix), and S^2 is independent of $Y_1, Y_2, \ldots Y_p$ and distributed as chi-square with v degrees of freedom.

The multivariate t-distribution is the distribution of the sample mean vector for random samples from a mixture of multinormal populations in which the mean vector has a multinormal distribution and the covariance matrix has an independent Wishart distribution. Among other applications, the multivariate t is used to construct simultaneous confidence intervals for means of a number of normal populations.

`MultivariateTDistribution[r, m]` MultinormalDistributions:MultivariateTDistribution represents the multivariate T distribution with correlation matrix `r` and degrees of freedom parameter `m`. For the random vector `{x1 Sqrt[m]/s, ..., xp Sqrt[m]/s}` to be distributed `MultivariateTDistribution[r, m]`, the random vector `{x1, ..., xp}` and the random value `s` must be independently distributed, with `{x1, ..., xp}` distributed `MultinormalDistribution[Table[0, {p}], r]` and `s` distributed `ChiSquareDistribution[m]`, respectively.

EXAMPLE 3: Plot the bivariate multivariate t pdf where the variables are independent and the degrees of freedom are 3.

SOLUTION: (a) Assuming that the **MultinormalDistributions** package has been loaded, we create the correlation matrix `corr`, which is the identity matrix because the variables are independent. Then we define the function `f[x,y]` to represent the multivariate t pdf with covariance matrix `corr` and degrees of freedom `df`. Finally, using `Plot3D` we display the pdf. The `PlotPoints` option is used to produce a smoother graph.

```
corr={{1,0},{0,1}};
df=3;
f[x_,y_]:=PDF[MultivariateTDistribution[corr,df],{x,y}]
```

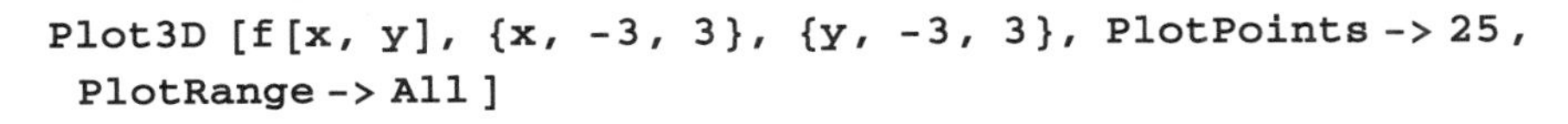

```
Plot3D [f[x, y], {x, -3, 3}, {y, -3, 3}, PlotPoints -> 25,
 PlotRange -> All ]
```

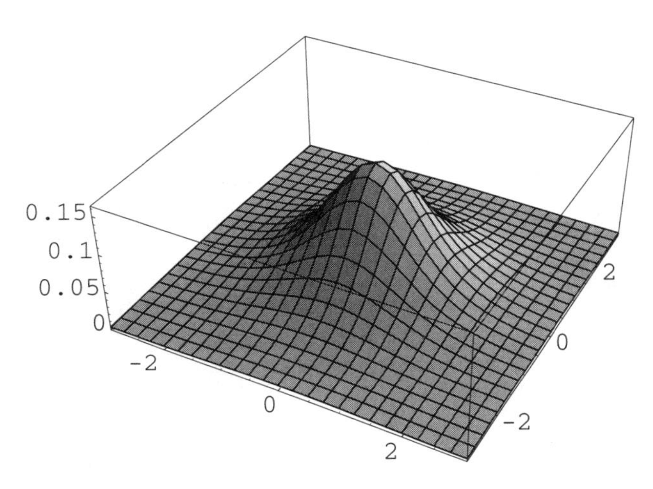

■

Wishart Distribution

The pdf for the Wishart distribution is complicated in form. (The density does not exist unless the sample size n is greater than the number of variables p.) The Wishart distribution is the multivariate analogue of the chi-square distribution. The univariate chi-square distribution is defined as the sum of the squares of independent standard normal random variables. In a similar manner, the Wishart distribution is the distribution of a sum of the form!

$$\sum_{i=1}^{p} (x_i - \mu)(x_i - \mu)$$

where $x_1, x_2, \ldots x_n$ are independently distributed as $N_p(\mu, \Sigma)$. The Wishart distribution is denoted by $W_p(\Sigma, n)$, in the *Mathematica* documentation, where Σ is called the scale matrix and n represents the degrees of freedom. In statistical inference, the Wishart distribution is the distribution of the sample covariance matrix from a random sample from a multinormal population with mean μ and covariance matrix Σ.

`WishartDistribution[sigma,m]` MultinormalDistributions:WishartDistribution represents the Wishart distribution with scale matrix `sigma` and degrees of freedom parameter `m`. For a $p \times p$ symmetric positive definite random matrix to be distributed

`WishartDistribution[sigma,m]`, `sigma` must be a $p \times p$ symmetric positive definite matrix, and `m` must be an integer satisfying $m \geq p + 1$.

Hotelling's T^2*-Distribution*

The Hotelling T^2-distribution is a univariate distribution. Let **d** be a p-dimensional vector of differences (distances) with multinomial distribution $N_p(\mathbf{0},\mathbf{I})$, where **0** is a vector of zeros and **I** is the identity matrix. Let **W** be a $p \times p$ matrix with Wishart distribution $W_p(I, n)$ which is independent of **d**. Then $\mathbf{T}^2 = n\mathbf{d}'\mathbf{W}^{-1}\mathbf{d}$ has the Hotelling T^2-distribution with dimensionality parameter p and degrees of freedom n, denoted by $T^2(p, n)$. In statistical inference, the T^2 random variable is the sample (Mahalanobis) distance between two vectors. For example, let $y_1, y_2, \ldots y_n$ be a random sample from a multinormal population $N_p(\mu,\Sigma)$, and let

$$T^2 = (\bar{y} - \mu_0)'\left(\frac{S}{n}\right)^{-1}(\bar{y} - \mu_0) = n(\bar{y} - \mu_0)'S^{-1}(\bar{y} - \mu_0),$$

where

$$\bar{y} = \frac{1}{n}\sum_{i=1}^{n} y_i,\ \mu_0 = \begin{pmatrix} \mu_{10} \\ \mu_{20} \\ \vdots \\ \mu_{p0} \end{pmatrix},\ \mathrm{S} = \frac{1}{n-1}\left(\sum_{i=1}^{n}(y_i - \bar{y})(y_i - \bar{y})',\right)$$

and $\frac{S}{n}$ is the estimated covariance matrix of $\bar{y}$.

Then T^2 has a Hotelling T^2-distribution with dimension p and degrees of freedom $n - 1$. Note that T^2 is the multivariate analog of the univariate test statistic

$$t^2 = \frac{(\bar{y} - \mu_0)^2}{s^2/n} = n(\bar{y} - \mu_0)(s^2)^{-1}(\bar{y} - \mu_0),$$

and that T^2 is distributed as

$$\frac{(n-1)p}{(n-p)}F_{(p,\, n-p)},$$

where F_p, n–p is a random variable having F-distribution with p and $(n - p)$ degrees of freedom.

`HotellingTSquareDistribution[p, m]` MultinormalDistributions:HotellingTSquareDistribution represents Hotelling's T^2-distribution with dimensionality parameter `p` and degrees of freedom parameter `m`. For the random variable `m* (d.Inverse[M].d)` to be distributed `HotellingTSquareDistribution[p, m]`, `d` and `M` must be independently distributed, with `d` distributed `MultinormalDistribution[Table[0, {p}], IdentityMatrix[p]]` and `M` distributed `WishartDistribution[IdentityMatrix[p], m]`, respectively.

EXAMPLE 4: Plot the pdfs for the Hotelling's T^2-distribution with dimensionality parameter `p` and degrees of freedom parameter `m` for (a) $m = 4$ and $p = 1, 2, 3, 4$ and (b) $p = 4$ and $m = 4, 5, 6, 7$.

SOLUTION: (a) Assuming that the **MultinormalDistributions** package has been loaded, we define the four pdfs for the four values of of the dimensionality parameter using the `Table` command and plot them on the same graph using the `Plot` command.

```
dimtab=Table[PDF[HotellingTSquareDistribution[j,4],x],
  {j,1,4}];
```

```
Plot[Evaluate[dimtab], {x, 0, 8}]
```

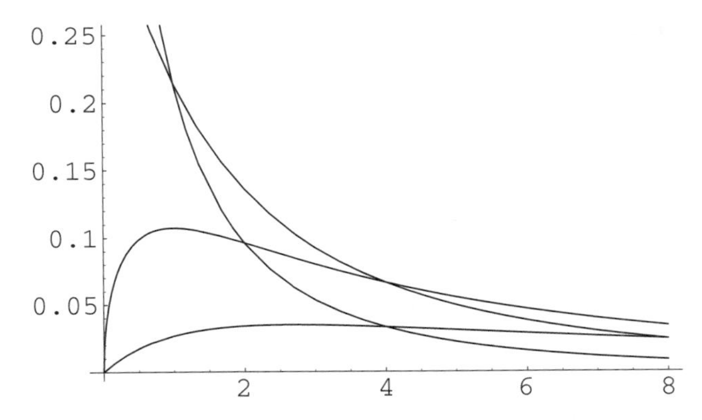

(b) We define the four pdfs for the four values of of the degrees of freedom parameter using the `Table` command and plot them with increasing length dashed lines on the same graph using the `Plot` command.

```
dftab=Table[PDF[HotellingTSquareDistribution[4,j],x],
  {j,4,8}];
```

```
Plot[Evaluate[dftab], {x, 0, 20},
 PlotStyle -> {Dashing[{0.01 }], Dashing[{0.02 }],
   Dashing[{0.03 }], Dashing[{0.04 }]}]
```

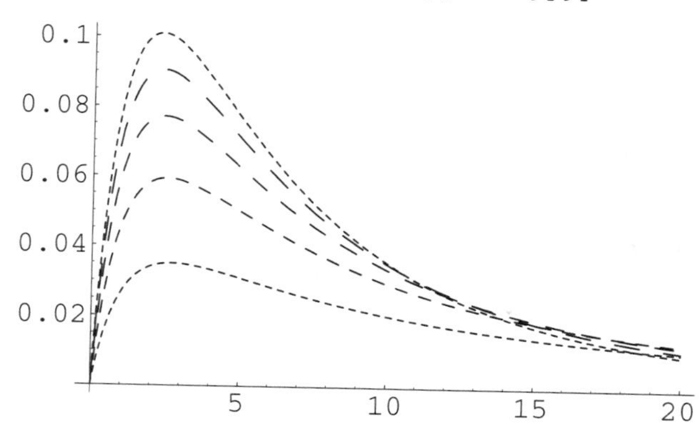

■

Quadratic Form Distribution

A quadratic form, $Q(x)$, in p variables $X_1, X_2, \ldots X_p$ is usually defined as $Q(\mathbf{x}) = \mathbf{x}'\mathbf{A}\mathbf{x}$, where $\mathbf{A}$ is a $p \times p$ symmetric matrix. The quadratic form can be written as

$$Q(\mathbf{x}) = \sum_{i=1}^{p} \sum_{j=1}^{p} a_{ij} X_i X_j.$$

In *Mathematica* the quadratic form is defined as $Q(\mathbf{x}) = \mathbf{x}'\mathbf{A}\mathbf{x} + \mathbf{b}'\mathbf{x} + c$, where $\mathbf{A}$ is a $p \times p$ symmetric matrix, $\mathbf{b}$ is a p-variate vector, and c is a scalar.

The univariate distribution of the quadratic form when $\mathbf{x}$ is a vector distributed as $N_p(\mu,\Sigma)$ is represented in *Mathematica* by `QuadraticFormDistribution[{A, b, c}, {μ, Σ}]`. When $\mu_i = 0$ for all $i = 1, 2, \ldots, p$, the quadratic from is called a central quadratic form. When one or more of $\mu_i \neq 0$, it is called a noncentral quadratic form. Simple cases of quadratic form distributions are central and noncentral chi-square distributions.

`QuadraticFormDistribution[{a, b, c}, {mu, sigma}]` MultinormalDistributions: QuadraticFormDistribution represents the (univariate) distribution of the quadratic form `z.a.z + b.z + c`, where `z` is distributed `MultinormalDistribution[mu, sigma]`. For the random variable `z.a.z + b.z + c` to be distributed `QuadraticFormDistribution[{a, b, c}, {mu, sigma}]`, `a` must be a $p \times p$ symmetric matrix, `b` must be a p-variate vector, and `c` must be a scalar.

CHAPTER 7

Random Number Generation and Simulation

There are two general methods of generating random numbers: physical methods and arithmetic methods. Examples of physical methods include drawing a card from a well-shuffled deck of cards, drawing a numbered ball from an urn, and rolling dice. Arithmetic methods are based on various mathematical algorithms. Random numbers generated in this way are often referred to as **pseudorandom numbers**, and the prerequisite to generating these random numbers is the ability to generate uniformly distributed random numbers over the interval between 0 and 1. In *Mathematica*, this basic capability is available through the built-in function `Random`. Nonuniform pseudorandom numbers are usually generated by starting with uniform pseudorandom numbers and transforming them to fit a given target distribution. The functions `Random` and `RandomArray` provide this capability in *Mathematica*. These functions are in the **ContinuousDistributions** and **DiscreteDistributions** packages, which are contained in the **Statistics** folder (or directory). To display information about these functions we use the ? command.

```
?Random
Random[ ] gives a uniformly distributed pseudorandom Real in
  the range 0 to 1. Random[type, range] gives a pseudorandom
  number of the specified type, lying in the specified
  range. Possible types are: Integer, Real and Complex. The
```

```
  default range is 0 to 1. You can give the range {min, max}
  explicitly; a range specification of max is equivalent to
  {0, max}. Random[distribution] gives a random number with
  the specified statistical distribution.
```

```
?RandomArray
RandomArray[distribution, n] generates a list of length n,
  where each element is a random number with the specified
  statistical distribution. RandomArray[distribution, {n1,
  n2, ...}] generates an n1 X n2 X ... array of nested lists
  of random numbers.
```

7.1 Simulating Simple Experiments

Simulation may be described as the process of designing a mathematical or logical model of a real system and then conducting computer-based experiments with the model to describe, explain, and predict the behavior of the real system. Often, in introductory probability, the experiments discussed include tossing coins and rolling dice. In the following examples, we will illustrate the use of `RandomArray` to simulate results from such simple experiments.

EXAMPLE 1: Use *Mathematica* to simulate tossing a fair coin 10 times and then 1000 times.

SOLUTION: Tossing a fair coin is an experiment consisting of repeated Bernoulli trials with probability of success 1/2. We can simulate tossing a fair coin 10 times using the Bernoulli distribution (Subsection 6.2.1) with probability of success $p = 1/2$. Assuming that **DiscreteDistributions** has been loaded, we use `RandomArray` to generate a list of length 10. Assuming zero represents tail (T) and one represents head (H), the simulation results six tails and four heads in the sequence T, T, T, T, T, H, H, T, H, H.

```
trial1=RandomArray[BernoulliDistribution[1/2],10]
{0,0,0,0,0,1,1,0,1,1}
```

Simulating tossing a coin 1000 times is done as easily by specifying the length of the list to be 1000. By using the semicolon we suppress the output.

```
trial2=RandomArray[BernoulliDistribution[1/2],1000];
```

Instead of displaying a list of 1000 zeros and ones, we load the **DataManipulation** package and then use `Frequencies` to see that 0 (tail) was obtained 486 times, and 1 (head) was obtained 514 times.

```
<<Statistics`DataManipulation`
Frequencies[trial2]
{{486,0},{514,1}}
```

■

EXAMPLE 2: Use *Mathematica* to simulate rolling a fair die 12 times and then rolling two dice 100 times.

SOLUTION: Rolling a fair die can be modeled using the discrete uniform distribution (Subsection 6.2.3) with the six outcomes, 1, 2, 3, 4, 5, and 6. Assuming that **DiscreteDistributions** has been loaded, the first experiment is simulated using `RandomArray` and specifying the length to be 12.

```
trial3=RandomArray[DiscreteUniformDistribution[6],12]
{3,3,3,6,3,5,3,5,6,4,2,4}
```

The second experiment is conducted similarly, except that we include 2 in the `RandomArray` command to indicate that two dice are rolled.

```
trial4=RandomArray[DiscreteUniformDistribution[6],
 {100,2}]
{{5,2},{6,6},{5,6},{1,2},{1,2},{2,6},{2,4},{4,2},{4,1},
 {2,1},{5,5},{4,3},{1,3},{3,5},{3,2},{6,6},{6,5},{4,6},
 {3,1},{4,2},{3,1},{3,4},{3,6},{3,4},{2,1},{4,4},{2,2},
 {2,1},{1,6},{4,6},{5,6},{3,4},{1,5},{2,5},{1,5},{1,1},
 {5,1},{3,3},{2,5},{1,2},{6,1},{6,2},{2,2},{5,1},{1,5},
 {6,6},{3,5},{4,3},{1,5},{3,4},{3,1},{3,2},{6,5},{2,5},
 {6,6},{5,5},{6,6},{2,2},{2,3},{6,6},{1,2},{1,6},{5,1},
 {5,5},{1,6},{4,6},{3,2},{3,3},{1,4},{6,2},{1,4},{4,4},
 {4,2},{1,5},{1,6},{6,1},{2,2},{2,3},{4,3},{1,5},{3,4},
 {1,2},{5,3},{6,2},{1,5},{1,5},{1,6},{6,1},{4,1},{4,2},
 {4,6},{3,1},{5,3},{1,1},{1,1},{2,4},{4,6},{3,5},{3,3},
 {6,2}}
```

Using `Map` together with `Apply` and `Plus` we find the sum of the dice for each trial.

```
sums=Map[Apply[Plus,#]&,trial4]
```

```
{7,12,11,3,3,8,6,6,5,3,10,7,4,8,5,12,11,10,4,6,4,7,9,
 7,3,8,4,3,7,10,11,7,6,7,6,2,6,6,7,3,7,8,4,6,6,12,8,
 7,6,7,4,5,11,7,12,10,12,4,5,12,3,7,6,10,7,10,5,6,5,
 8,5,8,6,6,7,7,4,5,7,6,7,3,8,8,6,6,7,7,5,6,10,4,8,2,
 2,6,10,8,6,8}
```

As in Example 1, we use `Frequencies` to see how often each sum occurred.

```
Frequencies[sums]
```

```
{{3,2},{8,3},{9,4},{9,5},{20,6},{20,7},{12,8},{1,9},
 {8,10},{4,11},{6,12}}
```

■

7.2 Simulation to Illustrate Concepts

In this section, we consider two important ideas. The Law of Large Numbers is important because it provides the base for the use of simulation in problem solving which is discussed in the next section. The Central Limit Theorem is important to statistical inference which we begin to discuss in the next chapter.

The Law of Large Numbers

In Section 6.1, we noted that the study of probability begins with an experiment, a sample space, and events. The probability of an event is then described as a measure of the likelihood that the event will occur. The traditional approach to expressing this likelihood is to use the relative frequency of the event when the experiment is repeated many times. We denote the relative frequency of the event E based on n repetitions of the experiment by $f_n(E) = n(E)/n$, where $n(E)$ is the number of times E occurred in the n repetitions. In Example 1 of Section 7.1, let E be the event "heads." Then the relative frequency of E in the 10 tosses of a fair coin was $f_{10}(E) = 4/10 = 0.4$. The relative frequency in 1000 tosses was $f_{1000}(E) = 514/1000 = 0.514$. Notice that the relative frequency does give an indication of the likelihood of E, but the result can vary. Also, it appears that the accuracy increases for large n. The (weak) **law of large numbers** verifies this observation. It says that the probability that $f_n(E)$ is close to $P(E)$ approaches 1 as n becomes large. Notice the law says that for large n it is almost certain that $f_n(E)$ will always be close to $P(E)$, but it provides no guarantee. It is after all possible (but unlikely) to have 1000 heads in 1000 tosses of a fair coin, in which case $f_n(E)$ would not be close to $P(E)$.

The law of large numbers applies to any finite sequence of independent, identically distributed random variables. For example, if $Y_1, Y_2, \ldots, Y_n$ is such a sequence where each Y_i has mean μ and variance σ^2, and $\bar{Y} = \frac{1}{n}\sum_{i=1}^{n} Y_i$ represents the sample mean, then the probability that $\bar{Y}$ is close to μ approaches 1 as n becomes large. Also, the probability that the sample variance, $S^2 = \frac{1}{n-1}\sum_{i=1}^{n}(Y_i - \bar{Y})^2$, is close to σ^2 approaches 1 as n becomes large.

EXAMPLE 1: Assume that two hospitals, Small Hospital and Big Hospital, have 20 and 200 births per day, respectively. Suppose that for one year, we count the number of days on which each hospital has over 60% boys in the day's births. Which hospital would be expected to have more such days, if we view the birth of a boy and the birth of a girl as equally likely events?

SOLUTION: The answer to the question lies with the law of large numbers, which says that the probability is high that on each day in Big Hospital the number of births of boys will be close to 50%. Thus, Small Hospital would be expected to have more days with over 60% boys in the day's births.

We will use simulation to verify this answer. Let 1 represent the birth of a boy and 0 represent the birth of a girl. To calculate the percent of boys in a list of the day's simulated births, we define the function `f`. Given a list of 0's and 1's, `f` returns the percent of 1's contained in the list.

```
f[x_]:=Module[{sum},
  sum=Apply[Plus,x];
  boys=sum/Length[x]//N]
```

We can model the birth of a child with a Bernoulli distribution with probability of success $p = 1/2$. Assuming that **DiscreteDistributions** has been loaded, we simulate the birth of 20 children a day for 365 days using `RandomArray` together with `BernoulliDistribution`. The result is named `small` and an abbreviated version of `small` is displayed using `Short`. We interpret the result to mean that on the first day, a girl is born, then two boys, followed by a girl, and so forth. On the second day, the first birth is a boy, the next three are girls, and so forth.

```
small=RandomArray[BernoulliDistribution[0.5],{365,20}];
Short[small,5]
```

```
{{0,1,1,0,1,1,0,1,1,1,0,1,0,1,1,0,0,0,1,0},{1,0,0,0,1,
  0,1,1,1,1,0,1,0,1,0,0,0,0,1,1},<<361>>,{1,0,1,0,1,1,
  1,1,1,1,0,1,1,0,1,1,0,0,0,0},{1,1,1,1,0,1,0,1,1,0,1,
  1,1,0,1,0,0,1,1,0}}
```

To determine the percent of each day's births that are boys, we use `Map` to apply `f` to `small`, and, after loading the **DataManipulation** package, we use `BinCounts` to learn that there were 57 days for which more than 60% of the day's births were boys.

```
smallboys=Map[f,small]; BinCounts[smallboys,{0,1,.1}]
```

```
{0,1,16,64,123,104,45,12,0,0}
```

```
BinCounts[smallboys,{.6,1,.4}]
```

```
{57}
```

We simulate the birth of 200 children a day for 365 days in the same way. The result is named `large` and then the percent of each day's births that are boys is computed by using `Map` to apply `f` to `large`. In this case, we see that there were no days in which more than 60% of the births were boys.

```
large=RandomArray[BernoulliDistribution[0.5],{365,200}];
```

```
largeboys=Map[f,large];BinCounts[largeboys,{0,1,.1}]
```

```
{0,0,0,2,187,176,0,0,0,0}
```

```
BinCounts[largeboys,{.6,1,.4}]
```

```
{0}
```

An alternative and simple approach to the simulation is to use `BinomialDistribution`. For this simulation we see there were 48 days for which more than 60% of the day's births were boys in Small Hospital and, again, no days for which more than 60% of the births were boys in Big Hospital.

```
n=20;
p=0.5;
q=0.5;
```

```
smallboys=RandomArray[BinomialDistribution[n,p],365]
/n//N;BinCounts[smallboys,{0,1,.1}]
```

```
{0,3,17,89,110,98,41,7,0,0}
```

```
BinCounts[smallboys,{.6,1,.4}]
```

```
{48}
```

```
n=200;
p=0.5;
q=0.5;

largeboys=RandomArray[BinomialDistribution[n,p],365]
/n//N;BinCounts[largeboys,{0,1,.1}]

{0,0,0,0,188,177,0,0,0,0}

BinCounts[largeboys,{.6,1,.4}]

{0}
```

■

The Central Limit Theorem

In most elementary statistics textbooks, the central limit theorem is used to describe the sampling distribution of the sample mean. Suppose a simple random sample (see Section 7.4) of size n is selected from a population, and a variable Y is measured on each sample unit. Then the data consist of observations on n random variables Y_1, Y_2, ..., Y_n, and the sample mean is a random variable having the form $\bar{Y} = \frac{1}{n}\sum_{i=1}^{n} Y_i$. Notice that the values that $\bar{Y}$ can take will vary depending on the sample selected. The **sampling distribution** of the sample mean $\bar{Y}$ is the distribution of these possible values. It has mean μ and standard deviation $\sigma/\sqrt{n}$, where μ is the mean and σ is the standard deviation of each Y_i. If each Y_i is normally distributed, denoted $N(\mu,\sigma)$, then the sampling distribution of the sample mean $\bar{Y}$ is normal as well. Thus, the sample mean based on a simple random sample of size n is $N(\mu, \sigma/\sqrt{n})$. Note that $\sigma/\sqrt{n}$ (or its point estimate from a sample $s/\sqrt{n}$) is often referred to as the **standard error**.

The central limit theorem guarantees that, if the sample size is large enough, the sampling distribution of the sample mean $\bar{Y}$ will be close to normal no matter what the distribution in the population may be, as long as the population has a finite standard deviation. More generally, the **central limit theorem** says that the distribution of a sum or an average of many small random quantities is close to normal. This is true even if these quantities are not independent, as long as they are not too closely related. Also, this remains true even if the quantities have different distributions, as long as none of them dominates the others. The central limit theorem has been referred to as the most important theorem of statistical inference, because it demonstrates why the normal distribution is such a common model for observed data.

We have written two procedures that will be useful in the examples which illustrate the central limit theorem. The first is called **scaledHistogram**, and the second is called **normalHistogram**. They can both be found on the disk that comes with this book. To use any of our procedures, first load the disk and click on its icon. Then locate the procedure name, click on the rightmost cell bracket, press **ENTER**, and return to the *Mathematica* work sheet.

Given a list of numbers, **scaledHistogram** generates a histogram with total area 1. That is, the sum of the areas of the rectangles which form the histogram is 1. The procedure **normalhHistogram** generates a scaled histogram representing the data using **scaled-Histogram** and displays this result together with a normal probability density function (pdf) with mean and standard deviation calculated from the list. Both procedures accept input of a data list, the number of classes (the default is 10), and any options that can be used with `GeneralizedBarChart`. Notice that **scaledHistogram** is automatically loaded with **normalHistogram**.

```
<<Statistics`ContinuousDistributions`
  Clear[normalHistogram]
  normalHistogram[data_, bars_: 10, opts___]:=Module
    [{p1, mu, sigma, p2},
p1 = scaledHistogram[data, bars, DisplayFunction ->
  Identity];mu = Mean[data];sigma = StandardDeviation[data];
p2 = Plot[PDF[NormalDistribution [mu, sigma], x],{x, mu-
  3sigma, mu+3sigma}, DisplayFunction -> Identity];
Show[p1, p2, PlotRange -> All, opts, DisplayFunction ->
  $DisplayFunction]]
```

EXAMPLE 2: Randomly select 10,000 samples of size 3 from the normal distribution having mean 100 and standard deviation 16. Use these samples to simulate the sampling distribution of (a) the sample mean and (b) the sample variance.

SOLUTION: Assuming that **ContinuousDistributions** has been loaded, we use `RandomArray` to select the samples and `Take` to display the first five of them.

```
dist=NormalDistribution[100,16];
randomsampleNormal=RandomArray[dist,{10000,3}];
```

```
Take[randomsampleNormal,5]
```

```
{{98.9361,77.0478,100.643},{108.767,92.587,93.566},
 {118.61,92.2593,112.405},{101.867,90.5351,75.426},
 {86.2466,114.014,86.14}}
```

(a) We use `Map` together with `Mean` to calculate the mean each sample and `Take` to display the first five. These sample means may be thought of as a random sample from the sampling distribution of the means of all possible samples of size 3 from the normal distribution having mean 100 and standard deviation 16. As such, they should represent the sampling distribution very closely. Notice that the mean of the 10,000 sample means is 100.058. The mean of the sampling distribution is 100, the mean of the population that was sampled. The standard deviation (standard error) of the 10,000 sample means is 9.22055. The standard deviation of the sampling distribution is $16/\sqrt{3n} = 9.23760$.

```
meanrandomsampleNormal=Map[Mean,randomsampleNormal];
Take[meanrandomsampleNormal,5]
```

```
{92.209,98.3066,107.758,89.276,95.4669}
```

```
Mean[meanrandomsampleNormal]
StandardDeviation[meanrandomsampleNormal]
```

```
100.058

9.22055
```

We next use `normalHistogram` to plot a histogram of the 10,000 sample means. Notice how closely this histogram resembles the normal pdf. As noted in the preceding discussion, the sampling distribution of the sample mean is normal (for any sample size) when the population sampled is normal.

```
normalHistogram[meanrandomsampleNormal, 40]
```

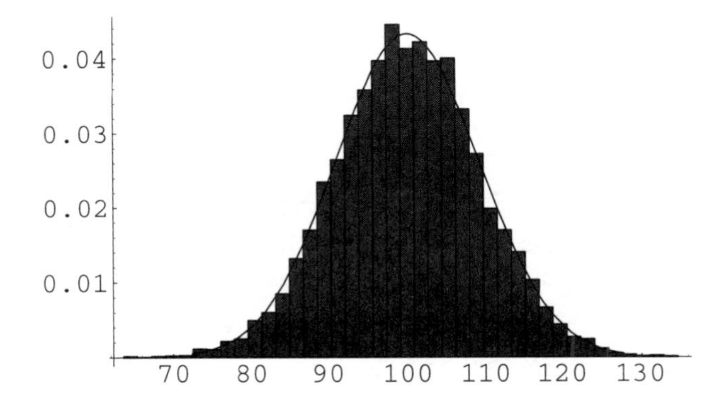

(b) Next we calculate the variance of each of the 10,000 random samples and use these variances to represent the sampling distribution of the variance of all possible samples of size 3 from the normal distribution having mean 100 and standard deviation 16. Note that this sampling distribution has mean $16^2 = 256$ and is proportional to the chi-square distribution with degrees of freedom 2. The 10,000 sample variances reflect the sampling distribution closely.

```
variancerandomsampleNormal=Map[Variance,
  randomsampleNormal];
Take[variancerandomsampleNormal,5]
```

```
{173.125,82.2996,189.783,175.969,258.002}
```

```
Mean[variancerandomsampleNormal]
StandardDeviation[variancerandomsampleNormal]
```

```
255.497
252.908
```

```
scaledHistogram[variancerandomsampleNormal, 40]
```

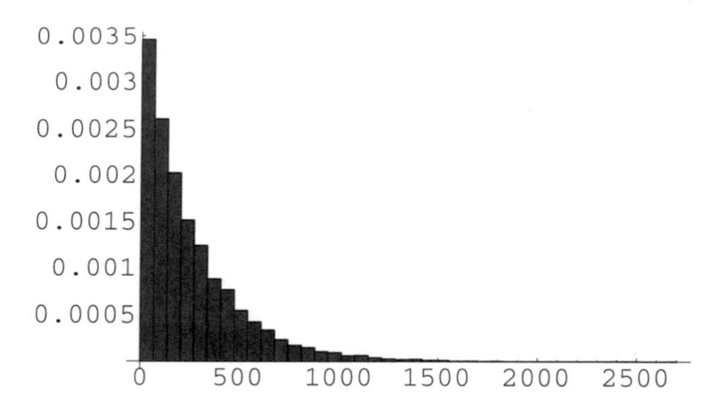

■

EXAMPLE 3: Randomly select 5000 samples of sizes 2, 3, 4, and 5, respectively, from a uniform distribution over the interval from 95 through 105. Plot histograms and assess the normality of the sample means for each sample size.

SOLUTION: Assuming that **ContinuousDistributions** has been loaded, we begin by defining `dist` to be a uniform distribution on the interval from 95 through 105.

```
dist=UniformDistribution[95,105];
```

We use `RandomArray` to select 5000 samples of size 2, 3, 4, and 5 from this distribution,

```
sample1=RandomArray[dist,{5000,2}];
sample2=RandomArray[dist,{5000,3}];
sample3=RandomArray[dist,{5000,4}];
sample4=RandomArray[dist,{5000,5}];
```

and then compute the mean of each sample with `Map` and `Mean`.

```
mean1=Map[Mean,sample1];
mean2=Map[Mean,sample2];
mean3=Map[Mean,sample3];
mean4=Map[Mean,sample4];
```

Finally, for each sample size, we use **normalHistogram** to plot a histogram representing the data together with a normal density function. The result is not displayed because we include `DisplayFunction->Identity` in the `normalHistogram` command. All four graphs are then displayed together using `Show` and `GraphicsArray`. Notice that starting with sample size 3 the simulated sampling distributions are close to normal.

```
g1=normalHistogram[mean1,40,DisplayFunction->Identity];
g2=normalHistogram[mean2,40,DisplayFunction->Identity];
g3=normalHistogram[mean3,40,DisplayFunction->Identity];
g4=normalHistogram[mean4,40,DisplayFunction->Identity];
```

```
Show[GraphicsArray[{{g1, g2}, {g3, g4}}]]
```

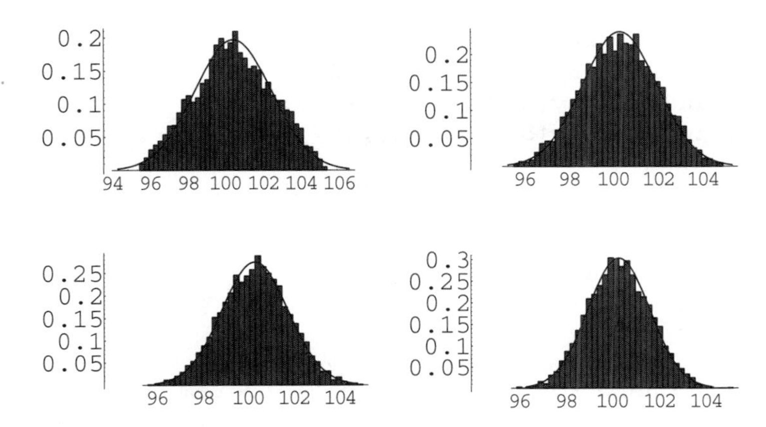

■

EXAMPLE 4: Randomly select 2000 samples of sizes 5, 10, 25, and 30, respectively, from an exponential distribution with mean 2 ($\lambda = 1/2$). Plot histograms and assess the normality of the sample means for each sample size.

SOLUTION: Assuming that **ContinuousDistributions** has been loaded, we begin by defining `dist` to be an exponential distribution with $\lambda = 1/2$ and plotting the probability density function.

```
dist = ExponentialDistribution[1 / 2];
Plot[PDF[dist, x], {x, 0, 8}]
```

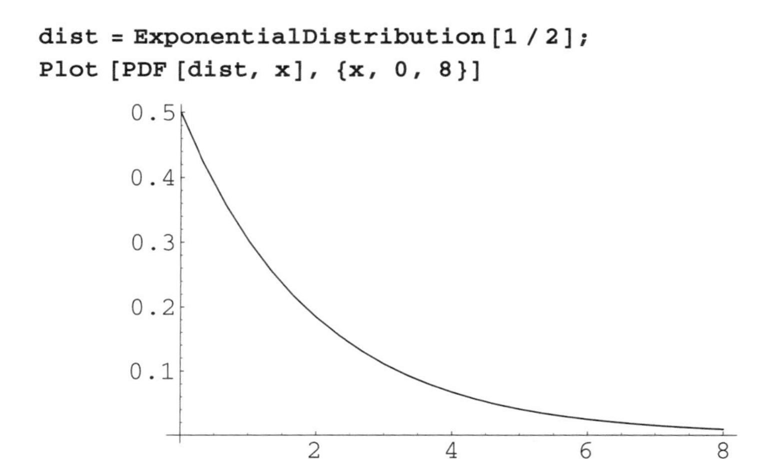

We use `RandomArray` to select 2000 samples of size 5, 10, 25, and 30 from this distribution,

```
sample1=RandomArray[dist,{2000,5}];
sample2=RandomArray[dist,{2000,10}];
sample3=RandomArray[dist,{2000,25}];
sample4=RandomArray[dist,{2000,30}];
```

and then compute the mean of each sample with `Map` and `Mean`.

```
mean1=Map[Mean,sample1];
mean2=Map[Mean,sample2];
mean3=Map[Mean,sample3];
mean4=Map[Mean,sample4];
```

Finally, for each sample size, we use **`normalHistogram`** to plot a histogram representing the data together with a normal density function. The result is not displayed because we include `DisplayFunction->Identity` in the **`normalHistogram`** command. All four graphs are then displayed together using `Show` and `GraphicsArray`. Notice that the exponential nature of the pdf is apparent for sample sizes 5 and 10 and still somewhat apparent even for sample size 25. On the other hand, it is clear that as the sample size increases the distribution becomes more normal.

```
g1=normalHistogram[mean1,50,DisplayFunction->Identity];
g2=normalHistogram[mean2,50,DisplayFunction->Identity];
g3=normalHistogram[mean3,50,DisplayFunction->Identity];
g4=normalHistogram[mean4,50,DisplayFunction->Identity];

Show[GraphicsArray[{{g1, g2}, {g3, g4}}]]
```

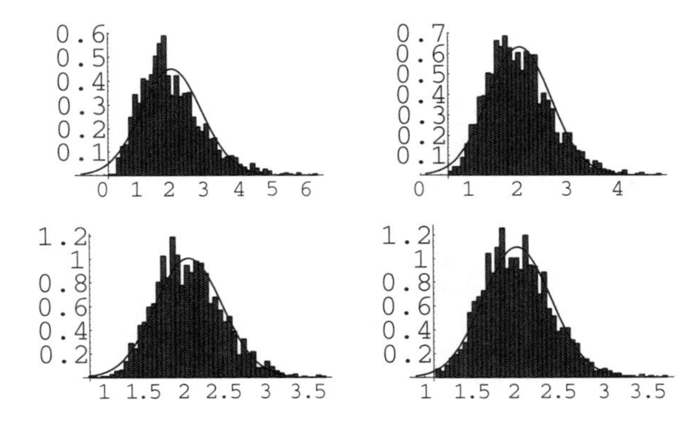

■

7.3 Simulation in Problem Solving (Monte Carlo Method)

Problem solving using simulation is often referred to as the **Monte Carlo method**. The technique uses artificial data generated through the use of a random number generator using the probability distribution appropriate to the problem. The first step is to identify an appropriate model and to decide how it will be used to solve the problem. Solutions are then estimated by random sampling. The theoretical basis for the method is the law of large numbers, which states that as the simulation is run a larger and larger number of times the probability is high that the simulated solutions will be close to the actual answer.

EXAMPLE 1: To estimate the proportion of brown candies in a small bag of M&M's, a random sample of 30 candies is selected and the number of brown candies determined. Because the number of brown candies can vary depending on the sample, the accuracy of the estimate obtained from one sample and the variability between possible estimates from other samples is of concern. (a) Develop a model to simulate the colors of candies in a bag of plain M&M's. (b) Use simulation to characterize the accuracy of and possible variability in potential estimates of the percent of brown candies based on a sample of size 30.

SOLUTION: (a) According to Mars, Inc., the proportions of the colors of Plain M&M's manufactured are orange 10%, green 10%, blue 10%, brown 30%, yellow 20%, and red 20%. To simulate these proportions, we divide the interval between 0 and 1 into appropriate length intervals and define the function `m`. Then using the built-in function `Random[]`, which will return a random value between 0 and 1, and `Table` we can generate a list of the colors of the 30 randomly selected candies and call the list `bag1`.

```
m[x_]:=o /; 0<x<=.1;
m[x_]:=g /; .1<x<=.2;
m[x_]:=bl /; .2<x<=.3;
m[x_]:=br /;.3<x<=.6;
m[x_]:=y /;.6<x<=.8;
m[x_]:=r/; .8<x<=1;
```

```
bag1=Table[m[Random[]],{30}]

{bl,g,br,y,br,y,r,y,o,r,r,br,bl,y,o,g,br,br,y,br,br,bl,
 r,r,br,g,y,g,o,br}
```

Using our procedure **countTable** we display the number of occurrences of each color. Then after loading **Graphics** we generate a bar graph where, on a color monitor, each bar is the color of the candy whose frequency it represents.

```
countTable[bag1]

   bl 3
   br 9
   g  4
   o  3
   r  5
   y  6
```

```
<<Graphics`Graphics`
```

```
freq1=Frequencies[bag1];BarChart[freq1,BarStyle->
 {{RGBColor[0,0,1]}, {RGBColor[.6,.4,.2]},
  {RGBColor[0,1,0]}, {RGBColor[.9,.2,0]},
  {RGBColor[1,0,0]}, {RGBColor[1,1,0]}}]
```

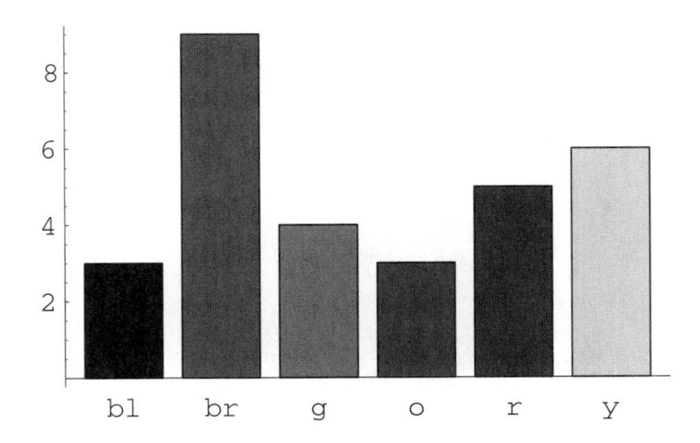

(b) We have written a procedure called `brpercent` to simulate a random sample of Plain M&M's using the function `m` and calculate the proportion of brown candies, where the sample size is entered by the user. Using `Table` with `brpercent` we compute the fraction of brown M&M's in 100 hypothetical bags of M&M's that contain 30 pieces of candy and store them in the list `brownsimulation1`. The first five results are displayed using `Take`. Each of the 100 values is an estimate of the true proportion (0.3) of manufactured brown candies.

```
Clear[brpercent]
brpercent[n_]:=Module[{bag},
   bag=Table[m[Random[]],{n}];
   Count[bag,br]/n//N]

brownsimulation1=Table[brpercent[30],{100}]
Take[brownsimulation1,5]

{0.333333,0.266667,0.3,0.266667,0.5}
```

Using our procedure **normalHistogram**, which is described in Section 7.2, we display a histogram of the 100 estimates. Notice that the estimates tend to center on the true proportion (0.3). The actual mean of the 100 estimates is 0.297333 and the standard deviation is 0.0891008.

```
normalHistogram[brownsimulation1]
```

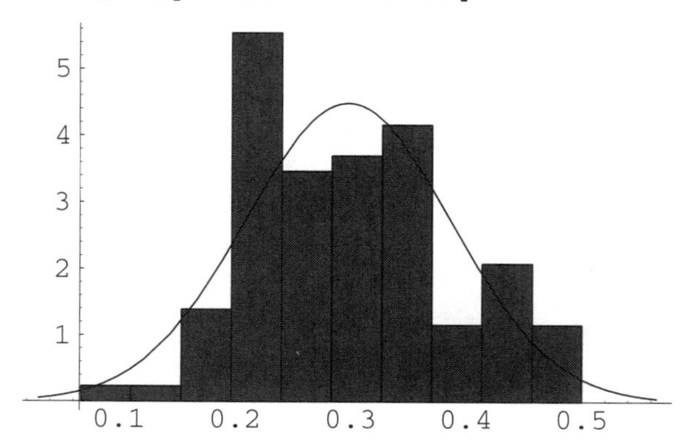


```
Mean[brownsimulation1]
StandardDeviation[brownsimulation1]
```

```
0.297333
0.0891008
```

Using the built-in functions `Select` and `Length`, we determine that 62% of the 100 estimates are within one standard deviation (about 0.09) of 0.3. Thus, if we denote an estimate by $\hat{p}$, for 62% of them, $\hat{p} \pm 0.09$ would cover the true value (0.3) of the proportion manufactured. For two standard deviations (about 0.18) the result is 97%.

```
onesigma1=Select[brownsimulation1,And[#<.389,#>.211]&];
Length[onesigma1]/100//N
```

```
0.62
```

```
twosigma1=Select[brownsimulation1,And[#<.478,#>.122]&];
Length[twosigma1]/100//N
```

```
0.97
```

Increasing the sample size from 30 to 75 and repeating these steps resulted in the following. Notice the apparent consistency of the centering of the estimates and the percent of the estimates within one and two standard deviations of the proportion (0.3). Also notice that the standard deviation is decreased (because of the increased sample size), so the estimates are all closer to the proportion (0.3).

```
brownsimulation2=Table[brpercent[75],{100}]
Take[brownsimulation2,5]
```

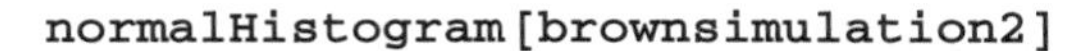

```
normalHistogram[brownsimulation2]
```

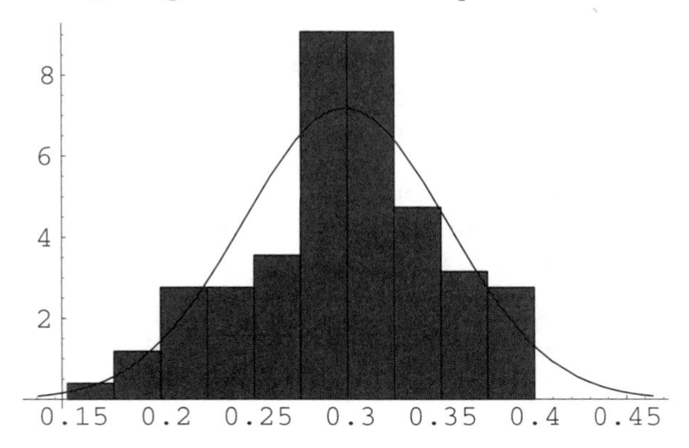


```
Mean[brownsimulation2]
StandardDeviation[brownsimulation2]
```

```
0.297467
0.0555659
```

```
onesigma2=Select[brownsimulation2,And[#<.356,#>.244]&];
Length[onesigma2]/100//N
```

```
0.67
```

```
twosigma2=Select[brownsimulation2,And[#<.411,#>.189]&];
Length[twosigma2]/100//N
```

```
0.96
```

■

EXAMPLE 2: A typical experiment to test extrasensory perception (ESP) is to ask a subject to identify which of four shapes appears on a card being studied by a second person but unseen by the subject. Develop and perform a simulation to characterize the distribution of the possible number of correct responses in 20 trials assuming neither participant has ESP.

SOLUTION: Assume one participant selects one of four cards: clubs, spades, hearts, and diamonds. Given that neither participant has ESP, the subject has probability 1/4 of guessing the correct shape. Assuming that 20 trials are performed, and the number of correct guesses is recorded, our model will be based on the binomial distribution. (Each trial is a Bernoulli trial with probability of success 1/4. The 20 trials are independent, and we are interested in the number of successes, which is a binomial random variable.) We can simulate the experiment with `Random` and `BinomialDistribution[20,1/4]`. After loading **DiscreteDistributions**, we use `Table` to simulate the number of correct responses given in 20 trials of the experiment conducted 10,000 times. We suppress the output of the resulting list, `simulate`, by using the semicolon.

```
<<Statistics`DiscreteDistributions`
dist=BinomialDistribution[20,1/4];
simulate=Table[Random[dist],{10000}];
```

Next we calculate the mean and standard deviation of the 10,000 results and plot a histogram with **normalHistogram**. We learn that five correct is expected and that it is unlikely to see more than nine correct. Having none correct is also unlikely.

```
Mean[simulate]//N
StandardDeviation[simulate]//N
```

```
5.0095
1.94334
```

```
normalHistogram[simulate, 12]
```

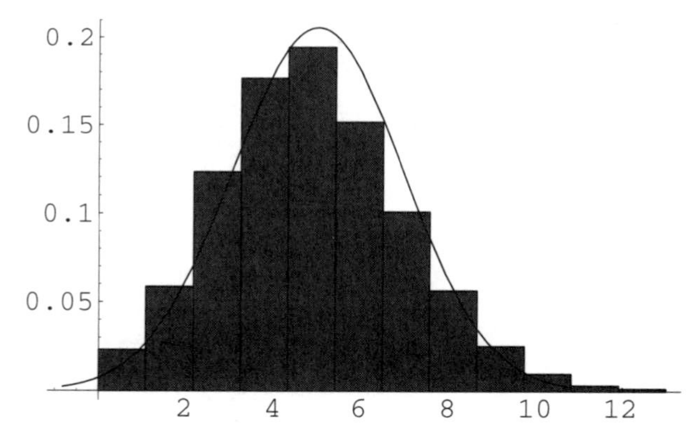

■

EXAMPLE 3: A **tontine** is an annuity or insurance plan whereby a group of participants hold shares in a common fund with right of survivorship, each participant's share being increased as one of the other dies, the final survivor receiving the whole. Develop and perform a simulation to characterize the behavior of a tontine.

SOLUTION: We first obtain mortality tables for males and females. The lists `femalemortalitytable` and `malemortalitytable` contain the *1995 U.S. Buck Mortality Tables* for females and males, respectively. The entries correspond to the age and probability of death. The first 10 entries in `femalemortalitytable` are listed using `Take`.

```
Take[femalemortalitytable,10]
{{1,0.000875},{2,0.000452},{3,0.000336},{4,0.00027},
 {5,0.000218},{6,0.000178},{7,0.000151},{8,0.000133},
 {9,0.000124},{10,0.000125}}
```

Next, we define the function `p`. Given a name, sex, and age, `p` returns the probability of death. To indicate sex use a 0 for male and a 1 (or any other number) for female. For example, we see that the probability of death occurring in a 31-year old male is approximately 0.00065.

```
p[{name_,sex_,age_}]:=
  If[sex==0,malemortalitytable[[age]][[2]],
    femalemortalitytable[[age]][[2]]]
p[{jim,0,31}]
0.00065
```

Further, we define the functions `f` and `increase`. Given an ordered triple containing name, sex, and age, `f` returns an ordered quadruple containing the name, sex, age, and 0 or 1 selected randomly using `BernoulliDistribution` with probability of success determined by `p`. (Note that "success" in this case corresponds to a 1 which indicates the person dies.) On the other hand, given an ordered quadruple containing name, sex, age, and 0 or 1, `increase` returns an ordered triple containing the name and sex as well as age increased by one year.

```
f[{name_,sex_,age_}]:={name,sex,age,
   Random[BernoulliDistribution[p[{name,sex,age}]]]};
increase[{name_,sex_,age_,z_}]:={name,sex,age+1};
```

Finally, we define the procedure `year` to run our simulation. Given a list of persons, each represented by name, sex, and age, it applies `f` to the list, selects those who have lived (that is, `Random` and `BernoulliDistribution` have returned 0), and applies `increase` to the resulting list. As an illustration, we define in `names` a list of five fictitious people. We then use a `While` loop to run `year` on `names` until only one name is left. (Note that for length considerations, only a portion of the result is displayed here.)

```
Clear[year]
year[list_]:=Module[{step1,step2},
   step1=Map[f,list];
```

```
    step2=Select[step1,#[[4]]==0&];
    step3=Map[increase,step2]]
names={{jim,0,30},{lori,1,35},{martha,1,40},
    {john,0,45},{ada,1,20}};
While[Length[names]>1,names=year[names];Print[names];]
{{jim,0,31},{lori,1,36},{martha,1,41},
 {john,0,46},{ada,1,21}}
{{jim,0,32},{lori,1,37},{martha,1,42},
 {john,0,47},{ada,1,22}}
{{jim,0,33},{lori,1,38},{martha,1,43},
 {john,0,48},{ada,1,23}}
{{jim,0,34},{lori,1,39},{martha,1,44},
 {john,0,49},{ada,1,24}}
```

We see that Lori "dies" at age 55.

```
{{jim,0,50},{lori,1,55},{martha,1,60},
 {john,0,65},{ada,1,40}}
{{jim,0,51},{martha,1,61},{john,0,66},{ada,1,41}}
```

Then John dies at age 86.

```
{{jim,0,71},{martha,1,81},{john,0,86},{ada,1,61}}
{{jim,0,72},{martha,1,82},{ada,1,62}}
```

Jim lives until the age of 80,

```
{{jim,0,80},{martha,1,90},{ada,1,70}}
{{martha,1,91},{ada,1,71}}
```

and Ada "wins" the game when Martha dies at age 105.

```
{{martha,1,105},{ada,1,85}}
{{ada,1,86}}
```

It is interesting to run the simulation several times. In this case, we use a Do loop to run the simulation 100 times. At the end of each trial, the "winner" is returned. Notice that the first line initializes the list `winner` to be empty, and at the end of the loop the list of the winners' names is displayed. For length considerations, only a portion of the intermediate results are displayed here.

```
winner={};
Do[names={{jim,0,30},{lori,1,35},{martha,1,40},
    {john,0,45},{ada,1,20}};
```

```
While[Length[names]>1,names=year[names]];
Print[names];
winner=Join[winner,{names[[1,1]]}],
{100}]
```

```
{{ada,1,69}}{{ada,1,70}}{{lori,1,92}}{{ada,1,81}}
{{ada,1,73}}{{ada,1,80}}{{lori,1,98}}{{jim,0,83}}
{{ada,1,80}}{{lori,1,101}}{{ada,1,86}}
.
.
.
{{ada,1,75}}{{ada,1,77}}{{ada,1,71}}{{lori,1,90}}
{{ada,1,80}}{{ada,1,78}}{{ada,1,78}}{{jim,0,88}}
{{ada,1,89}}
```

```
winner
```

```
{ada,ada,lori,ada,ada,ada,lori,jim,ada,lori,ada,ada,
 ada,lori,jim,ada,ada,ada,ada,ada,ada,jim,ada,ada,ada,
 ada,ada,ada,ada,ada,ada,ada,ada,jim,ada,ada,lori,ada,
 ada,ada,ada,ada,ada,ada,ada,ada,ada,ada,jim,ada,ada,
 ada,ada,ada,jim,ada,ada,lori,ada,ada,ada,ada,ada,ada,
 ada,ada,ada,ada,ada,ada,ada,ada,jim,ada,ada,ada,ada,
 jim,martha,ada,ada,ada,ada,ada,ada,lori,jim,jim,lori,
 ada,ada,ada,ada,ada,lori,ada,ada,ada,jim,ada}
```

We count the number of times that each person wins with `Frequencies`. Observe that after running the simulation 100 times, John (the oldest at the beginning) has never "won."

```
Frequencies[winner]
```

```
{{79,ada},{11,jim},{9,lori},{1,martha}}
```

To investigate if John might ever "win," we repeat the simulation 1000 times. Notice that *Mathematica* generates several error messages. These correspond to times when no one wins the game: that is, the two remaining people "die" in the same year.

```
winner={};
Do[names={{jim,0,30},{lori,1,35},
{martha,1,40},{john,0,45},{ada,1,20}};
While[Length[names]>1,names=year[names]];
winner=Join[winner,{names[[1,1]]}],{1000}]
```

```
Part::partw: Part 1 of {} does not exist.
Part::partw: Part 1 of {} does not exist.
```

```
Part::partw: Part 1 of {} does not exist.

General::stop: Further output of Part::partw will be
suppressed during this calculation.
```

In fact, when we look at how often each person "wins," it appears that it is more likely for no one to "win" the game than for John to "win" the game.

```
tograph=Frequencies[winner]

{{720,ada},{121,jim},{5,john},{97,lori},{31,martha},
 {26,{}[[1,1]]}}

<< Graphics`Graphics`
BarChart[tograph, PlotRange -> All ]
```

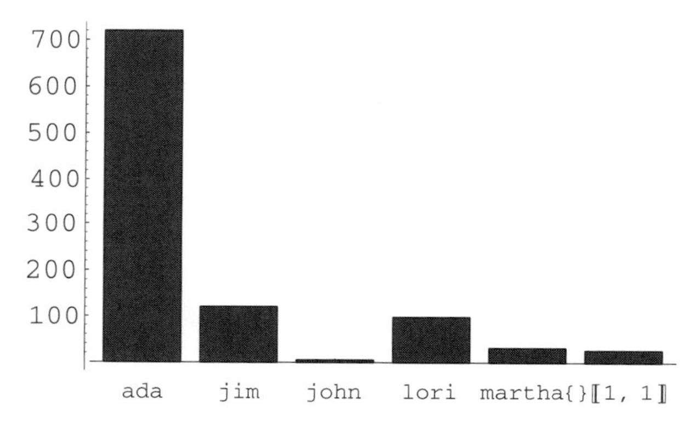

■

7.4 Random Sampling

The objective of inferential statistics is to make inferences about a population from information contained in a sample (subset) from the population. In doing this it is desirable to quantify the degree of uncertainty associated with the statistical inference. This requires that the sampling design be based on planned randomness, where the resulting samples are termed **probability samples**. The basic form of a probability sample is a **simple random sample**, which consists of a subset of n elements of the population selected in such a way that each subset of size n has the same chance of being selected. A consequence of this definition is that all individual elements in a population have the same chance of being selected. Note that the converse is not necessarily true. That is, giving each element of the population an equal chance of being selected does not imply that all samples of size n have the same chance of being selected.

When the population elements are divided into two or more groups, or strata, according to some relevant factor, and a simple random sample selected from each group, the resulting sample is called a **stratified random sample**. For example, a population of voters may contain individuals in two distinct income brackets (strata). Voters in the high bracket may have opinions on a bond issue that are quite different from the opinions of voters in the low bracket. To obtain accurate information about the population, voters from each bracket are sampled. When a simple random sample of groups of elements is selected and every element within a chosen group is included in the sample, the result is called a **cluster sample**. For example, in urban areas, it will be economical to sample families, whole apartment buildings, or whole city blocks rather than individual voters. Finally, if the names of persons in the population of interest are available in a list, such as a telephone directory, an economic technique for sampling is to select one name at random near the beginning of the list and then select names that follow at a fixed interval thereafter. The resulting sample is called a **systematic sample**.

We have written a procedure called **`simpleRandomSample`**. Given a list and a positive integer, n, it returns a simple random sample of size n from the list. It can be found on the disk that comes with this book. To use any of our procedures, first load the disk and click on its icon. Then locate the procedure name, click on the right most cell bracket, press **ENTER**, and return to the *Mathematica* work sheet.

EXAMPLE 1: The *Mathematica* list, `senators`, contains information about the members of the 1996 U.S. Senate. For each senator, name, party affiliation, and state are listed. Select a simple random sample of size 10 from this list. Give an example of the *Mathematica* code to perform a simulation using repeated samples from this list.

SOLUTION: Rather than displaying the entire list, we use `Short` to show the first five and the last five records in `senators`.

```
Short[senators,10]
```

```
{{Abraham, Spencer ,R,MI},{Akaka, Daniel K. ,D,HI},
 {Allard, Wayne ,R,CO},{Ashcroft, John ,R,MO},
 {Baucus, Max ,D,MT},{Bennett, Robert F. ,R,UT},
 {Biden, Joseph R., Jr. ,D,DE},{Bingaman, Jeff ,D,NM},
 {Bond, Christopher S. ,R,MO},{Boxer, Barbara ,D,CA},
 <<81>>,{Specter, Arlen ,R,PA},{Stevens, Ted ,R,AK},
```

```
{Thomas, Craig ,R,WY},{Thompson, Fred ,R,TN},
{Thurmond, Strom ,R,SC},{Torricelli, Robert ,D,NJ},
{Warner, John W. ,R,VA},{Wellstone, Paul D. ,D,MN},
{Wyden, Ron ,D,OR}}
```

Next, we use our procedure **simpleRandomSample** to select a simple random sample of size 10, naming the resulting list `srsample`.

```
srsample=simpleRandomSample[senators,10]
```

```
{{Boxer, Barbara ,D,CA},{Chafee, John H. ,R,RI},
 {Coverdell, Paul ,R,GA},{Faircloth, Lauch ,R,NC},
 {Hagel, Charles ,R,NE},{Jeffords, James M. ,R,VT},
 {Kerry, John F. ,D,MA},{Lugar, Richard G. ,R,IN},
 {Murray, Patty ,D,WA},{Reid, Harry ,D,NV}}
```

Note that the party affiliation of each senator can be extracted from `senators` with `Map`.

```
party=Map[#[[2]]&,senators]
```

```
{R,D,R,R,D,R,D,D,R,D,D,R,D,D,R,D,R,R,D,R,R,R,D,R,R,R,D,
 R,D,R,D,D,R,R,D,D,D,R,D,R,D,R,R,R,R,R,D,R,R,D,R,R,R,D,
 R,D,R,D,D,D,D,R,D,D,D,D,D,R,R,R,R,R,D,D,D,R,D,R,D,D,D,
 R,D,R,R,D,R,R,R,R,R,R,R,R,R,R,D,R,D,D}
```

Using `Frequencies`, we see that in 1996 there were 45 Democratic and 55 Republican senators.

```
<<Statistics`DataManipulation`
Frequencies[party]
```

```
{{45,D},{55,R}}
```

Repeating this process for our sample, we see that it contains four Democratic and six Republican senators.

```
step2=Map[#[[2]]&,srsample]
```

```
{D,R,R,R,R,R,D,R,D,D}
```

```
Frequencies[step2]
```

```
{{4,D},{6,R}}
```

Given `n`, `simulate[n]` randomly selects `n` senators from the list `senators` and then returns the number of Democratic senators in the list.

```
Clear[simulate,n]
simulate[n_]:=Module[{step1,step2,step3},
   step1=simpleRandomSample[senators,n];
```

```
step2=Map[#[[2]]&,step1];
step3=Frequencies[step2];
step3[[1,1]]]
```

As an illustration, we randomly select 10 groups of 5 senators each. The first group contains 3 Democrats, the second 2, the third 3, and so forth.

```
trial=Table[simulate[5],{10}]
```

```
{3,2,3,3,1,3,5,1,1,3}
```

We now use `simulate[10]` together with `Table` to simulate the selection of 500 committees of 10 senators and calculate the number of Democrats on each committee. Then we plot these frequencies using our procedure **normalHistogram**.

```
trial=Table[simulate[10],{500}];
```

```
normalHistogram [trial]
```

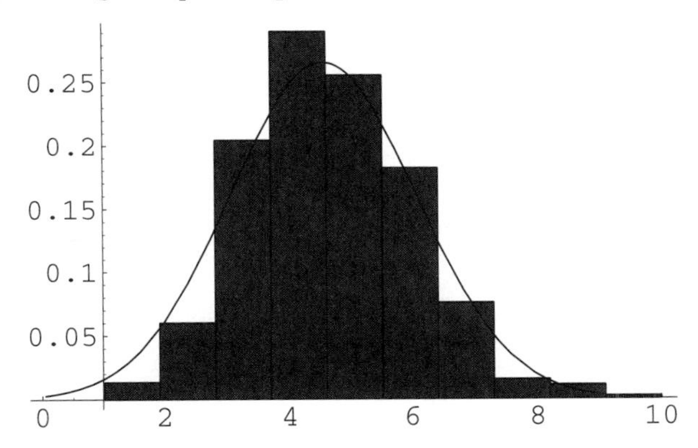

■

7.5 Randomization

Statistical studies can be classified as either observational or designed. In observational studies the experimenter has little or no control over the variables under study and only observes and records the values in the sample. In contrast, for designed experiments, the experimenter attempts to control one or more of the variables. In particular, the experimenter controls the specification of the treatments and the method of assigning the experimental units to treatments. Such specification should be based on some form of random assignment which is typically called a randomization scheme. Under some conditions the random assignment of experimental units to treatments should result in balanced groups (that is, groups of equal size). In other cases, it may be desirable to have one group with a larger number than other groups or all groups with different numbers. If the randomization is to be done over time, it may be desirable to randomize the total study in blocks or subunits that result in a fixed proportion in each group at the end of each block.

We have written a procedure called **randomPartition**. Given a list of individuals `list`, `randomPartition[list,n]` randomly divides list into `n` groups of equal size, while `randomPartition[list,{n1,n2,...,nm}]` randomly divides `list` into `m` groups of sizes `n1,n2,...,` and `nm`, respectively. We have written a second procedure called **randomSize**. Given a list of individuals `list`, `randomsize[list,n]` randomly divides the individuals into `n` groups whose sizes are randomly determined. Both procedures can be found on the disk that comes with this book. To use any of our procedures, first load the disk and click on its icon. Then locate the procedure name, click on the rightmost cell bracket, press ENTER, and return to the *Mathematica* work sheet.

EXAMPLE 1: Use the procedure **randomPartition** to create a random assignment (a) into balanced groups, (b) unbalanced groups, and (c) three blocks of balanced groups.

SOLUTION: We define `subjects` to be a list of 60 individuals using `Range`.

```
subjects=Range[60]
```

```
{1,2,3,4,5,6,7,8,9,10,11,12,13,14,15,16,17,18,19,20,21,
 22,23,24,25,26,27,28,29,30,31,32,33,34,35,36,37,38,39,
 40,41,42,43,44,45,46,47,48,49,50,51,52,53,54,55,56,57,
 58,59,60}
```

(a) The following command randomly partitions `subjects` into four groups of equal size.

```
randomPartition[subjects,4]
```

```
{{7,10,11,14,17,21,30,31,34,36,43,44,54,57,59},
 {3,6,13,18,19,25,29,32,33,40,47,49,51,55,58},
 {2,8,12,15,16,22,24,27,35,38,42,45,48,50,53},
 {1,4,5,9,20,23,26,28,37,39,41,46,52,56,60}}
```

(b) The following command randomly divides subjects into groups of sizes 6, 12, 18, and 24.

```
randomPartition[subjects,{6,12,18,24}]
```

```
{{2,3,5,9,10,14,17,21,24,29,31,35,36,38,40,42,46,47,
  48,50,52,57,58,59},{15,37,41,54,55,60},{7,8,11,12,18,
  20,22,25,32,39,43,56},{1,4,6,13,16,19,23,26,27,28,30,
  33,34,44,45,49,51,53}}
```

(c) The following commands randomly divides subjects into three blocks of balanced groups.

```
subjects1=Range[20];
subjects2=Range[21,40];
subjects3=Range[41,60];
```

```
randomPartition[subjects1,4]
randomPartition[subjects2,4]
randomPartition[subjects3,4]
```

```
{{4,5,15,18,20},{7,11,12,14,16},{2,8,13,17,19},
 {1,3,6,9,10}}
```

```
{{22,26,28,34,40},{29,30,31,35,38},{23,25,32,36,37},
 {21,24,27,33,39}}
```

```
{{44,47,48,50,51},{41,45,52,56,60},{42,43,46,53,58},
 {49,54,55,57,59}}
```

■

EXAMPLE 2: Use the procedure **randomSize** to create a random assignment.

SOLUTION: We again define subjects to be a list of 60 individuals using Range.

```
subjects=Range[60]
```

```
{1,2,3,4,5,6,7,8,9,10,11,12,13,14,15,16,17,18,19,20,21,
 22,23,24,25,26,27,28,29,30,31,32,33,34,35,36,37,38,39,
 40,41,42,43,44,45,46,47,48,49,50,51,52,53,54,55,56,57,
 58,59,60}
```

The procedure `randomSize[subjects,4]` randomly divides `subjects` into four groups.

```
randomSize[subjects,4]

{{4,6,10,13,17,30,35,50,51,52,55,56,59,60},{1,5,14,20,
  21,22,23,24,29,41,42,43,47,48,49,53,57},{3,11,12,15,
  18,26,28,31,34,36,37,39,45,46,58},{2,7,8,9,16,19,25,
  27,32,33,38,40,44,54}}
```

In this case, we see that the groups have sizes of 14, 17, 15, and 14.

```
Map[Length,%]

{14,17,15,14}
```

A second use of `randomSize[subjects,4]` gives groups of sizes 23, 20, 9, and 8.

```
randomSize[subjects,4]

{{1,2,4,5,7,9,10,19,22,23,24,26,28,36,38,39,40,42,46,
  54,56,57,59},{11,12,13,14,15,17,18,21,30,31,33,34,35,
  37,45,47,48,49,53,58},{3,16,25,27,29,50,52,55,60},
 {6,8,20,32,41,43,44,51}}
```

```
Map[Length,%]

{23,20,9,8}
```

■

7.6 Random Data Generation

Randomly generated data can be useful for testing procedures, constructing examples, and even for generating art. As an illustration of the last, consider the procedures contained in **`trees`**. We have written three procedures that can be found on the disk that comes with this book. First load the disk and click on its icon. Then locate the procedure **`trees`**, click on the rightmost cell bracket, press **ENTER**, and return to the *Mathematica* work sheet. This will load three procedures: **`winterTree`**, **`summerTree`**, and **`autumnTree`**. Each will randomly generate a tree that may be very healthy, somewhat healthy, or unhealthy. The argument of each procedure is the number of branchings.

EXAMPLE 1: Randomly select 1000 numbers from a normal distribution with mean 20 and standard deviation 2.

SOLUTION: Assuming that **ContinuousDistributions** has been loaded, we define `dist` to represent a normal distribution with mean 20 and standard deviation 2, use `RandomArray` to simulate the random selection of 1000 numbers from that distribution, and name the resulting list `data`. Several numbers from `data` are viewed using `Short`.

```
dist=NormalDistribution[20,2];

data=RandomArray[dist,1000];

Short[data,5]

{20.2321,18.0654,22.143,19.1031,16.0103,20.894,<<988>>,
 20.8372,21.2987,18.0782,21.5595,14.9567,17.7879}
```

We compute the mean and standard deviation of the list as well.

```
Mean[data]
StandardDeviation[data]

19.9884

2.07184
```

We then use `normalHistogram` to generate a histogram that represents the data.

```
normalHistogram[data, 20]
```

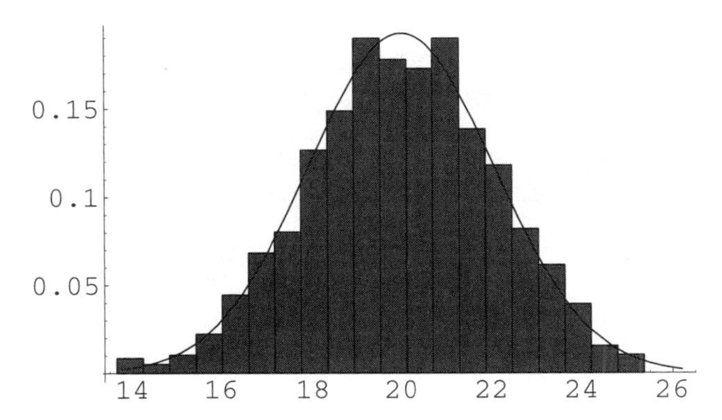

■

EXAMPLE 2: Randomly generate a winter tree, a summer tree, and an autumn tree.

SOLUTION: After loading **`trees`** we generate a tree for each season. The winter tree has no color; the summer tree has green leaves; and the autumn tree has brown leaves. These simulations use a large part of the memory. No more than six branchings are suggested.

```
winterTree[5]
```

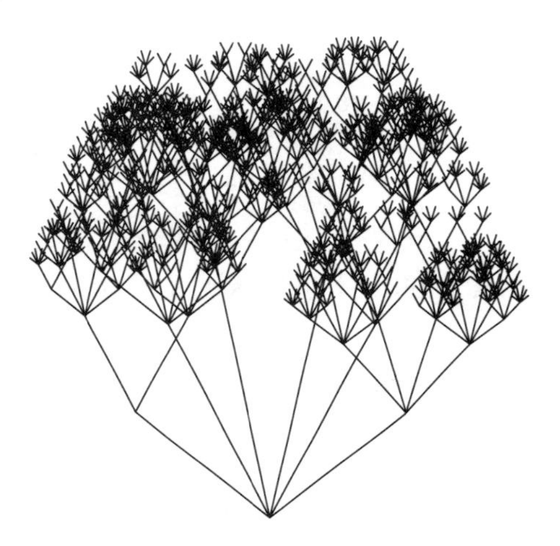

```
summerTree[6]
```

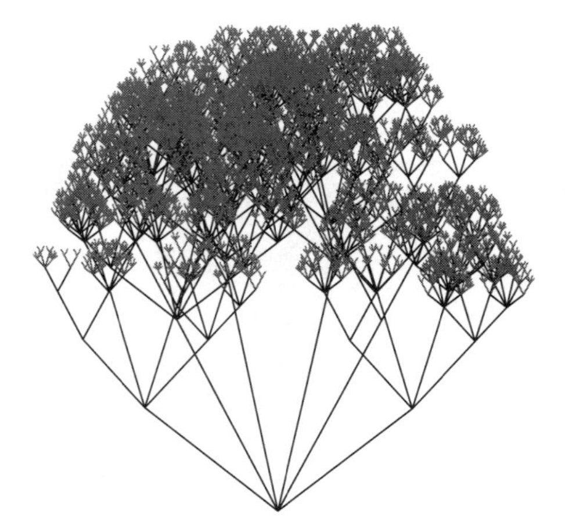

```
autumnTree[6]
```

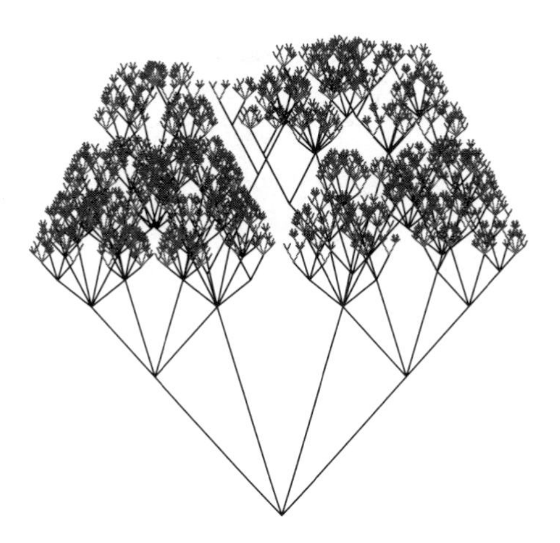

■

CHAPTER 8

One- and Two-Sample Inferential Procedures

The practice of the science of statistics is usually acknowledged to have two broad subdivisions: descriptive statistics, which is covered within the first seven chapters, and inferential statistics, which is the main topic of the last five chapters of this book. Statistical inference is usually divided into two basic types known as interval estimation and hypothesis testing. For the one- and two-sample procedures discussed in this chapter, interval estimation is in the form of confidence intervals. They are discussed in the first section of this chapter. Hypothesis testing is discussed in the second section, and the third section is devoted to a better understanding of these inferential procedures using simulation.

This chapter is concerned with inference about a single population mean, two population means, a single population variance, and two population variances. In *Mathematica*, the confidence interval procedures are contained in the package **ConfidenceIntervals**, and the hypothesis testing procedures are contained in the package **HypothesisTests**. Both packages are found in the **Statistics** folder (or directory). When either of them is loaded the **DescriptiveStatistics** package is also automatically loaded. The **Help Browser** can be used to obtain information about these packages. To display information about any of the procedures, use the ? command. Note that most introductory textbooks also cover inference about one- and two-sample proportions. Although the *Mathematica* standard package **Statistics** does not have any functions for this type of inference, we have written several procedures. These are discussed in Chapter 12.

8.1 Confidence Intervals

A confidence interval is used to estimate the value of a population parameter. Beginning with a simple random sample from the population, a point estimate of the parameter is calculated. For example, if the purpose is to estimate the population mean, the point estimate is the sample mean. Each possible sample has an associated possible point estimate, some of which are close to the parameter and some of which are farther away. The confidence interval provides a way of quantifying the accuracy of the point estimate from the sample that was taken. The point estimate is used to calculate two possible values of the parameter, a lower limit and an upper limit, which define the interval. Associated with the interval is the degree of confidence that the parameter is between the two limits; hence the name confidence interval. The level of confidence is usually signified by a percentage between 0 and 100. For example, it is common to calculate a 95% confidence interval or a 99% confidence interval. The width of the confidence interval (upper limit minus lower limit) provides a measure of the accuracy of the estimation process. A "narrow" interval is accurate in locating the parameter, whereas a "wide" interval has less accuracy.

Confidence Interval for the Population Mean: Population Variance Known

A common problem encountered in statistical inference is estimating the mean, μ, of a population. The best **estimator** of μ is the sample mean. An estimator is the rule which is applied to a set of sample data to calculate a **point estimate**. The estimator is in effect a random variable whose value is determined by the sample chosen. Recall that for a sample of n items $\{y_1, y_2, ..., y_n\}$ from a population, the sample mean is $\bar{y} = \frac{1}{n}\sum_{i=1}^{n} y_i$.

Assumptions

1. A random sample of size n is taken from the population of interest.
2. The population is normal. Note that if, based on the central limit theorem, the sampling distribution of the sample mean is approximately normal, then the method is still appropriate.
3. The population variance, σ^2, is known.

To determine whether assumptions 1 and 2 are reasonably met, there are diagnostic procedures that are discussed in Chapter 10. Also discussed are remedial procedures to transform the data set so that it better satisfies the assumptions.

Calculations

The $100(1-\alpha)\%$ confidence limits for μ are calculated using the formulas in the inequality

$$\bar{y} - z_{\alpha/2} \cdot \frac{\sigma}{\sqrt{n}} \leq \mu \leq \bar{y} + z_{\alpha/2} \cdot \frac{\sigma}{\sqrt{n}}.$$

The value $100(1-\alpha)\%$ is called the **confidence coefficient**. The value α is commonly no more than 0.1, so the resulting confidence interval will have a confidence coefficient of at least 90%. The value $z_{\alpha/2}$ is the $(1-\alpha/2)^{\text{th}}$ quantile of the standard normal distribution (the normal distribution with mean zero and standard deviation 1). That is, the area to the right of $z_{\alpha/2}$ under the standard normal probability density function (pdf) is $\alpha/2$. The confidence interval reflects the fact that the probability is $1-\alpha$ that the sample selected will result in an interval that covers (contains) μ.

EXAMPLE 1: Determine the values of $z_{\alpha/2}$ for $\alpha = 0.1, 0.05, 0.01, 0.001, 0.0001, 0.00001$.

SOLUTION: First, we load the **ContinuousDistributions** package. Note that the package **DescriptiveStatistics** is automatically loaded as well. After defining `zdist` to represent the standard normal distribution we compute several commonly used values of $z_{\alpha/2}$.

```
<<Statistics`ContinuousDistributions`
zdist=NormalDistribution[0,1];
```

Notice that in the table called `commonvalues`, we determine the list $\{x, (100-x)/100, (100-x)/200, z_{\alpha/2}\}$ for each of the values $x = 90, 95, 99, 99.9, 99.999$ where $(100-x)/100 = \alpha$ and $(100-x)/200 = \alpha/2$. We view this table using `TableForm` and `TableHeadings`.

```
commonvalues=Map[{#,(100-#)/100,
      (100-#)/200,Quantile[zdist,(100+#)/200]}&,
   {90,95,99,99.9,99.99,99.999}]//N;
TableForm[commonvalues,
 TableHeadings -> {{}, {"Confidence Level", α, α/2, z_{α/2}}}]
```

Confidence Level	α	$\alpha/2$	$z_{\alpha/2}$
90.	0.1	0.05	1.64485
95.	0.05	0.025	1.95996
99.	0.01	0.005	2.57583
99.9	0.001	0.0005	3.29053
99.99	0.0001	0.00005	3.89059
99.999	0.00001	$5.\times 10^{-6}$	4.41717

Procedure

The *Mathematica* procedures to calculate a confidence interval for μ when σ^2 is known are `MeanCI` and `NormalCI`. With `MeanCI` use either option setting `KnownVariance->` or `KnownStandardDeviation->`, where either the population variance or the population standard deviation is entered after the arrow. With `NormalCI` the sample mean and the population standard error $(\sigma/\sqrt{n})$ are entered. For both procedures the default setting for the confidence level is 95%, but any confidence level can be specified using the option `ConfidenceLevel->`.

> `MeanCI[list, options]` returns a list `{min,max}` representing a confidence interval for the population mean, using the entries in `list` as a sample drawn from the population. The options and default settings are `ConfidenceLevel->0.95`, `KnownStandardDeviation->None`, and `KnownVariance->None`.
>
> `NormalCI[mean, sd, ConfidenceLevel -> c]` returns a list `{min,max}` representing a confidence interval at confidence level `c` for the population mean, based on the sample mean and its standard deviation. The only option and its default setting for `NormalCI` is `ConfidenceLevel->0.95`. This function is used by `MeanCI` when the population variance is specified.

Confidence Interval for the Population Mean: Population Variance Unknown

Assumptions

1. A random sample of size n is taken from the population of interest.
2. The population is normal. Note that if, based on the central limit theorem, the sampling distribution of the sample mean is approximately normal, then the method is still appropriate.
3. The population variance, σ^2, is unknown.

To determine whether assumptions 1 and 2 are reasonably met, there are several diagnostic procedures that are discussed in Chapter 10. Also discussed are remedial procedures to transform the data set so that it better satisfies assumption 2. In case the data set cannot meet assumption 2, refer to the discussion in Chapter 12 of the sign test and the Wilcoxon signed-rank test.

Calculations

The $100(1-\alpha)\%$ confidence limits for μ are calculated using the formulas in the inequality

$$\bar{y} - t_{\alpha/2} \cdot \frac{s}{\sqrt{n}} \leq \mu \leq \bar{y} + t_{\alpha/2} \cdot \frac{s}{\sqrt{n}}.$$

The notation s represents the sample standard deviation, where

$$s = \sqrt{\frac{\sum_{i=1}^{n} (y_i - \bar{y})^2}{n-1}},$$

and the value $t_{\alpha/2}$ is the $(1-\alpha/2)^{\text{th}}$ quantile of the Student's t-distribution with degrees of freedom $\nu = n-1$. That is, the area to the right of $t_{\alpha/2}$ under the pdf of Student's t-distribution with $n-1$ degrees of freedom is $\alpha/2$. The confidence interval reflects the fact that the probability is $1-\alpha$ that the sample selected will result in an interval that covers μ.

EXAMPLE 2: Determine the value of $t_{\alpha/2}$ for $\alpha = 0.1, 0.05, 0.025, 0.01, 0.005, 0.001, 0.0005$ and degrees of freedom $n = 1, 2, \ldots, 15$.

SOLUTION: Assuming that the **ContinuousDistributions** package is loaded (from Example 1), we begin by defining the function `tvals[n]` which, when given a value of `n`, computes the value of t for $\alpha = 0.1, 0.05, 0.025, 0.01, 0.005, 0.001, 0.0005$. The command `Flatten` is used so that the result of `tvals[n]` is a list with no interior parentheses. As in Example 1, we view the table of t-values with `TableForm` and `TableHeadings`.

```
tvals[n_]:={n, Map[Quantile[StudentTDistribution[n],
   1-#]&,{.1,.05,.025,.01,.005,.001,.0005}]}//Flatten;

commonvalues=Table[tvals[n],{n,1,15}];

TableForm[commonvalues,
```

```
TableHeadings -> {{}, {"Degrees of Freedom",
  t.1, t.05, t.025, t.01, t.005, t.001, t.0005}}]
```

Degrees of Freedom	$t_{0.1}$	$t_{0.05}$	$t_{0.025}$	$t_{0.01}$	$t_{0.005}$	$t_{0.001}$	$t_{0.0005}$
1	3.07768	6.31375	12.7062	31.8205	63.6567	318.309	636.619
2	1.88562	2.91999	4.30265	6.96456	9.92484	22.3271	31.5991
3	1.63774	2.35336	3.18245	4.5407	5.84091	10.2145	12.924
4	1.53321	2.13185	2.77645	3.74695	4.60409	7.17318	8.6103
5	1.47588	2.01505	2.57058	3.36493	4.03214	5.89343	6.86883
6	1.43976	1.94318	2.44691	3.14267	3.70743	5.20763	5.95882
7	1.41492	1.89458	2.36462	2.99795	3.49948	4.78529	5.40788
8	1.39682	1.85955	2.306	2.89646	3.35539	4.50079	5.04131
9	1.38303	1.83311	2.26216	2.82144	3.24984	4.29681	4.78091
10	1.37218	1.81246	2.22814	2.76377	3.16927	4.1437	4.58689
11	1.36343	1.79588	2.20099	2.71808	3.10581	4.0247	4.43698
12	1.35622	1.78229	2.17881	2.681	3.05454	3.92963	4.31779
13	1.35017	1.77093	2.16037	2.65031	3.01228	3.85198	4.22083
14	1.34503	1.76131	2.14479	2.62449	2.97684	3.78739	4.14045
15	1.34061	1.75305	2.13145	2.60248	2.94671	3.73283	4.07277

■

Procedure

There are two *Mathematica* procedures to calculate a confidence interval for μ when σ^2 is unknown. One is `MeanCI`, using the default options of `KnownVariance->None` and `KnownStandardDeviation->None`. The other is `StudentTCI`, where the sample mean, sample standard error (standard deviation divided by the square root of sample size), and the degrees of freedom (sample size minus 1) must be specified. Note that the default setting for the confidence level is 95% for both procedures, but any confidence level can be specified using the option `ConfidenceLevel->`.

> `MeanCI[list, options]` returns a list `{min,max}` representing a confidence interval for the population mean, using the entries in `list` as a sample drawn from the population. The options and default settings are `ConfidenceLevel->0.95`, `KnownStandardDeviation->None`, and `KnownVariance->None`.
>
> `StudentTCI[mean,se,dof,ConfidenceLevel->c]` returns a list `{min,max}` representing a confidence interval at confidence level `c` for the population mean, based on a sample mean, `mean`, its standard error `se` given by `StandardErrorOfSampleMean`, and `dof` degrees of freedom. This function is used by `MeanCI` when the population variance is estimated from the sample. The only option and its default setting for `StudentTCI` is `ConfidenceLevel->0.95`.

EXAMPLE 3: (Population Standard Deviation Known): Table 8.1 lists the final average for a sample (collected over several terms) of 44 students taking multivariable calculus from a particular professor at Georgia Southern University. Assuming that the standard deviation for this professor is 10.5, determine 90%, 95%, and 99% confidence intervals for the mean average of all students taking multivariable calculus from this professor.

85.0	94.5	76.7	79.2	83.0	80.2	68.7	89.1	74.1	87.8	44.9
77.6	85.1	75.7	81.5	66.2	83.4	79.8	9.0	91.8	96.3	73.5
82.2	76.1	78.5	69.1	75.4	71.7	78.2	77.7	88.7	79.9	86.1
63.8	78.7	82.6	98.6	81.3	63.4	76.6	84.2	89.7	87.7	54.6

Table 8.1

SOLUTION: After loading the **ConfidenceIntervals** package, we enter the data, naming the data set `grades`. To compute the 90% confidence interval directly, we find the length of `grades`, `n`, and the mean `m`. We also find that $z_{\alpha/2} = 1.64485$ with `Quantile` and name this result `z`. According to our calculations, the 90% confidence interval is $76.7804 \leq \mu \leq 81.9878$.

```
grades={85,94.5,76.7,79.2,83,80.2,68.7,89.1,74.1,87.8,
 44.9,77.6,85.1,75.7,81.5,66.2,83.4,79.8,94,91.8,96.3,
 73.5,82.2,76.1,78.5,69.1,75.4,71.7,78.2,77.7,88.7,79.9,
 86.1,63.8,78.7,82.6,98.6,81.3,63.4,76.6,84.2,89.7,87.7,
 54.6};
```

```
n=Length[grades]
m=Mean[grades]
```

```
44
79.3841
```

```
z=Quantile[NormalDistribution[0,1],.95]
```

```
1.64485
```

```
m-z(10.5)/Sqrt[n]
m+z(10.5)/Sqrt[n]
```

```
76.7804
81.9878
```

When we use the package command `NormalTCI` with the `ConfidenceLevel->0.90` option, we find the same result for the 90% confidence interval. When we use the package command `MeanCI` with `KnownStandardDeviation->10.5` and `ConfidenceLevel->0.90` options, we again find the same result for the 90% confidence interval. Notice that the width of the intervals increases as the confidence coefficient increases.

```
NormalCI[79.3841,(10.5)/Sqrt[44],ConfidenceLevel->.90]
```

```
{76.7804,81.9878}
```

```
MeanCI[grades,KnownStandardDeviation->10.5,
 ConfidenceLevel->.90]
```

```
{76.7804,81.9878}
```

```
MeanCI[grades,KnownStandardDeviation->10.5]
```

```
{76.2816,82.4866}
```

```
MeanCI[grades,KnownStandardDeviation->10.5,
  ConfidenceLevel->.99]
```

```
{75.3067,83.4615}
```

■

EXAMPLE 4: (Population Standard Deviation Unknown): For the data in Table 8.1, assume that the population standard deviation is unknown and determine 90%, 95%, and 99% confidence intervals for the mean average of all students taking multivariable calculus from the professor.

SOLUTION: Assuming that we have the list `grades` from Example 3, we compute the 90% confidence interval directly using the length of `grades`, `n`, the mean `m`, and the standard deviation, `sig`. We also find $t_{\alpha/2} = 1.68107$ with `Quantile` and name this result `t`. According to our calculations, the 90% confidence interval is $76.7206 \leq \mu \leq 82.0476$.

```
n=Length[grades]
m=Mean[grades]
sig=StandardDeviation[grades]
```

```
44
79.3841
10.5096
```

```
t=Quantile[StudentTDistribution[43],.95]
1.68107

m-t sig/Sqrt[n]
m+t sig/Sqrt[n]
76.7206
82.0476
```

When we use the package command `StudentTCI` with the `ConfidenceLevel->0.90` option, we find the same result for the 90% confidence interval. We see the same result when using the command `MeanCI` as well. Notice that the intervals based on the Student's t-distribution are wider than the corresponding intervals based on the normal distribution. With less information available (not knowing the population standard deviation), there is more uncertainty, which is reflected by wider intervals.

```
StandardErrorOfSampleMean[grades]
1.58439

StudentTCI[79.3841,1.58439,43,ConfidenceLevel->0.90]
{76.7206,82.0476}

MeanCI[grades,ConfidenceLevel->0.90]
{76.7206,82.0476}

MeanCI[grades]
{76.1889,82.5793}

MeanCI[grades,ConfidenceLevel->0.99]
{75.114,83.6542}
```

■

Confidence Interval for the Variance of a Normal Population

Assumptions

1. A random sample of size n is taken from the population of interest.
2. The population is normal.

To determine whether these assumptions are reasonably met, there are diagnostic procedures that are discussed in Chapter 10. Also discussed are remedial procedures to transform the data set so that it better satisfies the assumptions.

Calculations

The 100(1 - α)% confidence limits for σ^2, the population variance, are calculated using the formulas in the inequality

$$\frac{(n-1)s^2}{\chi^2_{\alpha/2,\,n-1}} \le \sigma^2 \le \frac{(n-1)s^2}{\chi^2_{1-\alpha/2,\,n-1}}.$$

Note that s^2 represents the sample variance and $\chi^2_{\alpha,\,n-1}$ is the $(1-\alpha)^{\text{th}}$ quantile of the chi-square distribution with degrees of freedom $v = n - 1$. That is, the area to the right of $\chi^2_{\alpha,\,n-1}$ under the pdf of chi-square distribution with $n - 1$ degrees of freedom is α. The confidence interval reflects the fact that the probability is $1 - \alpha$ that the sample selected will result in an interval that covers σ^2.

EXAMPLE 5: Determine the value of $\chi^2_{\alpha,\,n}$ for α = 0.995, 0.99, 0.975, 0.95, 0.9, 0.005, 0.01, 0.025, 0.05, 0.1 and n = 1, 2, ..., 15.

SOLUTION: Assuming that the **ContinuousDistributions** package is loaded, we define the function `cvals[n]` so that when given a value of `n` it computes $\chi^2_{\alpha,\,n}$! for α = 0.995, 0.99, 0.975, 0.95, 0.9, 0.005, 0.01, 0.025, 0.05, 0.1. The result is a list with no internal parentheses (obtained with `Flatten`). We evaluate `cvals[n]` at the desired values of `n` to form the table `commonvalues`. As in previous examples, we view `commonvalues` with `TableForm` and `TableHeadings`.

```
cvals[n_]:={n,Map[Quantile[ChiSquareDistribution[n],#]&,{.005,.01,.
025,.05,.1}]}//Flatten;

commonvalues=Table[cvals[n],{n,1,15}];

(TableForm[commonvalues,TableHeadings -> {{}, {"Degrees
of Freedom", χ.995^2, χ.99^2, χ.975^2, χ.95^2, χ.9^2}}]
```

Degrees of Freedom	$\chi_{0.995}^2$	$\chi_{0.99}^2$	$\chi_{0.975}^2$	$\chi_{0.95}^2$	$\chi_{0.9}^2$
1	0.0000392704	0.000157088	0.000982069	0.00393214	0.0157908
2	0.0100251	0.0201007	0.0506356	0.102587	0.210721
3	0.0717218	0.114832	0.215795	0.351846	0.584374
4	0.206989	0.297109	0.484419	0.710723	1.06362
5	0.411742	0.554298	0.831212	1.14548	1.61031
6	0.675727	0.87209	1.23734	1.63538	2.20413
7	0.989256	1.23904	1.68987	2.16735	2.83311
8	1.34441	1.6465	2.17973	2.73264	3.48954
9	1.73493	2.0879	2.70039	3.32511	4.16816
10	2.15586	2.55821	3.24697	3.9403	4.86518

11	2.60322	3.05348	3.81575	4.57481	5.57778
12	3.07382	3.57057	4.40379	5.22603	6.3038
13	3.56503	4.10692	5.00875	5.89186	7.0415
14	4.07467	4.66043	5.62873	6.57063	7.78953
15	4.60092	5.22935	6.26214	7.26094	8.54676

Next, in `cvals2[n]`, we define a similar function to calculate the value of $\chi^2_{\alpha, n}$ for $\alpha = 0.995, 0.99, 0.975, 0.95, 0.9, 0.005, 0.01, 0.025, 0.05, 0.1$. We then compute the desired values of $\chi^2_{\alpha, n}$ and view the results in `TableForm`.

```
cvals2[n_]:={n,Map[Quantile[ChiSquareDistribution[n],1-
#]&,{.005,.01,.025,.05,.1}]}//Flatten;

commonvalues2=Table[cvals2[n],{n,1,15}];

TableForm[commonvalues2,TableHeadings -> {{},
{"Degrees of Freedom", χ.005^2, χ.01^2, χ.025^2, χ.05^2, χ.1^2}}]
```

Degrees of Freedom	$\chi_{0.005}^2$	$\chi_{0.01}^2$	$\chi_{0.025}^2$	$\chi_{0.05}^2$	$\chi_{0.1}^2$
1	7.87944	6.6349	5.02389	3.84146	2.70554
2	10.5966	9.21034	7.37776	5.99146	4.60517
3	12.8382	11.3449	9.3484	7.81473	6.25139
4	14.8603	13.2767	11.1433	9.48773	7.77944
5	16.7496	15.0863	12.8325	11.0705	9.23636
6	18.5476	16.8119	14.4494	12.5916	10.6446
7	20.2777	18.4753	16.0128	14.0671	12.017
8	21.955	20.0902	17.5345	15.5073	13.3616
9	23.5894	21.666	19.0228	16.919	14.6837
10	25.1882	23.2093	20.4832	18.307	15.9872
11	26.7568	24.725	21.92	19.6751	17.275
12	28.2995	26.217	23.3367	21.0261	18.5493
13	29.8194	27.6882	24.7356	22.362	19.8119
14	31.3193	29.1412	26.1189	23.6848	21.0641
15	32.8013	30.5779	27.4884	24.9958	22.3071

Procedure

`VarianceCI[list]` returns a list `{min,max}` representing a confidence interval for the population variance, using the entries in `list` as a sample drawn from the population. The option and default setting, `ConfidenceLevel->.95`, can be used with `VarianceCI`. This command uses `ChiSquareCI[var,dof,ConfidenceLevel->c]`, which returns a list {min,max} representing a confidence interval at confidence level c for the population variance, based on a sample with dof degrees of freedom and unbiased variance estimate var. The only option and its default setting is ConfidenceLevel->.95.

EXAMPLE 6: Determine 90%, 95%, and 99% confidence intervals for the population variance using the grades from Table 8.1.

SOLUTION: Assuming that we have loaded the **ConfidenceIntervals** package and entered the data in grades, we find that the 95% confidence interval is $75.3998 \leq \sigma^2 \leq 177.315$, Notice that the intervals increase as the confidence coefficient increases.

```
VarianceCI[grades,ConfidenceLevel->.90]
 {80.0873,163.974}

VarianceCI[grades]
 {75.3998,177.315}

VarianceCI[grades,ConfidenceLevel->.99]
 {67.2576,207.768}
```

■

Confidence Interval for the Difference between Two Population Means: Population Variances Known

Assumptions

1. Independent, random samples are taken from population 1 and population 2. The sample sizes are n^1 and n^2, respectively.
2. Both populations are normal. Note that if, based on the central limit theorem, the sampling distribution of each sample mean is approximately normal, then the method is still appropriate.
3. The population variances, σ_1^2 and σ_2^2 , are known.

To determine whether assumptions 1 and 2 are reasonably met, diagnostic procedures are discussed in Chapter 10. Also discussed are remedial procedures to transform the data set so that it better satisfies the assumptions.

Calculations

The $100(1-\alpha)\%$ confidence limits for $\mu_1 - \mu_2$ are calculated using the formulas in the inequality

$$\bar{y}_1 - \bar{y}_2 - z_{\alpha/2}\sqrt{\frac{\sigma_1^2}{n_1} + \frac{\sigma_2^2}{n_2}} \le \mu_1 - \mu_2 \le \bar{y}_1 - \bar{y}_2 + z_{\alpha/2}\sqrt{\frac{\sigma_1^2}{n_1} + \frac{\sigma_2^2}{n_2}}.$$

The value $z_{\alpha/2}$ is the $(1-\alpha/2)^{\text{th}}$ quantile of the standard normal distribution (the normal distribution with mean zero and standard deviation 1). That is, the area to the right of $z_{\alpha/2}$ under the standard normal probability density function (pdf) is $\alpha/2$. The confidence interval reflects the fact that the probability is $1-\alpha$ that the samples selected will result in an interval that covers $\mu_1 - \mu_2$.

Procedure

The *Mathematica* procedure to calculate a confidence interval for $\mu_1 - \mu_2$ when σ_1^2 and σ_2^2 are known is `MeanDifferenceCI` using either option `KnownVariance->{`σ_1^2, σ_2^2`}` or option `KnownStandardDeviation->{`σ_1, σ_2`}`. Note that the default setting for the confidence level is 95%, but any confidence level can be specified using the option `ConfidenceLevel->`.

> `MeanDifferenceCI[list1,list2,options]` returns a list `{min,max}` representing a confidence interval for the difference `Mean[list1]-Mean[list2]`, using the entries in `list1` as a sample from the first population and the entries in `list2` as a sample from the second population. Options and default settings for `MeanDifferenceCI` are `ConfidenceInterval->.95`, `KnownStandardDeviation->None`, `KnownVariance->None`, and `EqualVariances->False`.

Confidence Interval for the Difference between Two Population Means: Population Variances Unknown but Assumed Equal

Assumptions

1. Independent, random samples are taken from population 1 and population 2. The sample sizes are n_1 and n_2, respectively.
2. Both populations are normal. Note that if, based on the central limit theorem, the sampling distribution of each sample mean is approximately normal, then the method is still appropriate.
3. The population variances, σ_1^2 and σ_2^2 , are unknown but assumed equal

To determine whether these assumptions are reasonably met, diagnostic procedures are discussed in Chapter 10. Also discussed are remedial procedures to transform the data set so that it better satisfies the assumptions. In case the data cannot meet the assumptions, refer to the discussion of the Mann–Whitney test in Chapter 12.

Calculations

The $100(1-\alpha)\%$ confidence limits for $\mu_1-\mu_2$ are calculated using the formulas in the inequality

$$\bar{y}_1-\bar{y}_2-t_{\alpha/2}\cdot s_p\sqrt{\frac{1}{n_1}+\frac{1}{n_2}}\leq\mu_1-\mu_2\leq\bar{y}_1-\bar{y}_2+t_{\alpha/2}\cdot s_p\sqrt{\frac{1}{n_1}+\frac{1}{n_2}}.$$

Note that

$$s_p=\sqrt{\frac{(n_1-1)s_1^2+(n_2-1)s_2^2}{n_1+n_2-2}},$$

where the subscript p refers to the fact that the sample variances have been pooled to calculate an estimate of the common variance of the two populations. The value $t_{\alpha/2}$ is the $(1-\alpha/2)^{\text{th}}$ quantile of the Student's t-distribution with degrees of freedom $\nu = n_1 + n_2 - 2$. That is, the area to the right of $t_{\alpha/2}$ under the pdf of Student's t-distribution with $n_1 + n_2 - 2$ degrees of freedom is $\alpha/2$. The confidence interval reflects the fact that the probability is $1-\alpha$ that the samples selected will result in an interval that covers $\mu_1-\mu_2$.

Procedure

The *Mathematica* procedure to calculate a confidence interval for $\mu_1-\mu_2$ when σ_1^2 and σ_2^2 are unknown but assumed to be equal, is `MeanDifferenceCI` using the option `EqualVariances->True`. Note that the default setting for the confidence level is 95%, but any confidence level can be specified using the option `ConfidenceLevel->`.

`MeanDifferenceCI[list1,list2,options]` returns a list `{min,max}` representing a confidence interval for the difference `Mean[list1]-Mean[list2]`, using the entries in `list1` as a sample from the first population and the entries in `list2` as a sample from the second population. Options and default settings for `MeanDifferenceCI` are `ConfidenceInterval->.95`, `KnownStandardDeviation->None`, `KnownVariance->None`, and `EqualVariances->False`.

Confidence Interval for the Difference between Two Population Means: Population Variances Unknown and Assumed Unequal

Assumptions

1. Independent, random samples are taken from population 1 and population 2. The sample sizes are n_1 and n_2, respectively.
2. Both populations are normal. Note that if, based on the central limit theorem, the sampling distribution of each sample mean is approximately normal, then the method is still appropriate.
3. The population variances, σ_1^2 and σ_2^2, are unknown and assumed to be different.

To determine whether these assumptions are reasonably met, diagnostic procedures are discussed in Chapter 10. Also discussed are remedial procedures to transform the data set so that it better satisfies the assumptions. In case the data cannot meet the assumptions, refer to the discussion of the Mann–Whitney test in Chapter 12.

Calculations

The $100(1-\alpha)\%$ confidence limits for $\mu_1 - \mu_2$ are calculated using the formulas in the inequality

$$\bar{y}_1 - \bar{y}_2 - t_{\alpha/2}\sqrt{\frac{s_1^2}{n_1} + \frac{s_2^2}{n_2}} \le \mu - \bar{\mu}_2 \le \bar{y}_1 - \bar{y}_2 + t_{\alpha/2}\sqrt{\frac{s_1^2}{n_1} + \frac{s_2^2}{n_2}}.$$

The value $t_{\alpha/2}$ is the $(1-\alpha/2)^{\text{th}}$ quantile of Student's t-distribution with degrees of freedom

$$v = \frac{\left(\frac{s_1^2}{n_1} + \frac{s_2^2}{n_2}\right)^2}{\frac{1}{n_1-1}\left(\frac{s_1^2}{n_1}\right)^2 + \frac{1}{n_2-1}\left(\frac{s_2^2}{n_2}\right)^2}.$$

That is, the area to the right of $t_{\alpha/2}$ under the pdf of Student's t-distribution with ν degrees of freedom is $\alpha/2$. Note that the equation for the degrees of freedom is sometimes attributed to B. L. Welch, as is done in the *Mathematica* documentation. The confidence interval reflects the fact that the probability is approximately $1 - \alpha$ that the samples selected will result in an interval that covers $\mu_1 - \mu_2$.

Procedure

The *Mathematica* procedure to calculate a confidence interval for $\mu_1 - \mu_2$ when σ_1^2 and σ_2^2 are unknown and assumed to be different, is `MeanDifferenceCI` using the default options. Note that the default setting for the confidence level is 95%, but any confidence level can be specified using the option `ConfidenceLevel->`.

> `MeanDifferenceCI[list1,list2,options]` returns a list `{min,max}` representing a confidence interval for the difference `Mean[list1]-Mean[list2]`, using the entries in `list1` as a sample from the first population and the entries in `list2` as a sample from the second population. Options and default settings for `MeanDifferenceCI` are `ConfidenceInterval->.95`, `KnownStandardDeviation->None`, `KnownVariance->None`, and `EqualVariances->False`.

EXAMPLE 7: In Table 8.2, we list the average combined SAT score at eight public institutions in the southeastern part of the United States, while in Table 8.3, we list similar scores at seven public institutions in the western part of the United States. Determine a 95% confidence interval on the difference, $\mu_1 - \mu_2$, between the mean combined SAT scores for the southeastern institutions and those for the western institutions by assuming (a) equal standard deviations; (b) unequal standard deviations. (Note: Data are from the *1995 U.S. News and World Report's Guide to America's Best Colleges*.)

School	Average combined SAT score
Troy State University (AL)	980
Florida Atlantic University (FL)	990
Southern College of Technology (GA)	940
Louisiana State Univ. at Baton Rouge (LA)	997
University of Southern Mississippi (MS)	980
University of North Carolina at Asheville (NC)	1054

School	Average combined SAT score
College of Charleston (SC)	1019
Middle Tennessee State University (TN)	942

Table 8.2 Schools from the southeastern United States.

School	Average combined SAT score
Northern Arizona University (AZ)	939
San Jose State University (CA)	838
Colorado State University (CO)	1024
University of Nevada – Las Vegas (NV)	903
Oregon State University (OR)	965
University of Utah (UT)	1027
Western Washington University (WA)	1000

Table 8.3 Schools from the western United States.

SOLUTION: After loading the ConfidenceIntervals package, we enter the scores of the southeastern schools in `se` and those of the western schools in `w`. (a) We use the option `EqualVariances->True` to find that the 95% confidence interval assuming equal standard deviations is $-29.7877 \le \mu_1 - \mu_2 \le 92.1448$. (b) In this case, we find the confidence interval to be $-35.1969 \le \mu_1 - \mu_2 \le 97.554$.

```
<<Statistics`ConfidenceIntervals`
se={980,990,940,997,980,1054,1019,942};
w={939,838,1024,903,965,1027,1000};

MeanDifferenceCI[se,w,EqualVariances->True]
{-29.7877,92.1448}

MeanDifferenceCI[se,w]
{-35.1969,97.554}
```

■

Confidence Interval for the Ratio of Variances of Two Normal Populations

Assumptions

1. Independent, random samples are taken from population 1 and population 2. The sample sizes are n_1 and n_2, respectively.
2. Both populations are normal.

To determine whether these assumptions are reasonably met, diagnostic procedures are discussed in Chapter 10. Also discussed are remedial procedures to transform the data set so that it better satisfies the assumptions.

Calculations

The $100(1 - \alpha)\%$ confidence limits for the ratio $\frac{\sigma_1^2}{\sigma_2^2}$ of the population variances are calculated using the formulas in the inequality

$$\frac{s_1^2}{s_1^2} F_{1-\alpha/2, n_2-1, n_1-1} \leq \frac{\sigma_1^2}{\sigma_2^2} \leq \frac{s_1^2}{s_2^2} F_{\alpha/2, n_2-1, n_1-1}.$$

The value F_{α, v_2, v_1} is the $(1 - \alpha)^{\text{th}}$ quantile of the F (ratio) distribution with numerator degrees of freedom $v_2 = n_2 - 1$ and denominator degrees of freedom $v_1 = n_1 - 1$. That is, the area to the right of F_{α, v_2, v_1} under the pdf of an F-distribution with $v_2 = n_2 - 1$ and $v_1 = n_1 - 1$ degrees of freedom is α. The confidence interval reflects the fact that the probability is $1 - \alpha$ that the samples selected will result in an interval that covers $\frac{\sigma_1^2}{\sigma_2^2}$.

Procedure

`VarianceRatioCI[list1, list2, options]` returns a list `{min,max}` representing a confidence interval for the ratio of population variances, `Variance[list1]/Variance[list2]`, using the entries in `list1` as a sample from the first population and the entries in `list2` as a sample from the second population. The option for this command and the default setting is `ConfidenceLevel->.95`. This command uses `FRatioCI[ratio, numdof, dendof, ConfidenceLevel -> c]`, which returns a list `{min,max}` representing a confidence interval for the ratio of population variances at confidence level `c`, where `ratio` is a ratio of sample variances, `numdof` is the number of numerator degrees of freedom, and `dendof` is the number of denominator degrees of freedom. The only option for `FRatioCI` and the default setting is `ConfidenceLevel->.95`.

EXAMPLE 8: Use the data given in Table 8.2 and Table 8.3 to determine a 90% confidence interval for the ratio of the population variance for the Southeast over the population variance for the West.

SOLUTION: Assuming that we have loaded the **ConfidenceIntervals** package and entered the lists of scores in `se` and `w`, we find that this confidence interval is $0.0706333 \le \frac{\sigma_1^2}{\sigma_2^2} \le 1.1487$.

```
VarianceRatioCI[se,w,ConfidenceLevel->.90]
{0.0706333,1.1487}
```

■

8.2 Hypothesis Tests

A hypothesis is a statement about one or more populations. For example, as part of the scientific method, an experimental investigator formulates a research hypothesis. This is often based on a hunch or a suspicion resulting from extended observation. The experiment is then undertaken to confirm this hypothesis. In statistical inference, a hypothesis test begins with the designation of two hypotheses: the null hypothesis, which is usually designated by H_0, and the alternative hypothesis, which is often designated by H_a. Another common notation for the alternative hypothesis is H_1. The **null hypothesis** is the hypothesis to be tested. It is a statement of equality, no effect, or the status quo. For example, one may test the null hypothesis that two population means are equal, designated by the notation $H_0 : \mu_1 = \mu_2$. In hypothesis testing the null hypothesis is assumed to be true until sufficient sample evidence is amassed to reject it. The logic is equivalent to a criminal trial where the defendant is assumed innocent (the null hypothesis) until sufficient evidence is presented to reject this assumption. The test procedure, which is based on sample data, results in one of two statistical decisions: a decision to reject the null hypothesis as false, or a decision to fail to reject the null hypothesis because the sample data do not provide sufficient evidence to justify rejection. When the null hypothesis is rejected, the alternative hypothesis is accepted as being true. This is appropriate because the null hypothesis and the alternative hypothesis are developed to be mutually exclusive and exhaustive. That is, one or the other is true, but both cannot be true. Often the **alternative hypothesis** corresponds to the research hypothesis from the scientific method. Thus, the alternative hypothesis is often a statement of what we expect to be able to conclude.

The statistical hypothesis test is designed to make a decision regarding the truth of the *alternative* hypothesis: either that it is acceptable or that the data do not provide sufficient evidence to justify it. In order to make a decision to accept the *null* hypothesis, further consideration must be given and more calculations must be made. *Do not assume* that if you fail to reject the null hypothesis, then you can accept it. For example, a "not guilty" verdict in a criminal trial does not mean that the defendant is innocent.

A hypothesis test may be nondirectional and designated as two-sided or two-tailed, or it may be directional and designated as one-sided or one-tailed. For example, if we consider the means μ_1 and μ_2 of population 1 and population 2, respectively, then a two-side hypothesis test would have the form

$$H_0 : \mu_1 = \mu_2 \quad H_a : \mu_1 \neq \mu_2.$$

An equivalently representation would be

$$H_0 : \mu_1 - \mu_2 = 0 \quad H_a : \mu_1 - \mu_2 \neq 0.$$

The null hypothesis states that the means are equal and the alternative hypothesis states that they are different. In effect the investigator is asking whether, on the basis of the sample data, it can be concluded that the population means are different.

If, prior to data collection, the investigator can rule out one side, then a one-sided test is appropriate. A **left-sided hypothesis test** would be represented by

$$H_0 : \mu_1 = \mu_2 \quad H_a : \mu_1 < \mu_2$$

or by

$$H_0 : \mu_1 - \mu_2 = 0 \quad H_a : \mu_1 - \mu_2 < 0.$$

In this case, the investigator has good reason, based on considerations other than the sample data, to believe that, if the means are different, then μ_1 will be less than μ_2. Note that is has been traditional to include the side which was ruled out in the null hypothesis; for example, $H_0 : \mu_1 \geq \mu_2$. Regardless of the form, the test is performed assuming $\mu_1 = \mu_2$. A **right-sided hypothesis test** would be represented by

$$H_0 : \mu_1 = \mu_2 \quad H_a : \mu_1 > \mu_2$$

or by

$$H_0 : \mu_1 - \mu_2 = 0 \quad H_a : \mu_1 - \mu_2 > 0$$

To test a null hypothesis, an appropriate test statistic is selected and its distribution, assuming the null hypothesis is true, is determined. Because the test statistic is calculated from the sample data, it is a variable whose value depends on the data collected. Thus, there is a distribution of possible values. After the value of the test statistic is determined using the sample data, the analyst must determine whether it is surprisingly extreme given the assumption that the null hypothesis is true. If so, the null hypothesis is rejected. This determination has traditionally been based on specifying a small probability called the **level of significance** and denoted by α. The value specified for α is usually 0.05, often smaller (0.01, say), but rarely as large as 0.10. If the probability of obtaining a value of the test statistic which is equal to or more extreme than the calculated value, assuming the null hypothesis is true, is equal to or less than α, the null hypothesis is rejected. A critical value (or critical values) has been used to make the determination. A **critical value** of the test statistic is the value that is so extreme that the probability of getting it, or a more extreme value, when the null hypothesis is true is equal to α. In a one-sided test there is one critical value. It will typically be small in the case of a left-sided test and large in the case of a right-sided test. In a two-sided test there may be two critical values. The lower one is chosen so that the probability of getting it, or an even lower value, when the null hypothesis is true is equal to $\alpha/2$, and the upper one is chosen so that the probability of getting it, or an even higher value, when the null hypothesis is true is equal to $\alpha/2$. The interval(s) including the critical value(s) and values more extreme in the appropriate direction(s) is called the **rejection region** or the **critical region**.

In any form of inference there is uncertainty that the sample data are sufficiently representative of the population(s) to give a correct result. In hypothesis testing it is possible that, based on the data, a true null hypothesis is rejected. This mistake is called a **Type I error**. A **Type II error** occurs when a false null hypothesis is accepted. The probability of making a Type I error is denoted by α, and is another way to define the level of significance. Note that by specifying a value of α and rejecting the null hypothesis based on this value, the analyst controls the upper bound on the probability of being in error when the null hypothesis is rejected. The probability of making a Type II error is denoted by β. Before the null hypothesis can be accepted, it must be verified that this probability is small. Note that although there is no equation which relates the two, for a fixed sample size, decreasing the value of α leads to a corresponding increase in the value of β and increasing the value of α leads to a corresponding decrease in the value of β. Both probabilities are reduced when the sample size is increased. (The more information you have, the less likely you are to make an error.) The **power** of a hypothesis test is the probability of rejecting the null hypothesis when it is false. It may be calculated as $1 - \beta$. High power is always a desirable characteristic of a hypothesis test. Some tests are more powerful in a given situation than others.

Another way to decide whether the sample data contradict the null hypothesis is to calculate the ***p*-value**, also known as the **probability value**, the **observed level of significance**, and by other names. It is the probability of observing, when the null hypothesis is true, a value of the test statistic at least as extreme as the value actually observed. It is the smallest value that could be specified for α (the level of significance) and still result in the rejection of the null hypothesis. It has also been described as a measure of the degree of surprise which the observed value of the test statistic would cause a believer of the null hypothesis (where the smaller the p-value, the greater the surprise). Thus, a large p-value means that a test statistic as extreme (or more extreme) than the test statistic would not be a surprise if the null hypothesis were true. A small p-value means it is not likely that we would observe a test statistic as extreme or more extreme than the test statistic if the null hypothesis were true. When the hypothesis test is done using modern statistical software, the p-value is calculated and displayed. This is true for the functions in the *Mathematica*

Statistics folder (or directory), as well as for the procedures we have written that are available on the disk which comes with this book.

In the following tables we give some guidelines for interpreting p-values in hypothesis tests. Keep in mind when looking at these tables that evidence that supports the null hypothesis is not as convincing of the truth of the null hypothesis as evidence that is incompatible with the null hypothesis is convincing of the falsity of the null hypothesis. In Table 8.4 increasingly smaller p-values are seen to provide increasing evidence against the null hypothesis. In Table 8.5 increasingly larger p-values are seen to provide increasing evidence in support of the null hypothesis. This is true because of the inverse relationship between α (equivalently the p-value) and β (the probability of making an error when accepting the null hypothesis). An increase in the value of α leads to a corresponding decrease in the value of β.

p-value	**Evidence against H_0**
$P > 0.10$	weak
$0.05 < P \leq 0.10$	moderate
$0.01 < P \leq 0.05$	strong
$P \leq 0.10$	very strong

Table 8.4 Guidelines for using p-values as evidence against the null hypothesis.

p-value	**Evidence supporting H_0**
$P < 0.20$	none
$0.20 \leq P < 0.50$	weak
$0.50 \leq P < 0.75$	moderate
$0.75 \leq P < 1.0$	strong

Table 8.5 Guidelines for using p-values as evidence supporting the null hypothesis.

Hypothesis Test for the Population Mean: Population Variance Known

Recall that for a sample of n items $\{y_1, y_2, \ldots, y_n\}$ from a population, the sample mean is given by $\bar{y} = \frac{1}{n}\sum_{i=1}^{n} y_1$, and the sample standard deviation is given by $s = \sqrt{\frac{\sum_{i=1}^{n}(y_i - \bar{y})^2}{n-1}}$.

Assumptions

1. A random sample of size n is taken from the population of interest with unknown mean μ.
2. The population is normal. Note that if, based on the central limit theorem, the sampling distribution of the sample mean is approximately normal, then the method is still appropriate.
3. The population variance, σ^2, is known.

To determine whether assumptions 1 and 2 are reasonably met, diagnostic procedures are discussed in Chapter 10. Also discussed are remedial procedures to transform the data set so that it better satisfies the assumptions. In case the data cannot meet the assumptions, refer to the discussion of the sign test and the Wilcoxon signed-rank test in Chapter 12.

Hypotheses

The null hypothesis is $H_0 : \mu = \mu_0$. That is, the value of the population mean is μ_0. It is possible to test both the two-sided alternative H_a: $\mu \neq \mu_0$ and one-sided alternatives H_a: $\mu < \mu_0$ or H_a: $\mu > \mu_0$.

Test Statistic

The test statistic given by the following formula is assumed to have a standard normal distribution:

$$z = \frac{\bar{y} - \mu_0}{\sigma/\sqrt{n}}.$$

Procedure

The *Mathematica* procedures to perform a hypothesis test for μ when σ^2 is known are `MeanTest` and `NormalPValue`. With `MeanTest` use either option setting `KnownVariance->` or `KnownStandardDeviation->`, where either the population variance or the population standard deviation is entered after the arrow. The default output is a p-value, although more information can be displayed by using the option `FullReport->True`. With `NormalPValue` the value of the test statistic is entered and a p-value is returned. For both procedures the default setting is for a one-sided p-value, but a two-sided p-value can be specified using the option `TwoSided->True`. Also, both procedures have the option to specify a level of significance using the option `SignificanceLevel->`. In this case a decision to either reject or fail to reject the null hypothesis is also returned.

`MeanTest[list, mu0, options]` returns a probability estimate ($p$value) and other hypthesis test information for the relationship between the hypothesized population mean mu0 and `Mean[list]`. The options of `MeanTest` with default settings are `KnownStandardDeviation->None`, `FullReport->False`, `KnownVariance->None`, `SignificanceLevel->None`, and `TwoSided->False`.

`NormalPValue[teststat]` returns a one-sided p-value calculated using the standard normal distribution, `NormalDistribution[]`. The options of `NormalPValue` with default settings are `SignificanceLevel->None` and `TwoSided->False`.

Hypothesis Test for the Population Mean: Population Variance Unknown

Assumptions

1. A random sample of size n is taken from the population of interest with unknown mean μ.
2. The population is normal. Note that if, based on the central limit theorem, the sampling distribution of the sample mean is approximately normal, then the method is still appropriate.
3. The population variance, σ^2, is unknown.

To determine whether assumptions 1 and 2 are reasonably met, diagnostic procedures are discussed in Chapter 10. Also discussed are remedial procedures to transform the data set so that it better satisfies the assumptions. In case the data cannot meet the assumptions, refer to the discussion of the sign test and the Wilcoxon signed-rank test in Chapter 12.

Hypotheses

The null hypothesis is $H_0 : \mu = \mu_0$. That is, the value of the population mean is μ_0. It is possible to test both the two-sided alternative $H_a : \mu \neq \mu_0$ and one sided alternatives $H_a : \mu < \mu_0$ or $H_a : \mu > \mu_0$, where μ_0 is a given value.

Test Statistic

The test statistic given by the following formula is assumed to have a Student's t-distribution with $\nu = n - 1$ degrees of freedom:

$$t = \frac{\bar{y} - \mu_0}{s/\sqrt{n}}.$$

Procedure

The *Mathematica* procedures to perform a hypothesis test for μ when σ^2 is unknown are `MeanTest` and `StudentTPValue`. With `MeanTest` the default output is a p-value, although more information can be displayed by using the option `FullReport->True`. With `StudentTPValue` the value of the test statistic and the degrees of freedom (sample size minus 1) are entered and a p-value is returned. For both procedures the default setting is for a one-sided p-value, but a two-sided p-value can be specified using the option `TwoSided->True`. Also, both procedures have the option to specify a level of significance using the option `SignificanceLevel ->`. In this case a decision to either reject or fail to reject the null hypothesis is also returned.

> `MeanTest[list,` μ_0`, options]` returns a probability estimate ($p$value) and other hypothesis test information for the relationship between the hypothesized population mean μ_0 and `Mean[list]`. The options of `MeanTest` with default settings are `KnownStandardDeviation->None`, `FullReport->False`, `KnownVariance->None`, `SignificanceLevel->None`, and `TwoSided->False`.
>
> `StudentTPValue[teststat, dof]` returns a one-sided p-value calculated using the `StudentTDistribution`, with `dof` degrees of freedom. The options of `Student-TPValue` are `SignificanceLevel->None`, and `TwoSided->False`.

EXAMPLE 1: The mean final course average for students taking multivariable calculus courses in the 1980s was 80. Using the sample taken in 1996 given in Table 8.1, (a) perform a hypothesis to determine whether the mean final course average for students taking this course in the 1990s has changed from 80, $H_0 : \mu = 80$, $H_a : \mu \neq 80$. Assume the population standard deviation is 10.5. (b) Repeat the hypothesis test but use the one-sided alternative hypothesis $H_a : \mu < 80$ and assume the population standard deviation is unknown.

SOLUTION: First, we load the **HypothesisTests** package and enter the data into a list called `grades`. (a) We find that the p-value for the two-sided hypothesis test is $p = 0.697207$. Note that we include the options `TwoSided->True` and `KnownStandardDeviation->10.5.` to obtain this value. When we use the `FullReport->True` option, *Mathematica* gives more information, including the test statistic and the distribution used. If we also include the the option `SignificanceLevel->0.10`, the message `Fail to reject the null hypothesis` is also returned. Finally, note the example of using the function `NormalPValue`.

```
<<Statistics`HypothesisTests`

grades={85,94.5,76.7,79.2,83,80.2,68.7,89.1,74.1,87.8,
 44.9,77.6,85.1,75.7,81.5,66.2,83.4,79.8,94,91.8,96.3,
 73.5,82.2,76.1,78.5,69.1,75.4,71.7,78.2,77.7,88.7,79.9,
 86.1,63.8,78.7,82.6,98.6,81.3,63.4,76.6,84.2,89.7,87.7,
 54.6};
```

```
MeanTest[grades,80,TwoSided->True,
  KnownStandardDeviation->10.5]
```

```
TwoSidedPValue→0.697207
```

```
MeanTest[grades,80,TwoSided->True,
  KnownStandardDeviation->10.5,FullReport->True]
```

```
                Mean      TestStat   Distribution
{FullReport→  79.3841   -0.389093   NormalDistribution[]'
 TwoSidedPValue→0.697207}
```

```
MeanTest[grades,80,SignificanceLevel->0.10, TwoSided->
  True,KnownStandardDeviation->10.5,FullReport->True]
```

```
                Mean      TestStat   Distribution
{FullReport→  79.3841   -0.389093   NormalDistribution[]'
 TwoSidedPValue→0.697207,
 Fail to reject null hypothesis at significance level→
  0.1}
```

(b) Using MeanTest and the option FullReport->True, we find the one-sided p-value and learn that it was found using Student's t-distribution with 43 degrees of freedom.

```
MeanTest[grades,80,FullReport->True]
```

```
               Mean     TestStat  Distribution
{FullReport→ 79.3841  -0.388736  StudentTDistribution[43]'
 OneSidedPValue→0.349695}
```

■

Hypothesis Test for the Variance of a Normal Population

Assumptions

1. A random sample of size n is taken from the population of interest with unknown variance σ^2.
2. The population is normal.

To determine whether these assumptions are reasonably met, diagnostic procedures are discussed in Chapter 10. Also discussed are remedial procedures to transform the data set so that it better satisfies the assumptions.

Hypotheses

The null hypothesis is $H_0: \sigma^2 = \sigma_0^2$. That is, the value of the population variance is σ_0^2. It is possible to test both the two-sided alternative $H_a: \sigma^2 \neq \sigma_0^2$ and one-sided alternatives $H_a: \sigma^2 < \sigma_0^2$ or $H_a: \sigma^2 > \sigma_0^2$.

Test Statistic

The test statistic given by the following formula is assumed to have a chi-square distribution with $\nu = n - 1$ degrees of freedom:

$$\chi^2 = \frac{(n-1)s^2}{\sigma_0^2}.$$

Procedure

The *Mathematica* procedures to perform a hypothesis test for σ^2 are `VarianceTest` and `ChiSquarePValue`. The default output is a one-sided p-value, but a two-sided p-value can be specified using the option `TwoSided->True`. For `VarianceTest` more information can be displayed by using the option `FullReport->True`. There is also the option to specify a level of significance using the option `SignificanceLevel ->`. In this case a decision to either reject or fail to reject the null hypothesis is also returned.

> `VarianceTest[list, var0, options]` gives a probability estimate (p-value) and other hypothesis test information for the relationship between the hypothesized population variance `var0` and `Variance[list]`. The options of `VarianceTest` with default settings are `FullReport->False`, `SignificanceLevel->None`, and `TwoSided->False`.
>
> `ChiSquarePValue[teststat, dof]` returns a probability estimate (p-value) using the ChiSquareDistribution with `dof` degrees of freedom. The options of `ChiSquarePValue` with default settings are `SignificanceLevel->None` and `TwoSided->False`.

EXAMPLE 2: For the data in Table 8.1, test the hypothesis $H_0: \sigma^2 = 110.5$.

SOLUTION: Assuming that the data is entered in grades (see Example 1) and that we have loaded the **HypothesisTests** package, we use VarianceTest to obtain the p-value for a two-tailed alternative hypothesis. There is no basis in the data to reject our null hypothesis. We also show the result obtained when we use the option setting FullReport->True to observe that ChiSquareDistribution was employed in the computation of the test statistic as desired. Notice that when we use the SignificanceLevel option, *Mathematica* includes a statement to indicate that the null hypothesis is not rejected.

```
VarianceTest[grades,110.5,TwoSided->True]
TwoSidedPValue→0.85964

VarianceTest[grades,110.5,TwoSided->True,
  FullReport->True]
              Variance TestStat Distribution
{FullReport→ 110.453   43.9811  ChiSquareDistribution[43]'
 TwoSidedPValue→0.85964}

VarianceTest[grades,110.5,TwoSided->True,
  FullReport->True,SignificanceLevel->0.10]
              Variance TestStat Distribution
{FullReport→ 110.453   43.9811  ChiSquareDistribution[43]'
 TwoSidedPValue→0.85964,
 Fail to reject null hypothesis at significance level→
  0.1}
```

■

Hypothesis Tests for the Difference between Two Population Means: Population Variances Known

Assumptions

1. Independent, random samples are taken from population 1 with unknown mean μ_1 and population 2 with unknown mean μ_2. The sample sizes are n_1 and n_2, respectively.
2. Both populations are normal. Note that if, based on the central limit theorem, the sampling distribution of each sample mean is approximately normal, then the method is still appropriate.
3. The population variances, σ_1^2 and σ_2^2, are known.

To determine whether assumptions 1 and 2 are reasonably met, diagnostic procedures are discussed in Chapter 10. Also discussed are remedial procedures to transform the data set so that it better satisfies the assumptions. In case the data cannot meet the assumptions, refer to the discussion of the Mann–Whitney test in Chapter 12.

Hypotheses

The null hypothesis is $H_0: \mu_1 - \mu_2 = D_0$. That is, the value of the difference between the two population means is D_0. It is possible to test both the two-sided alternative $H_a: \mu_1 - \mu_2 \neq D_0$ and one-sided alternatives $H_a: \mu_1 - \mu_2 < D_0$ or $H_a: \mu_1 - \mu_2 > D_0$.

Test Statistic

The test statistic given by the following formula is assumed to have a standard normal distribution:

$$z = \frac{\bar{y}_1 - \bar{y}_2 - D_0}{\sqrt{\frac{\sigma_1^2}{n_1} + \frac{\sigma_2^2}{n_2}}}.$$

Procedure

The *Mathematica* procedures to perform a hypothesis test for the difference between the population means when σ_1^2 and σ_2^2 are known are `MeanDifferenceTest` and `NormalPValue`. With `MeanDifferenceTest` use either option setting `KnownVariance->` or `KnownStandardDeviation->`. The default output is a one-sided p-value, but a two-sided p-value can be specified using the option `TwoSided->True`. More information can be displayed by using the option `FullReport->True`. With `NormalPValue` the value of the test statistic is entered and a p-value is returned. Also, both procedures have the option to specify a level of significance using the option `SignificanceLevel->`. In this case a decision to either reject or fail to reject the null hypothesis is also returned.

> `MeanDifferenceTest[list1,list2,diff0,options]` gives a probability estimate (p-value) and other hypothesis test information for the relationship between the hypothesized population mean difference `diff0` and `Mean[list1]-Mean[list2]`. The options that can be used with `MeanDifferenceTest` with default settings are `KnownStandardDeviation->None`, `FullReport->False`, `KnownVariance->None`, `SignificanceLevel->None`, `TwoSided->False`, and `EqualVariances->False`.
>
> `NormalPValue[teststat]` returns a one-sided p-value calculated using the standard normal distribution, `NormalDistribution[]`. The options of `NormalPValue` with default settings are `SignificanceLevel->None` and `TwoSided->False`.

Hypothesis Test for the Difference between Two Population Means: Population Variances Unknown but Assumed Equal

Assumptions

1. Independent, random samples are taken from population 1 with unknown mean μ_1 and population 2 with unknown mean μ_2. The sample sizes are n_1 and n_2, respectively.
2. Both populations are normal. Note that if, based on the central limit theorem, the sampling distribution of each sample mean is approximately normal, then the method is still appropriate.
3. The population variances, σ_1^2 and σ_2^2 are unknown but assumed equal.

To determine whether these assumptions are reasonably met, diagnostic procedures are discussed in Chapter 10. Also discussed are remedial procedures to transform the data set so that it better satisfies the assumptions. In case the data cannot meet the assumptions, refer to the discussion of the Mann–Whitney test and the Wald–Wolfowitz test in Chapter 12.

Hypotheses

The **null hypothesis** is $H_0: \mu_1 - \mu_2 = D_0$. That is, the value of the difference between the two population means is D_0. It is possible to test both the two-sided alternative $H_a: \mu_1 - \mu_2 \neq D_0$ and one-sided alternatives $H_a: \mu_1 - \mu_2 < D_0$ or $H_a: \mu_1 - \mu_2 > D_0$.

Test Statistic

The test statistic given by the following formula is assumed to have a Student's t-distribution:

$$t = \frac{\bar{y}_1 - \bar{y}_2 - D_0}{s_p\sqrt{\frac{1}{n_1} + \frac{1}{n_2}}}.$$

Note that

$$s_p = \sqrt{\frac{(n_1 - 1)s_1^2 + (n_2 - 1)s_2^2}{n_1 + n_2 - 2}},$$

where the subscript p refers to the fact that the sample variances have been pooled to calculate an estimate of the common variance of the two populations. The number of degrees of freedom is given by $\nu = n_1 + n_2 - 2$.

Procedure

The *Mathematica* procedures to perform a hypothesis test for the difference between the means when σ_1^2 and σ_2^2 are unknown but assumed equal are `MeanDifferenceTest` and `StudentTPValue`. For `MeanDifferenceTest` the option `EqualVariances->True` must be specified. The default output is a one-sided p-value, but a two-sided p-value can be specified using the option `TwoSided->True`. More information can be displayed by using the option `FullReport->True`. With `StudentTPValue` the values of the test statistic and degrees of freedom are entered and a p-value is returned. Both procedures have the option to specify a level of significance using the option `SignificanceLevel ->`. In this case a decision to either reject or fail to reject the null hypothesis is also returned.

> `MeanDifferenceTest[list1,list2,diff0,options]` gives a probability estimate (p-value) and other hypothesis test information for the relationship between the hypothesized population mean difference `diff0` and `Mean[list1]-Mean[list2]`. The options that can be used with `MeanDifferenceTest` with default settings are `KnownStandardDeviation->None`, `FullReport->False`, `KnownVariance->None`, `SignificanceLevel->None`, `TwoSided->False`, and `EqualVariances->False`.
>
> `StudentTPValue[teststat, dof]` returns a one-sided p-value calculated using the `StudentTDistribution`, with `dof` degrees of freedom. The options of `StudentTPValue` are `SignificanceLevel->None` and `TwoSided->False`.

Hypothesis Test for the Difference between Two Population Means: Population Variances Unknown and Assumed Unequal

Assumptions

1. Independent, random samples are taken from population 1 with unknown mean μ_1 and population 2 with unknown mean μ_2. The sample sizes are n_1 and n_2, respectively.
2. Both populations are normal. Note that if, based on the central limit theorem, the sampling distribution of each sample mean is approximately normal, then the method is still appropriate.
3. The population variances, σ_1^2, and σ_2^2, are unknown and assumed to be different.

To determine whether these assumptions are reasonably met, diagnostic procedures are discussed in Chapter 10. Also discussed are remedial procedures to transform the data set so that it better satisfies the assumptions. In case the data cannot meet the assumptions, refer to the discussion of the Mann–Whitney test in Chapter 12.

Hypotheses

The null hypothesis is $H_0 : \mu_1 - \mu_2 = D_0$. That is, the value of the difference between the two population means is D_0. It is possible to test both the two-sided alternative $H_a : \mu_1 - \mu_2 \neq D_0$ and one-sided alternatives $H_a : \mu_1 - \mu_2 < D_0$ or $H_a : \mu_1 - \mu_2 > D_0$.

Test Statistic

The test statistic given by the following formula is assumed to have a Student's t-distribution:

$$t = \frac{\bar{y}_1 - \bar{y}_2 - D_0}{\sqrt{\frac{s_1^2}{n_1} + \frac{s_2^2}{n_2}}}.$$

The degrees of freedom is given by

$$\nu = \frac{\left(\frac{s_1^2}{n_1} + \frac{s_2^2}{n_2}\right)^2}{\frac{1}{n_1 - 1}\left(\frac{s_1^2}{n_1}\right)^2 + \frac{1}{n_1 - 1}\left(\frac{s_2^2}{n_2}\right)^2}.$$

Note that equation for the degrees of freedom is sometimes attributed to B. L. Welch, as is done in the *Mathematica* documentation.

Procedure

The *Mathematica* procedures to perform a hypothesis test for the difference between the means when σ_1^2 and σ_2^2 are unknown and assumed different are `MeanDifferenceTest` and `StudentTPValue`. The default output is a one-sided p-value, but a two-sided p-value can be specified using the option `TwoSided->True`. For `MeanDifferenceTest` more information can be displayed by using the option `FullReport->True`. With `StudentTPValue` the values of the test statistic and degrees of freedom are entered and a p-value is returned. Both procedures have the option to specify a level of significance using the option `SignificanceLevel->`. In this case a decision to either reject or fail to reject the null hypothesis is also returned.

> `MeanDifferenceTest[list1,list2,diff0,options]` gives a probability estimate (p-value) and other hypothesis test information for the relationship between the hypothesized population mean difference `diff0` and `Mean[list1]-Mean[list2]`. The options that can be used with `MeanDifferenceTest` with default settings are `KnownStandardDeviation->None`, `FullReport->False`, `KnownVariance->None`, `SignificanceLevel->None`, `TwoSided->False`, and `EqualVariances->False`.

`StudentTPValue[teststat, dof]` returns a one-sided p-value calculated using the `StudentTDistribution`, with `dof` degrees of freedom. The options of `StudentTPValue` are `SignificanceLevel->None` and `TwoSided->False`.

EXAMPLE 3: Table 8.6 lists the average combined SAT score at 12 public institutions, while Table 8.7 lists scores at 25 private institutions. Perform a hypothesis test to determine whether there is a difference between the mean score for public institutions and the mean score for private institutions. (*Note*: Data are from the *1995 U.S. News and World Report's Guide to America's Best Colleges*.)

Public institution	**Average SAT score**
North Carolina A & T State University	825
Appalachian State University	990
University of North Carolina at Asheville	1054
East Carolina University	921
Fayetteville State University	816
Pembroke State University	818
North Carolina State University	1071
University of North Carolina at Chapel Hill	1121
University of North Carolina at Charlotte	926
University of North Carolina at Greensboro	956
Western Carolina University	867
University of North Carolina at Wilmington	935

Table 8.6 Average SAT score at public institutions.

Private institution	**Average SAT score**
Barton College	840
Barber Scotia College	600
Belmont Abbey College	890
Bennett College	780
Campbell University	915
Catawba College	915
Davidson College	1230
Duke University	1302

Continued

Private institution	Average SAT score
Elon College	922
Gardner Webb University	845
Greensboro College	923
Guilford College	1030
High Point University	879
Lees-McRae College	757
Lenoir-Rhyne College	921
Mars Hill College	848
Montreat-Anderson College	870
North Carolina Wesleyan College	826
Pfeiffer College	831
Queens College	1005
Salem College	1002
St. Andrews Presbyterian College	915
St. Augustine's College	813
Wingate College	842
Winston-Salem State University	774

Table 8.7 Average SAT scores at private institutions.

SOLUTION: We begin by loading the **HypothesisTests** package. Then, we enter the data naming these lists `public` and `private`. We use `MeanDifferenceTest` to obtain a p-value where the value $D_0 = 0$. Assuming unknown and unequal population variances, the two-sided p-value is 0.30606, so we would not reject $H_0: \mu_1 = \mu_2$. If we assume that the variances of the two populations are equal, we would need to include the option `EqualVariances->True`. In this case, the two-sided p-value is 0.359269.

```
<<Statistics`HypothesisTests`

public={825,990,1054,921,816,818,1071,
  1121,926,956,867,935};

private={840,600,890,780,915,915,1230,
  1302,922,845,923,1030,879,757,921,848,
  870,826,831,1005,1002,915,813,842,774};
```

```
MeanDifferenceTest[public,private,0,TwoSided->True,
  FullReport->True]
```

```
                 MeanDiff   TestStat  Distribution
{FullReport→     42.6667    1.04174   StudentTDistribution[242702703932 03032/828966180838793]',
 TwoSidedPValue→0.30606}
```

```
MeanDifferenceTest[public,private,0,TwoSided->True,
  EqualVariances->True,FullReport->True]
```

```
                 MeanDiff   TestStat  Distribution
{FullReport→     42.6667    0.928967  StudentTDistribution[35]',
 TwoSidedPValue→0.359269}
```

■

Hypothesis Test for the Ratio of Variances of Two Normal Populations

Assumptions

1. Independent, random samples are taken from population 1 with unknown variance σ_1^2 and population 2 with unknown variance σ_2^2. The sample sizes are n_1 and n_2, respectively.
2. Both populations are normal.

To determine whether these assumptions are reasonably met, diagnostic procedures are discussed in Chapter 10. Also discussed are remedial procedures to transform the data set so that it better satisfies the assumptions. In case the data cannot meet the assumptions, refer to the discussion of the Siegel–Tukey test, the Moses test, and the Wald–Wolfowitz test in Chapter 12. Also see the discussion of the Bertlett variance test in Chapter 10.

Hypotheses

The null hypothesis is $H_0 : \frac{\sigma_1^2}{\sigma_2^2} = R_0$. That is, the ratio of the two population variances is R_0. It is possible to test both the two-sided alternative $H_a : \frac{\sigma_1^2}{\sigma_2^2} \neq R_0$ and one sided alternatives $H_a : \frac{\sigma_1^2}{\sigma_2^2} < R_0$ or $H_a : \frac{\sigma_1^2}{\sigma_2^2} > R_0$.

Test Statistic

The test statistic given by the following formula is assumed to have an F (ratio) distribution:

$$F = \frac{s_1^2}{R_0 s_2^2}.$$

The numerator degrees of freedom are $n_1 - 1$, and the denominator degrees of freedom are $n_2 - 1$.

Procedure

The *Mathematica* procedures to perform a hypothesis test of the ratio of two population variances are `VarianceRatioTest` and `FRatioPValue`. The default output is a one-sided p-value, but a two-sided p-value can be specified using the option `TwoSided->True`. For `VarianceRatioTest`, more information can be displayed by using the option `FullReport->True`. With `FRatioPValue` the values of the test statistic, numerator degrees of freedom and denominator degrees of freedom are entered and a p-value is returned. Also, both procedures have the option to specify a level of significance using the option `SignificanceLevel->`. In this case a decision to either reject or fail to reject the null hypothesis is also returned.

> `VarianceRatioTest[numlist,denlist,ratio0,options]` gives a probability estimate (p-value) and other hypothesis test information for the relationship between the hypothesized population variance ratio `ratio0` and the the ratio `Variance[numlist]/Variance[denlist]`. The options for use with `VarianceRatioTest` with default settings are `FullReport->False`, `SignificanceLevel->None`, and `TwoSided->False`.
>
> `FRatioPValue[teststat, numdof, dendof]` returns a probability estimate (p-value) using the FRatioDistribution with `numdof` numerator degrees of freedom and `dendof` denominator degrees of freedom. The options for use with `FRatioPValue` with default settings are `SignificanceLevel->None` and `TwoSided->False`.

EXAMPLE 4: For the data given in Tables 8.6 and 8.7, test the hypothesis that the ratio of the variances of SAT scores at public institutions and private institutions equals 1.

SOLUTION: Assuming that the **HypothesisTests** package is loaded and the data entered in `public` and `private`, we use `VarianceRatioTest` to obtain a p-value where the value $R_0 = 1$. The two-sided p-value is 0.257013, so we fail to reject the null hypothesis $H_0: \sigma_1 = \sigma_2$. Using the option `FullReport->True`, we observe that the `FRatioDistribution` is used in the calculation of the p-value.

```
VarianceRatioTest[public, private,1,TwoSided->True ]
 TwoSidedPValue→0.257013

VarianceRatioTest[public, private,1,TwoSided->True,
  FullReport->True]
                Ratio       TestStat     Distribution
{FullReport→    0.518999    0.518999     FRatioDistribution[11, 24]'
 TwoSidedPValue→0.257013}
```

■

8.3 Investigating Confidence Level and Significance Level with Simulation

In the usual applications of statistical methodology, sampling results are collected into a data set, which is then analyzed. It is rare for a population to be sampled repeatedly. Even if this were done, deciding which of the samples is really representative of the population is an impossible task. Using the simulation capability of *Mathematica*, on the other hand, a large number of samples can be drawn from a specified population and the quality of the samples can be evaluated. For example, a sample which results in a confidence interval that does not cover the population parameter it is intended to estimate is not representative. Likewise, a sample which results in a hypothesis test that rejects a true null hypothesis (Type I error) is not representative. We look into this capability in this section.

Confidence Level

We have represented the confidence level of a confidence interval using the notation $100(1-\alpha)\%$. Thus, a confidence coefficient of 95% would result from the value $\alpha = 0.05$. The confidence coefficient of 95% indicates that 95% of the possible samples of a given sample size from the population will produce an interval that covers the parameter being estimated. Thus, 5% of the samples will not result in a confidence interval that covers the parameter. To demonstrate the truth of this we can use simulated sampling.

EXAMPLE 1: Use *Mathematica* to select 1000 random samples of size 20 from a uniform population over the interval from 95 to 105. For each sample, calculate a 95% confidence interval for the population mean, and determine the number of intervals that actually cover the mean.

SOLUTION: After loading the **ContinuousDistributions** package, we use `dist` to stand for the uniform distribution over the interval from 95 to 105. Note that the the mean of this distribution is 100 and the variance is the difference between the upper and lower limits squared divided by 12. Using `RandomArray` we create the list `sample`, containing a random sample of size 20. After loading the **ConfidenceIntervals** package, we use the data in `sample` to calculate a 95% confidence interval for the population mean (100). This interval covers 100.

```
<<Statistics`ContinuousDistributions`
dist=UniformDistribution[95,105];
sample=RandomArray[dist,20];

<<Statistics`ConfidenceIntervals`
MeanCI[sample,KnownVariance->100/12]
{97.678,100.208}
```

Next, we define the function `confidence` that both selects a sample and calculates a 95% confidence interval. Then, using `Table`, we generate 1000 random samples and the corresponding confidence intervals. For length considerations, we do not display all of the results here, but instead display an abbreviated portion with `Short`.

```
confidence:=Module[{sample},
  sample=RandomArray[dist,20];
  MeanCI[sample,KnownVariance->100/12]]

confidence
{98.4329,100.963}

trials=Table[confidence,{1000}];

Short[trials,5]
{{99.3443,101.875},{99.3174,101.848},{99.063,101.593},
 «995»,{99.0164,101.547},{99.2475,101.778}}
```

Finally, we define the function `f` to returns `True` if 100 is not contained in the interval and `False` otherwise. We then use `Select` to extract those confidence intervals that do not contain 100 and `Length` to count them. Note that 51 of the 1000 do not contain 100. Thus, almost exactly 95% percent do cover 100.

```
f[op_]:=Not[op[[1]]<=18<=op[[2]]]
Clear[f]
  f[op_]:=Not[op[[1]]<=100<=op[[2]]]
```

```
Length[Select[trials,f[#]&]]
51
```

■

Significance Level

We have represented the significance level of a hypothesis using the notation α. A significance level of $\alpha = 0.05$ or 5% indicates that 5% of the possible samples of a given sample size from a population for which the null hypothesis is true will result it the erroneous rejection of the null hypothesis (a Type I error). To demonstrate the truth of this we can use simulated sampling.

EXAMPLE 2: Use *Mathematica* to select 1000 random samples of size 20 from a normal population with mean 10 and standard deviation 2. For each sample, perform a test at the 5% level of significance of the null hypothesis that the population mean is 10, and determine the number of tests that reject the null hypothesis.

SOLUTION: After loading the **ContinuousDistributions** package we use `dist` to stand for the normal distribution with mean 10 and standard deviation 2. Using `RandomArray` we create the list `sample`, containing a random sample of size 20. After loading the **HypothesisTests** package, we use the data in `sample` to perform a two-sided test of the null hypothesis that the population mean is 10. We fail to reject the null hypothesis, because the p-value is larger than 0.05. The p-value is the smallest level of significance for which the null hypothesis can be rejected.

```
<<Statistics`ContinuousDistributions`
dist=NormalDistribution[10,2];
sample=RandomArray[dist,20];

<<Statistics`HypothesisTests`
MeanTest[sample,10,TwoSided->True]
TwoSidedPValue→0.844502
```

Next, we define the function `hypothesis`, which both selects a sample and calculates a p-value. Then using `Table` we generate 1000 random samples and the corresponding p-values. For length considerations, we do not display the results.

```
hypothesis:=Module[{sample},
```

```
    sample=RandomArray[dist,20];
    MeanTest[sample,10,TwoSided->True][[2]]]
```

```
hypothesis
0.0527218
```

```
trials=Table[hypothesis,{1000}];
```

Finally, we use `Select` to extract the p-values which are less than or equal to 0.05 and `Length` to count them. If the p-value is equal to or less than the level of significance, then the null hypothesis is rejected at that level of significance. Note that 56 of the 1000 tests rejected the null hypothesis. Thus, a Type I error was committed in a little over 5% of the tests.

```
type1errors=Select[trials,#<=0.05&]
```

```
Length[type1errors]
56
```

■

Note that, in the version of *Mathematica* which was used for this book, the option `SignificanceLevel->` was not always correct when the decision to reject the null hypothesis was returned. This is seen for the following two tests.

```
MeanTest[test1,10,TwoSided->True,FullReport->True,
  SignificanceLevel->0.05]
```

```
                  Mean       TestStat   Distribution
{FullReport→      10.7608    1.97005    StudentTDistribution[19]'
 TwoSidedPValue→0.0635844,
 Reject null hypothesis at significance level→0.05}
```

```
MeanTest[test2,10,TwoSided->True,FullReport->True,
  SignificanceLevel->0.05]
```

```
                  Mean       TestStat   Distribution
{FullReport→      10.7648    1.77641    StudentTDistribution[19]'
 TwoSidedPValue→0.0916818,
 Reject null hypothesis at significance level→0.05}
```

In each case, the p-value is greater than the level of significance, which should be interpreted as "fail to reject," but the interpretation "reject" is returned. Based on simulation, it appears that the level of significance used for the decision was twice the value entered by the user.

CHAPTER 9

Analysis of Variance and Multiple Comparisons of Means

Analysis of variance is fundamental to a large area of statistical methodology. It is most widely used to test for differences among population means and linear combinations of population means. The name is derived from the fact that variance is used to quantify differences among the sample means, and the test statistic is a ratio of two measures of variability. The most common abbreviation for analysis of variance is ANOVA. Other abbreviations include ANOV and AOV.

There is no ANOVA package available with *Mathematica*, but we have written several procedures that may be used to evaluate differences among population means or linear combinations of population means. These include procedures to perform single-factor and two-factor ANOVA as well as several multiple comparison procedures. It is also possible to perform analysis of variance using linear regression. We do not discuss this in the book, although we do discuss the linear regression capabilities of *Mathematica* in detail in Chapter 11.

9.1 Single-Factor Analysis of Variance

Single-factor ANOVA, which is also known as **one way ANOVA**, is a method used to decide whether differences exist among several population means. It may be used to test hypotheses concerning two population means; however, one of the methods presented in Chapter 8 is usually more appropriate. When used to test the null hypothesis, $H_0 : \mu_1 = \mu_2$, the single-factor ANOVA is equivalent to a two-sided, pooled variance estimate t-test. See Subsection 8.2.5 in Chapter 8.

Assumptions

1. There are k groups that represent k independent, random samples, one from each of the k populations, where the population means are denoted by $\mu_1, \mu_2, \ldots, \mu_k$. The sample sizes are ,$n_1, n_2, \ldots, n_k$, respectively. Note that equal sample sizes are not required for the single-factor ANOVA, but the power of the test is increased by having sample sizes as nearly equal as possible.
2. Each population is normally distributed. Note that if, based on the central limit theorem, the sampling distribution of each sample mean is approximately normal, then the method is still appropriate.
3. The population variances are equal.

To determine whether these assumptions are reasonably met, there are several diagnostic procedures available. These are discussed in Chapter 10. Also discussed are remedial procedures to transform the data set so that it better satisfies the assumptions. It would then be appropriate to use the single-factor ANOVA on the transformed data. In case the data set can not meet the assumptions, an alternative method, the Kruskal–Wallis test, is discussed in Chapter 12.

Hypotheses

The null hypothesis is that all means are equal. $H_0 : \mu_1 = \mu_2 = \ldots = \mu_k$, where $k \geq 3$. The alternative hypothesis is that the means are not all equal. H_α: *The population means are not all equal.*

Test Statistic

We will represent the jth data value in the ith group by y_{ii}, where $j=1,2,\ldots n_i$ and $i=1,2,\ldots k$. Then the mean of the ith group is denoted by $\bar{y}_i = \frac{1}{n_i}\sum_{j=1}^{n_i} y_{ij}$, and the **grand mean** of all of the data is denoted by $\bar{y} = \frac{1}{n}\sum_{i=1}^{k}\sum_{j=1}^{n_i} y_{ij}$, where $n = \sum_{i=1}^{k} n_i$. The test statistic, F, is calculated as by dividing a measure of the variability among the sample means by a pooled estimate of the population variance (which is assumed to be the same in all k of the populations). The calculations are summarized in an ANOVA table.

Measures of variability are known as sums of squares and mean squares, where a mean square is a sum of squares divided by the corresponding degrees of freedom. For example, the sum of squares for the total data set, denoted by SST, is calculated by

$$SST = \sum_{i=1}^{k}\sum_{j=1}^{n_i} (y_{ij} - \bar{y})^2,$$

and the degrees of freedom for total is $DFT = n - 1$. Thus, the mean square for total would be calculated by

$$MST = \frac{SST}{DFT}.$$

Note that the MST is equivalent to the sample variance of a set of n values with sample mean $\bar{y}$. In single factor ANOVA the SST is divided into two parts. One part is due to the variability within the populations from which the samples are drawn and is called the **sum of squares within groups** or **sum of squares for error**. It is denoted by SSE. The second part is due to differences among the population means and is called the **sum of squares among groups** or simply **sum of squares for groups**. It is denoted by SSG. Note that $SST = SSG + SSE$. The sum of squares for error is calculated by

$$SSE = \sum_{i=1}^{k}\left[\sum_{j=1}^{n_i} (y_{ij} - \bar{y}_i)^2\right],$$

and the degrees of freedom for error is $DFE = n - k$. Thus, the mean square for error is

$$MSE = \frac{SSE}{DFE}.$$

It is a pooled estimate using all k sample variances of the comon population variance. The sum of squares for groups is calculated by

$$SSG = \sum_{i=1}^{k} n_i(\bar{y}_i - \bar{y})^2,$$

and the degrees of freedom for groups is $DFG = k - 1$. Note also that $DFT = DFG + DFE$. The the mean square for groups is

$$MSG = \frac{MSG}{DFG}.$$

Finally, the test statistic is the ratio of the mean square for groups to the mean square for error.

$$F = \frac{MSG}{MSE}.$$

It is assumed to have an F (ratio) distribution with numerator degrees of freedom $k - 1$, and denominator degrees of freedom $n - k$.

Procedure

We have written a *Mathematica* procedure called **oneWayAnova**. It is available on the disk that comes with this book. To use it, first load the disk and click on its icon. Next, locate the procedure name and click on the rightmost cell bracket as indicated in the following figure. Then press **ENTER** and return to the *Mathematica* work sheet.

▪ oneWayAnova

`oneWayAnova[dataset,catvar,quanvar,opts]` performs a single-factor ANOVA and displays an ANOVA table which includes sums of squares, mean squares, the value of the test statistic, and a *p*-value. It tests for a difference among the population means, where the groups (samples) are defined by the values of a categorical variable (assumed to be the first k integers). The groups are assumed to contain quantitative data. The name of the data set is entered first. The data set is assumed to be a rectangular array (a list containing n lists), where each record has at least one column containing a categorical variable and at least one column containing a quantitative variable. The location of the categorical variable (`catvar`) is entered next, followed by the location (`quanvar`) of the quantitative variable. The options that can be used with **oneWayAnova** with default settings are `categories->all` and `means->true`. The option `categories` is used to specify which values of the categorical variable to use in case an analysis for a subset of the population means is needed. The option `means` is used to specify whether the group means are displayed.

We illustrate this command using two simulated data sets in Example 1. In Example 2, we use a real data set. In the second example a process to transform the values of a categorial variable to the first k integers is demonstrated.

EXAMPLE 1: The list `dataset1` contains four categories where the data were sampled from populations with equal variances and equal means. The list `dataset2` contains four categories where the data were sampled from populations with equal variances but unequal means. For each data set, the categorical variable is the first entry in each record and the quantitative variable is the second entry in each record. Use **oneWayAnova** to perform a single-factor analysis of variance on each data set.

SOLUTION: After creating the first data set, we load **oneWayAnova** and use it on the data.

```
dataset1={{1,4.104},{1,4.159},{1,4.723},{1,3.757},
  {1,4.},{1,3.562},{1,4.069},{1,  3.701},{1,4.666},
  {1,3.429},{2,5.113},{2,3.361},{2,5.758},{2,3.822},
  {2,4.253},{2,3.287},{2,3.668},{2,4.697},{2,4.488},
  {2,3.485},{3,1.872},{3,4.548}, {3,3.367},{3,3.81},
  {3,3.008},{3,7.223},{3,3.754}, {3,2.802},{3,4.406},
  {3,3.216},{4,3.486},{4,4.211}, {4,3.483},{4,4.161},
  {4,4.338},{4,1.687},{4,2.906}, {4,1.561},
  {4,5.662},{4,4.491}};
```

Notice that the first column of the ANOVA table indicates the source of variation. The next three columns contain the sum of squares, degrees of freedom, and mean square for each source. There is no mean square for total, because it is not used in the calculation of the test statistic which is displayed in the fifth column. The p-value is displayed in the last column. It is the probability of observing a value at least as large as the observed Fratio ($P(F \geq 0.588457)$). Thes test does not detect a difference among the population means. Finally, notice that the group (sample) means are displayed below the ANOVA Table.

```
oneWayAnova[dataset1,1,2]
```

ANOVA Table

Source	Sum of Squares	DF	Mean Squares	Fratio	P-value
Groups	2.00355	3.	0.667852	0.588457	0.626532
Error	40.8571	36.	1.13492		
Total	42.8607	39.			

Means

Group	Number	Mean
1	10	4.017
2	10	4.1932
3	10	3.8006
4	10	3.5986

After generating `dataset2`, we run **oneWayAnova** on the data.

```
dataset2={{1,4.88},{1,4.294},{1,3.651},{1,5.368},
  {1,4.794},{1,5.234},{1,3.713},{1,2.581},{1,4.665},
  {1,5.374},{2,6.091},{2,5.198},{2,6.256},{2,6.132},
  {2,3.539},{2,6.947},{2,4.907},{2,5.765},{2,5.417},
  {2,3.617},{3,7.929},{3,3.503},{3,6.992},{3,5.98},
  {3,7.22},{3,5.955},{3,5.061},{3,5.779},{3,5.003},
  {3,5.992},{4,5.945},{4,6.88},{4,5.739},{4,6.923},
  {4,6.599},{4,8.821},{4,5.816},{4,6.963},{4,7.974},
  {4,6.024}};
```

Notice the small *p*-value which indicates a statistically significant difference among the means.

```
oneWayAnova[dataset2,1,2]
```

ANOVA Table

Source	Sum of Squares	DF	Mean Squares	Fratio	P-value
Groups	28.3145	3.	9.43817	8.11956	0.000293823
Error	41.8463	36.	1.1624		
Total	70.1608	39.			

Means

Group	Number	Mean
1	10	4.4554
2	10	5.3869
3	10	5.9414
4	10	6.7684

The following illustrates the use of the `categories` option.

```
oneWayAnova[dataset2,1,2,categories->{1,3,4}]
```

ANOVA Table

Source	Sum of Squares	DF	Mean Squares	Fratio	P-value
Groups	27.4736	2.	13.7368	12.1013	0.000176995
Error	30.649	27.	1.13515		

```
Total    58.1227         29.
Means

Group    Number    Mean
1        10        4.4554
3        10        5.9414
4        10        6.7684
```

■

EXAMPLE 2: The text file **Crash.txt** contains data from experiments in which automobiles with dummies in the driver and front passenger seat were crashed into walls at 35 miles per hour. The file contains 352 records with values of the 14 variables described in Table 9.1. Within each record the values are separated by tab marks. The data was generated by the National Transportation and Safety Administration and obtained from StatLib. Note that the symbols "-" and "nd" are used to represent missing data in the text file. (a) Use **oneWayAnova** to perform single factor analysis of variance to test for differences in "kinds of protection" with respect to "head injury." (b) Test for differences in a select group of "makes" with respect to "head injury."

Name	Description
Make	Car make
Model	Car model
CarID	(usually) Combination of make and model
CarID & Year	Full ID of the car
Head IC	Head injury criterion
Chest decel	Chest deceleration
L Leg	Left femur load
R Leg	Right femur load
D/P	Whether the dummy is in the driver or passenger seat
Protection	Kind of protection (seat belt, air bag, etc.)
Doors	Number of doors on the car
Year	Year of the car

Continued

Name	Description
Wt	Weight in pounds
Size	Categorical variable to classify the cars to a type (light, heavy, compact, etc.)

Table 9.1

SOLUTION: First, we use ReadList to load the file Crash.txt into *Mathematica*, naming the result cardata. Because Crash.txt contains both numerical and non-numerical data, we include Word in the ReadList command. Also, because the entries are separated by tab marks, we include the code {\t"} in the option WordSeparators. Our convention is to use the symbol null to represent missing data, so we replace all occurrences of "-" and "nd" with null using the symbol, (/.), representing ReplaceAll. Finally, after loading the **DataManipulation** package, we delete columns 2 through 4, which are redundant.

```
cardata=ReadList["Crash.txt",Word,
   RecordLists->True,WordSeparators->{"\t"}] /.
  {"-"->null,"nd"->null};
<<Statistics`DataManipulation`
cardata=ColumnDrop[cardata,{2,4}];
```

We use Length to see that the file contains 352 records, and view several records with Short.

```
Length[cardata]

352

Short[cardata,6]

{{Acura,599,35,791,262,Driver,manual belts,2,87,2350,lt

},{Acura,585,null,1545,

  1301,Driver,Motorized belts,4,90,2490,lt

},«349»,{Yugo,379,41,585,

  172,Passenger,manual belts,2,87,1850,mini

}}
```

Union[list] gives a sorted version of a list, in which all duplicated elements have been dropped.

```
makes=Union[Column[cardata,1]]
```

```
{Acura,Audi,BMW,Buick,Cadillac,
 Chevrolet,Chrysler,Daihatsu,Dodge,Eagle,
 Ford,Geo,Honda,Hyundai,Infiniti,Isuzu,Jeep,Lexus,
 Lincoln,Mazda,Mercedes,Mercury,Mitsubishi,Nissan,
 Oldsmobile,Peugeot,Plymouth,Pontiac,Renault,Saab,
 Saturn,Subaru,Suzuki,Toyota,Volkswagen,Volvo,Yugo}
```

We then define the rule, rule1, which replaces the various makes by digits.

```
rule1=Table[makes[[i]]->i,{i,1,Length[makes]}]
```

```
{Acura→1,Audi→2,BMW→3,Buick→4,Cadillac→5,
 Chevrolet→6,Chrysler→7,Daihatsu→8,Dodge→9,Eagle→10,
 Ford→11,Geo→12,Honda→13,Hyundai→14,Infiniti→15,
 Isuzu→16,Jeep→17,Lexus→18,Lincoln→19,Mazda→20,
 Mercedes→21,Mercury→22,Mitsubishi→23,Nissan→24,
 Oldsmobile→25,Peugeot→26,Plymouth→27,Pontiac→28,
 Renault→29,Saab→30,Saturn→31,Subaru→32,Suzuki→33,
 Toyota→34,Volkswagen→35,Volvo→36,Yugo→37}
```

Similarly, we define `rule2` to replace the various sizes by digits,

```
sizes=Union[Column[cardata,11]]
```

```
{comp
,hev
,lt
,med
,mini
,mpv
,pu
,van
}
```

```
rule2=Table[sizes[[i]]->i,{i,1,Length[sizes]}]
```

```
{comp
→1,hev
```

```
→2,lt
→3,med
→4,mini
→5,mpv
→6,pu
→7,van
→8}
```

and `rule3` to replace the various types of protection by digits. In this case, we assign the same code to more than one category.

```
protection=Union[Column[cardata,7]]

{d airbag,d&p airbags,manual belts,Motorized
belts,passive belts}

rule3={"d airbag"→1,"d&p airbags"→2,"manual belts"→3,
  "Motorized belts"3→,"passive belts"→3}

{d airbag→1,d&p airbags→2,manual belts→3,Motorized
belts→3,
 passive belts→3}
```

These three rules are then applied to `cardata`, and the resulting list is again named `cardata`. Although `cardata` is a valid *Mathematica* array, each entry is a string, including the numerical ones. Thus, we cannot perform arithmetic on the numerical entries at this point. Renaming the file using `ToExpression` converts the data from an array of strings to an array of numbers. The first 10 records in the new `cardata` are viewed with `Take`.

```
cardata=cardata /. rule1 /. rule2 /. rule3;

cardata=cardata//ToExpression;
Take[cardata,10]

{{1,599,35,791,262,Driver,3,2,87,2350,3},
 {1,585,null,1545,1301,Driver,3,4,90,2490,3},
 {1,435,50,926,708,Driver,1,4,88,3280,4},
 {2,600,49,168,1871,Driver,3,4,89,2790,1},
 {2,185,35,998,894,Driver,1,4,89,3100,4},
 {3,1036,56,865,null,Driver,1,2,90,2862,1},
 {4,815,47,1340,315,Driver,3,4,91,2992,1},
```

```
{4,1467,54,712,1366,Driver,3,4,88,3360,4},
{4,null,35,1049,908,Driver,3,2,90,3240,4},
{4,880,50,996,642,Driver,3,2,88,3210,4}}
```

(a) We will perform our analysis using the (quantitative) head injury data in the second column of `cardata`. To investigate this data, we create a file `headinjuries`, and, after loading our procedure **`normalHistogram`**, we plot the data.

```
headinjuries=Column[cardata,2]//DropNonNumeric;
normalHistogram[headinjuries,20]
```

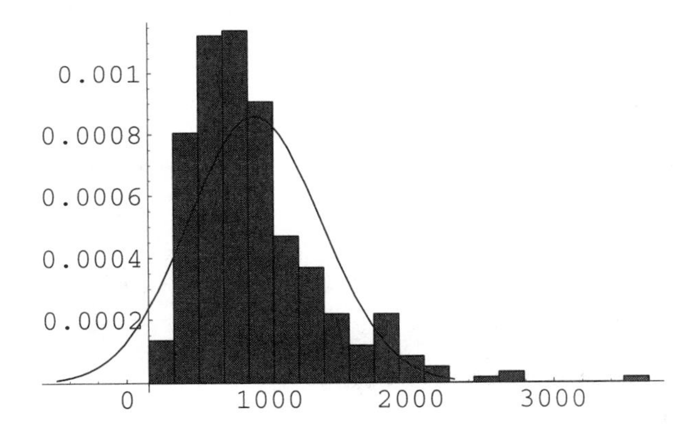

Because the data do not appear to have come from a normal distribution, we take the natural logarithm of each value (see Chapter 10). Again using **`normalHistogram`**, we see that the transformed data do appear to satisify the normally requirement.

```
logheadinjuries=Log[headinjuries]//N;
normalHistogram[logheadinjuries,20]
```

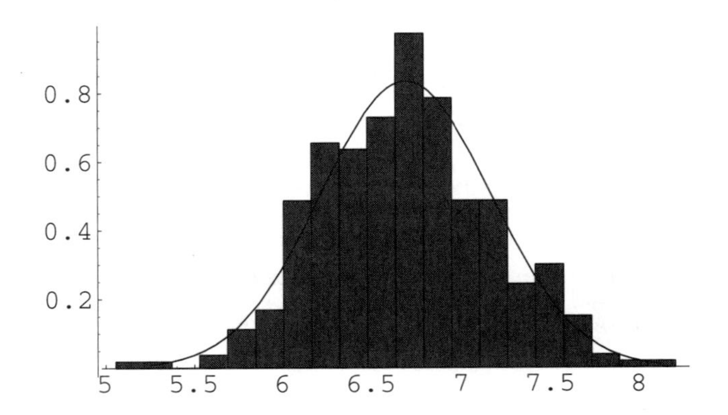

Now, we append the logarithm of the head injury criterion to each record by defining a function (f) to perform the transformation and performing the transformation using Map. We append the new column to cardata using RowJoin. The first record of the newest cardata is then viewed with Part (indicated by [[...]]).

```
f[list_]:={Log[list[[2]]]//N};

logs=Map[f,cardata];

cardata=RowJoin[cardata,logs];

cardata[[1]]

{1,599,35,791,262,Driver,3,2,87,2350,3,6.39526}
```

After loading our procedure **oneWayAnova**, we perform the analysis to look for differences in kinds of protection (column 7) with respect to the quantitative variable natural logarithm of the head injury (column 12).

```
oneWayAnova[cardata,7,12]
```

ANOVA Table

Source	SumofSquares	DF	MeanSquares	Fratio	P-value
Groups	6.18129	2.	3.09064	14.5475	8.7114×10^{-7}
Error	71.5963	337.	0.212452		
Total	77.7776	339.			

Means

```
Group    Number   Mean
1        55       6.39416
2        3        6.37125
3        282      6.75139
```

(b) In this case, we compare categories 2, 29, 33, and 36, which correspond to Audi, Renault, Suzuki, and Volvo.

```
oneWayAnova[cardata,1,12,categories->{2,29,33,36}]
```

```
ANOVA Table

Source   Sum of Squares   DF    Mean Squares   Fratio    P-value
Groups   4.04777          3.    1.34926        7.45791   0.0105179
Error    1.44733          8.    0.180916
Total    5.4951           11.
Means

Group    Number   Mean
2        4        6.10668
29       2        7.09205
33       4        7.36728
36       2        6.17499
```

■

9.2 Multiple Comparison Methods

Introduction

When the result of the ANOVA is to reject the null hypothesis, $H_0 : \mu_1 = \mu_2 = \dots = \mu_k$, the logical next step is to ask which of the population means are different, or more generally, which combinations of the population means are different. It is generally not valid to perform several two sample procedures of the type discussed in Chapter 8 to investigate possible differences among pairs of population means, because, with each test performed, the chance of finding at least one erroneous difference increases. For example, if three independent random samples are taken and three hypotheses, $H_0 : \mu_1 = \mu_2$, $H_0 : \mu_1 = \mu_3$, and $H_0 : \mu_2 = \mu_3$, are each tested at the 5% level of significance, then the chance of finding at least one erroneous difference is $1 - (1 - 0.05)^3 = 0.142625$ or over 14%. The formula in general for c comparisons made at level of significance α is $1 - (1 - \alpha)^c$. In this context, a **family** is a collection of inferences for which it is meaningful to take into account some combined measure of error. The family may be defined either by the collection of inferential problems or by the collection of parameters on which the inferences are to be made. However it is defined, the reason for its existance is to ensure the simultaneous correctness of a set of inferences and therby the correctness of the overall conclusion. Statistical procedures that take into account and properly control the combined measure of error for a given family are called **multiplication comparison methods**.

When using multiple comparison techniques, we must distinguish between the **per-comparison error rate**, which is 5% in the preceding example, and the **familywise** (also known as **experimentwise**) **error rate**, which would correspond to 14% in the example. If the purpose of the analysis is to distinguish among the three population means, then the familywise error rate must be controlled at a specified level which is often 5% but may be lower at the option of the experimenter. It is infrequent that the familywise error is set at over 10%. Many of the multiple comparison techniques discussed here can be formulated either in the form of hypothesis tests or in the form of confidence intervals. In the latter case, the family confidence coefficient must be controlled at a specified level, which is often 95%, as opposed to relying on per-comparison confidence levels. When appropriate, multiple comparisons are often displayed in confidence interval form because more information is provided. For example, simultaneous confidence intervals for $\mu_1 - \mu_2$, $\mu_1 - \mu_3$, and $\mu_1 - \mu_3$, with family confidence coefficient 95% can identify which of the population means are different (at 5% familywise level of significance) as well as provide bounds on the size of the difference between any two population means that are found to be significantly different. In general, two means, μ and μ_j, are significantly different at level of significance α if the confidence interval for $\mu_i - \mu_j$ with confidence coefficient $100(1-\alpha)\%$ does not cover (contain) zero.

Cautions

The subject of multiple comparisons is an active research field. More is being learned about existing methods and new methods are being developed. Whereas the user of these methods need not be an expert in the field, an understanding of some basic concepts is necessary to ensure appropriate use of a method and correct interpretation of the results. Multiple comparison methods are used in two situations: to confirm anticipated differences the study was designed to detect, and to identify differences which were not anticipated when the study was undertaken. The methods presented in this book are for population means and fit generally into the latter category, which has been called **data-snooping**. When deciding on a multiple comparison method, the analyst must be careful to choose one which guarantees (that is, protects) the specified familywise error rate or, equivalently, the specified family confidence level under all possible configurations of the unknown population means. This is known as **strongly protecting the familywise error rate**. Some methods only protect the familywise error rate when the the null hypothesis is true. For any other configuration of the population means, the familywise error rate would be, in actuality, higher than the specified value. Such a procedure is said to **protect the familywise error rate in a weak sense**. For example, for Fisher's least significant difference (LSD) test, the familywise error rate is α only when $H_0 : \mu_1 = \mu_2 = \ldots = \mu_k$ is true. But, the LSD test is only employed after the ANOVA rejects this hypothesis. This implies, of course, that the configuration of the population means is other than all equal. The result is, whenever this test is actually used, the familywise error rate exceeds α. Although they can be modified to do so, other multiple comparison methods that fail to strongly protect the familywise error rate are the Student–Newman–Keuls test and Duncan's multiple range test.

In choosing among the multiple comparison methods that do strongly protect the specified familywise error rate, a user should be aware of the following.

1. Some methods are exact in that the actual familywise error rate equals that specified. Others are based on approximations and are conservative in that the actual familywise error rate is at most (less than or equal to) that specified. Exact methods or those that are minimally conservative are often the best choice.
2. Depending on the type of comparisons in the family, one method may be more appropriate to use than others. For example, Tukey's procedure is most useful for a family of all pairwise comparisons; Scheffe's procedure is most useful when a large number of contrasts (not all of which are pairwise) is specified; Bonferroni and Dunn–Sidak procedures are most useful when small numbers of contrasts are specifies; and Dunnett's procedure is used for a family of comparisons of several experimental means to a control mean.
3. Some methods are accomplished in a single step, whereas others require multiple steps. The multiple step procedures are often more powerful than their single-step counterparts. However, they can usually only be used for hypothesis testing and not for confidence intervals, and it is not known whether directional decisions are protected under the familywise level of significance. With single-step procedures, confidence intervals are often possible. Also, if two population means are shown to be significantly different, the larger (or smaller) can be identified based on the larger (or smaller) sample mean. The chance of making an error in such **directional decisions** is protected at the specified familywise level of significance along with the chance of making an error in decisions concerning significant differences.
4. The multiple comparison methods for population means have the same underlying **assumptions** as does the ANOVA. However, whereas single inferential procedures based on the Student's t and F statistics are fairly robust to nonnormality, in the case of multiple comparisons, the departure from normality may become more serions. Its adverse effect on the level of significance may be magnified as the number of comparisons increases. That is, the actual level of significance may be a little higher because of a departure from normality for a single inferential procedure, but a lot higher for the multiple comparison procedure. Note that an inferential procedure is called **robust** if its

validity is not seriously affected by moderate deviations from an underlying assumption.

5. In practice, multiple comparison methods are most commonly performed only if the ANOVA rejects the null hypothesis, $H_0 : \mu_1 = \mu_2 = \ldots = \mu_k$. Although theoretically unnecessary for the methods discussed in this book, the practice has continued, because performing ANOVA first does provide numerical results which are used in the multiple comparison calculations. Many of the multiple comparison procedures which were developed for this book accept the same data input as **oneWayAnova**. In these cases it is not necessary to run it first.

Single-Step Procedures for Pairwise Comparisons

Perhaps the most common form of multiple comparisons, called pairwise comparisons, investigate each of the $\frac{1}{2}k(k-1)$ possible pairs of the k population means for significant differences. The null hypothesis to be tested by each comparison is of the form $H_0 : \mu_i - \mu_j = 0$. We have written procedures for several of these common pairwise comparison methods. The output of each is a table of confidence intervals for all pairwise differences, $\mu_i - \mu_j$. A significance difference is shown when the confidence interval does not cover zero. The table also includes an indication of which pairs of population means are significantly different.

mcmTukeyPairs

If the sample sizes are equal, this procedure performs **Tukey's honestly significant difference (HSD) test**, which is also known as **Tukey's wholly significant difference (WSD) test.** This test is exact and, in many respects, optimal. If the sample sizes are not all equal, the procedure performs a modification of the Tukey test, which is often called the **Tukey–Kramer test.** This is based on an approximation but is known to be minimally conservative. Both tests are robust to departures from normality. They are less robust to unequal variances and sensitive to violation of the assumption of independence.

`mcmTukeyPairs[dataset,catvar,quanvar,a,opts]` calculates and displays simultaneous confidence intervals for all pairwise differences, $\mu_i - \mu_j$. The table also includes an indication of which pairs of population means are significantly different. The groups are defined by the values (assumed to be the first k integers) of a categorical variable, but the groups are assumed to contain quantitative data. The name of the data set is entered first. The data set is assumed to be a rectangular array (a list containing n lists), where each record has at least one column containing a categorical variable and at least one column containing a quantitative variable. The location of the categorical variable (`catvar`) is entered next, followed by the location (`quanvar`) of the quantitative variable. Then the familywise level of significance (a) is entered. The option that can be used with **mcmTukey-Pairs** with default setting is `categories->all`. It is used to specify which values of the categorical variable to use in case an analysis for a subset of the k means is needed.

mcmBonferroniPairs

This procedure is based on an approximation known as the first-order Bonferroni inequality. When used to test all pairwise differences, it will result in wider confidence intervals than the Tukey procedure.

`mcmBonferroniPairs[dataset,catvar,quanvar,a,opts]` calculates and displays simultaneous confidence intervals for all pairwise differences, $\mu_i - \mu_j$. The table also includes an indication of which pairs of population means are significantly different. The groups are defined by the values (assumed to be the first k integers) of a categorical variable, but the groups are assumed to contain quantitative data. The name of the data set is entered first. The data set is assumed to be a rectangular array (a list containing n lists), where each record has at least one column containing a categorical variable and at least one column containing a quantitative variable. The location of the categorical variable (`catvar`) is entered next, followed by the location (`quanvar`) of the quantitative variable. Then the familywise level of significance (a) is entered. The option that can be used with **mcmBon-ferroniPairs** with default setting is `categories->all`. It is used to specify which values of the categorical variable to use in case an analysis for a subset of the k means is needed.

mcmDunnSidakPairs

This procedure is based on an inequality proved by Sidak. This inequality is a slight improvement on the Bonferroni inequality, so confidence intervals will be shorter, but still longer than for the Tukey procedure when all pairwise differences are tested.

`mcmDunnSidakPairs[dataset,catvar,quanvar,a,opts]` calculates and displays simultaneous confidence intervals for all pairwise differences, $\mu_i - \mu_j$. The table also includes an indication of which pairs of population means are significantly different. The groups are defined by the values (assumed to be the first k integers) of a categorical variable, but the groups are assumed to contain quantitative data. The name of the data set is entered first. The data set is assumed to be a rectangular array (a list containing n lists), where each record has at least one column containing a categorical variable and at least one column containing a quantitative variable. The location of the categorical variable (`catvar`) is entered next followed by the location (`quanvar`) of the quantitative variable. Then the familywise level of significance (a) is entered. The option that can be used with **mcmDunnSidakPairs** with the default setting is `categories->all`. It is used to specify which values of the categorical variable to use in case an analysis for a subset of the k means is needed.

mcmScheffePairs

This procedure is based on **Scheffe's fully significant difference (FSD) test,** which is also known as **Scheffe's globally significant difference (GSD) test.** The confidence intervals from this procedure will be longer than for the Tukey procedure.

`mcmScheffePairs[dataset,catvar,quanvar,a,opts]` calculates and displays simultaneous confidence intervals for all pairwise differences, $\mu_i - \mu_j$. The table also includes an indication of which pairs of population means are significantly different. The groups are defined by the values (assumed to be the first k integers) of a categorical variable, but the groups are assumed to contain quantitative data. The name of the data set is entered first. The data set is assumed to be a rectangular array (a list containing n lists), where each record has at least one column containing a categorical variable and at least one column containing a quantitative variable. The location of the categorical variable (`catvar`) is entered next, followed by the location (`quanvar`) of the quantitative variable. Then the familywise level of significance (`a`) is entered. The option that can be used with `mcmScheffePairs` with the default setting is `categories->all`. It is used to specify which values of the categorical variable to use in case an analysis for a subset of the k means is needed.

To use any of these pairwise procedures, first load the disk which comes with this book and click on its icon. Next, locate the procedure name and click on the rightmost cell bracket. Then press **ENTER** and return to the *Mathematica* work sheet.

We illustrate the use of these tests with a simulated data set.

EXAMPLE 1: The list `dataset1` contains four categories, where the data were sampled from populations with equal variances but unequal means except for categories 3 and 4. The categorical variable is the first entry in each record, and the quantitative variable is the second entry in each record. Test for differences among pairs of population means using a familywise level of significance of 5% and using (a) **`mcmTukeyPairs`**, (b) **`mcmBonferroniPairs`**, (c) **`mcmDunnSidakPairs`**, and (d) **`mcmScheffePairs`**.

SOLUTION: First, we simulate the data and then enter it into `dataset1`. The population means are 5, 4, 3, and 3, respectively. Note that the smaller the width of the confidence intervals the better for showing significance. For all multiple pairwise comparisons, Tukey intervals will be smallest followed by Dunn–Sidak, Bonferroni, and Scheffe intervals.

```
dataset1={{1,3.239},{1,5.972},{1,5.335},{1,6.556},
  {1,4.692},{1,4.439},{1,4.937},{1,6.02},
  {1,5.664},{1,4.529},{2,5.},{2,3.744},{2,2.8},
  {2,4.223},{2,2.901},{2,3.661},{2,4.378},{2,4.515},
  {2,2.665},{2,4.817},{3,3.051},{3,2.859},{3,3.303},
  {3,3.2},{3,2.714},{3,3.337},{3,3.434},{3,2.163},
  {3,4.049},{3,2.859},{4,2.053},{4,3.201},{4,1.737},
  {4,1.972},{4,3.74},{4,2.026},{4,1.19},{4,3.182},
  {4,2.622},{4,2.777}};
```

(a) Here are the results obtained using **mcmTukeyPairs**.

```
mcmTukeyPairs[dataset1,1,2,0.05]
```

Multiple Pairwise Comparisons using the Tukey-Kramer Procedure.

i,j			$\bar{Y}_i - \bar{Y}_j$	Confidence Interval					Significant Difference
1	,	2	1.2679	0.296532	≤	$\mu_1-\mu_2$	≤	2.23927	Yes
1	,	3	2.0414	1.07003	≤	$\mu_1-\mu_3$	≤	3.01277	Yes
2	,	3	0.7735	-0.197868	≤	$\mu_2-\mu_3$	≤	1.74487	No
1	,	4	2.6883	1.71693	≤	$\mu_1-\mu_4$	≤	3.65967	Yes
2	,	4	1.4204	0.449032	≤	$\mu_2-\mu_4$	≤	2.39177	Yes
3	,	4	0.6469	-0.324468	≤	$\mu_3-\mu_4$	≤	1.61827	No

The familywise confidence level is at least 95. percent.

The familywise level of significance is at most 5. percent.

(b) Here are the results obtained using **mcmBonferroniPairs**. The intervals are wider than those from the Tukey procedure.

```
mcmBonferroniPairs[dataset1,1,2,0.05]
```

Multiple Pairwise Comparisons using the Bonferroni inequality.

i,j$\bar{Y}_i - \bar{Y}_j$Confidence Interval Significant* Difference

1 , 2 1.26790.272058≤$\mu_1-\mu_2$≤2.26374Yes

1 , 3 2.04141.04556≤$\mu_1-\mu_3$≤3.03724Yes

2 , 3 0.7735-0.222342≤$\mu_2-\mu_3$≤1.76934No

1 , 4 2.68831.69246≤$\mu_1-\mu_4$≤3.68414Yes

2 , 4 1.42040.424558≤$\mu_2-\mu_4$≤2.41624Yes

3 , 4 0.6469-0.348942≤$\mu_3-\mu_4$≤1.64274No

The familywise confidence level is at least 95. percent.
The familywise level of significance is at most 5. percent.

(c) Here are the results obtained using **mcmDunnSidakPairs**. The intervals are wider than those from the Tukey procedure but not quite as wide as those from the Bonferroni procedure.

```
mcmDunnSidakPairs[dataset1,1,2,0.05]
```

Multiple Pairwise Comparisons using the Dunn-Sidak inequality.

i,j	$\bar{Y}_i - \bar{Y}_j$	Confidence Interval				Significant* Difference
1 , 2	1.2679	0.275057	≤	$\mu_1-\mu_2$	≤ 2.26074	Yes
1 , 3	2.0414	1.04856	≤	$\mu_1-\mu_3$	≤ 3.03424	Yes
2 , 3	0.7735	-0.219343	≤	$\mu_2-\mu_3$	≤ 1.76634	No
1 , 4	2.6883	1.69546	≤	$\mu_1-\mu_4$	≤ 3.68114	Yes
2 , 4	1.4204	0.427557	≤	$\mu_2-\mu_4$	≤ 2.41324	Yes
3 , 4	0.6469	-0.345943	≤	$\mu_3-\mu_4$	≤ 1.63974	No

The familywise confidence level is at least 95. percent.

The familywise level of significance is at most 5. percent.

(d) Here are the results obtained using **mcmScheffePairs**. The intervals are wider than from any of the three previous procedures.

```
mcmScheffePairs[dataset1,1,2,0.05]
```

Multiple Pairwise Comparisons using the Scheffe procedure.

i,j	$\bar{Y}_i - \bar{Y}_j$	Confidence Interval				Significant* Difference
1 , 2	1.2679	0.221981	≤	$\mu_1-\mu_2$	≤ 2.31382	Yes
1 , 3	2.0414	0.995481	≤	$\mu_1-\mu_3$	≤ 3.08732	Yes
2 , 3	0.7735	-0.272419	≤	$\mu_2-\mu_3$	≤ 1.81942	No
1 , 4	2.6883	1.64238	≤	$\mu_1-\mu_4$	≤ 3.73422	Yes
2 , 4	1.4204	0.374481	≤	$\mu_2-\mu_4$	≤ 2.46632	Yes
3 , 4	0.6469	-0.399019	≤	$\mu_3-\mu_4$	≤ 1.69282	No

The familywise confidence level is at least 95. percent.

The familywise level of significance is at most 5. percent.

■

EXAMPLE 2: The text file **Crash.txt** contains data from experiments in which automobiles with dummies in the driver and front passenger seats were crashed into walls at 35 miles per hour. The variables are described in Table 9.1, which is in Example 2 of Section 9.1. Use a pairwise multiple comparison procedure to determine differences (a) in "kinds of protection" with respect to "head injury" and (b) in a select group of "makes" with respect to "head injury."

SOLUTION: We assume that the final version of `cardata` is available from Example 2 of Section 9.1. In that example we decided to use its natural logarithm to represent "head injury." (a) After loading our procedure **`mcmTukeyPairs`** we perform pairwise comparisons at the 5% familywise level of significance among the categories of "kinds of protection" (column 7) with respect to the natural logarithm of "head injury" (column 12).

```
mcmTukeyPairs[cardata,7,12,0.05]
```

Multiple Pairwise Comparisons using the Tukey-Kramer Procedure.

i,j	$\bar{Y}_i - \bar{Y}_j$	Confidence Interval				Significant* Difference
1 , 2	0.0229101	-0.626239	≤	$\mu_1-\mu_2$	≤ 0.67206	No
1 , 3	-0.357231	-0.518623	≤	$\mu_1-\mu_3$	≤ -0.19584	Yes
2 , 3	-0.380142	-1.01563	≤	$\mu_2-\mu_3$	≤ 0.25535	No

The familywise confidence level is at least 95. percent.

The familywise level of significance is at most 5. percent.

(b) We perform pairwise comparisons at the 5% familywise level of significance among selected categories of "make" (column 1) with respect to the natural logarithm of "head injury" (column 12). The option `categories->{2,29,33,36}` is used to compare the makes corresponding to Audi, Renault, Suzuki, and Volvo.

```
mcmTukeyPairs[cardata,1,12,0.05,categories->{2,29,33,36}]
```

Multiple Pairwise Comparisons using the Tukey-Kramer Procedure.

i,j	$\bar{Y}_i - \bar{Y}_j$	Confidence Interval				Significant* Difference
2 , 29	-0.985369	-2.19186	≤	$\mu_2-\mu_{29}$	≤ 0.221127	No
2 , 33	-1.2606	-2.2457	≤	$\mu_2-\mu_{33}$	≤ -0.275501	Yes
29 , 33	-0.275232	-1.48173	≤	$\mu_{29}-\mu_{33}$	≤ 0.931264	No

```
2  ,  36 -0.0683097-1.27481    ≤    μ2-μ36    ≤ 1.13819    No
29 ,  36 0.917059 -0.476083    ≤    μ29-μ36   ≤ 2.3102     No
33 ,  36 1.19229  -0.0142052   ≤    μ33-μ36   ≤ 2.39879    No
```

■

Single-Step Procedures for Contrasts

A **contrast** is a linear combination whose coefficients sum to zero. We illustrate this concept in the following example.

EXAMPLE 3: Consider linear combinations of the sample means $\overline{Y}_1, \overline{Y}_2, \overline{Y}_3$, ,, and $\overline{Y}_4$, . Show that the following linear combinations of these sample means are contrasts: (a) $\overline{Y}_2 - \overline{Y}_4$, , (b) $(\overline{Y}_1 + \overline{Y}_3) - (\overline{Y}_2 + \overline{Y}_4)$, (c) $\frac{1}{2}(\overline{Y}_1 + \overline{Y}_3) - \frac{1}{2}(\overline{Y}_2 + \overline{Y}_4)$, and (d) $\overline{Y}_3 - \frac{1}{2}(\overline{Y}_1 + \overline{Y}_4)$.

SOLUTION:

(a) $\overline{Y}_2 - \overline{Y}_4$, can be written as $(0) \cdot \overline{Y}_1 + (1) \cdot \overline{Y}_2 + (0) \cdot \overline{Y}_3 + (-1) \cdot \overline{Y}_4$. The sum of the coefficients of this linear combination is 0 + 1 + 0 + (-1) = 0. Therefore, $\overline{Y}_2 - \overline{Y}_4$, is a contrast.

(b) $(\overline{Y}_1 + \overline{Y}_3) - (\overline{Y}_2 + \overline{Y}_4)$ can be written as $(1) \cdot \overline{Y}_1 + (-1) \cdot \overline{Y}_2 + (1) \cdot \overline{Y}_3 + (-1) \cdot \overline{Y}_4$. Because the sum of the coefficients is 1 + (−1) + 1 + (−1) = 0, it is a contrast.

(c) $\frac{1}{2}(\overline{Y}_1 + \overline{Y}_3) - \frac{1}{2}(\overline{Y}_2 + \overline{Y}_4)$ can be written as $\left(\frac{1}{2}\right) \cdot \overline{Y}_1 + \left(-\frac{1}{2}\right) \cdot \overline{Y}_2 + \left(\frac{1}{2}\right) \cdot \overline{Y}_3 + \left(-\frac{1}{2}\right) \cdot \overline{Y}_4$. Because the sum of the coefficients is $\frac{1}{2} + \left(-\frac{1}{2}\right) + \frac{1}{2} + \left(-\frac{1}{2}\right) = 0$, it is a contrast.

(d) $\overline{Y}_3 - \frac{1}{2}(\overline{Y}_1 + \overline{Y}_4)$ can be written as $\left(-\frac{1}{2}\right) \cdot \overline{Y}_1 + (0) \cdot \overline{Y}_2 + (1) \cdot \overline{Y}_3 + \left(-\frac{1}{2}\right) \cdot \overline{Y}_4$. Because the sum of the coefficients is $\left(-\frac{1}{2}\right) + 0 + 1 + \left(-\frac{1}{2}\right) = 0$, it is a contrast.

■

Pairwise differences in means that lead to the pairwise comparisons discussed previously are contrasts. More generally, contrasts are used to make more sophisticated comparisons. For example, a contrast can be used to test $H_0{:}\mu_1 - \frac{1}{2}(\mu_2 + \mu_3) = 0$. It has also been demonstrated that if $H_0 : \mu_1 = \mu_2 = \ldots = \mu_k$ is rejected using the ANOVA at level of significance α, then at least one contrast is significant at familywise level of significance α. That contrast may be a pairwise difference or a more sophisticated comparison.

We have written procedures for several of the common methods to test a family of contrasts specified by the user. The output is a table of confidence intervals, where a confidence interval of the form $L \le \mu_1 - \frac{1}{2}\mu_2 - \frac{1}{2}\mu_3 \le U$ is used to test the null hypothesis $H_0{:}\mu_1 - \frac{1}{2}(\mu_2 + \mu_3) = 0$. A significance difference is shown when the confidence interval does not cover zero. The table also contains an indication of which contrasts are significantly different from zero.

mcmScheffeContrasts

This procedure performs **Scheffe's test** on the specified family of contrasts. The width of the confidence intervals is not affected by the number of contrasts specified by the user, so Scheffe's test is most useful when a large number of contrasts is specified, not all of which are pairwise. Scheffe's test is exact and directional decisions are protected at the specified familywise level of significance. It is robust to departures from normality. It is less robust to unequal variances, and sensitive to violations of independence.

`mcmScheffeContrasts[dataset,catvar,quanvar,contrastvectors,a]` calculates and displays simultaneous confidence intervals for the contrasts specified in `contrastvectors`. The table also includes an indication of which of the contrasts of population means are significantly different from zero. The groups are defined by the values (assumed to be the first k integers) of a categorical variable, but the groups are assumed to contain quantitative data. The name of the data set is entered first. The data set is assumed to be a rectangular array (a list containing n lists), where each record has at least one column containing a categorical variable and at least one column containing a quantitative variable. The location of the categorical variable (`catvar`) is entered next, followed by the location (`quanvar`) of the quantitative variable. Then the family of contrasts (`contrastvectors`) and the familywise level of significance (`a`) are entered. A contrast is defined by a list of k coefficients, where the jth entry is the coefficient of μ_j in the null hypothesis.

mcmBonferroniContrasts

This procedure is based on the first-order Bonferroni inequality. The width of the confidence intervals is influenced by the number of contrasts in the family specified by the user. Thus, for small families, the confidence intervals may be shorter than for the Scheffe procedure. If the family is a subset of all pairwise comparisons, the confidence intervals may be shorter than for the Tukey pairwise procedure.

`mcmBonferroniContrasts[dataset,catvar,quanvar,contrastvectors,a]` calculates and displays simultaneous confidence intervals for the contrasts specified in `contrastvectors`. The table also includes an indication of which of the contrasts of population means are significantly different from zero. The groups are defined by the values (assumed to be the first k integers) of a categorical variable, but the groups are assumed to contain quantitative data. The name of the data set is entered first. The data set is assumed to be a rectangular array (a list containing n lists), where each record has at least one column containing a categorical variable and at least one column containing a quantitative variable. The location of the categorical variable (`catvar`) is entered next, followed by the location (`quanvar`) of the quantitative variable. Then the family of contrasts (`contrastvectors`) and the familywise level of significance (`a`) are entered. A contrast is defined by a list of k coefficients, where the j^{th}entry is the coefficient of μ_j *in* the null hypothesis in the null hypothesis.

mcmDunnSidakContrasts

This procedure is based on an inequality proved by Sidak, which is a slight improvement on the Bonferroni inequality. Thus, the confidence intervals will be shorter than for the Bonferroni inequality. The width of the intervals is influenced in the same way as for the Bonferroni inequality, so the previous discussion regarding the Scheffe and Tukey procedures also applies here.

`mcmDunnSidakContrasts[dataset,catvar,quanvar,contrastvectors, a]` calculates and displays simultaneous confidence intervals for the contrasts specified in `contrastvectors`. The table also includes an indication of which of the contrasts of population means are significantly different from zero. The groups are defined by the values (assumed to be the first k integers) of a categorical variable, but the groups are assumed to contain quantitative data. The name of the data set is entered first. The data set is assumed to be a rectangular array (a list containing n lists), where each record has at least one column containing a categorical variable and at least one column containing a quantitative variable. The location of the categorical variable (`catvar`) is entered next, followed by the location (`quanvar`) of the quantitative variable. Then the family of contrasts (`contrastvectors`) and the familywise level of significance (`a`) are entered. A contrast is defined by a list of k coefficients, where the jth entry is the coefficient of μ_j in the null hypothesis.

mcmTukeyContrasts

This procedure performs the Tukey test if sample sizes are equal or the Tukey–Kramer test if sample sizes are not all equal. The width of the confidence intervals is not influenced by the number of specified contrasts. When the contrasts involve two or three means, the confidence intervals may be shorter than for the Scheffe procedure. For contrasts involving more than three means, the intervals will be shorter for the Scheffe procedure.

`mcmTukeyContrasts[dataset,catvar,quanvar,contrastvectors,a]` calculates and displays simultaneous confidence intervals for the contrasts specified in `contrastvectors`. The table also includes an indication of which of the contrasts of population means are significantly different from zero. The groups are defined by the values (assumed to be the first k integers) of a categorical variable, but the groups are assumed to contain quantitative data. The name of the data set is entered first. The data set is assumed to be a rectangular array (a list containing n lists), where each record has at least one column containing a categorical variable and at least one column containing a quantitative variable. The location of the categorical variable (`catvar`) is entered next, followed by the location (`quanvar`) of the quantitative variable. Then the family of contrasts (`contrastvectors`) and the familywise level of significance (`a`) are entered. A contrast is defined by a list of k coefficients, where the jth entry is the coefficient of μ_j in the null hypothesis.

Note: Given a family of contrasts, it is appropriate to try several procedures and use the most informative. It cannot be overemphasized, however, that in a given application one must use only one procedure for all contrasts. If different procedures are used to find intervals for the contrasts in the family, then the specified familywise level of significance is not guaranteed.

To use any of these procedures, first load the disk which comes with this book and click on its icon. Next, locate the procedure name and click on the rightmost cell bracket. Then press **ENTER** and return to the *Mathematica* work sheet.

We illustrate the use of these procedures with a simulated data set.

EXAMPLE 4: The list `dataset1` contains four categories where the data were sampled from populations with equal variances, but with means of 4 (category 1), 6 (category 2), 2 (category 3), and 8 (category 4). The categorical variable is the first entry in each record, and the quantitative variable is the second entry in each record. Test for significance of the specified contrasts using a familywise level of significance of 5% and using (a) **`mcmScheffeContrasts`**, (b) **`mcmBonferroniContrasts`**, (c) **`mcmDunnSidakContrasts`**, and (d) **`mcmTukeyContrasts`**.

SOLUTION: First, we simulate the data and then enter it into `dataset1`.

```
dataset1={{1,4.806},{1,4.106},{1,4.33},{1,4.197},
  {1,3.653},{1,3.278},{1,3.54},{1,3.046},{1,2.929},
  {1,2.109},{2,4.948},{2,6.387},{2,5.582},{2,6.554},
  {2,6.538},{2,6.092},{2,5.775},{2,4.726},{2,7.797},
  {2,6.548},{3,1.444},{3,1.16},{3,3.227},{3,1.643},
  {3,1.77},{3,0.2487},{3,1.522},{3,0.6088},{3,2.384},
  {3,2.049},{4,8.262},{4,8.045},{4,7.827},{4,7.399},
  {4,7.1},{4,7.514},{4,8.774},{4,9.083},{4,7.889},
  {4,7.993}};
```

(a) The results obtained with **mcmScheffeContrasts** are shown below.

```
mcmScheffeContrasts[dataset1,1,2,{{1,-1,0,0},

{1/2,1/2,-1/2,-1/2},{0,1/2,1/2,-1},{1,-1/2,-1/2,0}},
  0.05]
```

Selected Multiple Comparisons using the Scheffe procedure.

		Confidence Interval			Significant* Difference
-3.57116	≤	$\mu_1-\mu_2$	≤	-1.41944	Yes
-0.685952	≤	$\mu_1/2+\mu_2/2-\mu_3/2-\mu_4/2$	≤	0.835542	No
-5.09502	≤	$\mu_2/2+\mu_3/2-\mu_4$	≤	-3.23157	Yes
-1.15763	≤	$\mu_1-\mu_2/2-\mu_3/2$	≤	0.705817	No

The familywise confidence level is at least 95. percent.

The familywise level of significance is at most 5. percent.

(b) The results obtained with **mcmBonferroniContrasts** are shown below.

```
mcmBonferroniContrasts[dataset1,1,2,{{1,-1,0,0},
  {1/2,1/2,-1/2,-1/2},{0,1/2,1/2,-1},{1,-1/2,-1/2,0}},
 0.05]
```

Selected Multiple Comparisons using the Bonferroni inequality.

		Confidence Interval			Significant* Difference
-3.46002	≤	$\mu_1-\mu_2$	≤	-1.53058	Yes
-0.607366	≤	$\mu_1/2+\mu_2/2-\mu_3/2-\mu_4/2$	≤	0.756956	No
-4.99877	≤	$\mu_2/2+\mu_3/2-\mu_4$	≤	-3.32782	Yes
-1.06138	≤	$\mu_1-\mu_2/2-\mu_3/2$	≤	0.609569	No

```
The familywise confidence level is at least 95. percent.

The familywise level of significance is at most 5.
percent.
```

(c) Next, we perform the procedure with **mcmDunnSidakContrasts**.

```
mcmDunnSidakContrasts[dataset1,1,2,{{1,-1,0,0},
  {1/2,1/2,-1/2,-1/2},{0,1/2,1/2,-1},{1,-1/2,-1/2,0}},
 0.05]
```

```
Selected Multiple Comparisons using the Dunn-Sidak
Inequality.
```

		Confidence Interval		Significant*	Difference
-3.45716	≤	$\mu_1-\mu_2$	≤	-1.53344	Yes
-0.605345	≤	$\mu_1/2+\mu_2/2-\mu_3/2-\mu_4/2$	≤	0.754935	No
-4.99629	≤	$\mu_2/2+\mu_3/2-\mu_4$	≤	-3.3303	Yes
-1.0589	≤	$\mu_1-\mu_2/2-\mu_3/2$	≤	0.607093	No

```
The familywise confidence level is at least 95. percent.

The familywise level of significance is at most 5.
percent.
```

(d) Finally, we use **mcmTukeyContrasts**.

```
mcmTukeyContrasts[dataset1,1,2,{{1,-1,0,0},
  {1/2,1/2,-1/2,-1/2},{0,1/2,1/2,-1},{1,-1/2,-1/2,0}},
 0.05]
```

```
Selected Multiple Comparisons using the Tukey-Kramer
procedure.
```

		Confidence Interval		Significant*	Difference
-3.90835	≤	$\mu_1-\mu_2$	≤	-1.08225	Yes
-1.33825	≤	$\mu_1/2+\mu_2/2-\mu_3/2-\mu_4/2$	≤	1.48784	No
-5.57634	≤	$\mu_2/2+\mu_3/2-\mu_4$	≤	-2.75025	Yes
-1.63895	≤	$\mu_1-\mu_2/2-\mu_3/2$	≤	1.18714	No

```
The familywise confidence level is at least 95. percent.

The familywise level of significance is at most 5.
percent.
```

Notice that (based on the first contrast) the intervals from the Dunn–Sidak procedure are marginally the best (narrowest) followed in order by the intervals from the Bonferroni, Scheffe, and Tukey–Kramer procedures.

■

Single-Step Procedures for Many-to-One Comparisons

The common form of many-to-one comparisons arises when the means for different experimental treatments are compared to the mean for a control treatment that was designated prior to data collection. Often, the goal is to identify the experimental treatments that have a significantly "better" outcome than the control. This requires a one-sided procedure. If the goal is to identify treatments with significantly "better" or "worse" outcomes, then a two-sided procedure is required. The procedure used for this is the **Dunnett test.** It is an exact test. Because the control group is of such importance, it has been suggested that it contain more values than the other groups. The following rule of thumb tends to optimize the procedure. If each of the experimental groups has n values, then the control group should have $n\sqrt{k-1}$ values. If the experimental groups do not have equal numbers of values, calculate the mean number and multiply by $\sqrt{k-1}$.

We have written a procedure to perform either a one-sided or two-sided Dunnett test. The output is a table of confidence intervals for all pairwise differences of the form $\mu_i - \mu_c$ where μ_i is the mean of an experimental population, and μ_c is the mean of the control population. For a two-sided test, a significant difference is shown when the confidence interval does not cover zero. For a lower one-sided test with corresponding alternative hypothesis $H_a : \mu_i - \mu_c > 0$, reject the null hypothesis when the upper one-sided confidence limit is less than zero. For an upper one-sided test with corresponding alternative hypothesis $H_a : \mu_i - \mu_c > 0$, reject the null hypothesis when the lower one-sided confidence limit is greater than zero. The output table also includes an indication of which pairs of population means are significantly different.

mcmDunnettPairs

`mcmDunnettPairs[yVec,nVec,s,nuValue,opts]` calculates and displays simultaneous confidence intervals for all pairwise differences of the form $u_i - u_c$, where u_i is the mean of an experimental population, and u_c is the mean of the control population. The table also includes an indication of which pairs of population means are significantly different. A list (with length 3 to 20) of sample means (`yVec`) is entered first followed by a list of corresponding sample sizes (`nVec`). These values are available in the output of **`oneWayAnova`**. The mean square for error (`s`) from the ANOVA table is entered next, followed by the degrees of freedom for error (`nuValue`) from the ANOVA table. The options (`opts`) that can be used with **`mcmDunnettPairs`** with default settings are `sided->two` and `confidence->0.95`. The possible settings for `sided->` are `LowerOneSided` and `UpperOneSided`. The possible settings for `confidence->` include all values between 0.80 and 0.99.

To use this procedure, first load the disk which comes with this book and click on its icon. Next, locate the procedure name and click on the rightmost cell bracket. Then press **ENTER** and return to the *Mathematica* work sheet.

EXAMPLE 5: Consider five samples having sample means $\bar{Y}_1 = 10.0$, $\bar{Y}_2 = 13.5$, $\bar{Y}_3 = 20.0$, $\bar{Y}_4 = 14.0$, and $\bar{Y}_5 = 17.5$ with corresponding sample sizes $n_1 = 9$, $n_2 = 9$, $n_3 = 9$, $n_4 = 18$, and $n_5 = 9$. Assuming a familywise confidence coefficient of 0.95 and $MSE = \sqrt{12}$ and $DFE = 40$ from the ANOVA table, use the Dunnett test to determine which experimental means differ from the control mean, μ_4.

SOLUTION: First we enter the sample means in the list `ybar`; the sample sizes in the list `n`, the MSE from the ANOVA table as `s`; and the DFE from the ANOVA table as `nu`. **Note:** The control sample mean must be last in the list, `ybar`. Also, the sample size which corresponds to a sample mean must have the same relative position in `n` as that sample mean has in `ybar`.

```
Clear[n,ybar,s,nu]
ybar={10.0,13.5,20.0,17.5,14.0};
n={9,9,9,9,18};
s=Sqrt[12]//N;
nu=40;
```

We use **`mcmDunnettPairs`** to investigate the pairwise differences of each experimental mean with the control mean. We also show the results obtained with the option settings `sided->UpperOneSided` and `sided->LowerOneSided`. Finally, we demonstrate the use of the option `confidence->` and the output if one or more of the limits on the procedure is exceeded.

```
mcmDunnettPairs[ybar,n,s,nu]
```

```
MULTIPLE PAIRWISE COMPARISONS WITH A CONTROL USING
DUNNETT'S PROCEDURE
```

i	c	Confidence Interval	Significant Difference
1	5	$-7.6416 \leq \mu_1 - \mu_5 \leq -0.3584$	Yes
2	5	$-4.1416 \leq \mu_2 - \mu_5 \leq 3.1416$	No
3	5	$2.3584 \leq \mu_3 - \mu_5 \leq 9.6416$	Yes
4	5	$-0.1416 \leq \mu_4 - \mu_5 \leq 7.1416$	No

The familywise confidence coefficient is at least 95. percent.

The familywise level of significance is at most 5. percent.

mcmDunnettPairs[ybar,n,s,nu,sided->UpperOneSided]

MULTIPLE PAIRWISE COMPARISONS WITH A CONTROL USING DUNNETT'S PROCEDURE

i	c	Confidence Interval	Significant Difference
1	5	$-7.21557 \le \mu_1 - \mu_5$	No
2	5	$-3.71557 \le \mu_2 - \mu_5$	No
3	5	$2.78443 \le \mu_3 - \mu_5$	Yes
4	5	$0.284432 \le \mu_4 - \mu_5$	Yes

The familywise confidence coefficient is at least 95. percent.

The familywise level of significance is at most 5. percent.

mcmDunnettPairs[ybar,n,s,nu,sided->LowerOneSided]

MULTIPLE PAIRWISE COMPARISONS WITH A CONTROL USING DUNNETT'S PROCEDURE

i	c	Confidence Interval	Significant Difference
1	5	$\mu_1 - \mu_5 \le -0.784432$	Yes
2	5	$\mu_2 - \mu_5 \le 2.71557$	No
3	5	$\mu_3 - \mu_5 \le 9.21557$	No
4	5	$\mu_4 - \mu_5 \le 6.71557$	No

The familywise confidence coefficient is at least 95. percent. The familywise level of significance is at most 5. percent.

mcmDunnettPairs[ybar,n,s,nu,sided->UpperOneSided,confidence->0.79]

The mcmDunnettPairs test calculations are valid only when the familywise confidence coefficient is between 80% and 99% (familywise level of significance between 0.01 and 0.20), the number of groups is between 3 and 20, and an average correlation coefficient calculated

```
from the data is between 0.1 and 0.7.  One or more of
these conditions is not satisfied.
```

■

Step-Down Procedures

Step-down procedures are multiple step procedures that begin by testing all k means. If significance is found, then all subsets consisting of $k - 1$ means are tested. In general, if a set consisting of p means is significant, then all of its subsets of size $p - 1$ are tested. The purpose is to identify the maximal nonsignificant subsets of the k means. A set is **maximal nonsignificant** if it is not significant and not a proper subset of another nonsignificant set. Significance or nonsignificance of any set of means can be ascertained from the list of maximal nonsignificant subsets. All of the means which occupy the same maximal nonsignificant subset are not significantly different, and means which do not occupy the same maximal nonsignificant subset are significantly different. The comparisons may be extended to sets of means in a like manner.

Two well-known step-down procedures are the Student–Newmann–Keuls and Duncan multiple range tests. As they were originally presented and are commonly used, these tests do not guarantee that the familywise level of significance is not exceeded for every possible configuration of the unknown population means. In other words, the chance of being wrong about one or more of the differences identified by these procedures will likely be larger than specified by the familywise level of significance. The procedures presented here do protect the specified familywise level of significance under all possible configurations of the population means. They are generally more powerful than comparable single-step procedures; however, methods for developing corresponding confidence intervals are not available, and it is not known whether the chance of making an error in directional decisions is protected at the specified familywise level of significance. For example, there is uncertainty in deciding μ_1 is different from μ_2, and, knowing the difference exists, there is uncertainty in deciding that $\mu_1 < \mu_2$ (a directional decision) based on $y_1 < y_2$. For single-step procedures, both inferences are protected at the specified familywise level of significance. Whether the second case is protected for the step-down procedures is not known.

We have written procedures for four step-down multiple range procedures. The first two follow the same algorithm and differ only with respect to the probability distribution used to determine statistical significance. Their development is attributed to various authors, and they may be called by different names depending on the reference consulted. The authors credited include Einot, Gabriel, Ryan, Shaffer, and Welsch. The second two procedures incorporate what is generally called the Peritz method. This method takes the algorithm from the first two procedures and extends it in a way that makes the Peritz procedure more powerful. That is, the Peritz method finds at least as many significant differences. The two Peritz procedures differ only with respect to the probability distribution used to determine statistical significance. The F in the names refers to the *F* (ratio) distribution, and the Q refers to the **studentized range** distribution. The latter is the distribution used to perform the Tukey test. Although these procedures will work with unbalanced groups (unequal sample sizes), some references suggest that they are most accurate when all sample sizes are equal. The output of each of the procedures is a table which classifies the pairwise comparisons as significant or not and displays the maximal nonsignificant subsets.

mcmMultipleF

`mcmMultipleF[ybar,n,s,nu,a]` performs a step-down multiple range procedure, displays a table which classifies the pairwise comparisons as significant or not, and displays the maximal nonsignificant subsets. The procedure uses the *F* (ratio) distribution. A list of sample means *ordered from smallest to largest* (`yVec`) is entered first, followed by a list of the corresponding sample sizes (`nVec`). All of these values are available in the output of **`oneWayAnova`**. The mean square for error (`s`) from the ANOVA table is entered next, followed by the degrees of freedom for error (`nu`) from the ANOVA table. The final entry (`a`) is the familywise level of significance.

mcmMultipleQ

`mcmMultipleQ[ybar,n,s,nu,a]` performs a step-down multiple range procedure, displays a table which classifies the pairwise comparisons as significant or not, and displays the maximal nonsignificant subsets. The procedure uses the studentized range distribution. A list of sample means *ordered from smallest to largest* (`yVec`) is entered first, followed by a list of corresponding sample sizes (`nVec`). All of these values are available in the output of **`oneWayAnova`**. The mean square for error (`s`) from the ANOVA table is entered next, followed by the degrees of freedom for error (`nu`) from the ANOVA table. The final entry (`a`) is the familywise level of significance.

mcmPeritzF

`mcmPeritzF[ybar,n,s,nu,a]` performs the Peritz step-down multiple range procedure, displays a table which classifies the pairwise comparisons as significant or not, and displays the maximal nonsignificant subsets. The procedure uses the *F* (ratio) distribution. A list of sample means *ordered from smallest to largest* (`yVec`) is entered first, followed by a list of corresponding sample sizes (`nVec`). All of these values are available in the output of **`oneWayAnova`**. The mean square for error (`s`) from the ANOVA table is entered next, followed by the degrees of freedom for error (`nu`) from the ANOVA table. The final entry (`a`) is the familywise level of significance.

mcmPeritzQ

`mcmPeritzQ[ybar,n,s,nu,a]` performs the Peritz step-down multiple range procedure, displays a table which classifies the pairwise comparisons as significant or not, and displays the maximal nonsignificant subsets. The procedure uses the studentized range distribution. A list of sample means *ordered from smallest to largest* (`yVec`) is entered first, followed by a list of corresponding sample sizes (`nVec`). All of these values are available in the output of **`oneWayAnova`**. The mean square for error (`s`) from the ANOVA table is entered next, followed by the degrees of freedom for error (`nu`) from the ANOVA table. The final entry (`a`) is the familywise level of significance.

To use any of these procedures, first load the disk which comes with this book and click on its icon. Next, locate the procedure name and click on the rightmost cell bracket. Then press **ENTER** and return to the *Mathematica* work sheet.

EXAMPLE 6: Consider five samples having sample means $\overline{Y}_1 = 10.0$, $\overline{Y}_2 = 13.5$, $\overline{Y}_3 = 20.0$, $\overline{Y}_4 = 14.0$, and $\overline{Y}_5 = 17.5$ with corresponding sample sizes $n_1 = 9$, $n_2 = 9$, $n_3 = 9$, $n_4 = 9$, and $n_5 = 9$. Assuming a familywise level of significance of 0.05 and $MSE = \sqrt{12}$ and $DFE = 40$ from the ANOVA table, use each step-down procedure to find the maximal nonsignificant subsets.

SOLUTION: First we enter the sample means in assending order in the list `ybar`. If the sample sizes were not all equal, we would be careful to list them in the same position as the corresponding sample mean. Even though they are all equal, it is necessary to enter them all in `n`. We then enter the mean square for error in `s`, the degrees of freedom in `nu`, and the familywise level of significance in `a`. Next we load and run our procedure **`mcmMultipleF`**. In addition to indicating the pairs that differ significantly, the output list also indicates the maximal nonsignificant subsets. The list {0, 1, 1, 1, 0} means that a significant difference cannot be shown among μ_2, μ_3,, and μ_4. The list {1, 1, 1, 0, 0} indicates that a significant difference cannot be shown among μ_1, μ_2, and μ_3. The list {0, 0, 0, 1, 1} indicates that a significant difference cannot be shown between μ_4 and μ_5.

```
Clear[n,ybar,s,nu,a]
ybar={10.0,13.5,14.0,17.5,20.0};
n={9,9,9,9,9};
s=Sqrt[12]//N;
nu=40;
a=0.05;
```

mcmMultipleF[ybar,n,s,nu,a]

i	j	$\bar{Y}_i \bar{Y}_j$	Significant Difference
4	5	-2.5	No
3	5	-6.	Yes
3	4	-3.5	No
2	5	-6.5	Yes
2	4	-4.	No
2	3	-0.5	No
1	5	-10.	Yes
1	4	-7.5	Yes
1	3	-4.	No
1	2	-3.5	No

The Maximal Nonsignificant Subsets are

{{0,1,1,1,0},{1,1,1,0,0},{0,0,0,1,1}}

Next we load and the run our procedure **mcmMultipleQ**.

mcmMultipleQ[ybar,n,s,nu,a]

i	j	$\bar{Y}_i \bar{Y}_j$	Significant Difference
4	5	-2.5	No
3	5	-6.	Yes
3	4	-3.5	No
2	5	-6.5	Yes
2	4	-4.	No
2	3	-0.5	No
1	5	-10.	Yes
1	4	-7.5	Yes
1	3	-4.	No
1	2	-3.5	No

The Maximal Nonsignificant Subsets are
{{0,1,1,1,0},{1,1,1,0,0},{0,0,0,1,1}}

Next, we have the results obtained from loading and running our procedure **mcmPeritzF** follow by the output from loading and running our procedure **mcmPeritzQ**. Notice that the Peritz method identifies an additional significant difference.

i	j	$\bar{Y}_i \bar{Y}_j$	Significant Difference
4	5	-2.5	No
3	5	-6.	Yes

```
3        4        -3.5        No
2        5        -6.5        Yes
2        4        -4.         No
2        3        -0.5        No
1        5        -10.        Yes
1        4        -7.5        Yes
1        3        -4.         No
1        2        -3.5        No

The Maximal Nonsignificant Subsets are
 {{1,1,1,0,0},{0,0,1,1,0},{0,0,0,1,1}}
```

mcmPeritzQ[ybar,n,s,nu,a]

i	j	$\bar{Y}_i \bar{Y}_j$	Significant Difference
4	5	-2.5	No
3	5	-6.	Yes
3	4	-3.5	No
2	5	-6.5	Yes
2	4	-4.	No
2	3	-0.5	No
1	5	-10.	Yes
1	4	-7.5	Yes
1	3	-4.	No
1	2	-3.5	No

```
The Maximal Nonsignificant Subsets are
 {{1,1,1,0,0},{0,0,1,1,0},{0,0,0,1,1}}
```

■

9.3 Two-Factor Analysis of Variance

A simultaneous analysis of the effect of more than one factor on the population means is known as a **factorial analysis of variance.** The advantage of the simultaneous analysis is that, in addition to testing each factor, the interactions among factors can be tested. For the two-factor ANOVA, we designate one factor as A and the other as B. We assume that both are categorical and that A has a categories (called levels) and B has b categories. Thus, using the vocabulary of the statistician, we would say that factor A has a levels and factor B has b levels. There is assumed to be a population corresponding to each factor level combination, and samples from these populations form the $a \times b$ groups in the analysis. For example, we may classify cars according to size (factor A) as light, intermediate, or heavy and according to whether or not they have air bags (factor B). Then factor A has three levels and factor B has two levels, and there are six groups to include in the analysis: light with air bag, light without air bag, intermediate with air bag, intermediate without air bag, heavy with air bag, and heavy without air bag. To test the effect of each of the factors on head injuries, we assume that each of the six groups contains a random sample of head injury data from a population of cars having the appropriate size and air bag configuration. In general, it is advantageous to have equal group sizes, which is often described by calling the groups **balanced**. In this section we will assume that each group contains n (the same number of) observations. Note that two-factor ANOVA is often called **two-way ANOVA**. There is a possible source of confusion in using this designation, because in some introductory textbooks and references, two-way ANOVA refers to the analysis of a particular case of having one observation per group. This case is called a (complete) randomized block design.

An **interaction** between two factors means that the effect of one factor is not independent of the presence of a particular level of the second factor. For example, if there is no interaction between car size and air bags, then one would expect the difference in the means for the groups "light with air bag" and "light without air bag" to be the same as the difference in the means for the groups "intermediate with air bag" and "intermediate without air bag" and the groups "heavy with air bag" and "heavy without air bag." If, on the other hand, the difference for light cars is very large, but the difference for heavy cars is quite small, then an interaction is present. This interaction might be described by saying, "Light cars benefit more from air bags than do heavy cars."

Assumptions

1. There are two factors: factor A with a levels and factor B with b levels. There are $k = a \times b$ groups that represent k independent, random samples, one from each of k populations, where the population means are denoted by μ_{ij}, $i = 1, \ldots, a$, and $j = 1, \ldots, b$. The sample sizes are all equal to,n
2. Each population is normally distributed. Note that if, based on the central limit theorem, the sampling distribution of each sample mean is approximately normal, then the method is still appropriate.
3. The population variances are equal.

To determine whether these assumptions are reasonably met, there are several diagnostic procedures available. These are discussed in Chapter 10. Also discussed are remedial procedures to transform the data set so that it better satisfies the assumptions. It would then be appropriate to use the two-factor ANOVA on the transformed data. In case the data set cannot meet the assumptions, an alternative method, the Friedman test, may help. It is discussed in Chapter 12.

Hypotheses

There are three sets of hypotheses to be tested. We use the following notation to express the hypotheses symbolically. Let μ_i represent the mean effect at the ith level of factor A, $i = 1, \ldots, a$. Let μ_i represent the mean effect at the jth level of factor B, $j = 1, \ldots, b$. Let $\mu_{..}$ represent the overall mean for all levels of both factors. As noted previously, μ_{ij} represents the mean effect at combined i^{th} level of factor A and j^{th} level of factor B. Then the hypotheses to be tested are as follows.

1. The null hypothesis is that there is no interaction: $H_0 : \mu_{ij} - \mu_{1.} - \mu_{.j} + \mu_{..} = 0$ for all i, j. The alternative hypothesis is that there is an interaction: $H_a : \mu_{ij} - \mu_{1.} - \mu_{.j} + \mu_{..} \neq 0$ for some i, j.
2. The null hypothesis is that there is no effect of factor A: $A : H_0 : \mu_{1.} = \ldots = \mu_{a}$. The alternative hypothesis is that there is an effect of factor A: H_a : *not all* $\mu_{i.}$ *are equal.*
3. The null hypothesis is that there is no effect of factor B: $B : H_0 : \mu_{1.} = \ldots = \mu_{.b}$. The alternative hypothesis is that there is an effect of factor B: H_a : *not all* $\mu_{.j}$ *are equal.*

ANOVA Table Calculations

We will represent the lth data value in the group corresponding to the ith level of factor A and the jth level of factor B by y_{ijt}, where $l = 1, \ldots, n$, $i = 1 \ldots, a$, and $j = 1, \ldots, b$. The mean of the group corresponding to the ith level of factor A and the jth level of factor B is denoted by

$$\bar{y}_{ij.} = \frac{1}{n} \sum_{t=1}^{n} \bar{y}_{ijt}.$$

The mean of the data corresponding to the ith level of factor A is denoted by

$$\bar{y}_{i..} = \frac{1}{bn} \sum_{j=1}^{b} \sum_{t=1}^{n} \bar{y}_{ijt}.$$

The mean of the data corresponding to the jth level of factor B is denoted by.

$$\bar{y}_{.j.} = \frac{1}{an} \sum_{i=1}^{a} \sum_{t=1}^{n} \bar{y}_{ijt}.$$

The grand mean of all of the data is denoted by .

$$\bar{y}_{...} = \frac{1}{abn}\sum_{i=1}^{a}\sum_{j=1}^{b}\sum_{t=1}^{n}\bar{y}_{ijt}.$$

The sum of squares and degrees of freedom for total are calculated by

$$SST = \sum_{i=1}^{a}\sum_{j=1}^{b}\sum_{t=1}^{n}(\bar{y}_{ijt}-\bar{y}_{...})^2 \text{ and } DFT = abn-1.$$

The sum of squares and degrees of freedom for groups are calculated by

$$SSG = n\sum_{i=1}^{a}\sum_{j=1}^{b}(\bar{y}_{ij.}-\bar{y}_{...})^2 \text{ and } DFG = ab-1.$$

The sum of squares and degrees of freedom for error are calculated by

$$SSE = \sum_{i=1}^{a}\sum_{j=1}^{b}\sum_{t=1}^{n}(\bar{y}_{ijt}-\bar{y}_{ij.})^2 \text{ and } DFE = ab(n-1).$$

Note that $SST = SSG + SSE$ and $DFT = DFG + DFE$. In order to test the hypotheses, the sum of squares for groups is divided (partitioned) into three parts corresponding to factor A, factor B, and the interaction. The degrees of freedom for groups is divided similarly. The sum of squares and degrees of freedom for factor A are calculated by

$$SSA = bn\sum_{i=1}^{a}(\bar{y}_{i..}-\bar{y}_{...})^2 \text{ and } DFA = a-1.$$

The sum of squares and degrees of freedom for factor B are calculated by

$$SSB = an\sum_{j=1}^{b}(\bar{y}_{.j.}-\bar{y}_{...})^2 \text{ and } DFB = b-1.$$

The sum of squares and degrees of freedom for the A by B interaction are calculated by

$$SSAB = n\sum_{i=1}^{a}\sum_{j=1}^{b}(\bar{y}_{ij.}-\bar{y}_{i..}-\bar{y}_{.j.}+\bar{y}_{...})^2 \text{ and } DFAB = (n-1)(b-1).$$

Note that $SSAB = SSG - SSA - SSB$ and $DFAB = DFG - DFA - DFB$.

Test Statistic

As with single-factor ANOVA, the test statistics are the ratio of two mean squares, where the mean square in the numerator corresponds to the factor or interaction being tested. To determine the mean square for the denominator, the analyst needs to understand the difference between a fixed factor and a random factor and determine which case best describes the factors of the data being analyzed. When the levels of a factor are specifically chosen and interest is centered only on them, the factor is called a **fixed factor**. When the levels of a factor are not of intrinsic interest in themselves but constitute a sample from a larger population of possible levels, the factor is call a **random factor**.

Model I ANOVA: Both Factors Fixed

The test statistic for factor A is

$$F = \frac{MSA}{MSE},$$

which is assumed to have an F (ratio) distribution with numerator degrees of freedom $a - 1$ and denominator degrees of freedom $ab(n - 1)$.

The test statistic for factor B is

$$F = \frac{MSB}{MSE},$$

which is assumed to have an F (ratio) distribution with numerator degrees of freedom $b - 1$ and denominator degrees of freedom $ab(n - 1)$.

The test statistic for the A by B interaction is

$$F = \frac{MSAB}{MSE},$$

which is assumed to have an F (ratio) distribution with numerator degrees of freedom $(a - 1)(b - 1)$ and denominator degrees of freedom $ab(n - 1)$.

Model II ANOVA: Both Factors Random

The test statistic for factor A is

$$F = \frac{MSA}{MSAB},$$

which is assumed to have an F (ratio) distribution with numerator degrees of freedom $(a - 1)$ and denominator degrees of freedom $(a - 1)(b - 1)$.

The test statistic for factor B is

$$F = \frac{MSB}{MSAB},$$

which is assumed to have an F (ratio) distribution with numerator degrees of freedom (b – 1), and denominator degrees of freedom (a – 1)(b – 1).

The test statistic for the A by B interaction is,

$$F = \frac{MSAB}{MSE},$$

which is assumed to have an F (ratio) distribution with numerator degrees of freedom (a – 1)(b – 1), and denominator degrees of freedom $ab(n - 1)$.

Model III ANOVA: Factor *A* Fixed and Factor *B* Random

The test statistic for factor A is,

$$F = \frac{MSA}{MSAB},$$

which is assumed to have an F (ratio) distribution with numerator degrees of freedom $a - 1$ and denominator degrees of freedom (a – 1)(b – 1.

The test statistic for factor B is,

$$F = \frac{MSB}{MSE},$$

which is assumed to have an F (ratio) distribution with numerator degrees of freedom $b - 1$ and denominator degrees of freedom $ab(n - 1)$.

The test statistic for the A by B interaction is,

$$F = \frac{MSAB}{MSE},$$

which is assumed to have an F (ratio) distribution with numerator degrees of freedom (a – 1)

(b – 1) and denominator degrees of freedom $ab(n - 1)$.

Procedure

We have written two *Mathematica* procedures to perform two-factor analysis of variance which differ only in the form of data entry. If the data are presented in a table as, for example, one might find in a textbook, then **`twoWayANOVA`** is used. If the data are in the form of an array where the first two columns of each record contain values of categorical variables and the third column contains a quantitative variable, then **`twoWayANOVAarray`** is used. Both procedures produce a two-way ANOVA table which includes sums of squares, mean squares, test statistics, and p-values. Both procedures have options to permit the analysis of Model I, Model II, and

Model III data. We have also written procedures to plot the group means. The procedure **groupMeanPlot** accepts the data in the same format as is used with **twoWayANOVA**, and **groupMeanPlotArray** accepts the data in the format used with **twoWayANOVAarray**.

twoWayANOVA[data,opts] performs a two-factor ANOVA and displays an ANOVA table which includes sums of squares, mean squares, test statistics, and *p*-values to test for an interaction and for the effect of each factor. The data are entered as a list of groups where each group has the same number of values. The first group in the list corresponds to the first level of factor *A* and the first level of factor *B*. The second group in the list corresponds to the first level of factor *A* and the second level of factor *B*. When all groups corresponding to the first level of factor *A* have been entered in order corresponding to the levels of factor *B*, the groups corresponding to the second level of factor *A* are entered in the same order corresponding to the levels of factor *B*. The same orderings are followed until all of the data has been entered. The default option is model->fixed, resulting in a Model I (both factors fixed) ANOVA. Other options are random, resulting in a Model II (both factors random) ANOVA; mixed, resulting in a Model III (first factor fixed; second factor random) ANOVA; and single- ValuePerCell, when the data set has exactly one data value per cell.

twoWayANOVAarray[data,opts] performs a two-factor ANOVA and displays an ANOVA table which includes sums of squares, mean squares, test statistics, and *p*-values to test for an interaction and for the effect of each factor. The data are entered in the form of an array where for each record the first column contains the level (assumed to be from the first *a* integers) of factor *A*, the second column contains the level (assumed to be from the first *b* integers) of factor *B*, and the third column contains a corresponding data value. The default option is model->fixed, resulting in a Model I (both factors fixed) ANOVA. Other options are random, resulting in a Model II (both factors random) ANOVA; mixed, resulting in a Model III (first factor fixed; second factor random) ANOVA; and singleValuePerCell when the data set has exactly one data value per cell.

groupMeanPlotArray[data,{factornum,symbol},{factornum,symbol},opts] produces group mean plots for the given data which must be in the form described for **twoWayANOVA**. The first ordered pair specifies which factor will be displayed on the horizontal axis. Enter either 1 or 2 for factornum. The default for symbol is *A*, but a word or symbol can be specified. The second ordered pair will correspond to the factor which is not defined on the horizontal axis. The default for symbol is *B*. The vertical axis is defined by the values of the group means. Any option for the built-in function Show may be specified in the opts position.

groupMeanPlotArray [data, { factornum, symbol }, { factornum, symbol }, quanvar, opts] produces group mean plots for the given data which must be in the form described for **twoWayANOVAarray**. The first ordered pair specifies which factor will be displayed on the horizontal axis. Enter either 1 or 2 for factornum. The default for symbol is *A*, but a word or symbol can be specified. The second ordered pair will correspond to the factor which is not defined on the horizontal axis. The default for symbol is *B*. The column location of the quantitative variable is entered next for quanvar. The vertical axis is defined by the values of the group means. Any option for the built-in function Show may be specified in the opts position.

To use any of these procedures, first load the disk which comes with this book and click on its icon. Next, locate the procedure name and click on the rightmost cell bracket. Then press **ENTER** and return to the *Mathematica* work sheet.

EXAMPLE 7: In a test to study the effects of technician and brand of disk drive on service time, each of three technicians was required to complete five repair jobs on each of three brands of disk drive. The number of minutes needed to complete each task is shown in Table 9.2. (The problem and data were taken from the textbook *Applied Linear Statistical Models* by Netter *et al.*) Generate a two-factor analysis of variance table assuming (a) fixed factors, (b) random factors, and (c) factor A fixed and factor B random. (d) Produce group means plots for the data.

	Factor B **(brand of disk drive)**		
Factor A **(technician)**	**Brand 1**	**Brand 2**	**Brand 3**
Technician 1	62.0	57.0	59.0
	48.0	45.0	53.0
	63.0	39.0	67.0
	57.0	54.0	66.0
	69.0	44.0	47.0
Technician 2	51.0	61.0	55.0
	57.0	58.0	58.0
	45.0	70.0	50.0
	50.0	66.0	69.0
	39.0	51.0	49.0
Technician 3	59.0	58.0	47.0
	65.0	63.0	56.0
	55.0	70.0	51.0
	52.0	53.0	44.0
	70.0	60.0	50.0

Table 9.2 Repair Time (Minutes)

SOLUTION: First, in `diskDrive`, we enter the data for Technician 1 (Brand 1 followed by Brand 2 followed by Brand 3), then Technician 2 (brands in the same order), and then Technician 3 (brands in the same order). (a) Having loaded **`twoWayANOVA`**, we use it to produce the two-factor analysis of variance table. Note that the default setting of **`twoWayANOVA`** is `model->fixed`. The value $F \approx 5.84$ with p-value under 0.001 suggests that there is a significant interaction. The value of $F \approx 0.24$ with approximate p-value 0.79 indicates that we fail to reject the null hypothesis for technicians. We cannot find a difference among technicians in mean service time. Similarly, for $F \approx 0.27$ and p-value approximately 0.76, we fail to reject the null hypothesis for brands. We cannot find a difference among brands in mean service time.

```
diskDrive={{{62.0,48.0,63.0,57.0,69.0},{57.0,45.0,39.0,54.0,44.0},
  {59.0,53.0,67.0,66.0,47.0}},{{51.0 ,57.0,45.0,50.0,39.0},
  {61.0,58.0,70.0,66.0,51.0},{55.0,58.0,50.0,69.0,49.0}},
 {{59.0,65.0,55.0,52.0,70.0},{58.0,63.0,70.0,53.0,60.0},
  {47.0,56.0,51.0,44.0,50.0}}};

twoWayAnova[diskDrive]
```

Analysis of Variance Table

Source	Sum of Squares	DF	Mean Squares	Fratio	P-value
Factor A	24.5778	2	12.2889	0.236274	0.790779
Factor B	28.3111	2	14.1556	0.272164	0.763283
A B Interaction	1215.29	4	303.822	5.84149	0.000994107
Error	1872.4	36	52.0111		
Total	3140.58	44			

(b) Next, we use the `model->random` option to produce the two-factor analysis of variance assuming Model II.

```
twoWayAnova[diskDrive,model->random]
```

Analysis of Variance Table

Source	Sum of Squares	DF	Mean Squares	Fratio	P-value
Factor A	24.5778	2	12.2889	0.236274	0.790779
Factor B	28.3111	2	14.1556	0.272164	0.763283
A B Interaction	1215.29	4	303.822	5.84149	0.000994107
Error	1872.4	36	52.0111		
Total	3140.58	44			

(c) With the option setting, `model->mixed`, we assume the mixed model (Model III), where the procedure assumes the first factor (factor A) is fixed and the second factor (factor B) is random.

```
twoWayAnova[diskDrive,model->mixed]
```

Analysis of Variance Table

Source	Sum of Squares	DF	Mean Squares	Fratio	P-value
Factor A	24.5778	2	12.2889	0.236274	0.790779
Factor B	28.3111	2	14.1556	0.272164	0.763283
A B Interaction	1215.29	4	303.822	5.84149	0.000994107
Error	1872.4	36	52.0111		
Total	3140.58	44			

(d) After loading our procedure, `groupMeanPlot`, we use it to plot the means of each group. In the first case, we assign technicians to the horizontal axis. In the second plot we assign brands to the horizontal axis. The first plot helps to describe the interaction between technicians and brand. Each technician is most efficient on a different brand.

```
groupMeanPlot [diskDrive , {{1, Tech }, {2, Brand }}]
```

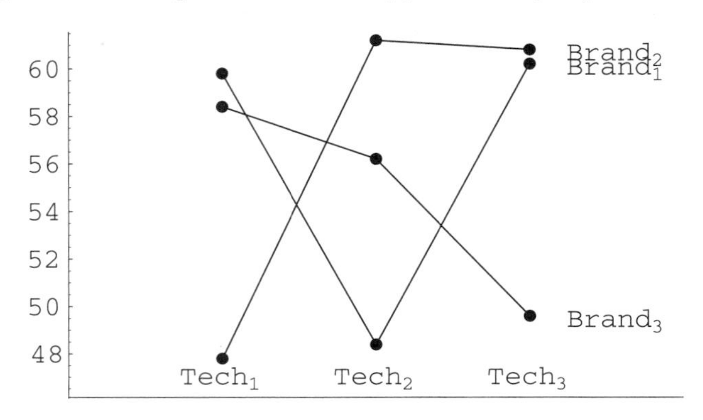

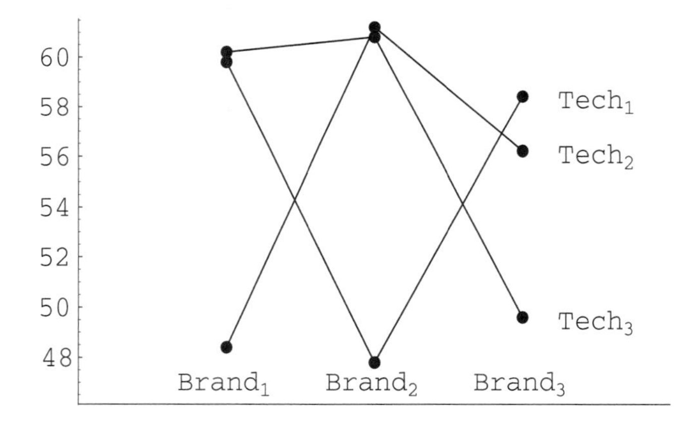

EXAMPLE 8: The text file **Crash.txt** contains data from experiments in which automobiles with dummies in the driver and front passenger seats were crashed into walls at 35 miles per hour. The variables are described in Table 9.1, which is in Example 2 of Section 9.1. Use **twoWayAnova** to perform two factor analysis of variance to test for differences in automobile size (light, intermediate, or heavy) and whether or not they have air bags (yes or no) with respect to "head injury."

SOLUTION: We begin by entering the data in Crash.txt into the list `cardata`. Following a procedure similar to that in Example 2 of Section 9.1, we code the variables size and protection. For size, 1 represents light, 2 represents intermediate, and 3 represents heavy. For protection, 1 represents the presence of airbags and 2 represents the absence of airbags.

```
cardata=ReadList["Crash.txt",Word,
  RecordLists->True,WordSeparators->{"\t"}] /.
  {"-"->null,"nd"->null};

<<Statistics`DataManipulation`

cardata=ColumnDrop[cardata,{2,4}];

sizes=Union[Column[cardata,11]];

rule2={sizes[[1]]->1,sizes[[2]]->3,
  sizes[[3]]->1,sizes[[4]]->2,sizes[[5]]->2,
  sizes[[6]]->3,sizes[[7]]->2,sizes[[8]]->3};

rule3={"d airbag"\->1,"d&p airbags"\->1,
  "manual belts"\->2,"Motorized belts"\->2,
  "passive belts"\->2};

cardata=cardata  /. rule2 /. rule3;

cardata=cardata//ToExpression;
```

In Example 2 of Section 9.1, we decided to use its natural logarithm to represent "head injury." In the next commands we create a new list called `injury` and display the first record. Notice that for each record the first column represents size, the second column represents airbag, and the third column represents the logarithm of "head injury."

```
f[list_]:={Log[list[[2]]]//N};
logs=Map[f,cardata];
size=ColumnTake[cardata,{11}];
airbag=ColumnTake[cardata,{7}];
injury=RowJoin[size,airbag,logs];
injury[[1]]
```

```
{1,2,6.39526}
```

Next we delete all records that have missing data and rename the file `headinjury`. Because **twoWayANOVAarray** expects the data to be sorted on column 1 and on column 2 within column 1 and expects equal numbers in each group, we sort `headinjury` and then determine the number of observations in each group.

```
headinjury=DropNonNumeric[injury];
Length[headinjury]
```

```
340
```

```
headinjury=Sort[headinjury];
try2=Flatten[Table[Select[headinjury,
  #[[1]]===i&&#[[2]]==j&],
  {i,1,3},{j,1,2}],1]
Map[Length,try2]
```

```
{20,136,28,80,10,66}
```

The smallest group has 10, so we randomly choose 10 records from each group using our procedure **simpleRandomSample**. The resulting list is called `try3`.

```
try3=Flatten[Map[simpleRandomSample[#,10]&,try2],1]
```

```
{{1,1,5.69373},{1,1,5.83773},{1,1,6.10032},{1,1,6.12468},
 {1,1,6.20051},{1,1,6.22059},{1,1,6.6107},{1,1,6.63595},
 {1,1,6.72623},{1,1,6.94312},{1,2,6.03548},{1,2,6.05209},
 {1,2,6.30262},{1,2,6.40192},{1,2,6.8222},{1,2,6.86693},
 {1,2,6.9256},{1,2,6.93147},{1,2,6.95368},{1,2,7.25912},
 {2,1,5.55296},{2,1,6.07535},{2,1,6.23637},{2,1,6.28972},
 {2,1,6.36819},{2,1,6.56526},{2,1,6.59987},{2,1,6.63988},
 {2,1,6.68461},{2,1,7.62217},{2,2,5.60947},{2,2,6.31173},
 {2,2,6.52942},{2,2,6.56808},{2,2,6.62539},{2,2,6.85857},
 {2,2,7.14677},{2,2,7.21377},{2,2,7.32647},{2,2,7.47534},
 {3,1,6.04737},{3,1,6.15486},{3,1,6.19848},{3,1,6.24804},
```

```
{3,1,6.27852},{3,1,6.52503},{3,1,6.66696},{3,1,6.76041},
{3,1,7.00397},{3,1,7.66482},{3,2,6.61204},{3,2,6.73459},
{3,2,6.85646},{3,2,7.14598},{3,2,7.29777},{3,2,7.355},
{3,2,7.35756},{3,2,7.4495},{3,2,7.49332},{3,2,7.60937}}
```

Finally, using **twoWayANOVAarray** and **groupMeanPlotArray**, we find significant differences for both size and airbags and display the group means. Although the plots suggest a possible interection, it is not statistically significant.

```
twoWayANOVAarray[try3]
```

Analysis of Variance Table

Source	Sum of Squares	DF	Mean Squares	Fratio	P-value
Factor A	1.57934	2	0.789669	3.64539	0.0327313
Factor B	2.75256	1	2.7526	12.707	0.000771634
A B Interaction	0.328836	2	0.164418	0.759011	0.473057
Error	11.6975	54	0.216621		
Total	16.3583	59			

```
groupMeanPlotArray[try3,{{1,Size},{2,Airbag}},3,
  AxesOrigin->{0,6.2}]
```

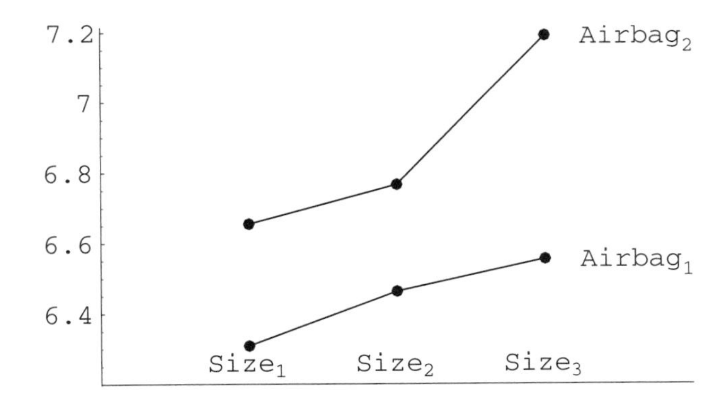

```
groupMeanPlotArray[try3,{{2,Airbag},{1,Size}},3,
  AxesOrigin->{0,6.2}]
```

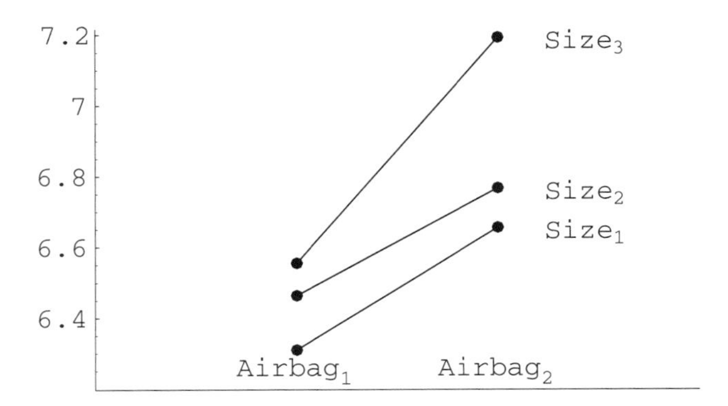

■

EXAMPLE 9: Two weeks prior to final exams, 10 students took part in an experiment to determine what effect the presence of a live plant, a photo of a plant, or the absence of a plant has on the student's ability to relax. Each student participated in three sessions: with a live plant, with a photo of a plant, and with no plant. Finger temperature was used as an indication of stress, where higher temperature indicates an increased level of relaxation. The data are summarized in Table 9.3. (a) Use two-factor ANOVA to determine if there is an effect of the presence of a live plant on stress. (b) Graph the data.

Student	Live plant	Plant photo	No plant
1	91.4	93.5	96.6
2	94.9	96.6	90.5
3	97.0	95.8	95.4
4	93.7	96.2	96.7
5	96.0	96.6	93.5
6	96.7	95.5	94.8
7	95.2	94.6	95.7
8	96.0	97.2	96.2
9	95.6	94.8	96.0
10	95.6	92.6	96.6

Table 9.3 Finger temperature in degrees

SOLUTION: The data in Table 9.3 are typical of a (complete) randomized block design, where one of the factors, students in this example, is of little interest and is often a random factor. This factor is refered to as a blocking variable. The reason to include it in the analysis is to account for variability that would otherwise be included in the mean square for error in a one-factor ANOVA.

We begin by entering the data in `temperature`. Our convention for Model III ANOVA is for factor A to be fixed and factor B to be random, so we enter the first column, followed by the second, and then the third column. Factor A is the presence of a plant, and factor B is students. We then load and run **twoWayANOVA** with the option setting `model->singleValuePerCell`. When there is only one observation per cell, the mean square for the A by B interaction is the only available measure of error. Thus, the appropriate Model III test statistic is calculated for factor A. No effect on stress due to the presence of a live plant can be shown with the data.

```
temperature={{{91.4},{94.9},{97.0},{93.7},{96.0},{96.7},
  {95.2},{96.0},{95.6},{95.6}},{{93.5},{96.6},{95.8},
  {96.2},{96.6},{95.5},{94.6},{97.2},{94.8},{92.6}},
 {{96.6},{90.5},{95.4},{96.7},{93.5},{94.8},{95.7},
  {96.2},{96.0},{96.6}}};

twoWayAnova[temperature,model->singleValuePerCell]
```

Analysis of Variance Table

Source	Sum of Squares	DF	Mean Squares	Fratio	P-value
Factor A	0.122	2	0.061	0.0189186	0.981279
Factor B	18.415	9	2.04611	0.634584	0.753712
Error	58.038	18	3.22433		
Total	76.575	29			

(b) We have written two functions to produce graphs for two-factor, single value per cell data. One, called `pointsPlotAbyB`, displays the levels of factor A on the horizontal axis. The other, called `pointsPlotBbyA`, displays the levels of factor B on the horizontal axis. These procedures are both on the disk that comes with this book. They are both loaded by clicking on the rightmost cell bracket of `Code for Single Cell Data Plot`. Then press **ENTER** and return to the *Mathematica* work sheet. After doing this, we produce the following plots.

```
pointsPlotAbyB[temperature]
```

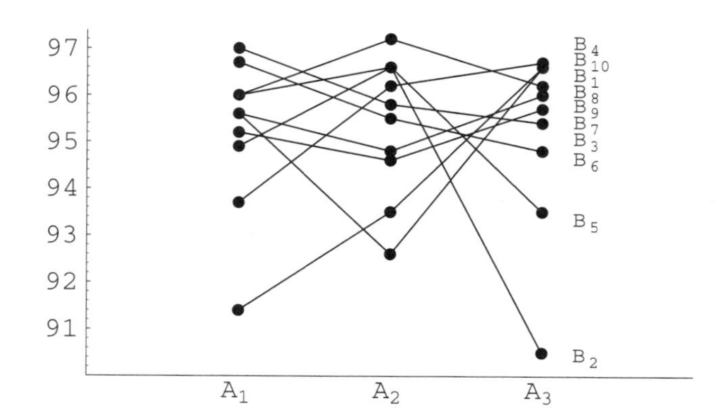


```
pointsPlotBbyA[temperature]
```

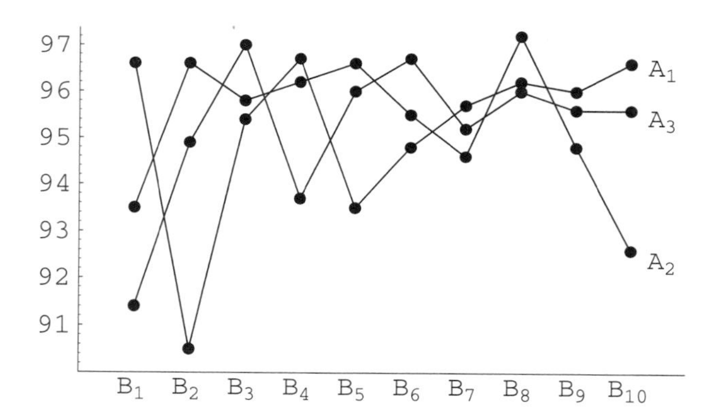

■

CHAPTER 10

Diagnostic Procedures and Transformations

10.1 Introduction

In earlier chapters we have discussed various methods for statistical inference and indicated the underlying assumptions. In this chapter, we will examine these assumptions together with diagnostic procedures for determining whether these assumptions are valid for a given set of data. The consequences, if the assumptions are violated, and remedial methods to be employed, if the assumptions cannot be met, are discussed. Before using an inferential method, the analyst should determine that the assumptions seem reasonable for the data set and, if not, carry out the steps necessary to remedy the situation.

The assumptions to be discussed are random sampling, independence, equality of variance, and normality. In many cases, departure from the assumptions can be rectified by a transformation of the original data into a new scale. In case transformations are not able to make the data conform to the assumptions, other techniques known as nonparametric or distribution free methods are used. These methods are discussed in Chapter 12. When the above assumptions are met, the procedure based on these assumptions should be used, because it is more powerful than the corresponding nonparametric procedure.

10.2 Diagnostic Procedures

Residuals

Residuals form the basis for many of the diagnostic methods used to check assumptions. Given $k \geq 2$ lists of sample data that may correspond to k levels of a single factor or k cells from a multifactor experiment, a residual is calculated for each data value. Let y_{ij} denote the jth value in the ith list. Then, the **residual** is

$$e_{ij} = y_{ij} - \bar{y}_{i\cdot},$$

where $\bar{y}_{i\cdot}$ is the mean of the ith list. Once calculated, the residuals are considered to be one data set. The mean of this set is zero. The variance is the **mean square for error**

$$MSE = \frac{SSE}{n-k},$$

where

$$n = \sum_{i=1}^{k} n_i, \; n_i$$

is the number in list i, and the sum of squares for error (SSE) is the sum of the squared residuals. Thus

$$MSE = \frac{1}{n-k} \sum_{i=1}^{k} \sum_{j=1}^{n_i} e_{ij}^2 .$$

Standardized Residuals

It is often useful to standardize the residuals before performing any analysis. For example, residuals are often plotted, and standardizing usually means that the scale on the residual axis can extend from around −3 to +3. A commonly used form of standardization is to divide each residual by the square root of the MSE. For convenience, we will call $e_{ij}/\sqrt{MSE}$ the **standardized residual**. Because not all residuals have the same variance, it is appropriate to consider standardizing each residual using an estimate of its standard deviation. The calculation

$$\frac{e_{ij}}{MSE(n_i - 1)/n_i}$$

results in the **studentized residual**. In some applications, the **studentized separate variance residual** is used, as opposed to the studentized residual. This residual is given by

$$e_{ij}\sqrt{\frac{s_i^2\,(n_i - 1)}{n_i}};$$

where s_i^2 is the sample variance of the ith list. When residuals are used to identify outliers, **deleted residuals** are often used. These residuals are calculated by first omitting the value in question, say y_{ij}, from the calculations for the mean, $\bar{y}_i$, and the mean square for error. The calculation

$$e_{ij}\sqrt{\frac{n - k - 1}{SSE(1 - 1/n_i) - e_{ij}^2}}$$

results in the **studentized deleted residual**. For convenience, we have written a procedure called **`anovaResiduals`** that will calculate residuals and the four types of standardized residuals just described. The procedure **`anovaResiduals`** has an option that results in a list of ordered pairs of the form $(\bar{y}_i, r_{ij})$ where r_{ij} represents a residual for a value from list i, and $\bar{y}_i$ represents the mean of the values in list i. The list of ordered pairs is used to make a scatter plot using the built-in function `ListPlot`, as discussed in the next section.

The **LinearRegression** package that is located in the **Statistics** folder (or directory) contains three procedures that perform similar calculations. `FitResiduals` calculates residuals for regression calculations. `StandardizedResiduals` calculates what we have termed **studentized residuals,** and `StudentizedResiduals` calculates what we have called **studentized deleted residuals**. Our names are more consistent with both the calculations performed and commonly recognized terminology.

Caution

Residuals are not independent. Consequently, statistical methods that require independent observations are not exactly appropriate for residuals. However, any adverse effect will be slight, provided that the number of residuals for each cell is not too small. Also, plots of residuals are little affected by the lack of independence, because the plots contain individual residuals, not functions of them.

Outliers

An **outlier** is an individual value located outside the overall pattern of the data. It may result from a data handling error, a miscalculation, or some other identifiable cause and, if so, it should be corrected or discarded. Otherwise, outliers should remain with the data and receive special consideration when necessary. Outliers can be identified based on the absolute value of the corresponding studentized deleted residuals, by using plots, and/or by using formal tests of significance. When the data set is large, a rough rule of thumb is to identify as outliers values whose studentized deleted residual is over 3.5 in absolute value. A scatter plot of residuals (vertical axis) against the means used in their calculation is particularly useful to identify points outside the overall pattern of the data. In *Mathematica,* use the built-in function `ListPlot` with the optional output from our procedure **`anovaResiduals`** to generate scatter plots. A box-and-whiskers plot of the studentized deleted residuals provides another means of outlier identification. (Our procedure to display box-and-whiskers plots is presented and discussed in Chapter 4.)

A variety of formal test for outliers have been developed. One, which is easy to perform using *Mathematica,* is based on the Bonferroni inequality. Calculate a critical value, *cv,* using `StudentTDistribution` which is found in the **ContinuousDistribution** package. Note that

cv = `Quantile[StudentTDistribution[n-k-1],1-α/(2n)],`

where

$$n = \sum_{i=1}^{k} n_i.$$

Any studentized deleted residuals larger than *cv* in absolute value correspond to outliers using familywise level of significance α. It should be noted that this test assumes that the data reasonably satisfy the assumptions discussed in the introduction, because, if so, each studentized deleted residual will follow the Student's *t*-distribution. It should also be noted that, if there are outliers that turn out to be valid data values, their presence may indicate departures from the assumptions. In particular, the outliers may help establish departure from normality and/or constant variance. Therefore, even though there are no specific restrictions about outliers, their existence in a data set can affect the validity of the assumptions for that data set.

> `anovaResiduals[data,opts]`, given two or more lists of sample data, returns a list of the residuals of the lists. The default option, `option->normal`, returns a list of the residuals. The option `residuals->standardized` returns a list of the standardized residuals; `residuals->studentized` returns a list of studentized residuals; `residuals->studentizedseparatevariance` returns a list of studentized separate variance residuals; and `residuals->studentizeddeleted` returns a list of studentized deleted residuals. The option `output->orderedpairs` results in a list of ordered pairs of the form $\bar{y}_{i.}, r_{ij}$ where r_{ij} represents a residual for a value from list *i*, and r_{ij} represents the mean of the values in list *i*.

The procedure **`anovaResiduals`** is available on the disk that comes with this book. To use it, first load the disk and click on its icon. Next, locate the procedure name and click on the rightmost cell bracket. Then press **ENTER** and return to the *Mathematica* work sheet.

EXAMPLE 1: Each of four groups of pigs was given a different feed. The mass of each of the 19 pigs is given (in kilograms) in Table 10.1. (a) Determine the residuals, the standardized residuals, the studentized residuals, the studentized separate variance residuals, and the studentized deleted residuals. (b) Graph a scatter plot of the studentized deleted residuals. (c) Using the Student's *t*-distribution, calculate a critical value and determine if any of the data values is an outlier. (**Note:** These values were randomly generated using the normal distribution.)

Group 1	65.2	58.4	49.0	70.1	54.4
Group 2	74.2	68.0	61.1	61.7	62.6
Group 3	95.0	91.2	98.6	97.7	
Group 4	75.1	91.8	79.4	91.0	93.7

Table 10.1 Mass (Kilograms) of 19 Pigs

SOLUTION: After entering our procedure **anovaResiduals** we enter the weights in Table 10.1, naming the list `data`. Next, we compute the residuals (in `res`), the standardized residuals (in `st`), the studentized residuals (in `stu`), the studentized separate variance residuals (in `stsv`), and the studentized deleted residuals (in `stdel`). Notice that all of the studentized deleted residuals lie between –3.5 and 3.5.

```
data={{65.2,58.4,49.0,70.1,54.4},
 {74.2,68.0,61.1,61.7,62.6},{95.0,91.2,98.6,97.7},
 {75.1,91.8,79.4,91.0,93.7}};

res=anovaResiduals[data]

{{5.78,-1.02,-10.42,10.68,-5.02},
 {8.68,2.48,-4.42,-3.82,-2.92},
 {-0.625,-4.425,2.975,2.075},
 {-11.1,5.6,-6.8,4.8,7.5}}

st=anovaResiduals[data,residuals->standardized]

{{0.834466,-0.147259,-1.50435,1.54189,-0.724744},
 {1.25314,0.358041,-0.638121,-0.551498,-0.421564},
 {-0.0902321,-0.638843,0.429505,0.29957},
 {-1.60252,0.808479,-0.981725,0.692982,1.08278}}
```

```
stu=anovaResiduals[data,residuals->studentized]
{{0.932962,-0.16464,-1.68191,1.72388,-0.810288},
 {1.40106,0.400302,-0.713441,-0.616594,-0.471323},
 {-0.104191,-0.737672,0.495949,0.345914},
 {-1.79167,0.903907,-1.0976,0.774778,1.21059}}
```

```
stsv=anovaResiduals
 [data,residuals>studentizedseparatevariance]
{{0.860046,-0.151773,-1.55046,1.58915,-0.74696},
 {1.94756,0.556444,-0.991728,-0.857104,-0.655168},
 {-0.250775,-1.77549,1.19369,0.832575},
 {-1.65808,0.836511,-1.01576,0.717009,1.12033}}
```

```
stdel=anovaResiduals[data,residuals->studentizeddeleted]
{{0.14921,-0.0248482,-0.312274,0.323869,-0.127659},
 {0.242851,0.0609572,-0.111256,-0.0953063,-0.0720728},
 {-0.016223,-0.119096,0.0784521,0.0542542},
 {-0.34341,0.144019,-0.179795,0.121587,0.202125}}
```

(b) Using the option settings `residuals->studentizeddeleted` and `output->orderedpairs`, we obtain a list of ordered pairs of the form $(\bar{y}_i, r_{ij})$ where r_{ij} represents the studentized deleted residual. We name this list `rvals`. Notice that `rvals` contains sublists (i.e., the ordered pairs for each group are written as a list), so we use `Flatten` (at level 1) to eliminate the sublists. The new list is called `rpoints`. Finally, we use `ListPlot` to graph the ordered pairs in `rpoints`.

```
rvals=anovaResiduals
 [data,residuals->studentizeddeleted,output>orderedpairs]
{{{59.42,0.14921},{59.42,-0.0248482},{59.42,-0.312274},
  {59.42,0.323869},{59.42,-0.127659}},{{65.52,0.242851},
  {65.52,0.0609572},{65.52,-0.111256},{65.52,-0.0953063},
  {65.52,-0.0720728}},{{95.625,-0.016223},{95.625,-0.119096},
  {95.625,0.0784521},{95.625,0.0542542}},{{86.2,-0.34341},
  {86.2,0.144019},{86.2,-0.179795},{86.2,0.121587},
  {86.2,0.202125}}}
```

```
rpoints=Flatten[rvals,1]
{{59.42,0.14921},{59.42,-0.0248482},{59.42,-0.312274},
 {59.42,0.323869},{59.42,-0.127659},{65.52,0.242851},
 {65.52,0.0609572},{65.52,-0.111256},{65.52,-0.0953063},
 {65.52,-0.0720728},{95.625,-0.016223},
```

```
{95.625,-0.119096},{95.625,0.0784521},
{95.625,0.0542542},{86.2,-0.34341},
{86.2,0.144019},{86.2,-0.179795},
{86.2,0.121587},{86.2,0.202125}}
```

```
ListPlot[rpoints,PlotRange->{{50,100},Automatic}]
```

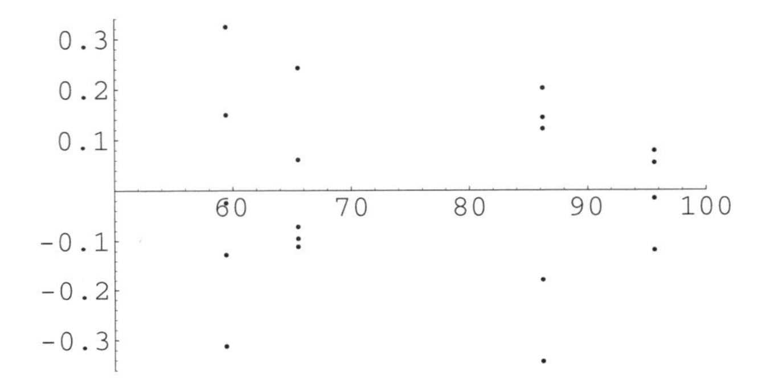

(c) To find the total number of data values n, we first use `Flatten` to remove all inner pairs of parentheses in the list `data`, naming the new list `fldata`, and then find the length of the list `fldata`. We then compute the length of `data` to define k. Finally, we enter the value $\alpha = 0.05$ and find that the critical value is approximately 3.6481. Notice that none of the studentized deleted residuals in the list `stdel` are larger (in absolute value) than 3.6481. Therefore, at the 5% familywise level of significance, we can find no outliers.

```
fldata=Flatten[data]
```

```
{65.2,58.4,49.,70.1,54.4,74.2,68.,61.1,61.7,62.6,95.,
 91.2,98.6,97.7,75.1,91.8,79.4,91.,93.7}
```

```
n=Length[fldata]
```

```
19
```

```
k=Length[data]
```

```
4
```

```
α=0.05;
Quantile[StudentTDistribution[n-k-1],1-α/(2n)]//N
```

```
3.64871
```

■

Random Sampling

The fundamental requirement of all inferential methods is that sampling of individuals be random. Most random sampling procedures employ probability sampling, where a random device is used to decide which members of the population will constitute the sample. This process does not guarantee that the sample will accurately represent the population, but it does eliminate unintentional bias in the selection process and permits the experimenter to determine the chance of obtaining a nonrepresentative sample. Nonrandomness of sample selection may well be reflected by a data set that demonstrates departures from one or more of the other assumptions. Our procedures for random sampling are discussed in Chapter 7.

Independence

A requirement for many statistical methods is that the data values be obtained independently. For a designed experiment, the experimenter can secure independence by appropriate use of randomization (see Chapter 7). When data are collected in a time sequence or in some other logical sequence, such as in a geographical sequence, a plot of residuals against the ordering variable is useful. Under independence, a long sequence of positive values followed by an equally long sequence of negative values is quite unlikely. Positive and negative values which alternate regularly would also not be expected. One of these or another apparent pattern in the plot indicates a departure from independence. A simple, although not very powerful, test for a pattern in sequentially collected data is the Runs test that is discussed in Chapter 12. It requires that the sequence be reduced to one containing two types. Whether the residual is positive or negative may be the appropriate dichotomy to use.

Equality of Variance

Equality of variance refers to the requirement of several statistical methods that the populations from which samples are drawn all have the same variance. For example, procedures based on the Student's t-distribution to test hypotheses concerning two population means or estimate the difference between two population means (see Chapter 8) may require equality of variance. This is also a requirement for single and multifactor analysis of variance. In the latter case, the data in each group is assumed to be taken from a population and equality of variance requires that all "group" populations have equal variances. Synonyms for equality of variance include **homogeneity** of variance and **homoscedasticity** (literally "equal scatter"). The converse of equality of variance is called **heterogeneity** of variance or **heteroscedasticity**.

Heterogeneity of variance can be investigated using a scatter plot of studentized residuals (vertical axis) against the means used in their calculations. In *Mathematica*, use `ListPlot` with the optional output from our procedure **anovaResiduals**. In effect, the plot is a set of simultaneous scatter plots, one for each sample. Heteroscedasticity is indicated if the extent of the scatter differs excessively among two or more samples. Several formal statistical hypothesis tests are available for studying heterogeneity of variance. In each case, the null hypothesis is homoscedasticity (that is, equal population variances) and the alternative hypothesis is heteroscedasticity (not all population variances are equal). The test which is most often discussed in introductory textbooks is Bartlett's test, also called the Bartlett–Box test.

Bartlett–Box Test of Heterogeneity of Variance

Assumptions

1. There are k groups that represent k independent, random samples, one from each of the k populations, where the population variances are denoted by $\sigma_1^2, \ldots, \sigma_k^2$. The sample sizes are $n_1, \ldots, n_k$, respectively.
2. Each population is normally distributed.

Bartlett's test is very sensitive to departures from normality, which is discussed in the next section, so a significant result may indicate nonnormality rather than heteroscedasticity and a nonsignificant result may fail to detect an important difference. Thus, Bartlett's test should be used only when there is no doubt of the normality of the populations from which the samples were selected. Under this circumstance, it has reasonable power.

Hypotheses

The null hypothesis is that all variances are equal: $H_0\colon \sigma_1^2 = \ldots = \sigma_k^2$, where $k \geq 2$. The alternative hypothesis is that the variances are not all equal: H_α : *The population variances are not all equal.*

If $k = 2$ and $n1 = n2$, Bartlett's test is equivalent to the two-tailed variance ratio test discussed in Chapter 8. With two samples of unequal size, the two procedures may yield different results with sometimes one and sometimes the other being better (more powerful). For one-tailed tests or to check Bartlett's test when $n1 \neq n_2$, use `VarianceRatioTest` in the **HypothesisTests** package (see Chapter 8).

Test Statistic

Let s_i^2 represent the sample variance of the ith list, and let

$$s_p^2 = \left(\sum_{i=1}^{k} (n_i - 1)s_i^2\right) \Big/ \left(\sum_{i=1}^{k} (n_i - 1)\right),$$

which is a pooled estimate of variability. The test statistic is $B_C = B/C$, where B is given by

$$B = (\ln s_p^2)\left(\sum_{i=1}^{k} n_i - 1\right) - \sum_{i=1}^{k} (n_i - 1)\ln s_i^2,$$

and C is given by

$$C = 1 + \frac{1}{3(k-1)}\left(\sum_{i=1}^{k} \frac{1}{n_i - 1} - \frac{1}{\sum_{i=1}^{k} (n_i - 1)}\right).$$

The distribution of B_C is approximated by the Chi-Square Distribution with $k - 1$ degrees of freedom. Because this approximation may not be good for $n_i < 5$, a further calculation is done that results in a test statistic whose distribution is approximated by the F-distribution with no restrictions on n_i. The modified test statistic is

$$B'_C = \frac{f_2 B_C C}{f_1(A - B_C C)},$$

where $f_1 = k - 1$, $f_2 = (k + 1)/(C - 1)^2$, and $A = f_2/(2 - C - 2/f_2)$. The distribution of B'_C is approximated by an F (ratio) distribution with numerator degrees of freedom f_1 and denominator degrees of freedom f_2. The modified test is often called the **Bartlett–Box test.**

Procedure

We have written a *Mathematica* procedure called **`bartlettVariance`**. It is available on the disk that comes with this book. To use it, first load the disk and click on its icon. Next, locate the procedure name and click on the rightmost cell bracket. Then press **ENTER** and return to the *Mathematica* work sheet.

`bartlettVariance[dataset]` performs the Bartlett–Box test and returns the sample variances, the test statistic, B'_C, a p-value, and a description of the distribution used to calculate the p-value.

Modified Levene's Test of Heterogeneity of Variance

An alternative test for heteroscedasticity that is less sensitive to departures from normality is called the **Modified Levene's test** or **Levene's Median test**.

Assumptions

1. There are k groups that represent k independent, random samples, one from each of the k populations, where the population variances are denoted by $\sigma_1^2, \ldots, \sigma_k^2$. The sample sizes are $n_1, \ldots, n_k$, respectively.
2. Each population is normally distributed.

Hypotheses

The null hypothesis is that all variances are equal: H_0: $\sigma_1^2 = \ldots = \sigma_k^2$, where $k \geq 2$. The alternative hypothesis is that the variances are not all equal: H_α: *The population variances are not all equal.*

Test Statistic

Let y_{ij} be the jth value in list i, and m_1 the median of list i. For each y_{ij} calculate

$$d_{ij} = |y_{ij} - m_i|.$$

The test statistic is the ratio F from a single factor analysis of variance on the d_{ij} values.

It has been suggested that if the sample size is an odd number and less than 20 for one or more lists, the modified Levene's test is too likely to find heteroscedasticity. This peculiarity disappears if values of $d_{ij} = 0$ are removed from groups with an odd number less than 20 before doing the analysis of variance.

We have written a procedure to perform the modified Levene's test called `leveneVariance`. It is available on the disk that comes with this book. To use it, first load the disk and click on its icon. Next, locate the procedure name and click on the rightmost cell bracket. Then press **ENTER** and return to the *Mathematica* work sheet.

`leveneVariance[dataset]` performs the modified Levene test and returns the sample variances, the test statistic, F, a p-value, and a description of the distribution used to calculate the p-value.

Caution

Some authors in the statistical literature have recommended that formal tests for heterogeneity of variance should not be used prior to procedures based on Student's t-distribution or analysis of variance. Their reasoning is based on what they term the relatively poor performance for tests of heterogeneity of variances and the robustness of the procedures based on Student's t-distribution and analysis of variance. We strongly recommend investigating the equal variance assumption. This should be done using Bartlett's test if normality is not in question. Regarding performance of the tests for heterogeneity of variance, it is known that Bartlett's test has reasonable power when normality holds and that the modified Levene's test is not overly sensitive to nonnormality but slightly less powerful than Bartlett's test when normality holds. Regarding the robust properties of the procedures based on Student's t-distribution and the analysis of variance, it is known that if sample sizes are equal or approximately equal, they are robust against heterogeneity of variance. The only effect is to raise the actual level of significance only slightly higher than the specified level. If the factors in the analysis of variance are random, unequal variances can have a pronounced adverse effect on inferences, even with equal sample sizes.

Test of Heterogeneity of Coefficients of Variation

The tests just described can also be used to investigate heteroscedasticity when that is the primary focus of the data analysis. Because variability in a data set is influenced by the magnitude of the values in the set, to compare variability between populations of measures with different magnitudes, the coefficient of variation is often used. A test for heterogeneity of coefficients has also been included on the disk that accompanies this text, called **`coefficientsOfVariation`**. The test statistic, CV, has a distribution that is approximated by the chi-square distribution with $k - 1$ degrees of freedom:

$$CV = \frac{1}{V_p^2\,(0.5 + V_p^2)}\left(\sum_{i=1}^{k}(n_i - 1)V_i^2 - \frac{\left(\sum_{i=1}^{k}(n_i - 1)V_i\right)^2}{\sum_{i=1}^{k}(n_i - 1)}\right),$$

where n_i is the sample size of the ith list, V_i is the coefficient of variation of the ith list, and

$$V_p = \frac{\sum_{i=1}^{k}(n_i - 1)V_i}{\sum_{i=1}^{k}(n_i - 1)}.$$

`coefficientsOfVariation[dataset]` returns the sample coefficients of variation, the test statistic, CV, a p-value, and a description of the distribution used to calculate the p-value. To use the procedure, first load the disk and click on its icon. Next, locate the procedure name and click on the rightmost cell bracket. Then press **ENTER** and return to the *Mathematica* work sheet.

EXAMPLE 2: (a) Perform tests for heterogeneity of variance using the data in Groups 1, 2, 3, and 4 given in Table 10.1. Use the Bartlett-Box test and the Modified Levene's test. Also test for heterogeneity of coefficients of variation. (b) Repeat for the data in Table 10.2.

Group 1:

0.390569	1.19312	0.375175	-0.943365	1.30642
0.208836	-0.324407	0.97112	0.302299	1.02387

Group 2:

0.630718	-0.994795	-5.45224	-0.723693	0.318204
-2.14474	-1.21454	-2.40184	0.0917247	-0.152679
-2.20593	-2.54901			

Group 3:

-9.88767	-1.66042	-1.91557	0.803153	-0.8476
-3.93527	-3.59389	2.23085	-1.65416	0.979322
0.524861	1.90809	-0.617991	-0.314847	-0.820731
-1.93908	1.26113	-3.75642	4.17884	0.283673

Table 10.2

SOLUTION: (a) First, we consider Table 10.1 and begin by entering the values in groups 1, 2, 3, and 4 in the list pig. When we perform the Bartlett–Box test, we find such a large *p*-value that we cannot dispute the null hypothesis that the variances are equal.

```
pig={{60.8,57.0,65.0,58.6,61.7},{68.7,67.7,74.0,66.3,69.8},
  {102.6,102.1,100.2,96.5},{87.9,84.2,83.1,85.7,90.3}};

bartlettVariance[pig]
Bartlett/Box Test for Heterogeneity of Variance
 SampleVariances→9.392,8.565,7.65667,8.388}
 TestStatistic→0.0108709
 PValue→0.998446
 Distribution→FRatioDistribution[3,391.83]
```

Next, we use the modified Levene's test and then investigate the heterogeneity of the coefficients of variation with similar results.

```
leveneVariance[pig]
Modified Levene's Test for Heterogeneity of Variance
SampleVariances→9.392,8.565,7.65667,8.388}
TestStatistic→0.0238055
PValue→0.99481
Distribution→FRatioDistribution[3, 15]
```

```
coefficientsOfVariation[pig]
Test for Heterogeneity of Coefficients of Variation
CoeffsOfVar→0.0505549,0.0422309,0.0275742,0.0335831
TestStatistic→1.4058
PValue→0.704176
Distribution→ChiSquareDistribution[3]
```

(b) We create the list `data` using the values in Table 10.2. When we use the Bartlett–Box test, we obtain a small p-value (0.0001958) which indicates a difference among the population variances.

```
data={{0.390569,1.19312,0.375175,-0.943365,1.30642,0.208836,
 -0.324407,0.97112,0.302299,1.02387},{0.630718,-0.994795,
 -5.45224,-0.723693,0.318204,-2.14474,-1.21454,-2.40184,
  0.0917247,-0.152679,-2.20593,-2.54901},{-9.88767,-1.66042,
 -1.91557,0.803153,-0.8476,-3.93527,-3.59389,2.23085,
 -1.65416,0.979322,0.524861,1.90809,-0.617991,-0.314847,
 -0.820731,-1.93908,1.26113,-3.75642,4.17884,0.283673}}
```

```
bartlettVariance[data]
Bartlett/Box Test for Heterogeneity of Variance
SampleVariances→{0.501872,2.86017,8.79384}
TestStatistic→8.56502
PValue→0.0001958
Distribution→FRatioDistribution[2,2745.69]
```

We find similar results with the modified Levene's test and the coefficients of variation test. The larger p-value for the modified Levene test in comparison to the Bartlett–Box test is explained by knowing that the data were randomly generated from normal populations and recalling that the Bartlett–Box test is more powerful in this case.

```
leveneVariance[data]
Modified Levene's Test for Heterogeneity of Variance
SampleVariances→{0.501872,2.86017,8.79384}
TestStatistic→3.2652
PValue→0.0488521
Distribution→FRatioDistribution[2,39]
```

```
coefficientsOfVariation[data]
Test for Heterogeneity of Coefficients of Variation
CoeffsOfVar→{1.57302,-1.20809,-3.15914}
TestStatistic→21.4507
PValue→0.0000219807
Distribution→ChiSquareDistribution[2]
```

■

Normality

Normality refers to the requirement of several statistical methods that the populations from which samples are drawn be all normally distributed. For example, procedures based on the Student's *t*-distribution and analysis of variance all require normality. Normality should be investigated for each individual sample if the sample size is large enough or, if not, by combining the residuals for all samples into one set. For a two-factor experimental design with four cells, for example, it would be preferred to evaluate each cell for normality. If the sample sizes in the cells are small, then residuals from all four cells would be combined to investigate normality. The residuals are calculated using our `anovaResiduals` procedure.

For procedures based on Student's *t*-distribution and analysis of variance procedures, the consequences of nonnormality are not too serious, because the sample means will follow the normal distribution more closely than the populations sampled (a consequence of the central limit theorem). Consequently, both sets of procedures are said to be robust against departures from normality. Typically, the achieved level of significance in the presence of nonnormality is slightly higher that the specified one. If the factors in the analysis of variance are random factors, lack of normality can have a pronounced adverse effect on inferences. It should be noted that if unequal variances are indicated and normality is examined for the combined data using residuals, studentized residuals based on separate variance estimates should be used. Refer to the earlier discussion on studentized separate variance residuals.

Normality of residuals or for each list of data can be studied graphically by using a histogram, a normal probability plot, and measures of skewness and kurtosis.

Histogram and Normal Density Plot

We have written a procedure called `normalHistogram` to display a histogram of a list of residuals or data values with a normal density curve superimposed on it (see Chapter 7).

EXAMPLE 3: Use a **normalHistogram** to investigate the normality of the population from which the data in Table 10.3 was taken. Similarly investigate the population from which the data in Table 10.4 was taken. (**Note:** The values in Table 10.3 were randomly generated from the standard normal distribution. The values in Table 10.4 were randomly generated from the chi-square distribution with two degrees of freedom.)

0.584876	0.164822	0.449716	1.28953	0.520173
-1.69339	-0.225564	0.462183	-0.446375	0.226037
0.570508	0.735886	2.59758	0.231171	1.0225
-1.31356	2.60839	1.31471	0.893	2.17977
0.184372	-0.661991	-0.066023	0.140468	-0.000891814
-0.291963	0.728657	0.713564	0.724345	-0.224926
0.256552	-0.771846	-0.525295	-0.0889038	-1.28344
-1.91511	-0.599998	-0.0474356	0.197584	-0.598371
-0.582649	0.238121	-0.566242	-0.484493	-1.61822
-2.08677	-0.191946	0.697967	1.81547	-1.39897

Table 10.3

3.48566	3.55124	2.26459	0.116474	1.56762
1.99768	1.57667	1.75057	1.90107	0.678585
0.955201	0.20144	1.13616	0.185295	0.44635
1.52742	0.423731	2.41548	0.524364	0.394676
6.45974	10.1795	0.721821	0.155046	0.291119
0.356386	1.963	0.0363703	1.79359	1.51655
0.166446	1.14102	7.69427	0.558897	2.40877
0.827852	1.5761	0.33758	1.38669	3.26832
0.874976	1.21096	0.627969	1.96591	1.00144
1.23365	6.79307	1.60237	0.598011	0.705157

Table 10.4

SOLUTION: First, we enter the numbers in Table 10.3, naming the list `data1`. We then plot the normal histogram by specifying that 20 bars be used in the histogram, but we suppress the output initially with the option setting `DisplayFunction->Identity`. (If this option setting is not included in the **normalHistogram** command, then the graph is shown.) We name this plot `p1`. Next, we follow similar steps for the values in Table 10.4. In this case, we name the values `data2` and the normal histogram `p2`. Finally, we use `GraphicsArray` to display the two graphs side-by-side. Notice that the first data set suggests the normal distribution while the second does not.

```
data1={0.584876,0.164822,0.449716,1.28953,0.520173,
 -1.69339,-0.225564,0.462183,-0.446375,0.226037,0.570508,
 0.735886,2.59758,0.231171,1.0225,-1.31356,2.60839,1.31471,
 0.893,2.17977,0.184372,-0.661991,-0.066023,0.140468,
 -0.000891814,-0.291963,0.728657,0.713564,0.724345,
 -0.224926,0.256552,-0.771846,-0.525295,-0.0889038,
 -1.28344,-1.91511,-0.599998,-0.0474356,0.197584,-0.598371,
 -0.582649,0.238121,-0.566242,-0.484493,-1.61822,-2.08677,
 -0.191946,0.697967,1.81547,-1.39897}
```

```
p1=normalHistogram[data1,20,DisplayFunction->Identity];
```

```
data2={3.48566,3.55124,2.26459,0.116474,1.56762,
 1.99768,1.57667,1.75057,1.90107,0.678585,0.955201,
 0.20144,1.13616,0.185295,0.44635,1.52742,0.423731,
 2.41548,0.524364,0.394676,6.45974,10.1795,0.721821,
 0.155046,0.291119,0.356386,1.963,0.0363703,1.79359,
 1.51655,0.166446,1.14102,7.69427,0.558897,2.40877,
 0.827852,1.5761,0.33758,1.38669,3.26832,0.874976,
 1.21096,0.627969,1.96591,1.00144,1.23365,6.79307,
 1.60237,0.598011,0.705157}
```

```
p2=normalHistogram[data2,20,DisplayFunction->Identity];
```

```
Show[GraphicsArray[{p1,p2}]]
```

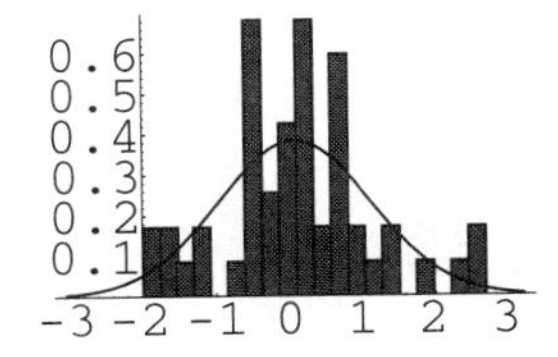

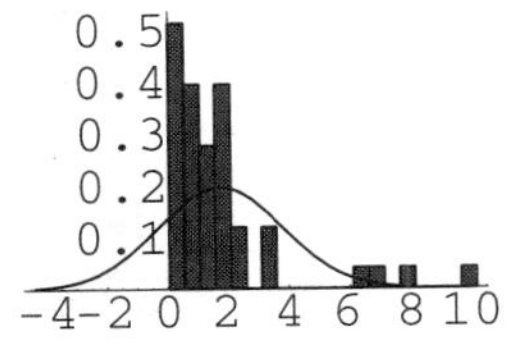

■

Normal Probability Plot

A normal probability plot is a scatter plot where each value in a list is plotted against a corresponding value (quantile) from a normal distribution. A plot that is linear, or nearly so, indicates agreement with the normality assumption. Departure from linearity indicates nonnormality. The list may contain a sample from a single population or the residuals for two or more samples. Often, to make the plot more compact, the values in the list are standardized. In case of a single sample, each value is coded by subtracting the sample mean and dividing by the sample standard deviation. For a list, this is done with the command `Standardize[list]` located in the **DescriptiveStatistics** folder (or directory). For residuals, our **`anovaResiduals`** procedure can be used to calculate a variety of standardized residuals. The values or coded values are then ranked and corresponding quantiles from the standard normal distribution are calculated. For the value with rank r, the standard normal quantile for

$$\frac{r - \frac{3}{8}}{n + \frac{1}{4}}$$

is used, where n is the number of values in the list. If the values in the list are not standardized, it is possible, but not necessary, to code the quantiles by multiplying each one by the sample standard deviation and adding the sample mean. Then, both axes would be on the same scale.

For a distribution that is skewed right, the normal probability plot will be concave downward. It will be concave upward for a distribution that is skewed to the left. If a distribution is symmetric but with high probability in the tails, such as a uniform distribution or a platykurtic distribution, the normal probability plot is concave downward at the left end and concave upward at the right end. The opposite pattern is seen for a leptokurtic distribution.

There are formal statistical hypothesis tests that can be used to investigate the normality assumption. One simple-to-use test in association with a normal probability plot involves the Pearson's correlation coefficient. Because this coefficient is equal to 1 if the plot is exactly linear, it can be used to measure the linearity of the plot and to make inference about the normality of the sampled population(s). The null hypothesis is that the population(s) is normal. The alternative hypothesis is the population(s) is not normal. The null hypothesis is rejected if the Pearson correlation coefficient r is too small (in absolute value).

We have written a procedure called **`normalProbability`** to display a normal probability plot of a list of residuals or data values. The Pearson's correlation coefficient and a corresponding approximate p-value (assuming normality) is displayed on the normal probability plot. Reject the null hypothesis and conclude a departure from normality if the p-value is smaller than the level of significance. To use the procedure, first load the disk that comes with this book and click on its icon. Next, locate the procedure name and click on the rightmost cell bracket. Then press **ENTER** and return to the *Mathematica* work sheet.

`normalProbability[list,opts]` displays a normal probability plot including the value of the Pearson's correlation coefficient and an approximate p-value for testing the null hypothesis of normality. Any option available for the built-in function `Show` may be specified.

EXAMPLE 4: Generate 50 random numbers each from the standard normal distribution, the uniform distribution on the interval [0,1], the chi-square distribution with four degrees of freedom, and the Student's t-distribution with five degrees of freedom. Use a normal probability plot to test for normality of the respective populations using each data set.

SOLUTION: We begin by loading the commands needed to execute **`normalProbability`**. In `plot1`, `plot2`, `plot3`, and `plot4`, respectively, we graph the pdf for the standard normal distribution, the uniform distribution on the interval [0,1], the chi-square distribution with four degrees of freedom, and the Student's t-distribution with one degree of freedom. (**Note:** We have selected the Student's t-distribution with one degree of freedom instead of five degrees of freedom to illustrate the general differences between the shapes of the pdfs for the standard normal distribution and the Student's t-distribution.) The output for each `Plot` command is suppressed initially with the option setting `DisplayFunction->Identity`. We then use `Show` to view the graphs as a `GraphicsArray`. The graphs are shown in the order indicated in the `GraphicsArray` with the normal and uniform distributions on the first row and the chi-square and Student's t-distributions on the second row.

```
plot1=Plot[PDF[-NormalDistribution[0,1],x],
  {x,-3,3},DisplayFunction->Identity];
plot2=Plot[PDF[UniformDistribution[0,1],x],
  {x,0,1},DisplayFunction->Identity];
plot3=Plot[PDF[ChiSquareDistribution[4],x],
  {x,0,14},DisplayFunction->Identity];
plot4=Plot[PDF[StudentTDistribution[1],x],
  {x,-3,3},DisplayFunction->Identity];

Show[GraphicsArray[{{plot1,plot2},{plot3,plot4}}]]
```

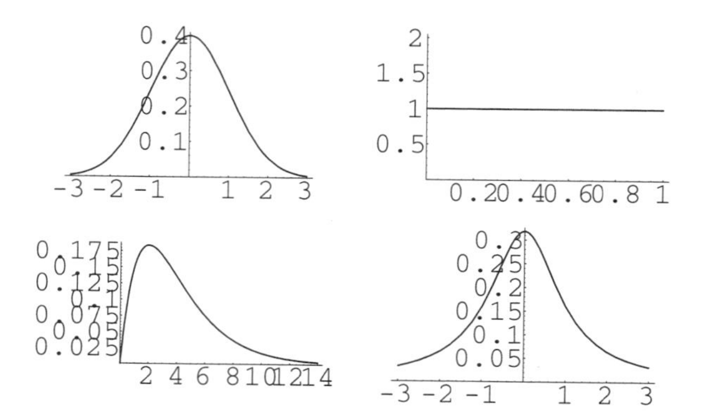

In the list named `normal`, we randomly generate 50 values from the standard normal distribution. Then, we construct the normal probability plot for these numbers. We name this graph `n1`. Pearson's correlation coefficient is approximately 0.993382 in this case, and the p-value exceeds 0.250. As expected, the normal probability plot for the standard normal distribution is almost linear.

```
normal=RandomArray[NormalDistribution[0,1],50]
```

```
{0.280585,-0.515308,-1.20967,-1.71066,-0.315709,-1.45923,
 0.648545,-0.356955,-0.241109,1.2189,1.1893,-0.144886,
 0.874057,0.0760477,0.374537,1.78797,-0.75668,-0.939977,
 0.74049,-1.1628,-0.31726,-0.0658284,-0.145251,-1.04731,
 -0.612713,-2.02349,1.10608,0.572592,0.704576,-0.057163,
 0.46979,0.701704,-0.413544,-1.45876,0.79543,
 1.57084,-0.337686,-0.888965,2.6422,-0.709234,1.84116,
 0.599437,0.700319,0.641387,-1.77072,-0.1958 -1.79975,1.48989,
 0.848462,-1.9836}
```

```
n1=normalProbability[normal]
```

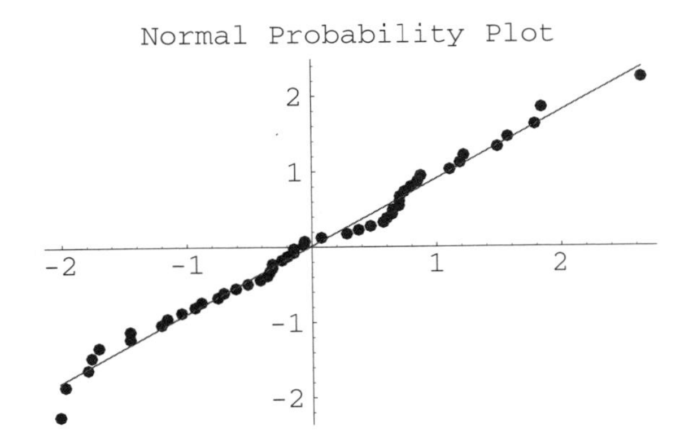


```
Pearson'sCorrelationCoefficient ->0.993382

PValue -> > 0.250
```

In the list we name `uniform`, we randomly generate 50 numbers using the uniform distribution on the interval [0,1]. As with the remaining two lists of randomly generated numbers, we use `Short` to view only a portion of the list. We name the normal probability plot of this distribution `n2`. In this case, Pearson's correlation coefficient is approximately 0.960812 with p-value $< .001$. We reject the null hypothesis that the data are normal. Notice that the normal probability plot for the uniform distribution is concave downward at the left end and concave upward at the right end.

```
uniform=RandomArray[UniformDistribution[0,1],50];
Short[uniform]
 {0.604844,«48»,0.41328}

n2=normalProbability[uniform]
```

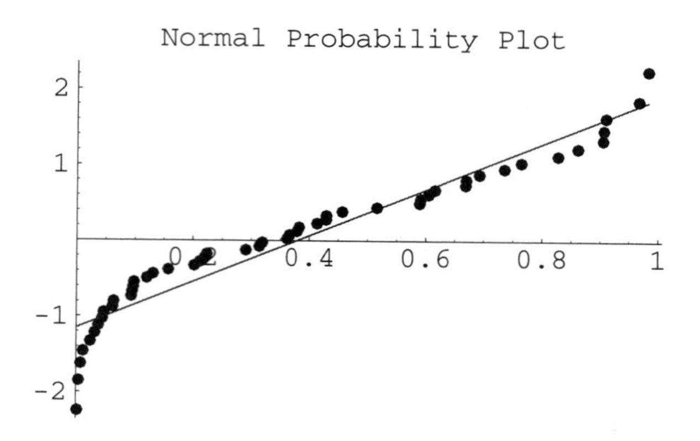


```
Pearson'sCorrelationCoefficient ->0.960812

PValue -> 0
```

Using the chi-square distribution with four degrees of freedom, we obtain the list of 50 randomly generated numbers that we name `chi`. We name the normal probability plot of these values `n3` and find Pearson's correlation coefficient to be approximately 0.959434 with p-value $< .001$. We reject the null hypothesis. As expected for the chi-square distribution which is skewed to the right, the plot is concave downward.

```
chi=RandomArray[ChiSquareDistribution[4],50];
Short[chi]
 {11.8465,2.89625,«47»,2.12715}

n3=normalProbability[chi]
```

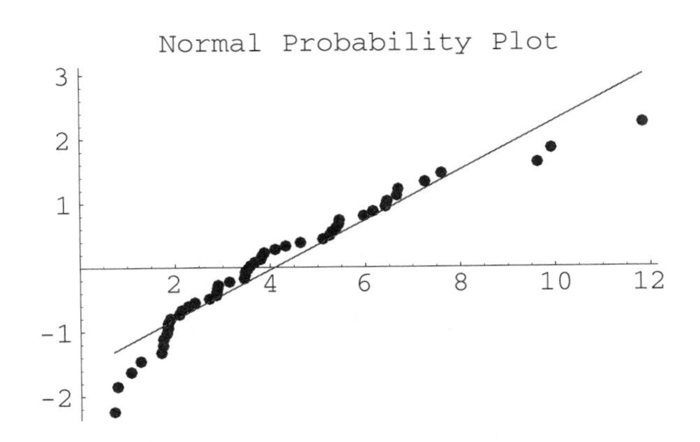


```
Pearson'sCorrelationCoefficient ->0.959434

PValue -> 0
```

In the list named `student`, we use the Student's t-distribution with five degrees of freedom to generate 50 values at random. The corresponding normal probability plot is called `n4`, and Pearson's correlation coefficient is found to be approximately 0.980997. Because the p-value is 0.0999402, we may not reject the null hypothesis at the 5% level, but may reject it at the 10% level of significance.

```
student=RandomArray[StudentTDistribution[5],50];

Short[student]

 {0.491388,«48»,-1.04938}

n4=normalProbability[student]
```

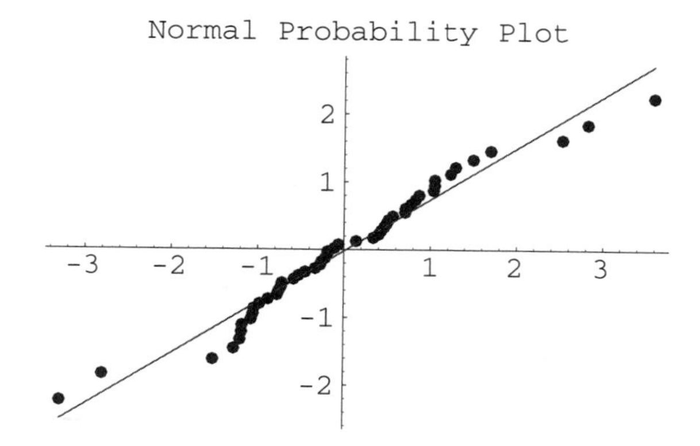


```
Pearson'sCorrelationCoefficient ->0.980997
PValue -> 0.0999402
```

■

D'Agostino-Pearson Test for Nonnormality

Another test that can be done easily using *Mathematica* involves the use of measures of skewness and kurtosis. We have written a procedure called **`normalAssumption`** that may be used for single samples or a set of residuals. The null hypothesis is that the population(s) is normal, and the alternative hypothesis is that the population(s) is not normally distributed. For this test, rejection of the null hypothesis implies departure from symmetry, or from kurtosis, or both. It has been suggested that this test works best if the number of input values is 20 or more.

`normalAssumption[list]` performs the D'Agostino–Pearson test for departure from normality and outputs the skewness coefficient, the kurtosis coefficient, the test statistic, and a p-value calculated using the chi-square distribution with two degrees of freedom. The test statistic is a function of the sum of the squares of the two coefficients.

To use the procedure, first load the disk that comes with this book and click on its icon. Next, locate the procedure name and click on the rightmost cell bracket. Then press **ENTER** and return to the *Mathematica* work sheet.

EXAMPLE 5: Use the function **normalAssumption** to test the data sets in Tables 10.5 and 10.6 for normality based on skewness and kurtosis. (**Note:** The 30 numbers in Table 10.5 were randomly generated from the standard normal distribution. The numbers in Table 10.6 were randomly generated from the chi-square distribution with four degrees of freedom.)

77.7331	86.7357	82.9031	59.6612	81.2474	65.1297	81.0629	104.793	82.4776	82.2584
88.1696	61.4329	69.1927	67.6756	80.8268	55.6229	68.5386	83.3563	54.261	62.4938
88.6138	67.5976	79.1755	84.5068	73.3097	68.8738	65.9694	86.1401	85.2654	63.6961

Table 10.5

9.49794	3.76105	2.58316	3.1983	.417378	2.6296	.980215	2.26886	12.7192	.639708
1.25589	2.6607	2.17027	2.5262	3.95442	4.39354	3.1799	2.50287	.803577	.557738
3.43366	2.8698	7.39507	3.12274	1.19529	2.981	8.70338	9.71008	5.96435	2.99968

Table 10.6

SOLUTION: After loading our procedure **normalAssumption**, we enter the values in Table 10.5 and name this list data1. We then evaluate **normalAssumption** for these values and obtain a *p*-value of approximately 0.83196, so we have no reason to dispute the null hypothesis that the population is normal.

```
data1={77.7331,86.7357,82.9031,59.6612,81.2474,65.1297,
 81.0629,104.793,82.4776,82.2584,88.1696,61.4329,69.1927,
 67.6756,80.8268,55.6229,68.5386,83.3563,54.261,62.4938,
 88.6138,67.5976,79.1755,84.5068,73.3097,68.8738,65.9694,
 86.1401,85.2654,63.6961}

normalAssumption[data1]
D'Agnostino-Pearson Test for Departure from Normality
 {Skewness→0.147652,Kurtosis→-0.204058,
  TestStatistic→0.367942,PValue→0.83196}
```

Next, we enter the data in Table 10.6, naming the list `data2`. When we use **normalAssumption** with this data, we obtain a p-value of approximately 0.00133468, so we reject the null hypothesis that the population is normal.

```
data2={9.49794,3.76105,2.58316,3.1983,0.417378,2.6296,
 0.980215,2.26886,12.7192,0.639708,1.25589,2.6607,
 2.17027,2.5262,3.95442,4.39354,3.1799,2.50287,0.803577,
 0.557738,3.43366,2.8698,7.39507,3.12274,1.19529,2.981,
 8.70338,9.71008,5.96435,2.99968}
```

```
normalAssumption[data2]
```

```
D'Agnostino-Pearson Test for Departure from Normality

{Skewness→1.5176,Kurtosis→1.87542,TestStatistic→13.2381,
 PValue→0.00133468}
```

■

10.3 Transformations

Transformations are most useful for departures from the assumption of homogeneous variance and normality. However, because there is good evidence that inferential methods based on the Student's t-distribution and analysis of variance are robust enough to perform well when these assumptions are violated, such remedial methods may not be needed unless the departures are severe. When the appropriate plot(s) indicates an obvious departure which is statistically significant using appropriate hypothesis testing, the data set may be adjusted by transforming each value from its original form (Y) to a different form (Y'). Common transformations involve logarithms, square roots, and reciprocals. If the transformed data can reasonably satisfy the assumptions, then the analysis is carried out on the adjusted data set. Heterogeneity of variance and nonnormality frequently appear together and the same transformation can often correct both departures. On the other hand, a transformation used to remedy one departure may cause a different departure in the transformed data. The transformed data should be carefully checked for all assumptions. We have stored several functions to be used in making transformations on the disk which comes with this book. To access them, first load the disk and click on its icon. Next, locate the procedure called **transformations** and click on the rightmost cell bracket. Then press **ENTER** and return to the *Mathematica* work sheet.

An alternative to transforming the data is to use an analysis technique not dependent on the assumption(s) that is violated. Such methods are called nonparametric or distribution-free, and they are discussed in Chapter 12. When the assumptions are met, procedures based on those assumptions are more powerful that the more generally applicable nonparametric procedures.

Outliers

If the presence of valid outliers does not result in departure from normality or constant variance, then no remedial action is necessary. Otherwise, remedial procedures appropriate for departure from normality and/or constant variance can be used. If there are only a few outliers and no problems with the data otherwise, each can be replaced with its closest neighbor that is not an outlier. This procedure is said to **winsorize** the data. See also the discussion of weighted least squares in Chapter 11.

Random Sampling and Independence

There is no simple adjustment for lack of random sampling or lack of independence. The basic design of the experiment or the way in which the data are collected must be changed. If the sampling technique contributes to departure from normality or constant variance, remedial procedures for these departures can be used. If lack of independence is related to a sequence variable such as time, including the sequence variable in the statistical analysis along with the original variable(s) of interest may remedy the problem.

Equality of Variance

The transformation that is most useful in stabilizing variance will depend on the pattern of variability in the data and/or the type of data. To examine the pattern of variability, sample means, variances, and standard deviations can be calculated for each sample or each cell in a multifactor experiment. Ratios of the form $s_i^2/\overline{Y}_i$, $s_i/\overline{Y}_i$, and $s_i/\overline{Y}^2_i$ can then be evaluated. Approximate constancy of one of the three for all samples (cells) would suggest the corresponding transformation (square root, logarithmic, or reciprocal, respectively) as useful. An inverse relationship between standard deviations and means would suggest an appropriate (the square) transformation.

We will list several situations and suggest the appropriate transformations to try.

Square Root

Pattern of Variability: Variance proportional to the mean; that is, large sample means are accompanied by large sample variances.

Type of Data: Counts of randomly occurring objects or events; that is, data which follows the Poisson Distribution.

Transformation: Square root:

$$Y' = \sqrt{Y}$$

$$Y' = \sqrt{Y + \frac{3}{8}}$$

$$Y' = \sqrt{Y} + \sqrt{Y + 1}.$$

Note: If there are negative data values, the square root may not be the correct transformation. If you try it, first add a constant to each data value that is large enough to avoid taking the square root of a negative number.

EXAMPLE 1: Use a transformation to stabilize the variance of the data in Table 10.7.

Group 1	2	0	2	3	0
Group 2	6	4	8	2	4
Group 3	9	5	6	5	11
Group 4	2	4	1	0	2

Table 10.7

SOLUTION: Note that loading our procedure **`transformations`** will automatically load the packages **DescriptiveStatistics** and **ContinuousDistributions**. Having done this, we enter the data in Table 10.7 into a list named `data`. Next, we define a function `groupmeansVars` to calculate the group mean and variance and use it on `data`. Notice that large sample means correspond to large sample variances. Therefore, we will try a square root transformation, and, because some of the values are zero, we use the transformation

$$Y' = \sqrt{Y + \frac{3}{8}},$$

which is called `transSqrt` in **`transformations`**. We also define the function `transData` to apply the transformation to each group in `data`. Finally, we apply the transformation and show the results in `newData`. Notice in groupmeansVars2 that when we determine the means and variances of the transformed values, the variance is stabilized.

```
data={{2,0,2,3,0},{6,4,8,2,4},{9,5,6,5,11},{2,4,1,0,2}};

groupmeansVars=
 Table[{Mean[data[[i]]],Variance[data[[i]]]}//N,
  {i,1,Length[data]}]
{{1.4,1.8},{4.8,5.2},{7.2,7.2},{1.8,2.2}}

transSqrt[y_]:=Sqrt[y+3/8]

transData[i_]:=transSqrt[data[[i]]]//N

newData=Table[transData[i],{i,1,Length[data]}]
{{1.5411,0.612372,1.5411,1.83712,0.612372},{2.52488,
  2.09165,2.89396,1.5411,2.09165},{3.06186,2.3184,
  2.52488,2.3184,3.37268},{1.5411,2.09165,1.1726,
  0.612372,1.5411}}

groupmeansVars2=
 Table[{Mean[newData[[i]]],Variance[newData[[i]]]}//N,
  {i,1,Length[newData]}]
{{1.22881,0.331271},{2.22865,0.260161},{2.71925,
  0.225874},{1.39177,0.297482}}
```

■

Logarithm

Pattern of Variability: Standard deviation proportional to the mean; that is, the coefficient of variation is constant (the same) for each list of data values.

Transformation: Logarithmic:

$$Y' = \log Y$$
$$Y' = \log Y(Y+1).$$

Note: Logarithms with base 10 are frequently used, but any base is satisfactory. It is preferable to take logarithms of numbers greater than or equal to 1. Adding a constant to each value before taking the logarithm may be useful. If all the data values are between 0 and 1, multiplying each by a constant (usually a power of 10) may be more appropriate.

Reciprocal

Pattern of Variability: Standard deviation proportional to the square of the mean.

Transformation: Reciprocal:

$$Y' = \frac{1}{Y}$$

$$Y' = \frac{1}{Y+1}$$

Note: If there are zeros in the data, add a constant to each data value to avoid taking the reciprocal of zero.

Square

Pattern of Variability: Standard deviation inversely proportional to the mean; that is, large sample means are accompanied by small standard deviations.

Transformation: Square.

$$Y' = Y^2$$

EXAMPLE 2: Determine which transformation is applicable for the values in Table 10.8.

Group 1	3.5	3.6	3.3	2.9	3.1
Group 2	7.5	6.4	7.6	6.3	6.9

Table 10.8

SOLUTION: Following a procedure like that in Example 1, we load **`transformations`**, enter the data, and for each group calculate the group mean, variance, and standard deviation using `groupmeansVarsStd`. Next, we define `entry[i]` to calculate $s_i^2/\overline{Y}_i$, $s_i/\overline{Y}_i$, and $s_i/\overline{Y}_i^2$ for group i. We compute the values in the table named `tb`, and display the results with `TableForm`. Because the values of $s_i/\overline{Y}_i$ are almost equal, we use the transformation $Y' = \log(Y+1)$. We define this transformation in `transLog` and write the function `transData2` to apply the transformation to the elements of each group. Notice that `Log[y,10]` gives the logarithm with base 10. The transformed data is shown in `newData2`. In `groupmeansVars2`, we find the mean and variance of each group of transformed data. Notice that the variance is stabilized with this transformation.

```
data2=
 {{3.5,3.6,3.3,2.9,3.1},{7.5,6.4,7.6,6.3,6.9}};

groupmeansVarsStd=Table[{Mean[data2[[i]]],
    Variance[data2[[i]]],Sqrt[Variance[data2[[i]]]]}//N,
  {i,1,Length[data2]}]
{{3.28,0.082,0.286356},{6.94,0.363,0.602495}}

entry[i_]:=
 {i,groupmeansVarsStd[[i,2]]/groupmeansVarsStd[[i,1]],
  groupmeansVarsStd[[i,3]]/groupmeansVarsStd[[i,1]],
  groupmeansVarsStd[[i,3]]/groupmeansVarsStd[[i,1]]^2}

tb=Table[entry[i],{i,1,Length[groupmeansVarsStd]}];

TableForm[tb,TableAlignments -> {Center},
 TableHeadings -> {None, {"i",
"si^2/Ȳi","si/Ȳi","si/Ȳi^2"}}]
i  si^2/Ȳi si/Ȳi si/Ȳi^2
1  0.0250.08730380.026617
2  0.05230550.08681480.0125093

transLog[y_]:=Log[y+1,10]

transData2[i_]:=transLog[data2[[i]]]//N

newData2=Table[transData2[i],{i,1,Length[data2]}]
{{1.5309,1.50885,1.57861,1.69186,1.6319},
  {1.07594,1.15044,1.07009,1.15832,1.11405}}

groupmeansVars2=Table[
  {Mean[newData2[[i]]],Variance[newData2[[i]]]}//N,
  {i,1,Length[newData2]}]
{{1.58842,0.00558194},{1.11377,0.00166697}}
```

■

EXAMPLE 3: Generate a collection of data using the chi-square distribution and determine an appropriate transformation to stabilize the variance.

SOLUTION: We begin by loading the necessary packages. Next, we generate three sets of 10 numbers using the chi square distribution with three, two, and four degrees of freedom. We name this collection of data `dataRandom3`. In `groupMeansVarsStd`, we find the mean, variance, and standard deviation of each group. The variances are 2.42985, 4.53176, and 12.5422.

```
dataRandom3={RandomArray[ChiSquareDistribution[3],10],
  RandomArray[ChiSquareDistribution[2],10],
  RandomArray[ChiSquareDistribution[4],10]}
```

```
{{3.64735,4.06056,1.08711,2.21093,4.46602,1.50053,
  1.54348,0.438803,0.75392,4.25178},{0.983526,0.319084,
  2.99698,3.21586,5.07494,1.43377,7.08766,0.704366,
  3.95643,2.83079},{7.21526,1.04185,2.23396,1.01254,
  11.686,0.903573,1.17295,4.53716,3.85088,6.18729}}
```

```
groupmeansVarsStd=Table[
  {Mean[dataRandom3[[i]]],Variance[dataRandom3[[i]]],
    Sqrt[Variance[dataRandom3[[i]]]]}//N,
  {i,1,Length[dataRandom3]}]
```

```
{{2.39605,2.42985,1.5588},{2.86034,4.53176,2.12879},
 {3.98415,12.5422,3.5415}}
```

We compute $s_i^2/\overline{Y}_i$, $s_i/\overline{Y}_i$, and $s_i/\overline{Y}_i^2$ for each group; we see that $s_i/\overline{Y}_i^2$ is most stable. Therefore, we use the reciprocal transformation. When we use $Y' = 1/Y$, we find that the variance is stabilized somewhat.

```
entry[i_]:=
 {i,groupmeansVarsStd[[i,2]]/groupmeansVarsStd[[i,1]],
  groupmeansVarsStd[[i,3]]/groupmeansVarsStd[[i,1]],
  groupmeansVarsStd[[i,3]]/groupmeansVarsStd[[i,1]]^2}

tb=Table[entry[i],{i,1,Length[groupmeansVarsStd]}];

TableForm[tb,TableAlignments -> {Center},
 TableHeadings -> {None, {"i",
"si2/Ȳi","si/Ȳi","si/Ȳi2"}}]
```

i	$s_i^2/\bar{Y}_i$	$s_i/\bar{Y}_i$	$s_i/\bar{Y}_i^2$
1	1.01411	0.65057	0.271518
2	1.58434	0.744244	0.260194
3	3.14803	0.888898	0.223109

```
transReciprocal[y_]:=1/y

transData3[i_]:=transReciprocal[dataRandom3[[i]]]//N

newData3=Table[transData3[i],{i,1,Length[dataRandom3]}]
{{0.274172,0.246271,0.919867,0.452299,0.223913,
  0.66643,0.647886,2.27893,1.3264,0.235196},
 {1.01675,3.13397,0.333669,0.310959,0.197047,
  0.697462,0.14109,1.41972,0.252753,0.353258},
 {0.138595,0.959827,0.447636,0.987613,0.0855722,
  1.10672,0.852554,0.220402,0.259681,0.161622}}

groupmeansVars3=
 Table[{Mean[newData3[[i]]],Variance[newData3[[i]]]}//N,
  {i,1,Length[newData3]}]
{{0.727136,0.424608},{0.785668,0.848924},
 {0.522022,0.16581}}
```

When we use $Y' = 1/(Y + 1)$, we obtain more favorable results, as the variance of the three groups of transformed data is 0.031442, 0.04314376, and 0.0312142.

```
transReciprocal2[y_]:=1/(y+1)

transData32[i_]:=transReciprocal2[dataRandom3[[i]]]//N

newData32=
 Table[transData32[i],{i,1,Length[dataRandom3]}]
{{0.215177,0.197606,0.47913,0.311437,0.182949,
  0.399915,0.393162,0.695022,0.570151,0.190412},
 {0.504153,0.758102,0.250189,0.237199,0.164611,
  0.410885,0.123645,0.586728,0.201758,0.261043},
 {0.121725,0.489751,0.309219,0.496884,0.0788269,
  0.525328,0.460205,0.180598,0.206148,0.139135}}
```

```
groupmeansVars32=Table[
  {Mean[newData32[[i]]],Variance[newData32[[i]]]}//N,
  {i,1,Length[newData32]}]
{{0.363496,0.031442},{0.349831,0.0431376},
 {0.300782,0.0312142}}
```

■

EXAMPLE 4: Generate a collection of data using the uniform distribution and determine which transformation should be used to stabilize the variance.

SOLUTION: After loading the packages needed to execute the commands, we randomly generate three groups of 10 numbers using the uniform distribution over the intervals [1, 4], [2, 4], and [1, 5 ,4], respectively. This collection of data is called randData. When we determine the mean, variance, and standard deviation for each group, we observe that the standard deviation decreases as the mean increases. Therefore, we use the transformation $Y' = Y^2$ that is defined in transSquare. When given a group, the function transData applies $Y' = Y^2$ to each member of the group. The set of transformed data is listed in newData. As we observe in groupmeansVars, the variance is almost the same for each group in newData.

```
randData={RandomArray[UniformDistribution[1,4],10],
  RandomArray[UniformDistribution[2,4],10],
  RandomArray[UniformDistribution[2.5,4],10]}
{{2.16977,2.50442,2.97788,2.58228,3.23156,3.76594,
  3.07865,2.48051,2.63454,1.24597},{2.84166,2.87938,
  3.17009,3.39635,3.80056,3.53005,2.31078,2.03487,2.2698,
  2.56909},{3.71231,2.89319,3.82092,2.85131,2.59421,
  3.34337,2.79327,2.54576,3.9904,3.01174}}

groupmeansVarsStd=Table[{Mean[randData[[i]]],

Variance[randData[[i]]],Sqrt[Variance[randData[[i]]]]}//
N,

{i,1,Length[randData]}]
```

```
{{2.66715,0.458723,0.677291},{2.88026,0.34772,0.589678},
 {3.15565,0.276134,0.525485}}

transSquare[y_]:=y^2

transData[i_]:=transSquare[randData[[i]]]//N

newData=Table[transData[i],{i,1,Length[randData]}]

{{4.70791,6.27214,8.86779,6.66817,10.443,14.1823,9.47809,
  6.15291,6.94081,1.55243},{8.07502,8.29085,10.0495,11.5352,
  14.4442,12.4612,5.3397,4.14071,5.15201,6.60022},{13.7813,
  8.37057,14.5994,8.12998,6.72994,11.1781,7.80236,6.48088,
  15.9233,9.07058}}

groupmeansVars=
 Table[{Mean[newData[[i]]],Variance[newData[[i]]]}//N,
  {i,1,Length[newData]}]
{{7.52655,11.843},{8.60886,11.8393},{10.2066,11.8246}}
```

Arcsine

Type of Data: Percentages from 0 to 100% or proportions from 0 to 1 based on differing numbers of cases (that is, with differing denominators).

Transformation: Arcsine:

$$Y' = 2\arcsin\sqrt{Y}.$$

Note: If you know the actual proportions *Y/n,* then the arcsine transformation may be improved by replacing *0/n* by *1/(4n)* and *n/n* by 1 – 1/(4*n*) before making the transformation. Other transformations to try if the data are of the form *Y/n* are

$$Y' = \arcsin\sqrt{\frac{Y+\frac{3}{8}}{n+\frac{3}{4}}}$$

$$Y' = \frac{1}{2}\left(\arcsin\sqrt{\frac{Y}{n+1}} + \arcsin\sqrt{\frac{Y+1}{n+1}}\right).$$

EXAMPLE 5: Table 10.9 gives two groups of percentage data. Use the arcsine transformation to stabilize the sample variance.

Group 1	84.2	88.9	89.2	83.4	80.1	81.3	85.8
Group 2	92.3	95.1	90.3	88.6	92.6	96.0	93.7

Table 10.9

SOLUTION: After loading the necessary package, we enter the data in Table 10.9, naming it `data4`. We then calculate the mean, variance, and standard deviation for each group. We observe that the variances differ substantially. Next, we define the function `convertToProportions` to convert each group from percentages to proportions. In `proportions`, we convert the data using this function.

```
data4={{84.2,88.9,89.2,83.4,80.1,81.3,85.8},
  {92.3,95.1,90.3,88.6,92.6,96.0,93.7}};

groupmeansVars=
 Table[{i,Mean[data4[[i]]]//N,Variance[data4[[i]]]//N,
   Sqrt[Variance[data4[[i]]]]//N},
  {i,1,Length[data4]}];

TableForm[groupmeansVars,TableAlignments -> {Center},
 TableHeadings -> {None, {"i", "p̄_i ", "s_i^2", "s_i"}}]
i      p̄_i          s_i^2         s_i
1      84.7         12.2933       3.50619
2      92.6571       6.72952      2.59413

convertToProportions[i_]:=data4[[i]]/100//N

proportions=
 Table[convertToProportions[i],{i,1,Length[data4]}]
{{0.842,0.889,0.892,0.834,0.801,0.813,0.858},
 {0.923,0.951,0.903,0.886,0.926,0.96,0.937}}
```

Next, we define the transformation $Y' = 2\arcsin\sqrt{Y}$ in `transArcSinRadians` so that the result is given in degrees instead of radians. (This is accomplished by multiplying by $180/\pi$.) The function `transData4` applies this transformation to each group in `proportions`, so the transformed data are obtained in `newData4`.

In groupmeansVars2, we find the mean, variance, and standard deviation of each group of transformed data. The results show that the variance is stabilized.

```
transArcSinRadians[y_]:=2 (ArcSin[Sqrt[y]]*180/Pi)

transData4[i_]:=transArcSinRadians[proportions[[i]]]//N

newData4=Table[transData4[i],{i,1,Length[proportions]}]

{{133.157,141.078,141.628,131.913,127.013,128.756,
  135.725},{147.779,154.422,143.707,140.534,148.43,
  156.926,150.927}}

groupmeansVars2=Table[
  {i,Mean[newData4[[i]]]//N,Variance[newData4[[i]]]//N,
   Sqrt[Variance[newData4[[i]]]]//N},
  {i,1,Length[newData4]}];

TableForm[groupmeansVars2,TableAlignments -> {Center},
 TableHeadings -> {None, {"i", "p̄_i ", "s_i^2", "s_i"}}]
```

i	$\bar{p}_i$	s_i^2	s_i
1	134.181	32.0694	5.66299
2	148.961	32.9055	5.73633

If we wish to consider the other two arcsine transformations, we need to determine the sample size, n. This value is 14 in this case. Following a procedure similar to that above, we define the transformation function:

$$Y' = \arcsin\sqrt{\left(Y + \frac{3}{8}\right)/\left(n + \frac{3}{4}\right)}$$

in transArcSinRadians2. (Again, the result is given in degrees.) We see that the variance is stabilized with this transformation as well.

```
n=Length[Flatten[proportions]]

14

transArcSinRadians2[y_]:=
 ArcSin[Sqrt[(y+3/8)/(n+3/4)]]*180/Pi

transData42[i_
]:=transArcSinRadians2[proportions[[i]]]//N
```

```
newData42=
 Table[transData42[i],{i,1,Length[proportions]}]
{{16.693,17.0219,17.0427,16.6364,16.4013,16.4871,16.805
6},
 {17.2564,17.4474,17.1188,
  17.0011,17.2769,17.5084,17.3521}}
```

```
groupmeansVars42=Table[{i,
   Mean[newData42[[i]]]//N,Variance[newData42[[i]]]//N,
   Sqrt[Variance[newData42[[i]]]]//N},
  {i,1,Length[newData42]}];
```

```
TableForm[groupmeansVars42,TableAlignments -> {Center},
 TableHeadings -> {None, {"i", "p̄i ", "si2", "si"}}]
```

i	$\bar{p}_i$	s_i^2	s_i
1	16.7268	0.0609646	0.24691
2	17.2802	0.0316307	0.17785

Finally, we consider the transformation $Y' = \frac{1}{2}(\arcsin\sqrt{Y/(n+1)} + \arcsin\sqrt{(Y+1)/(n+1)})$, defined on `transArcSinRadians3`. As in the other cases, the transformed values are given in degrees. The results show a stabilization of the variance.

```
transArcSinRadians3[y_]:=
 (ArcSin[Sqrt[y/(n+1)]]+ArcSin[Sqrt[(y+1)/(n+1)]])*
  180/Pi
```

```
transData43[i_
]:=transArcSinRadians3[proportions[[i]]]//N
```

```
newData43=
 Table[transData43[i],{i,1,Length[proportions]}]
{{34.2186,34.8756,34.9171,34.1055,33.6347,33.8067,34.4437},
 {35.3432,35.7236,35.0689,34.8341,35.3842,35.8451,35.5339}}
```

```
groupmeansVars43=Table[{i,
   Mean[newData43[[i]]]//N,Variance[newData43[[i]]]//N,
   Sqrt[Variance[newData43[[i]]]]//N},
```

```
{i,1,Length[newData43]}];

TableForm[groupmeansVars43,TableAlignments -> {Center},
 TableHeadings -> {None, {"i", "p̄i ", "si2", "si"}}]
```

i	$\bar{p}_i$	s_i^2	s_i
1	34.286	0.243656	0.493615
2	35.3904	0.125583	0.354377

■

EXAMPLE 6: Use the arcsine transformation to stabilize the variance in a collection of data generated using the beta distribution.

SOLUTION: After loading the necessary packages, we randomly generate three groups of data, each containing 10 elements, using the beta distribution with parameters $p = 2$ and $q = 2$; $p = 3$ and $q = 3$; $p = 4$ and $q = 4$. We name this collection `dataRandom4` and determine the variance of each group in `groupVars`.

```
dataRandom4={RandomArray[BetaDistribution[2,2],10],
  RandomArray[BetaDistribution[3,3],10],
  RandomArray[BetaDistribution[4,4],10]}
{{0.53239,0.802439,0.869672,0.290151,0.704132,0.450182,
  0.565891,0.140704,0.390127,0.255477},{0.394409,
  0.578338,0.255149,0.682648,0.281775,0.656174,0.28842,
  0.671808,0.326259,0.148435},{0.535306,0.608895,0.175994,
  0.0844495,0.384303,0.263363,0.336048,0.588978,0.403997,
  0.400093}}

groupVars=Table[Variance[dataRandom4[[i]]]//N,
  {i,1,Length[dataRandom4]}]
{0.0580736,0.0399255,0.029517}
```

To attempt to stabilize these values, we introduce the function `transArcSin`, which represents $Y' = 2\arcsin\sqrt{Y}$. In this case, we leave the result in radians. We also define `transData4` to make the transformation of each member of the ith group `dataRandom4[[i]]`, and we apply this function to obtain the transformed values in `newData4`. The variance of each of the three groups of transformed data is shown in `transgroupVars`.

```
transArcSin[y_]:=2 (ArcSin[Sqrt[y]])
```

```
transData4[i_]:=transArcSin[dataRandom4[[i]]]//N

newData4=Table[transData4[i],{i,1,Length[dataRandom4]}]

{{1.63562,2.22041,2.40289,1.13768,1.99135,1.471,1.70296,
  0.769021,1.34924,1.0598},{1.35801,1.72812,1.05905,1.94475,
  1.11915,1.88846,1.13387,1.92156,1.21591,0.791007},
 {1.64147,1.79035,0.865826,0.589712,1.33729,1.07779,1.23671,
  1.74971,1.37759,1.36963}}

transgroupVars=Table[Variance[newData4[[i]]]//N,
  {i,1,Length[newData4]}]

{0.273711,0.176106,0.146422}
```

Next, we define the sample size so that we can apply the transformation $Y' = \arcsin\sqrt{\left(Y+\frac{3}{8}\right)/\left(n+\frac{3}{4}\right)}$, which is given in `transArcSin2`. We follow a similar procedure as that used above to transform the data and to determine the variance of each group of transformed data in `transgroupVars2`.

```
n=Length[Flatten[dataRandom4]]

30

transArcSin2[y_]:=ArcSin[Sqrt[(y+3/8)/(n+3/4)]]

transData42[i_]:=transArcSin2[dataRandom4[[i]]]//N

newData42=
 Table[transData42[i],{i,1,Length[dataRandom4]}]

{{0.172637,0.196951,0.202572,0.14761,0.188447,0.164556,
  0.175828,0.129867,0.158402,0.143684},{0.158849,0.176999,
  0.143646,0.186539,0.146671,0.184162,0.147416,0.185569,
  0.151594,0.130842},{0.172917,0.179844,0.134263,0.122542,
  0.157793,0.144586,0.152656,0.177995,0.159844,0.159439}}

transgroupVars2=Table[Variance[newData42[[i]]]//N,
  {i,1,Length[newData42]}]

{0.000566589,0.000415713,0.000344595}
```

Finally, we make use of $Y' = \frac{1}{2}(\arcsin\sqrt{Y/(n+1)} + \arcsin\sqrt{(Y+1)/(n+1)})$, to transform the data. In this case, the list of variances is given in `transgroupVars3`.

```
transArcSin3[y_]:=
 (ArcSin[Sqrt[y/(n+1)]]+ArcSin[Sqrt[(y+1)/(n+1)]])

transData43[i_]:=transArcSin3[dataRandom4[[i]]]//N

newData43=
 Table[transData43[i],{i,1,Length[dataRandom4]}]
{{0.355634,0.40512,0.416409,0.302344,0.387952,0.338811,
  0.36221,0.260444,0.325795,0.293534},{0.326746,0.364617,
  0.293449,0.384084,0.300253,0.379254,0.301913,0.382115,
  0.311113,0.262935},{0.356212,0.370446,0.271442,0.240361,
  0.324495,0.295575,0.313428,0.366659,0.328863,0.328003}}

transgroupVars3=Table[Variance[newData43[[i]]]//N,
  {i,1,Length[newData43]}]
{0.00255475,0.00190399,0.001726}
```

■

Normality

The transformation that is most useful will depend on the shape of the distribution of residuals and/or the type of data. Because each of the data values must be transformed in the same way, a suggested technique for determining an appropriate transformation is to generate the set of residuals using our function `anovaResiduals` and then plot them using our function `normalHistogram`.

Logarithm

Shape: Positively skewed.

Transformation: Logarithm:

$$Y' = \log Y$$

$$Y' = \log(Y+1).$$

Square

Shape: Negatively skewed.

Transformation: Square:

$$Y' = Y^2.$$

Arcsine

Type of Data: Percentages from 0 to 100% or proportions from 0 to 1.

Transformation: Arcsine:

$$Y' = 2\arcsin\sqrt{Y}.$$

Note: When percentages in the original data fall between 30 and 70%, it is generally not necessary to apply the arcsine transformation.

Box–Cox Transformations

If there is not a good reason for selecting a transformation or one wants to confirm a selection, the Box–Cox procedure provides a systematic way to search for the best transformation of the form $Y' = Y^\lambda$. The values of λ that are evaluated are usually in the range of –2 to 2, which includes all of the following transformations.

λ	Y^λ
−2	$1/Y^2$
−1	$1/Y$
−1/2	$1/\sqrt{Y}$
0	log Y (by definition)
1/2	$\sqrt{Y}$
1	Y
2	Y^2

The user may wish to alter the range depending on the data. For example, for data that are positively skewed, values of $\lambda \leq 0$ may be emphasized, while for negatively skewed data, values of $\lambda \geq 2$ may be investigated. The best transformation is the one that minimizes the variance, $\hat{\sigma}^2(\lambda)$ among the transformed values. For a single sample, $\hat{\sigma}^2(\lambda)$ would be the sample variance of the transformed values. For more than one sample, $\hat{\sigma}^2(\lambda)$ would be the mean square for error (*MSE*). There is often little difference in values of $\hat{\sigma}^2(\lambda)$ near the minimum, so several values of λ are possible for being best (or very close). In this case, a "meaningful" value of λ in the neighborhood of best is chosen. For example, the smallest value of $\hat{\sigma}^2(\lambda)$ might be $\hat{\sigma}^2(0.1)$; that is, the transformation $Y' = Y^{1/10}$ is the best transformation. But, the values of $\hat{\sigma}^2(\lambda)$ are very close for values of λ between -0.1 and 0.2. Thus, we might decide to use the familiar transformation $Y' = \log Y$, which corresponds to $\lambda = 0$, this being a "meaningful" value for λ.

We have written a procedure called **boxCox** to find the best transformation using the Box–Cox approach. It will determine the value of λ, called the **Box–Cox lambda**, that defines the best transformation of the form

$$Y' = Y^{\lambda}.$$

To use **boxCox**, first load the disk that comes with this book and click on its icon. Next, locate the procedure name and click on the rightmost cell bracket. Then press **ENTER** and return to the *Mathematica* work sheet.

`boxCox[list,opts]` displays a value for the Box–Cox lambda, a confidence coefficient, and a confidence interval. The options and their default values are `limits->{-2,2}`, `gridsize->20`, and `confidence->0.95`. The limits determine the range of possible values for the Box–Cox lambda. The grid size determines the number of values that are considered at equal intervals along the range of possible values.

EXAMPLE 7: Use the Box-Cox procedure with the data in Table 10.10 to determine the best transformation of the form $Y' = Y^{\lambda}$. (**Note:** The values in Groups 1, 2, and 3 were generated at random from a chi-square distribution with 10, 15, and 20 degrees of freedom, respectively.)

Group 1	9.25667	5.0732	0.66261	6.78225	6.68503	19.6864	10.8548
Group 2	28.0191	13.6191	5.82406	7.72194	9.99296	20.0193	10.417
Group 3	27.3743	9.36934	21.9378	34.8322	19.7593	23.7833	17.9993

Table 10.10

SOLUTION: We enter the three groups of numbers and name the list `data`. When we perform the Box–Cox procedure with **boxCox**, we obtain the value $\lambda = 10/19$ and the confidence interval {6/19, 18/19}. Therefore, a reasonable choice is $\lambda = 1/2$, so we will use the transformation $Y' = Y^{1/2} = \sqrt{Y}$.

```
data={
{9.25667,5.0732,0.66261,6.78225,6.68503,19.6864,10.8548},
 {28.0191,13.6191,5.82406,7.72194,9.99296,20.0193,10.417},
 {27.3743,9.36934,21.9378,34.8322,19.7593,23.7833,17.9993}};

boxCox[data]

Box-Cox Transformation

{BoxCoxλ→10/19, Confidence→95. %,
ConfidenceInterval→{6/19,18/19}}
```

Next, we will compare the normal probability plot of the original numbers and that of the transformed numbers. We begin by making sure that the functions needed to execute **anovaResiduals** and **normalProbability** are loaded. We then use this function to calculate the residuals for the values in `data`. We name this list `reslist` and then graph the normal probability plot for the set of residuals in `reslist`.

```
reslist=anovaResiduals[data]//Flatten

{0.827961,-3.35551,-7.7661,-1.64646,-1.74368,11.2577,
 2.42609,14.36,-0.0399657,-7.83501,-5.93713,-3.66611,
 6.36023,-3.24207,5.22351,-12.7815,-0.212991,12.6814,
 -2.39149,1.63251,-4.15149}
```

```
normalProbability[reslist]
```

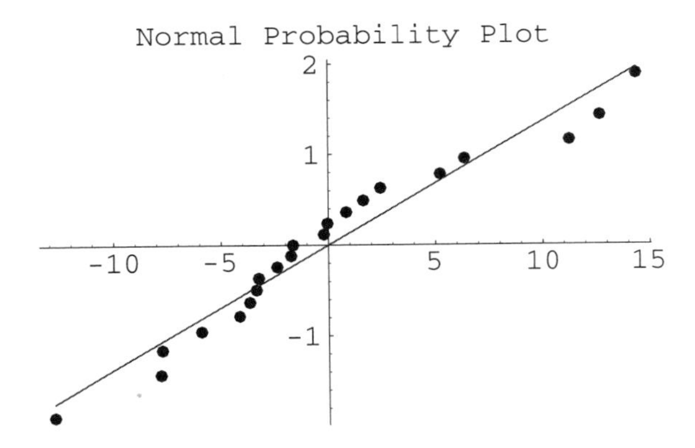


```
Pearson'sCorrelationCoefficient ->0.976158
```

Next, we apply the square root function to the values in `data` and name this list `newvals`. Following a similar procedure as before, we obtain the associated residuals for the numbers in `newvals`, naming this list `reslistNew` and graphing the normal probability plot for these residuals. The two plots indicate that the values in the transformed set of data in `newvals` produce a more linear normal probability plot (for the associated residuals).

```
newvals=Map[Sqrt,data]

{{3.04248,2.25238,0.814009,2.60428,2.58554,4.43694,3.29466},
 {5.29331,3.69041,2.41331,2.77884,3.16116,4.47429,3.22754},
 {5.23205,3.06094,4.68378,5.90188,4.44514,4.87681,4.24256}}

reslistNew=anovaResiduals[newvals]//Flatten

{0.323866,-0.466235,-1.9046,-0.114336,-0.133069,
 1.71832,0.576051,1.71633,0.113427,-1.16367,-0.798142,
 -0.415815,0.897314,-0.349442,0.597308,-1.5738,
 0.0490435,1.26714,-0.189594,0.242075,-0.392179}
```

```
normalProbability[reslistNew]
```

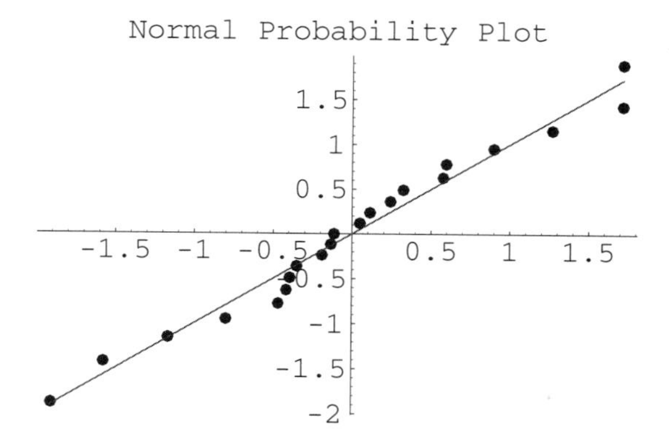


```
Pearson'sCorrelationCoefficient ->0.988125
```

■

CHAPTER 11

Regression and Correlation

11.1 Introduction

Regression

In this chapter we discuss methods used to express the relationship between quantitative variables. Simple regression and correlation are used when there are two variables, and multiple regression is used to express the relationship between one variable and two or more other variables. In general, regression is useful when the relationship between variables is of functional dependence. If so, the magnitude of one variable, called the dependent variable, is assumed to be determined by the other variables, called the independent variables. In regression, it is assumed that the functional relationship is less than a perfect one. That is, the value of the dependent variable is not completely determined by the independent variables. As a result, the dependent variable is often renamed the **response variable**, and the independent variables are referred to as either **explanatory**, **predictor**, or **regressor** variables. When there is only one independent variable in addition to the dependent variable, the term **simple regression** is used. When there are two or more independent variables, the term **multiple regression** is used. Note that a regression relationship does not necessarily imply a cause-and-effect relationship between the independent variable(s) and the dependent variable.

Correlation

Correlation is a term used to refer to a relationship between two variables which is not a dependence relationship. In such cases the magnitude of one variable changes as the magnitude of the other variable changes, but it is not reasonable to consider one to be dependent on the other. Although correlation among several variables, called **multiple correlation,** can be defined, correlation is most often used with two variables.

The standard *Mathematica* packages **Statistics`LinearRegression`** and **Statistics`NonLinearFit`**, as well as the built-in function `Fit`, provide the *Mathematica* user with a set of useful tools to perform regression analysis. We begin our discussion with simple linear regression, then discuss multiple linear regression, and nonlinear regression. We finish the chapter with a discussion of correlation.

Caution

The term *linear* can be a source of confusion when the applications of linear regression methodology are discussed. Consider simple regression. If the independent variable is represented by x and the dependent variable by y, then the simplest form of a model to represent the regression relationship is given by

$$y = \beta_0 + \beta_1 x + \varepsilon,$$

where ε, known as the **error term**, is included to account for addition (or subtraction) to the value of y due to other sources apart from the value of x. The values β_0 and β_1 are called **regression coefficients**. Because the graph of the equation

$$y = \beta_0 + \beta_1 x$$

is a line, users wrongly assume that y must have a linear relationship with x in order to use linear regression methodology. What is required is that the model, which represents the relationship, be additive (an equation with plus or minus signs between all of the terms) like the above model. Another way to describe the additive model is as a linear combination of the parameters, β_0 and β_1, and ε. Consequently, the additive model is also known as a linear model even though the dependent variable is not necessarily linearly related to the independent variable. Linear regression can be used as long as the model is additive. For example, the model

$$y = \beta_0 + \beta_1 x + \beta_2 x^2 + \varepsilon$$

is additive and can be fit using linear regression, but the relationship between x and y is not linear.

11.2 Simple Linear Regression

Least Squares Regression Equation

Suppose that we have a collection of n data points $\{(x_1, y_1), (x_2, y_2), \ldots, (x_n, y_n)\}$ which constitute a sample from a population of points, and we would like to investigate the relationship between the independent variable x and the dependent variable y in the population. If we assume that the relationship is linear and additive, then a model of the form

$$y_1 = \beta_0 + \beta_1 x_1 + \varepsilon,\ i = 1, 2, \ldots n,$$

can represent the relationship, where β_0 and β_1 are parameters. Because there is a unique value ε_1 corresponding to each value of y_i, the n data points will not all lie on a line. (They would if ε_1 was always zero.) The n data points can be used, however, to estimate β_0 and β_1 using both the built-in function `Fit` and the standard package **LinearRegression** found in the **Statistics** folder (or directory. The estimates called b_0 and b_1, respectively, are calculated using the method of least squares, which requires the estimates to be the values of β_0 and β_1 that minimize

$$\sum_{i=1}^{n} (y_i - \beta_0 - \beta_1 x_i)^2.$$

The following equations are used to calculate the least square estimates of β_0 and β_1 :

$$b_1 = \frac{\sum_{i=1}^{n} (x_1 - \bar{x})(y_i - \bar{y})}{\sum_{i=1}^{n} (x_i - \bar{x})^2} \text{ and } b_0 = \frac{1}{n}\left(\sum_{i=1}^{n} y_i - b_1 \sum_{i=1}^{n} x_i\right) = \bar{y} - b_1\bar{x}.$$

The built-in function `Fit` returns an expression of the form $b_0 + b_1 x$, where $\hat{y} = b_0 + b_1 x$ is the least squares regression equation (also called the **prediction equation**). Given a value of x, this equation can be used to calculate a value of y which is called a **predicted value** or a **fitted value**. The notation $\hat{y}$ is used to distinguish the predicted value $\hat{y}_i$ from the observed value y_i corresponding to a particular x-value x_i in the data. The standard package **LinearRegression** calculates the estimates b_0 and b_1 and an extensive list of other results. The following is the information displayed by *Mathematica* when the command `?LinearRegression` is used. The general description will be discussed in the section on multiple regression. The various option settings will be discussed throughout the chapter. In particular, the `Weights` and `Tolerance` settings will be discussed in the section on diagnostic procedures and remedial methods.

`Regress[data, funs, vars]` finds a least-squares fit to a list of data as a linear combination of the functions funs of variables `vars`. The data can have the form `{{x1, y1, ..., f1}, {x2, y2, ..., f2}, ...}`, where the number of coordinates `x`, `y`, ... is equal to the number of variables in the list `vars`. The data can also be written as a list of the response variables `{y1, y2, ...}`, where a single independent variable is assumed to take the values 1, 2, ... The argument `funs` can be any list of functions that depend only on the objects `vars`. The result is a list identifying summary statistics about the fit, or other statistics as specified by the option `RegressionReport`.

Options for `Regress` with corresponding default settings are `RegressionReport->SummaryReport`, `IncludeConstant->True`, `BasisNames->Automatic`, `Weights->Equal`, `Tolerance->Automatic`, and `ConfidenceLevel->0.95`.

Other settings for `RegressionReport` are `BestFit`, `BestFitCoefficients`, `ANOVATable`, `ConfidenceIntervalTable`, `CovarianceMatrix`, `CorrelationMatrix`, `EstimatedVariance`, `FitResiduals`, `ParameterTable`, `ParameterCITable`, `ParameterConfidenceRegion`, `SinglePredictionCITable`, `MeanPredictionCITable`, `CoefficientOfVariation`, `PredictedResponse`, `RSquared`, and `AdjustedRSquared`.

EXAMPLE 1: Table 11.1 gives the batting average of opposing teams and winning percentage of teams in the National League of Major League Baseball. Plot the data, determine a linear regression function to predict the winning percentage of a team based on the opposing team batting average, and plot the corresponding line on the scatter plot of the data.

Team	Opponent batting average	Winning percentage
Atlanta	0.240	0.625
Los Angeles	0.254	0.512
Florida	0.249	0.488
San Diego	0.245	0.524
Montreal	0.250	0.588
Chicago	0.252	0.475
St Louis	0.254	0.513
New York	0.270	0.463
Houston	0.274	0.512
Philadelphia	0.264	0.405
Pittsburgh	0.280	0.450
Cincinnati	0.266	0.480

Team	Opponent batting average	Winning percentage
San Francisco	0.268	0.456
Colorado	0.286	0.506

Table 11.1 National League: Opponent Batting Average and Winning Percentage

SOLUTION: First, we load the **LinearRegression** package for later use. Next we enter the data in `oppavg` and `winpct`, respectively, and then use `Table` to create a list of data points in the form (x_1, y_1) called `dpoints`. We then plot the data points with `ListPlot` using the option `PointSize[{0.02}]` to increase the size of each point when graphed. Note that points which are close together may appear to be one point. The graph is named `dots`.

```
<<Statistics`LinearRegression`

oppbavg={0.240,0.254,0.249,0.245,0.250,0.252,0.254,
  0.270,0.274,0.264,0.280,0.266,0.268,0.286};
winpct={0.625,0.512,0.488,0.524,0.588,0.475,0.513,
  0.463,0.512,0.405,0.450,0.480,0.456,0.506};

dpoints=Table[{oppbavg[[i]],winpct[[i]]},
  {i,1,Length[winpct]}]

{{0.24,0.625},{0.254,0.512},{0.249,0.488},{0.245,0.524},
 {0.25,0.588},{0.252,0.475},{0.254,0.513},{0.27,0.463},
 {0.274,0.512},{0.264,0.405},{0.28,0.45},{0.266,0.48},
 {0.268,0.456},{0.286,0.506}}

Clear[dots]
dots=ListPlot[dpoints,Prolog->{PointSize[0.02]}]
```

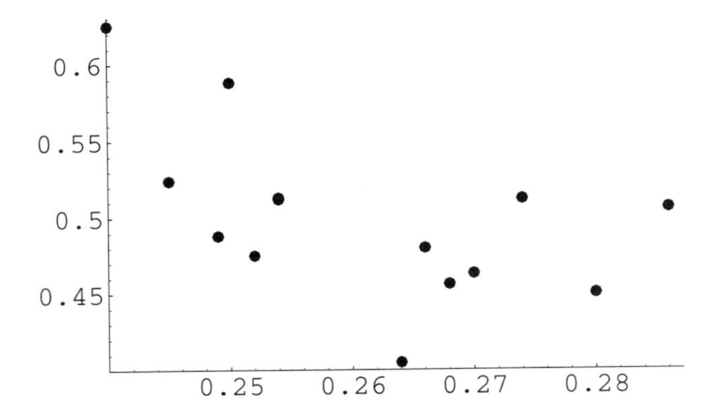

Assuming winning percentage y and opposing team batting average x are functionally related, we fit the model

$$y = \beta_0 + \beta_1 x + \varepsilon$$

using Regress. The first argument of Regress specifies the data, where the value of the dependent variable is always the second (last) entry in each ordered pair. Note that if the data are single values, the corresponding values of the independent variable are assumed to be 1, 2, 3, ... The second argument specifies the model, where the value 1 indicates that the model contains a constant term β_0. The third argument specifies the independent variable. In this example, the arguments are somewhat redundant, but they are all needed to use Regress in more sophisticated applications.

```
Regress[dpoints,{1,x},x]
```

```
                       Estimate      SE          TStat       PValue
{ParameterTable→  1    1.07813       0.25596     4.21211     0.00120568,
                  x    -2.2171       0.979963    -2.26243    0.0430218
 RSquared→0.299008, AdjustedRSquared→0.240592, EstimatedVariance→0.00236213,
                        DF    SumOfSq      MeanSq        FRatio     PValue
 ANOVATable→  Model     1     0.0120908    0.0120908     5.11859    0.0430218}
              Error     12    0.0283456    0.00236213
              Total     13    0.0404364
```

Regress has several options as listed above. If an option is not specifically included in the command, then the default applies. For example, the option IncludeConstant has default value True. Even if the value 1 had not been included in the second argument of Regress, the prediction equation would have a constant term. The option RegressionReport determines most of the output of Regress. The default is SummaryReport, which gives an output list that includes ParameterTable, RSquared, AdjustedRSquared, EstimatedVariance, and ANOVATable.

From ParameterTable we see that the least squares estimates are $b_0 = 1.07813$ and $b_0 = -2.2171$.. ParameterTable also provides estimates of the variability of the least squares estimates (in the column SE) and test statistics and associated p-values for hypothesis tests to determine whether β_0 and β_1 are different from zero. RSquared and AdjustedRSquared give the values of the **coefficient of determination** and the **adjusted coefficient of determination**, respectively. Both measure the proportionate reduction in variability in the values of the dependent variable y associated with the use of the independent variable x. Values close to one typically indicate strong regression relationships. AdjustedRSquared is most useful in multiple regression. EstimatedVariance gives the value of the mean square for error (*MSE*) which is also given in the ANOVATable output. The analysis of variance (ANOVA) table also provides a hypothesis test to determine whether β_1 is different from zero which is equivalent (in this application) to the test done in `ParameterTable`. Note that the *p*-values are identical.

In the ANOVA table, the sum of squares for total (*SST*) is calculated using

$$SST = \sum_{i=1}^{n} (y_i - \bar{y}_i)^2$$

and the sum of squares for error (*SSE*) is calculated using

$$SSE = \sum_{i=1}^{n} (y_i - \hat{y}_i).$$

The degrees of freedom for total (*DFT*) is $n-1$ and the degrees of freedom for error (*DFE*) is $n-2$. A mean square is a sum of squares divided by the corresponding degrees of freedom. Thus, $MSE = \frac{SSE}{DFE}$. Note that the row for `Model` in the ANOVA table is often labeled **Regression** by others. Likewise, `Error` is often called **Residual.**

The coefficient of determination, `RSquared`, is calculated from the ANOVA table. It is $r^2 = 1 - \frac{\text{SSE}}{\text{SSTO}}$. The adjusted coefficient of determination is

$$adj\ r^2 = 1 - \frac{MSE}{SSTO/(n-1)}.$$

Next, we illustrate the *`RegressionReport`* option *`BestFit`* to yield the least squares regression line for the data. We show that this is the same formula obtained with *`Fit`* and graph the least squares line simultaneously with the data using *`Show`*. Notice that initially in *`plotline`*, we suppress the output of the *`Plot`* command with the option setting *`DisplayFunction->Identity`*. In the *`Show`* command, we use *`DisplayFunction->$DisplayFunction`* in order to view the graph.

```
Regress[dpoints,{1,x},x,RegressionReport->BestFit]
```

```
{BestFit→1.07813-2.2171 x}
```

```
lsq[x_]=Fit[dpoints,{1,x},x]
```

```
1.07813-2.2171 x
```

```
plotline=Plot[lsq[x],{x,0.24,0.29},
  DisplayFunction->Identity];
```

```
Show[dots,plotline,DisplayFunction->$DisplayFunction]
```

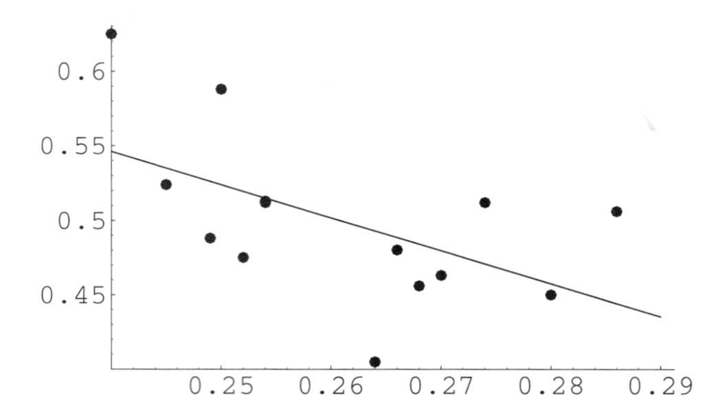

Another option for `Regress` is `BasisNames` which permits the labeling of the rows of `ParameterTable` and `ParameterCITable`.

```
Regress[dpoints, {1, x}, x, RegressionReport ->
  ParameterTable, BasisNames -> {b0, b1}]
```

{ParameterTable→

	Estimate	SE	Tstat	PValue
b_0	1.07813	0.25596	4.21211	0.00120568
b_1	-2.2171	0.9799623	-2.26243	0.0430218

}

■

Inferences in Regression Analysis

Assumptions

For valid inferences to be made, the data must be a randomly selected sample from a population of points which reflect an additive model of the form $y = \beta_0 + \beta_1 x + \varepsilon$, where the value of y is dependent on x as well as other influences represented by the error term ε. We assume that for a given x', call it $x_{i,}$, there is a whole distribution of possible y's (each with a corresponding value of ε). The mean of this distribution is assumed to be $\beta_0 + \beta_1 x_{i'}$, and we assume any y's selected from this distribution are independent. We assume that x_i is measured without error, and that the corresponding distribution of y's has a normal probability distribution. Furthermore, we assume that every such distribution has the same variance. We denote the common variance by σ^2. It is estimated by the mean square for error (*MSE*). A common terminology for σ is the **standard error of the regression**. It is of course estimated by $\sqrt{MSE}$.

Confidence Intervals and Hypothesis Tests for the Regression Coefficients

The $100(1-\alpha)\%$ confidence limits for β_1 are calculated using $b_1 \pm t_{\alpha/2,\, n-2}\, s_{b_1}$, where

$$s^2_{b_1} = \frac{MSE}{\sum_{i=1}^{n} (x_i - \bar{x})^2}.$$

The $100(1-\alpha)\%$ confidence limits for β_0 are calculated using $b_0 \pm t_{\alpha/2,\, n-2}\, s_{b_0}$, where

$$s^2_{b_0} = MSE\left[\frac{1}{n} + \frac{\bar{x}^2}{\sum_{i=1}^{n} (x_i - \bar{x})^2}\right].$$

The notation $t_{\alpha/2,\, n-2}$ represents the $\left(1 - \frac{\alpha}{2}\right)$th quantile of the Student's t-distribution with $n-2$ degrees of freedom. The `RegressionReport` option setting `ParameterCITable` yields the confidence intervals for β_1 and β_0. Note that the `Regress` option `ConfidenceLevel` has default setting 0.95, so the confidence level will be 95% unless the command specifies another value.

Often, there is interest in drawing inferences about the slope of the regression line, β_1. For example, we may test

$$\begin{cases} H_0: \beta_1 = 0 \\ H_a: \beta_1 \neq 0 \end{cases}$$

to investigate if there is a regression relationship between x and y. If $\beta_1 = 0$, there is none, because the model becomes $y = \beta_0 + \varepsilon$, which is no longer influenced by the value of x. We test H_0 with the test statistic

$$t = \frac{b_1}{s_{b_1}}.$$

When the model of the regression relationship is $y = \beta_0 + \beta_1 x + \varepsilon$, this hypothesis test can be equivalently done using analysis of variance. Referring back to the output from the `RegressionReport` option ANOVA table, the value of the test statistic is given in the column headed `FRatio`. Notice that in this table the first two rows are labeled `Model` and

`Error`. If we represent the sum of squares for `Model` as *SSM* and note that the value of degrees of freedom for `Model` is 1, then the mean square for `Model` (*MSM*) is the same as the sum of squares, and the test statistic is $F = \frac{MSM}{MSE}$. The *p*-value is calculated from the *F* (ratio) distribution with degrees of freedom 1 for the numerator and $n-2$ for the denominator.

It is not frequently of interest to draw inference about β_0, the intercept of the regression line. Testing

$$\begin{cases} H_0: \beta_0 = 0 \\ H_a: \beta_0 \neq 0 \end{cases}$$

can only show that the regression does not pass through the origin. Knowing this can only be useful if the range of the *x*-values in the data includes zero. We test H_0 with the test statistic

$$t = \frac{b_0}{s_{b_0}}.$$

As noted earlier, the `RegressionReport` option setting `ParameterTable` yields values of b_0, b_1, s_{b_0}, s_{b_1}, the test statistics, and their corresponding *p*-values.

Confidence Interval for the Mean of the Distribution of y's

Let x_i denote a particular value of the independent variable. Then, according to the assumptions, the corresponding distribution of possible *y*-values has mean $\beta_0 + \beta_1 x_i$. The point estimate of this mean is calculated using $\hat{y}_i = b_0 + b_x x_i$. The $100(1-\alpha)\%$ confidence limits for the mean $\beta_0 + \beta_1 x_i$ are calculated using $\hat{y}_i \pm t_{\alpha/2, n-2}\, s_{\hat{y}_i}$, where

$$s^2_{\hat{y}_i} = MSE\left[1 + \frac{1}{n} + \frac{(x_i - \bar{x})^2}{\sum_{i=1}^{n} (x_i - \bar{x})^2}\right].$$

Note that these calculations are only meaningful for values x_i in the range of the values of the independent variable in the data set.

The setting `RegressionReport->MeanPredictionCITable` gives a table of confidence intervals for estimating the mean (also called the expected value) of the possible *y*-values corresponding to each x_i in the data set.

Prediction Interval for a New Observation Corresponding to x_i

Given a value of x, call it x_i, which is within the range of the values of the independent variable in the data set, we can predict a corresponding value of y which is assumed to be randomly chosen from the distribution of y's corresponding to x_i. The predicted value is $\hat{y}_{i(new)} = b_0 + b_1 x_i$. (Note that both $\hat{y}_i$ and $\hat{y}_{i(new)}$ are calculated using the same equation.) An interval which covers the range of possible y-values corresponding to x_i with confidence coefficient $100(1-\alpha)\%$ is called a **prediction interval.** The limits are calculated using $\hat{y}_{i(new)} \pm t_{\alpha/2,\, n-2}\, s_{\hat{y}_{i(new)}}$, where

$$s^2_{\hat{y}_{i(new)}} = MSE\left[1 + \frac{1}{n} + \frac{(x_i - \bar{x})^2}{\sum_{i=1}^{n}(x_i - \bar{x})^2}\right].$$

The setting `RegressionReport->SinglePredictionCITable` gives a table of prediction intervals corresponding to each x_i in the data set.

EXAMPLE 2: Using the data in Table 11.1, determine: (a) 95% confidence intervals for β_0 and β_1; (b) 90% confidence intervals for β_0 and β_1; (c) 95% confidence intervals for the mean of the distribution of y's for each x_i in the data set; (d) 95% prediction intervals for each x_i in the data set. (e) Graph the 95% confidence band for the regression line.

SOLUTION: We begin by making sure that the **LinearRegression** package and the data are loaded. (a) We can obtain confidence intervals for the parameters β_0 and β_1 with the option setting `RegressionReport->ParameterCITable`. This command yields the 95% confidence intervals $0.520442 \le \beta_0 \le 1.63482$. and $-4.35225 \le \beta_1 \le -0.0819426$. (b) With the additional option setting `ConfidenceLevel->0.90`, we find that the 90% confidence intervals are $0.621937 \le \beta_0 \le 1.53433$ and $-3.96367 \le \beta_1 \le -0.470523$.

```
Regress[dpoints,{1,x},x,RegressionReport->
  ParameterCITable]
```

{ParameterCITable→

	Estimate	SE	CI
1	1.07813	0.25596	{0.520442, 1.63582}
x	-2.2171	0.9799623	{-4.35225, -0.0819426}

}

```
Regress[dpoints,{1,x},x,RegressionReport->
  ParameterCITable, ConfidenceLevel->0.90]
```

{ParameterCITable→

	Estimate	SE	CI
1	1.07813	0.25596	{0.621937, 1.53433}
x	-2.2171	0.979963	{-3.96367, -0.470523}

}

(c) With `RegressionReport->MeanPredictionCITable`, we obtain a table containing each of the observed values y_i, the corresponding predicted values $\hat{y}_i$, their standard errors $s_{\hat{y}_i}$, and the 95% confidence intervals for the means. (d) We obtain a similar table for prediction intervals with `RegressionReport->SinglePredictionCITable`.

```
Regress[dpoints,{1,x},x,RegressionReport->
  MeanPredictionCITable->
```

{MeanPredictionCITable→

Observed	Predicted	SE	CI
0.625	0.546028	0.0242175	{0.493263, 0.598793}
0.512	0.514989	0.0146246	{0.483124, 0.546853}
0.488	0.526074	0.0174281	{0.488102, 0.564047}
0.524	0.534943	0.0202533	{0.490814, 0.579071}
0.588	0.523857	0.0167906	{0.487273, 0.560441}
0.475	0.519423	0.0156224	{0.485385, 0.553461}
0.513	0.514989	0.0146246	{0.483124, 0.546853}
0.463	0.479515	0.0157797	{0.445134, 0.513896}
0.512	0.470647	0.0182922	{0.430791, 0.510502}
0.405	0.492818	0.0133495	{0.463732, 0.521904}
0.45	0.457344	0.0228174	{0.407629, 0.507059}
0.48	0.488383	0.0139328	{0.458027, 0.51874}
0.456	0.483949	0.0147553	{0.4518, 0.516098}
0.506	0.444042	0.0278533	{0.383354, 0.504729}

}

```
Regress[dpoints,{1,x},x, RegressionReport->
  SinglePredictionCITable->
```

{SinglePredictionCITable→

Observed	Predicted	SE	CI
0.625	0.546028	0.0543012	{0.427716, 0.66434}
0.512	0.514989	0.0507544	{0.404404, 0.625573}
0.488	0.526074	0.0516321	{0.413578, 0.638571}
0.524	0.534943	0.0526529	{0.420222, 0.649663}
0.588	0.523857	0.0514204	{0.411822, 0.635892}
0.475	0.519423	0.0510509	{0.408193, 0.630653}
0.513	0.514989	0.0507544	{0.404404, 0.625573}
0.463	0.479515	0.0510992	{0.368179, 0.590851}
0.512	0.470647	0.0519301	{0.357501, 0.583793}
0.405	0.492818	0.0504018	{0.383002, 0.602634}
0.45	0.457344	0.0536914	{0.340361, 0.574328}
0.48	0.488383	0.0505594	{0.378224, 0.598543}
0.456	0.483949	0.0507922	{0.373283, 0.594616}
0.506	0.444042	0.0560173	{0.32199, 0.566093}

}

(e) We begin by generating a `MeanPredictionCITable` and naming the result `rtable`. We then extract the elements (columns) of the table, naming them `obs`, `pred`, `se`, and `ci`, respectively. In `predpts`, we create a collection of points with the first coordinate representing the opponent batting average and the second the predicted winning percentage. In `lowerCI`, we use the same first coordinates as those in `predpts`, but we use `First` to extract the lower endpoint of the confidence intervals in `ci` to form the second coordinate. Similarly in `upperCI`, we use `Last` to generate points by extracting the upper endpoint of the confidence intervals.

```
rtable=Regress[dpoints,{1,x},x, RegressionReport->
  MeanPredictionCITable];{obs,pred,se,ci}=
  Transpose[(MeanPredictionCITable/.rtable)[[1]]]
```

```
{{0.625,0.512,0.488,0.524,0.588,0.475,0.513,
  0.463,0.512,0.405,0.45,0.48,0.456,0.506},
 {0.546028,0.514989,0.526074,0.534943,
  0.523857,0.519423,0.514989,0.479515,
  0.470647,0.492818,0.457344,0.488383,
  0.483949,0.444042},{0.0242175,0.0146246,
  0.0174281,0.0202533,0.0167906,0.0156224,
  0.0146246,0.0157797,0.0182922,0.0133495,
  0.0228174,0.0139328,0.0147553,0.0278533},
{{0.493263,0.598793},{0.483124,0.546853},
 {0.488102,0.564047},{0.490814,0.579071},
 {0.487273,0.560441},{0.485385,0.553461},
 {0.483124,0.546853},{0.445134,0.513896},
 {0.430791,0.510502},{0.463732,0.521904},
 {0.407629,0.507059},{0.458027,0.51874},
 {0.4518,0.516098},{0.383354,0.504729}}}
```

```
predpts=Transpose[{oppbavg,pred}]
```

```
{{0.24,0.546028},{0.254,0.514989},{0.249,0.526074},
 {0.245,0.534943},{0.25,0.523857},{0.252,0.519423},
 {0.254,0.514989},{0.27,0.479515},{0.274,0.470647},
 {0.264,0.492818},{0.28,0.457344},{0.266,0.488383},
 {0.268,0.483949},{0.286,0.444042}}
```

```
lowerCI=Transpose[{oppbavg,Map[First,ci]}]
```

```
{{0.24,0.493263},{0.254,0.483124},{0.249,0.488102},
 {0.245,0.490814},{0.25,0.487273},{0.252,0.485385},
 {0.254,0.483124},{0.27,0.445134},{0.274,0.430791},
 {0.264,0.463732},{0.28,0.407629},{0.266,0.458027},
 {0.268,0.4518},{0.286,0.383354}}
```

```
upperCI=Transpose[{oppbavg,Map[Last,ci]}]
```

```
{{0.24,0.598793},{0.254,0.546853},{0.249,0.564047},
 {0.245,0.579071},{0.25,0.560441},{0.252,0.553461},
 {0.254,0.546853},{0.27,0.513896},{0.274,0.510502},
 {0.264,0.521904},{0.28,0.507059},{0.266,0.51874},
 {0.268,0.516098},{0.286,0.504729}}
```

After loading **MultipleListPlot**, found in the **Graphics** folder (or directory), we graph simultaneously the points in `dpoints`, `predpts`, `lowerCI`, and `upperCI`. Note that we use the `Diamond` `SymbolShape` to plot the points in `dpoints`. Because we use `PlotJoined->True` with the last three graphs, we specify that the points in `predpts` be joined with line segments graphed with the `GrayLevel[0.5]` option. The points in `lowerCI` and `upperCI` are joined using `Dashing[{0.003}]` (automatically defined with `Dashing[Dot]`). Note that another automatic dashing setting is `Dashing[Dash]` corresponding to the setting `Dashing[{0.04}]`.)

```
<<Graphics`MultipleListPlot`

MultipleListPlot[dpoints,predpts,lowerCI,upperCI,
  SymbolShape->{PlotSymbol[Diamond],None,None,None},
  PlotJoined->{False,True,True,True},PlotStyle->
  {Automatic,GrayLevel[0.5],Dashing[Dot],Dashing[Dot]}]
```

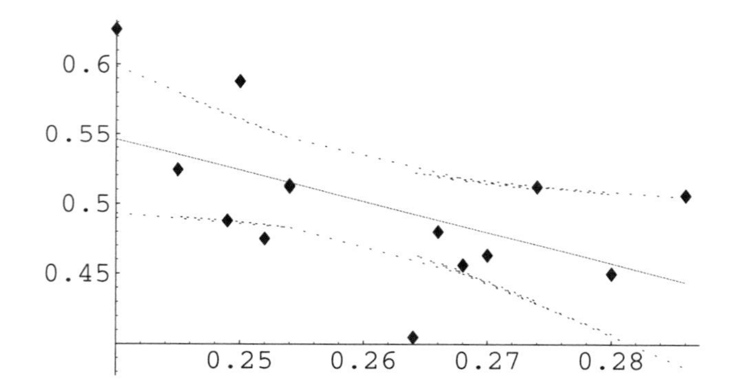

■

11.3 Polynomial Regression

An instance of simple regression where the independent variable and the dependent variable do not have a linear relationship is polynomial regression. Consider the model with the form

$$y = \beta_0 + \beta_1 x + \beta_2 x^2 + \varepsilon.$$

This model is known as a second-order or quadratic model. Because it is additive (i.e., linear in the coefficients), it can be fit using **LinearRegression**. The coefficients β_1 and β_2 are called the linear effect coefficient and the quadratic effect coefficient, respectively. The three parameters β_0, β_1, and β_2 are estimated using the method of least squares by the statistics b_0, b_1, and b_2.

In general, the polynomial regression model of order m is

$$y = \beta_0 + \beta_1 x + \beta_2 x^2 + \ldots + \beta_m x^m + \varepsilon.$$

Note that in fitting data with a polynomial model, the maximum power m in the polynomial cannot exceed $n-1$ and more practically cannot exceed $n-2$ if statistical analysis is to be performed. To determine if the polynomial model has a maximum power that is larger than necessary, we may use a stepwise procedure and examine the significance of β_m associated with the highest power in the model. If this term is significantly different from zero, the analysis is complete. If not, the highest-order term is dropped from the model and a lower-order model is fitted. The procedure is repeated until the β_j associated with the highest power in the model is significantly different from zero. Another, more commonly used, approach is to begin with a lower-order model and add higher power terms following a stepwise procedure. For example, we may begin with the linear model, then move to the quadratic model, then on to higher-order models. To determine if the addition of the quadratic term (to the linear model) significantly improves the accuracy of the fit, test $H_0\beta_2 = 0$ where failure to reject H_0 indicates that the linear model describes the dependence of y on x sufficiently and the contribution of the quadratic term is insignificant. If we conclude that $\beta_2 = 0$, then we may fit the data with a polynomial of degree 3, and test $H_0\beta_3 = 0$, and so forth. Note that whenever another term is added to the model, r^2 increases. In order to compare the usefulness of models with different numbers of terms, *adj* r^2 is useful. The model with the larger *adj* r^2 is preferred.

Note further that $x, x^2, x^3, \ldots$ may be so closely correlated that computational difficulties arise in fitting a polynomial model. If so, then replacing each x_i in the data with $x_i - \bar{x}$ and fitting the model using the adjusted data can remedy the problem. The predicted values for the regression function in terms of x_i are the same as for the regression function in terms of $x_i - \bar{x}$. This is not true, however, of the standard deviations (standard errors) of the estimates of the regression coefficients.

EXAMPLE 1: Determine whether the data in Table 11.1 are modeled better with a linear equation, a quadratic equation, or a cubic equation.

SOLUTION: We begin by loading the **LinearRegression** package and entering the data. When we fit the data with a linear equation, we find that each coefficient is significant. We also observe that r^2=0.299008 and *adj* r^2 = 0.240592. We will use these values for comparison with the other models.

```
<<Statistics`LinearRegression`

oppbavg={0.240,0.254,0.249,0.245,0.250,0.252,0.254,
  0.270,0.274,0.264,0.280,0.266,0.268,0.286};
winpct={0.625,0.512,0.488,0.524,0.588,0.475,0.513,
  0.463,0.512,0.405,0.450,0.480,0.456,0.506};
```

```
dpoints=Table[{oppbavg[[i]],winpct[[i]]},
  {i,1,Length[winpct]}];
```

```
{{0.24,0.625},{0.254,0.512},{0.249,0.488},{0.245,0.524},
{0.25,0.588},{0.252,0.475},{0.254,0.513},{0.27,0.463},
 {0.274,0.512},{0.264,0.405},{0.28,0.45},{0.266,0.48},
 {0.268,0.456},{0.286,0.506}}
```

```
Regress[dpoints,{1,x},x]
```

```
                     Estimate    SE          TStat       PValue
{ParameterTable→ 1    1.07813    0.25596     4.21211     0.00120568
                 x   -2.2171     0.979963   -2.26243     0.0430218
 RSquared→0.299008, AdjustedRSquared→0.240592,EstimatedVariance→0.00236213,
                     DF     SumOfSq      MeanSq       FRatio     PValue
 ANOVATable→  Model   1     0.0120908    0.0120908    5.11859    0.0430218 }
              Error  12     0.0283456    0.00236213
              Total  13     0.0404364
```

For the quadratic model, we again find that each coefficient is significant. We also notice that the values of r^2 and *adj* r^2 increase. Finally, when we use a cubic model, we find all three coefficients have large *p*-values and, thus, are not significant. We also see that *adj* r^2 decreases, which indicates that the cubic model is less useful than the quadratic model. Based on our findings, we conclude that the best fit is found using the quadratic model.

```
Regress[dpoints,{1,x,x^2},x]
```

```
                     Estimate    SE          TStat       PValue
{ParameterTable→ 1    12.9504    4.17042     3.10529     0.0100092
                 x   -92.8343    31.8034    -2.91901     0.0139622
                 x^2  172.464    60.5108     2.85014     0.0157927
 RSquared→0.59678, AdjustedRSquared→0.523467, EstimatedVariance→0.00148225,
                     DF     SumOfSq      MeanSq       FRatio     PValue
 ANOVATable→  Model   2     0.0241316    0.0120658    8.14018    0.00676837 }
              Error  11     0.0163048    0.00148225
              Total  13     0.0404364
```

```
Regress[dpoints,{1,x,x^2,x^3},x]
```

```
                     Estimate    SE          TStat       PValue
{ParameterTable→ 1    48.0489    82.9408     0.579316    0.575193
                 x   -494.226    947.803    -0.521444    0.613409
                 x^2  1700.06    3605.45     0.471525    0.647385
                 x^3 -1934.66    4565.52    -0.423755    0.680715
 RSquared→0.603892, AdjustedRSquared→0.48506, EstimatedVariance→0.00160171,
                     DF     SumOfSq      MeanSq       FRatio     PValue
 ANOVATable→  Model   3     0.0244192    0.00813974   5.08189    0.0215877 }
              Error  10     0.0160171    0.00160171
              Total  13     0.0404364
```

■

11.4 Multiple Regression

Basis Functions

The Wolfram Research documentation for **LinearRegression** refers to "basis functions" defined as functions of the independent variables which specify the predictors. (Thus, a predictor is an independent variable or a function of an independent variable.) The documentation represents the ith response y_i in general as,

$$y_i = \beta_1 f_{1i} + \beta_2 f_{2i} + \ldots + \beta_p f_{pi} + \varepsilon_i ,$$

where f_{ji} is the jth basis function evaluated at the ith case and ε_i is the error for the ith case. Note that we have altered the Wolfram Research notation, using ε_i in place of e_i to represent the error term. Recall that the error term is included in the model to account for addition to (or subtraction from) the value of y_i due to other sources apart from the predictors. If one has not thought in terms of functions of independent variables, the use of "basis functions" in the documentation can be a sourse of confusion. Two examples will be used to illustrate the usefulness of "basis functions."

For simple linear regression, the Wolfram Research notation could be

$$y_i = \beta_1 f_{1i} + \beta_2 f_{2i} + \varepsilon_i$$

where $f_{1i} = 1$ and $f_{2i} = x_i$ for all i. Thus, their model would reduce to

$$y_i = \beta_1 + \beta_2 x_i + \varepsilon_i.$$

For quadratic regression, the Wolfram Research notation could be

$$y_i = \beta_1 f_{1i} + \beta_2 f_{2i} + \beta_3 f_{3i} + \varepsilon_i,$$

where $f_{1i} = 1$, $f_{2i} = x_i$, and $f_{3i} = x_i^2$. Note that the function f_{ji} can represent a constant, an independent variable, or a function of one or more independent variables. In this example, f_{3i} is a function that squares the value of the independent variable.

Matrix Approach

Note that the Wolfram notation uses "p" to represent the number of "basis functions" or, equivalently, the number of parameters β_i's to be estimated. It is possible to have either more or fewer independent variables than "basis functions," so we use (a different notation) m-1 to represent the number of independent variables. The number of variables including the dependent variable is, therefore, m. The general linear regression model is usually given in the form

$$y = \beta_0 + \beta_1 x + \ldots + \beta_{m-1} x_{m-1} + \varepsilon,$$

where the $\beta_0, \beta_1, \ldots, \beta_{m-1}$ are called regression coefficients or **partial regression coefficients**. Note that m is the number of partial regression coefficients to be estimated. The model including the data can be represented in matrix notation as

$$Y = X\beta + \varepsilon,$$

where **Y** is the $n \times 1$ vector of responses, β is the $m \times 1$ vector of parameters, **X** is the $n \times m$ matrix of constants (independent variables), and ε is the $n \times 1$ vector. In the general linear regression model, the least squares estimators are those values of $\beta_0, \beta_1, \ldots, \beta_{m-1}$ that minimize

$$\sum_{i=1}^{n} (y_i - \beta_0 - \beta_1 x_{i1} - \beta_2 x_{i2} - \ldots - \beta_{m-1} x_{i,m-1})^2.$$

We represent the vector of the least squares estimated regression coefficients $b_0, b_1, \ldots, b_{m-1}$ by

$$\mathbf{b} = \begin{pmatrix} b_0 \\ b_1 \\ \vdots \\ b_{m-1} \end{pmatrix},$$

where $\mathbf{b} = (\mathbf{X'X})^{-1}(\mathbf{X'Y})$. The vector of fitted values $\hat{y}_i$ and the vector of the residual terms $e_i = y_i - \hat{y}_i$ are given by

$$\hat{\mathbf{Y}} = \begin{pmatrix} \hat{y}_1 \\ \hat{y}_2 \\ \vdots \\ \hat{y}_n \end{pmatrix} \text{ and } \mathbf{e} = \begin{pmatrix} e_1 \\ e_2 \\ \vdots \\ e_n \end{pmatrix},$$

respectively. The fitted values are found with $\hat{\mathbf{Y}} = \mathbf{Xb}$ and the residual terms with $\mathbf{e} = \mathbf{Y} - \hat{\mathbf{Y}} = \mathbf{Y} - \mathbf{Xb}$.

Assumptions

The assumptions required to make valid inferences presented earlier for simple regression generalize for multiple regression. The data vectors, which are of the form $\{x_{i1}, x_{i2}, \ldots, x_{1,m-1}, y_i\}$ in *Mathematica*, must be a randomly selected sample from a population of vectors which reflect an additive model. We assume that for a given set $\{x_{i1}, x_{i2}, \ldots, x_{1,m-1}\}$ there is a distribution of possible y's (each with a corresponding value of ε). The mean of this distribution is assumed to be $\beta_0 + \beta_1 x_1 + \ldots + \beta_{m-1} x_{m-1}$, and we assume any y's selected from this distribution are independent. We assume that

$\{x_{i1}, x_{i2}, \ldots, x_{1, m-1}\}$ are measured without error, and that the corresponding distribution of y's has a normal probability distribution. Furthermore, we assume that every such distribution has the same variance. We denote the common variance by σ^2. It is estimated by the mean square for error (*MSE*). The assumptions of independence and constant variance are represented by a matrix of the form

$$\sigma^2 = \begin{pmatrix} \sigma^2 & 0 & \cdots & 0 \\ 0 & \sigma^2 & \cdots & 0 \\ \vdots & \vdots & \ddots & \vdots \\ 0 & 0 & \cdots & \sigma^2 \end{pmatrix},$$

where an entry on the main diagonal (i^{th} row, j^{th} column) is the variance of the distribution corresponding to y_i, and an off-diagonal entry is the covariance between distributions for y_i and y_j. The off-diagonal elements are zero because of independence.

The matrix $\mathbf{H} = \mathbf{X}(\mathbf{X}'\mathbf{X})^{-1}\mathbf{X}'$, called the **hat matrix**, can be used to represent $\hat{\mathbf{Y}}$ as $\hat{\mathbf{Y}} = \mathbf{HY}$ as well as $\mathbf{e}$ and σ^2 as $\mathbf{e} = (\mathbf{I}-\mathbf{H})\mathbf{Y}$ and $\sigma^2 = \sigma^2(\mathbf{I}-\mathbf{H})$, respectively. Note that σ^2 is estimated with $s^2 = MSE(\mathbf{I}-\mathbf{H})$.

Statistical Inference

We will now revert to the notation in the Wolfram documentation and use "p" to represent the number of predictors in the regression model to be fit using least squares. Their notation specifies the partial regression coefficients as $\beta_1, \beta_2, \cdots \beta_p$.

The entries in the ANOVA Table for the general linear regression model are given in Table 11.2, where **J** is the $n \times n$ matrix in which each entry is 1 given by.

$$\mathbf{J} = \begin{pmatrix} 1 & \cdots & 1 \\ \vdots & & \vdots \\ 1 & \cdots & 1 \end{pmatrix}.$$

In terms of the hat matrix **H**, $SSE = \mathbf{Y}'(\mathbf{I}-\mathbf{H})\mathbf{Y}$ and $SSM = \mathbf{Y}'\left(\mathbf{H} - \frac{1}{n}\mathbf{J}\right)\mathbf{Y}$.

Source of variation	*SS*	*df*	*MS*
Model	$SSM = \mathbf{b}'\mathbf{X}'\mathbf{Y} - \frac{1}{n}\mathbf{Y}'\mathbf{J}\mathbf{Y}$	$p-1$	$MSM = \frac{SSM}{p-1}$
Error	$SSE = \mathbf{Y}'\mathbf{Y} - \mathbf{b}'\mathbf{X}'\mathbf{Y}$	$n-p$	$MSE = \frac{SSE}{n-p}$
Total	$SSTO = \mathbf{Y}'\mathbf{Y} - \frac{1}{n}\mathbf{Y}'\mathbf{J}\mathbf{Y}$	$n-1$	

Table 11.2

We use the test statistic $F = \frac{MSM}{MSE}$ to determine if there is a regression relationship between the response variable y and the predictors based on the hypothesis test

$$H_0{:}\beta_1 = \beta_2 = \cdots = \beta_p = 0$$
$$H_a{:}\text{ not all } \beta_k = 0 \quad k = 1,2,\ldots,p.$$

The coefficient of multiple determination,

$$R^2 = 1 - \frac{SSE}{SSTO},$$

measures the proportionate reduction of total variation in y associated with the use of the predictors. The adjusted coefficient of multiple determination, $adj\ R^2$, adjusts R^2 by dividing each sum of squares by the associated degrees of freedom. Therefore,

$$adj\ R^2 = 1 - \frac{SSE/(n-p)}{(SSTO)/(n-1)} = 1 - \left(\frac{n-1}{n-p}\right)\frac{SSE}{SSTO}.$$

Notice that the Wolfram notation for the adjusted coefficient of multiple determination is $\bar{R}^2$, whereas we have used $adj\ R^2$. The coefficient of multiple correlation, R, is given by $R = \sqrt{R^2}$. The variance–covariance matrix for the estimated partial regression coefficients is

$$\sigma^2(\mathbf{b}) = \begin{pmatrix} \sigma_{b_1}^2 & \sigma_{b_1,b_2} & \cdots & \sigma_{b_1,b_p} \\ \sigma_{b_2,b_1} & \sigma_{b_2}^2 & \cdots & \sigma_{b_2,b_p} \\ \vdots & \vdots & \ddots & \vdots \\ \sigma_{b_p,b_1} & \sigma_{b_p,b_2} & \cdots & \sigma_{b_p}^2 \end{pmatrix},$$

where this $p \times p$ matix is found with $\sigma^2(\mathbf{b}) = \sigma^2(\mathbf{X}'\mathbf{X})^{-1}$. The notation $\sigma_{b_i}^2$ represents the variance of the estimator b_i, and σ_{b_i,b_j} represents the covariance between the estimators b_i and b_j. The estimated variance–covariance matrix is given by $\mathbf{s}^2(\mathbf{b}) = MSE(\mathbf{X}'\mathbf{X})^{-1}$.

The $100(1-\alpha)\%$ confidence limits for β_k are $b_k \pm t_{\alpha/2} s_{b_k}$. The hypothesis tests for the β_k are

$$H_0{:}\ \beta_k = 0$$
$$H_a{:}\beta_k \neq 0,$$

tested using the test statistic $t = \frac{b_k}{s_{b_k}}$.

EXAMPLE 1: Table 11.3 gives the 1996 team earned run average (ERA), team batting average, opposing team batting average, and winning percentage for each of the 14 teams in the National League of Major League Baseball. Determine the least squares regression equation to predict the team winning percentage (dependent variable) based on the three independent variables team ERA, team batting average, and opposing team batting average.

Team	Team earned run average (ERA)	Team batting average	Opposing team batting average	Team winning percentage
Atlanta	3.33	0.276	0.240	0.625
Los Angeles	3.51	0.249	0.254	0.512
Florida	3.55	0.249	0.249	0.488
San Diego	3.65	0.260	0.245	0.524
Montreal	3.80	0.271	0.250	0.588
Chicago	4.20	0.241	0.252	0.475
St. Louis	4.22	0.269	0.254	0.513
New York	4.27	0.264	0.270	0.463
Houston	4.31	0.270	0.274	0.512
Philadelphia	4.48	0.240	0.264	0.405
Pittsburgh	4.53	0.259	0.280	0.450
Cincinnati	4.55	0.252	0.266	0.480
San Francisco	4.62	0.258	0.268	0.456
Colorado	5.86	0.293	0.286	0.506

Table 11.3 National League: Team ERA, Team Batting Average, Opponent Batting Average and Winning Percentage (1996 Season)

SOLUTION: We load the **LinearRegression** package (if we are beginning a new *Mathematica* session) and enter the data in `teamera`, `ownbavg`, `oppbavg`, and `winpct`.

```
<<Statistics`LinearRegression`

teamera={3.33,3.51,3.55,3.65,3.80,4.20,4.22,4.27,
  4.31,4.48,4.53,4.55,4.62,5.86};
```

```
ownbavg={0.276,0.249,0.249,0.260,0.271,0.241,0.269,
  0.264,0.270,0.240,0.259,0.252,0.258,0.293};
oppbavg={0.240,0.254,0.249,0.245,0.250,0.252,0.254,
  0.270,0.274,0.264,0.280,0.266,0.268,0.286};
winpct={0.625,0.512,0.488,0.524,0.588,0.475,0.513,
  0.463,0.512,0.405,0.450,0.480,0.456,0.506};
```

In dpoints, we place the data in matrix form, and use this matrix to perform the regression analysis. We use scatter plots to investigate whether there appears to be a linear relationship between the response variable and each of the predictors.

```
Clear[dpoints]

dpoints=Table[{teamera[[i]],ownbavg[[i]],
  oppbavg[[i]],winpct[[i]]},{i,1,Length[winpct]}]

{{3.33,0.276,0.24,0.625},{3.51,0.249,0.254,0.512},
 {3.55,0.249,0.249,0.488},{3.65,0.26,0.245,0.524},
 {3.8,0.271,0.25,0.588},{4.2,0.241,0.252,0.475},
 {4.22,0.269,0.254,0.513},{4.27,0.264,0.27,0.463},
 {4.31,0.27,0.274,0.512},{4.48,0.24,0.264,0.405},
 {4.53,0.259,0.28,0.45},{4.55,0.252,0.266,0.48},
 {4.62,0.258,0.268,0.456},{5.86,0.293,0.286,0.506}}

ListPlot[Transpose[{teamera,winpct}],
  Prolog->{PointSize[0.02]},AxesOrigin->
  {Min[teamera],Min[winpct]},
  PlotLabel->"Team ERA vs. Winning Percentage"]
```

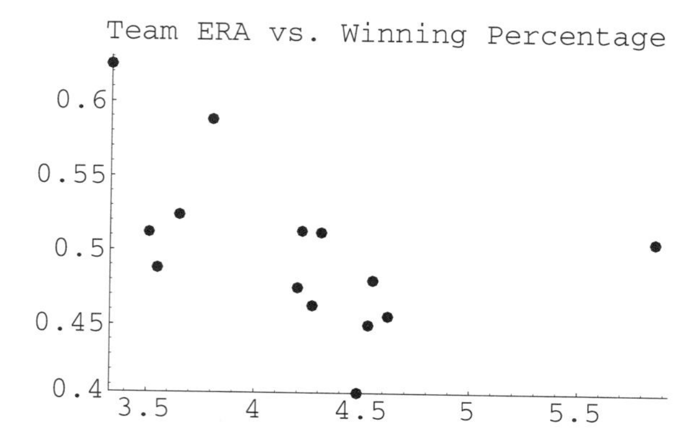


```
ListPlot[Transpose[{ownbavg,winpct}],
  Prolog->{PointSize[0.02]},AxesOrigin->
  {Min[ownbavg],Min[winpct]},PlotLabel->
  "Own Batting Average vs. Winning Percentage"]
```

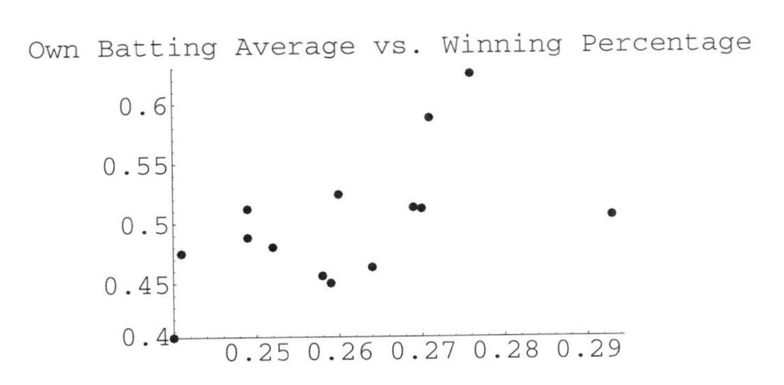


```
ListPlot[Transpose[{oppbavg,winpct}],
  Prolog->{PointSize[0.02]},AxesOrigin->
  {Min[oppbavg],Min[winpct]},PlotLabel->
  "Opponent Batting Average vs. Winning Percentage"]
```

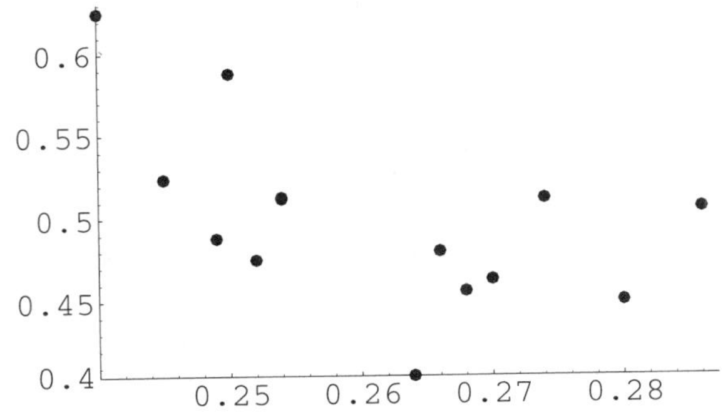

Assuming a linear relationship of each predictor with `winpct`, we fit a model of the form

$$y = \beta_1 + \beta_2 x_1 + \beta_3 x_2 + \beta_4 x_3 + \varepsilon.$$

Note that in the following commands `x1` represents the data in `teamera`, `x2` that in `ownbavg`, and `x3` that in `oppbavg`. In this case, the response variable is contained in `winpct`. We can generate the prediction equation either using the `RegressionReport` option setting `BestFit` or with `Fit` (where `Fit` has the same arguments as `Regress`). The prediction equation is given by

$$\hat{y} = 0.302668 - 0.0303858x_1 + 3.16123x_2 + -1.91482x_3.$$

```
Regress[dpoints,{1,x1,x2,x3},{x1,x2,x3},
  RegressionReport->BestFit]
```

{BestFit→0.302668-0.0303858 x1+3.16123 x2-1.91482 x3}

```
Fit[dpoints,{1,x1,x2,x3},{x1,x2,x3}]
```

0.302668-0.0303858 x1+3.16123 x2-1.91482 x3

More generally, the least squares estimates and the other default output is displayed with the command `Regress[dpoints,{1,x1,x2,x3},{x1,x2,x3}]` where we can read the coefficients of the prediction equation from the column in the `ParameterTable` labeled `Estimate`.

```
Regress[dpoints,{1,x1,x2,x3},x1,x2,x3]
```

{ParameterTable→

	Estimate	SE	TStat	PValue
1	0.302668	0.196793	1.538	0.155066
x1	-0.0303858	0.0190012	-1.59915	0.14087
x2	3.16123	0.442659	7.14147	0.0000313646
x3	-1.91482	0.864137	-2.21587	0.0510505,

RSquared→0.885111, AdjustedRSquared→0.850644, EstimatedVariance→0.00046457,

ANOVATable→

	DF	SumOfSq	MeanSq	FRatio	PValue
Model	3	0.0357907	0.0119302	25.6801	0.000051525 }
Error	10	0.0046457	0.00046457		
Total	13	0.0404364			

From the ANOVA table we learn there is a highly significant (p-value = 0.00005) regression relationship. The value of *adj* R^2 = 0.850 suggests that 85% of the variability in `winpct` is explained by the predictors. From the parameter table, we see that x2 (`ownbavg`) with p-value = 0.00003 and perhaps x3 (`oppbavg`) with p-value = 0.051 are the statistically significant predictors of `winpct`. The material in the next section on diagnostic procedures and remedial methods can help with questions such as whether the relationship between `winpct` and the significant predictors is best described by a linear equation.

As an illustration of the use of "basis functions" of more than one variable we include predictors x_1x_2, x_2x_3, and x_1x_3 in the model. Such cross-product terms are called **interaction terms**, because they help model situations where one of the variables behaves differently depending on the value of a second variable.

```
Regress[dpoints,{1,x1,x2,x3,x1 x2,x2 x3,x1x3},
  {x1,x2,x3}]
```

{ParameterTable→

	Estimate	SE	TStat	PValue
1	4.3393	6.70683	0.646997	0.538264
x1	0.146461	0.528633	0.277055	0.789741
x2	-8.45276	23.6866	-0.356858	0.731713
x3	-24.2133	35.3193	-0.685554	0.515049
x1x2	-1.64926	2.42043	-0.68139	0.517525
x2x3	70.7988	126.636	0.559074	0.593538
x1x3	0.959189	1.41262	0.679013	0.518942,

```
RSquared→0.89415,AdjustedRSquared→0.803413,EstimatedVariance→0.000611483,

                       DF    SumOfSq       MeanSq         FRatio    PValue
ANOVATable→  Model      6    0.036156      0.006026       9.85473   0.00402332  }
             Error      7    0.00428038    0.000611483
             Total     13    0.0404364
```

Although the ANOVA table shows a significant regression relationship (p-value = 0.004), the *adj* R^2 = 0.803 is lower than for the original model. Also, the parameter table indicates that none of the predictors is statistically significant. Thus, the new model is not as desirable as the original model. One possible explanation for the situation where the ANOVA indicates significance but none of the partial regression coefficients is significant is multicolinearity, which is discussed in the next section.

Two other `RegressionReport` option settings, `CovarianceMatrix` and `CorrelationMatrix`, are demonstrated next. `CovarianceMatrix` displays the variance–covariance matrix for the estimated partial regression coefficients, $\mathbf{s}^2(\mathbf{b})$. Values in the display resulting from `CorrelationMatrix` can be calculated from corresponding values from the `CovarianceMatrix` display with the formula

$$r_{b_i,b_j} = \frac{s_{b_i,b_j}}{s_{b_i}s_{b_j}},$$

where s_{b_i} and s_{b_j} are from the column `SE` in parameter table.

The results obtained with these options are named `cov` and `corr`, respectively. Notice that we extract the correlation matrix from the output list with `corr[[1,2]]`. However, this output is given in the form of a matrix. With `corr[[1,2,1]]`, we obtain the matrix in `StandardForm` (as a list of lists) and name it `corrmat`. With the correlation matrix in this form, we can use subscript notation to extract elements of the correlation matrix. For example, `corrmat[[1,3]]` gives the member of `corrmat` in the first row, third column. The covariance matrix and its elements can be extracted from `cov` in a similar way.

```
cov=Regress[dpoints,{1,x1,x2,x3,x1 x2,x2 x3,x1 x3},
  {x1,x2,x3},RegressionReport->CovarianceMatrix]
```

```
{CovarianceMatrix→
 ( 44.9816   2.73914    -155.174  -234.987  -14.7961   836.777   4.139      )
 ( 2.73914   0.279453   -10.5961  -15.0672  -1.03209   58.082    -0.0360002 )
 ( -155.174  -10.5961   561.055   804.323   49.0586    2961.71   -7.89514   )
 ( -234.987  -15.0672   804.322   1247.45   81.6642    4420.59   -23.2543   ) }
 ( -14.7961  -1.03209   49.0586   81.6642   5.8585     283.878   -1.88114   )
 ( 836.777   58.082     -2961.71  -4420.59  -283.878   16036.7   58.7984    )
 ( 4.13907   0.0360002  -7.89514  -23.2543  -1.88114   58.7984   1.9955     )
```

```
corr=Regress[dpoints,{1,x1,x2,x3,x1 x2,x2 x3,x1 x3},
  {x1,x2,x3},RegressionReport->CorrelationMatrix];
```

```
corr[[1,2]]
```

```
1.          0.772577    0.976785   -0.992006    0.911457    0.985224    0.436877
0.772577    1.          0.84623    -0.806985    0.806625    0.867621    0.0482087
-0.976785   -0.84623\   1.          0.961426    0.855696    0.987375   -0.235956
-0.992006   -0.806985   0.961426    1.          0.955271    0.988351   -0.466086
-0.911457   -0.806625   0.855696    0.955271    1.          0.926149    0.550177
0.985224    0.867621    0.987375   -0.988351    0.926149    1.          0.328687
0.436877    0.0482087   0.235956   -0.466086    0.550177    0.328687    1.
```

```
corrmat=corr[[1,2,1]]
```

```
{{1.,0.772577,-0.976785,-0.992006,-0.911457,0.985224,0.436877},
 {0.772577,1.,-0.84623,-0.806985,-0.806625,0.867621,-0.0482087},
 {-0.976785,-0.84623,1.,0.961426,0.855696,-0.987375,-0.235956},
 {-0.992006,-0.806985,0.961426,1.,0.955271,-0.988351,-0.466086},
 {-0.911457,-0.806625,0.855696,0.955271,1.,-0.926149,-0.550177},
 {0.985224,0.867621,-0.987375,-0.988351,-0.926149,1.,0.328687},
 {0.436877,-0.0482087,-0.235956,-0.466086,-0.550177,0.328687,1.}}
```

```
corrmat[[1,3]]
```

```
-0.976785
```

■

11.5 Diagnostic Procedures

Introduction

In Chapter 10, we discussed procedures used to identify outliers and to investigate whether a data set satisfies the assumptions that underlie the statistical methods presented to that point. We called these diagnostic procedures. We also discussed methods for correcting the problems that were diagnosed. With only minor modifications, these procedures and methods apply for regression. In addition, we will discuss some other procedures and methods. The diagnostic topics to be discussed here include outliers, random sampling, independence, linearity, equality of variance, normality, and multicolinearity. We will discuss remedial methods in the next section.

Residuals

Residuals are key to many of the diagnostic procedures commonly employed. The residual corresponding to the ith case is calculated by $e_i = y_i - \hat{y}_i$, where the ith case is represented by a vector of the form $\{x_{i1}, x_{i2}, \ldots, y_i\}$. The ith value of the kth independent variable is x_{ik}; the ith value of the dependent variable is y_i. The predicted (or fitted) value found by substituting the values of $\{x_{i1}, x_{i2}, \ldots\}$ into the least squares regression (or prediction) equation is $\hat{y}_i$. In the **LinearRegression** package, the residuals are displayed by specifying the

RegressionReport option setting FitResiduals, and the predicted values ($\hat{y}_i$'s) are displayed by specifying the RegressionReport option setting PredictedResponse.

Standardized residuals are often used in diagnostic procedures. The most common form of the standardization is to divide each residual by the square root of the mean square for error (*MSE*):

$$\text{Standardized Residual} = \frac{e_i}{\sqrt{MSE}}.$$

It is better, however, to standardize each residual using an estimate of its standard deviation. The resulting value is called a **studentized residual,**

$$\text{Studentized Residual} = \frac{e_i}{\sqrt{MSE(1-h_{ii})}},$$

where h_{ii} is a value called the **leverage** and corresponds to the ith element in the main diagonal of the hat matrix. These residuals are found with the RegressionReport option setting StandardizedResidual. (Note the difference between our terminology and that employed by Wolfram Research.) When residuals are used to identify outliers, deleted residuals are often used. These are calculated in theory by first omitting the case in question, call it $\{x_{i1}, x_{i2}, \ldots, y_i\}$, from the regression calculations, and then calculating the studentized residual. In this way the case influences the value of e_i but not its standard deviation. In practice, deleted residuals can be calculated without repeating the regression calculations.

$$\text{Studentized Deleted Residual} = e_i\left[\frac{n-p-1}{SSE(1-h_{ii})-e_i^2}\right]^{1/2},$$

where n is the number of cases and $p-1$ and *SSE* are respectively the degrees of freedom for Model and the sum of the squares for error from the analysis of variance table. We use the RegressionReport option setting StudentizedResidual. (Again, note the difference between our terminology and that used in by Wolfram Research.) Residual plots against the independent variable and against the predicted values that have the points randomly scattered in a band about the zero line can indicate the data are consistent with some of the underlying assumptions. A particular pattern to the scatter can indicate a departure from a corresponding assumption. This is discussed again in the subsection on linearity and additivity.

For simple linear regression, each case is of the form $\{x_i, y_i\}$ and the leverage, h_i, is calculated with

$$h_i = \frac{1}{n} + \frac{(x_i-\bar{x})^2}{\sum_{i=1}^{n} x_i^2 - \frac{1}{n}\left(\sum_{i=1}^{n} x_i\right)^2}.$$

EXAMPLE 1: Calculate the residuals, predicted values, studentized residuals, and studentized deleted residuals for the data in Table 11.1. In addition, plot the residuals against the independent variable (opponent batting average) and the residuals against the predicted values.

SOLUTION: After making sure that the package **LinearRegression** is loaded, we enter the data in `oppbavg` and `winpct`. We then place this information in ordered pairs of the form `{oppbavg[[i]],winpct[[i]]}` in `dpoints`, and determine the residuals with the option setting `RegressionReport->FitResiduals` and the predicted values with `RegressionReport->PredictedResponse`. We name these lists `res` and `pr`, respectively.

```
<<Statistics`LinearRegression`

oppbavg={0.240,0.254,0.249,0.245,0.250,0.252,0.254,
  0.270,0.274,0.264,0.280,0.266,0.268,0.286};
winpct={0.625,0.512,0.488,0.524,0.588,0.475,0.513,
  0.463,0.512,0.405,0.450,0.480,0.456,0.506};

dpoints=Table[{oppbavg[[i]],winpct[[i]]},
  {i,1,Length[winpct]}];

res=Regress[dpoints,{1,x},x,RegressionReport->
  FitResiduals]

{FitResiduals→{0.0789719,-0.00298867,-0.0380742,
  -0.0109426,0.0641429,-0.0444229,-0.00198867,
  -0.0165151,0.0413533,-0.0878177,-0.00734412,
  -0.00838349,-0.0279493,0.0619585}}

pr=Regress[dpoints,{1,x},x,RegressionReport->
  PredictedResponse]

{PredictedResponse→{0.546028,0.514989,0.526074,
  0.534943,0.523857,0.519423,0.514989,0.479515,
  0.470647,0.492818,0.457344,0.488383,0.483949,
  0.444042}}
```

To illustrate that the values just obtained are to be expected, we find the least squares line with `Fit` and define it as the function `lsq[x]`. We then use `Map` to evaluate this function at each value in `oppbavg`. This gives the predicted values, the same as those found earlier. In addition, if we calculate `winpct-Map[lsq,oppbavg]`, we find the residuals. Note that `Map` can be used to predict a value of `winpct` for any value of `oppbavg` which is within the range of the `oppbavg` values in the data.

```
lsq[x_]=Fit[dpoints,{1,x},x]

1.07813-2.2171 x
```

```
Map[lsq,oppbavg]
```

```
{0.546028,0.514989,0.526074,0.534943,0.523857,
 0.519423,0.514989,0.479515,0.470647,0.492818,
 0.457344,0.488383,0.483949,0.444042}
```

```
winpct-Map[lsq,oppbavg]
```

```
{0.0789719,-0.00298867,-0.0380742,-0.0109426,
 0.0641429,-0.0444229,-0.00198867,-0.0165151,
 0.0413533,-0.0878177,-0.00734412,-0.00838349,
 -0.0279493,0.0619585}
```

We determine the studentized residuals and studentized deleted residuals in the same command with `RegressionReport->{StandardizedResiduals, StudentizedResiduals}`. First, the studentized residuals are given. The studentized deleted residuals follow. (Recall the difference between the names we use in this book and those used by Wolfram Research.)

```
Regress[dpoints,{1,x},x,RegressionReport->
  {StandardizedResiduals,StudentizedResiduals}]
```

```
{StandardizedResiduals→{1.87411,-0.0644816,
  -0.839202,-0.247677,1.40636,-0.965242,-0.0429063,
  -0.359268,0.91839,-1.87916,-0.171141,-0.180051,
  -0.603555,1.55562},
 StudentizedResiduals→{2.13351,-0.0617472,
  -0.828143,-0.237742,1.47337,-0.962259,-0.0410828,
  -0.345837,0.911923,-2.14166,-0.164055,-0.172619,
  -0.586836,1.66693}}
```

To generate the residual plots, we extract the predicted values from the result named `pr` earlier and the residuals from `res`. (Of course, we could have "copied and pasted" these results and assigned the given names.) After names have been assigned, we use `ListPlot` to first plot the residuals in `errvals` against the independent variable in `oppbavg` and then to plot the residuals against the predicted values in `predvals`. Notice that we use several `Plot` options so that the residual plots are shown in the appropriate "window."

```
predvals=pr[[1,2]]
```

```
{0.546028,0.514989,0.526074,0.534943,0.523857,
 0.519423,0.514989,0.479515,0.470647,0.492818,
 0.457344,0.488383,0.483949,0.444042}
```

```
errvals=res[[1,2]]
```

```
{0.0789719,-0.00298867,-0.0380742,-0.0109426,0.0641429,
 -0.0444229,-0.00198867,-0.0165151,0.0413533,-0.0878177,
 -0.00734412,-0.00838349,-0.0279493,0.0619585}
```

```
eps=(Max[oppbavg]-Min[oppbavg])/Length[oppbavg];
ListPlot[Transpose[{oppbavg,errvals}],
  Prolog->{PointSize[0.025]},
  AxesOrigin->{Min[oppbavg]-eps,0},
  PlotRange->{{Min[oppbavg]-eps,Max[oppbavg]+eps},
   {Min[errvals]-eps,Max[errvals]+eps}}]
```

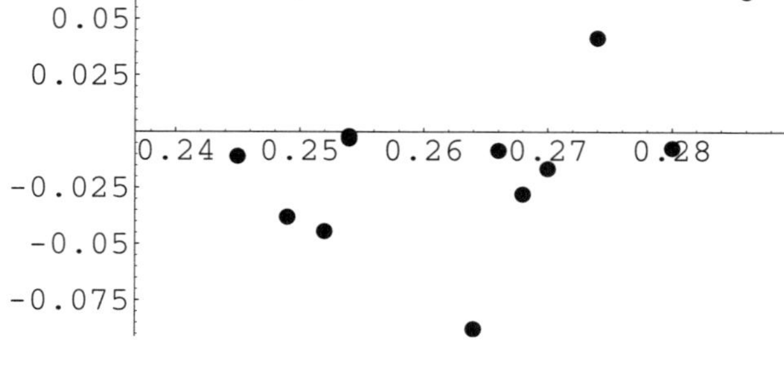


```
eps=(Max[predvals]-Min[predvals])/Length[predvals];
ListPlot[Transpose[{predvals,errvals}],
  Prolog->{PointSize[0.03]},
  AxesOrigin->{Min[predvals]-eps,0},
  PlotRange->{{Min[predvals]-eps,Max[predvals]+eps},
   {Min[errvals]-eps,Max[errvals]+eps}}]
```

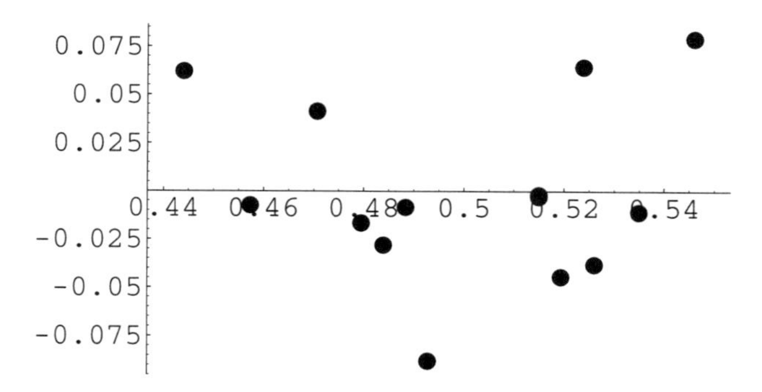

We can use studentized residuals in these plots which are found with the setting `RegressionReport->StandardizedResidual`. In `errSTvals`, we calculate the studentized residuals. Then, we create the residual plot of studentized residuals against predicted values in a similar manner as those shown previously.

```
errSTvals=Regress[dpoints,{1,x},x,
  RegressionReport->StandardizedResiduals][[1,2]]
```

```
{1.87411,-0.0644816,-0.839202,-0.247677,
 1.40636,-0.965242,-0.0429063,-0.359268,0.91839,
 -1.87916,-0.171141,-0.180051,-0.603555,1.55562}
```

```
eps=(Max[predvals]-Min[predvals])/Length[predvals];
ListPlot[Transpose[{predvals,errSTvals}],
  Prolog->{PointSize[0.02]},
  AxesOrigin->{Min[predvals]-eps,0},
  PlotRange->{{Min[predvals]-eps,Max[predvals]+eps},
   {Min[errSTvals]-eps,Max[errSTvals]+eps}}]
```

1.5
1
0.5
0.44 0.46 0.48 0.5 0.52 0.54
-0.5
-1
-1.5

■

Outliers

To identify outliers, we look for those cases with large studentized deleted residuals. We informally classify as an outlier any case for which the absolute value of the studentized deleted residual is greater than 3.5. A formal test for outliers based on the Bonferroni inequality, which was discussed in Chapter 10, also applies here. We calculate a two-sided critical value $t_{1-\alpha/2n,\, n-p-1}$. Any studentized deleted residual larger than the critical value in absolute value corresponds to an outlier using a familywise level of significance, α.

EXAMPLE 2: Use residuals to determine if there are outlying observations in the data given in Table 11.3.

SOLUTION: First we load the **LinearRegression** package. Then we enter the lists of data in `teamera`, `ownbavg`, `oppbavg`, and `winpct` and define the matrix `dpoints`. Next, we perform a linear regression using the information in `teamera`, `ownbavg`, `oppbavg` as the independent variables to predict the winning percentage of each team (the dependent variable). By including the `RegressionReport->StudentizedResiduals` option setting, we obtain a list of the studentized deleted residuals. Notice that none of these values is greater than 3.5 in absolute value. Therefore, according to our first criterion, there are no outliers.

```
<<Statistics`LinearRegression`

teamera={3.33,3.51,3.55,3.65,3.80,4.20,4.22,
  4.27,4.31,4.48,4.53,4.55,4.62,5.86};
ownbavg={0.276,0.249,0.249,0.260,0.271,0.241,0.269,
  0.264,0.270,0.240,0.259,0.252,0.258,0.293};
oppbavg={0.240,0.254,0.249,0.245,0.250,0.252,0.254,
  0.270,0.274,0.264,0.280,0.266,0.268,0.286};
winpct={0.625,0.512,0.488,0.524,0.588,0.475,0.513,
  0.463,0.512,0.405,0.450,0.480,0.456,0.506};

Clear[dpoints]
  dpoints=Table[{teamera[[i]],ownbavg[[i]],
  oppbavg[[i]],winpct[[i]]},{i,1,Length[winpct]}];

Regress[dpoints,{1,x1,x2,x3},{x1,x2,x3},
  RegressionReport->StudentizedResiduals]

{StudentizedResiduals→{0.639771,0.789271,-0.862064,
  -1.06129,1.20324,1.24538,-1.35245,-1.49127,0.620229,
  -0.813636,0.131561,1.51698,-0.412657,0.25586}}
```

In order to calculate the Bonferroni critical value, we load the **NormalDistribution** (or **ContinuousDistributions**) package. After entering the information $n = 14$, $p = 4$, and $\alpha = 0.05$, we use `Quantile` to find that the critical value is 3.90877. Notice that none of the residuals in the list above are larger (in absolute value) than 3.90877. Therefore, we cannot detect any outliers in the data set.

```
<<Statistics`NormalDistribution`
n=14;
p=4;
α=0.05;
Quantile[StudentTDistribution[n-p-1],1-α/(2n)]
3.90877
```

■

Influential Cases

An influential case is one which has a greater effect on the prediction equation than other cases. This effect on the prediction equation is measured using a quantity called **leverage,** where for the ith case the notation is h_{ii}. Leverage values are the diagonal elements of the hat matrix where $0 \le h_{ii} \le 1$ and $\sum_{i=1}^{n} h_{ii} = p$. We can interpret h_{ii} as a measure of the distance between the x values for the ith case and the means of the x values for all n cases. Therefore, a large value of h_{ii} implies that the ith case is at a large distance from the center of all x observations. A leverage value is usually considered large if it is more than twice as large as the mean leverage value,

$$\bar{h} = \frac{1}{n}\sum_{i=1}^{n} h_{ii} = \frac{p}{n}.$$

Another rule of thumb suggests that values greater than 0.5 indicate very high leverage while values between 0.2 and 0.5 indicate moderate leverage. In practice, an influential case is identified if its omission results in a major change in the prediction equation. We will present three often-used measures of influence.

DFFITS

$(DFFITS)_i$ represents the standardized difference between the predicted value for the ith case when all n cases are used in fitting the regression equation and the predicted value for the ith case obtained when the ith case is excluded in fitting the regression equation. It can be calculated as follows:

$$(DFFITS)_i = e_i\left[\frac{n-p-1}{SSE(1-h_{ii})-e_i^2}\right]^{1/2}\left(\frac{h_{ii}}{1-h_{ii}}\right)^{1/2} = t_i\left(\frac{h_{ii}}{1-h_{ii}}\right)^{1/2}.$$

For small to medium data sets, a case is considered influential if $(DFFITS)_i > 1$. For large data sets, a case is influential if $(DFFITS)_i > 2\sqrt{p/n}$. The `RegressionReport` option setting `PredictedResponseDelta` yields a list of values of $(DFFITS)_i$.

Cook's Distance

Cook's distance measure considers the influence of the ith case on all n fitted values. It can be calculated as follows:.

$$d_i = \frac{e_i^2}{pMSE}\left(\frac{h_{ii}}{(1-h_{ii})^2}\right).$$

Because d_i depends on e_i and h_{ii}, the ith case can be influential by having a large residual e_i and only a moderate leverage value h_{ii}, or by having a large leverage value and only a moderately sized residual, or by having both a large residual and a large leverage value. To interpret Cook's distance measure use the F (ratio) distribution with numerator degrees of freedom p and denominator degrees of freedom n - p. If d_i is less than the 20th percentile of this distribution, the ith case has little influence. If d_i is around the 50th percentile or more, the ith case has significant influence on the fit of the regression function. The `RegressionReport` option setting `CookD` yields the list $\{d_1, d_2, \ldots, d_n\}$ of the Cook's distance measures for the n cases.

DFBETAS

$(DFBETAS)_{i,k}$ is a measure of the influence of the ith case on the regression coefficient b_k. It is the difference between the estimated partial regression coefficient b_k based on all n cases and the partial regression coefficient obtained when the ith case is excluded, $b_{k(i)}$, divide by the standard deviation of b_k. It is calculated using

$$(DFBETAS)_{i,k} = \frac{b_k - b_{k(i)}}{\sqrt{MSE_{(i)}c_{kk}}},$$

where c_{kk} is the kth diagonal element of $(\mathbf{X}'\mathbf{X})^{-1}$ and $MSE_{(i)}$ is the mean square for error from the regression where the ith case is deleted. Notice that for each of the n cases there are p corresponding values of *DFBETAS*. The sign of the *DFBETAS* value shows whether inclusion of the case causes an increase or a decrease in the estimated regression coefficient, whereas the magnitude (in absolute value) of the *DFBETAS* value indicates the size of the difference relative to the estimated standard deviation of the regression coefficient. A large value of $|(DFBETAS)_{i,k}|$ indicates that the ith case has a large impact on the kth regression coefficient. We typically consider a case i influential if $|(DFBETAS)_{i,k}| > 1$ for small to medium data sets or if $|(DFBETAS)_{i,k}| > 2/\sqrt{n}$ for large data sets. Using the `RegressionReport` option setting `BestFitParametersDelta` outputs the list $\{\{d_{11}, \ldots, d_{1p}\}, \ldots, \{d_{n1}, \ldots, d_{np}\}\}$, where d_{ik} is $(DFBETAS)_{i,k}$.

EXAMPLE 3: Investigate the data given in Table 11.3 to determine if there are any influential cases.

SOLUTION: After loading the **LinearRegression** package and entering the information in `teamera`, `ownbavg`, `oppbavg`, and `winpct`, we form the matrix `dpoints`. We find the diagonal elements of the hat matrix with `HatDiagonal` and name the result `hd` so that we can extract the list of diagonal elements with `hd[[1,2]]`. We name this expression `hdlist`, so we can extract elements easily. For example, `hdlist[[1]]` gives us the first member of the list, h_{11}. We also verify that $\sum_{i=1}^{n} h_{ii} = p$, where $n = 14$ and $p = 4$ in this case. Notice that $2p/n \approx 0.571429$, and the only element in the list that is larger than this value is the last diagonal element, which corresponds to the Colorado Rockies. This case has high leverage.

```
<<Statistics`LinearRegression`

teamera={3.33,3.51,3.55,3.65,3.80,4.20,4.22,4.27,
  4.31,4.48,4.53,4.55,4.62,5.86};
ownbavg={0.276,0.249,0.249,0.260,0.271,0.241,0.269,
  0.264,0.270,0.240,0.259,0.252,0.258,0.293};
oppbavg={0.240,0.254,0.249,0.245,0.250,0.252,0.254,
  0.270,0.274,0.264,0.280,0.266,0.268,0.286};
winpct={0.625,0.512,0.488,0.524,0.588,0.475,0.513,
  0.463,0.512,0.405,0.450,0.480,0.456,0.506};

Clear[dpoints]
  dpoints=Table[{teamera[[i]],ownbavg[[i]],
  oppbavg[[i]],winpct[[i]]},{i,1,Length[winpct]}];

hd=Regress[dpoints,{1,x1,x2,x3},{x1,x2,x3},

RegressionReport->HatDiagonal]
{HatDiagonal→{0.446993,0.231595,0.169291,0.182651,
  0.192002,0.3769,0.174635,0.179229,0.315062,
  0.318829,0.371122,0.153562,0.119834,0.768295}}
```

```
hdlist=hd[[1,2]]
```

```
{0.446993,0.231595,0.169291,0.182651,0.192002,
  0.3769,0.174635,0.179229,0.315062,0.318829,
  0.371122,0.153562,0.119834,0.768295}
```

```
hdlist[[1]]
```

```
0.446993
```

```
Sum[hdlist[[i]],{i,1,Length[hdlist]}]
```

```
4.
```

```
p=4;
n=14;
2p/n//N
```

```
0.571429
```

We use the `CookD` option setting to find the Cook's distance measures for the data. Next, we load the **NormalDistribution** package (if we have not already done so) to determine the 20th percentile for the F (ratio) distribution with 4 and 10 degrees of freedom, which is approximately 0.406574. Notice that all of the distance measures for our data set are less than this value, so we cannot detect any influential cases. In fact, when we calculate the 50th percentile (about 0.898817), we see that none of the distance measures are close to the level necessary to have a large influence on the regression function.

```
Regress[dpoints,{1,x1,x2,x3},{x1,x2,x3},
  RegressionReport->CookD]
```

```
{CookD→{0.0879027,0.0487779,0.0388601,0.0621402,
  0.0823222,0.222289,0.0893455,0.108167,0.0471378,
  0.0801743,0.00283181,0.0923556,0.00632049,0.0598614}}
```

```
<<Statistics`NormalDistribution`
```

```
n=14;
p=4;
Quantile[FRatioDistribution[p,n-p],0.2]
```

```
0.406574
```

```
Quantile[FRatioDistribution[p,n-p],0.50]
```

```
0.898817
```

With the `RegressionReport` option `PredictedResponseDelta`, we obtain the list of *DFFITS* values. No influential cases are identified because none of the values exceeds 1 in absolute value. The closest value to 1 in the list (0.96858) corresponds to the Chicago Cubs.

```
Regress[dpoints,{1,x1,x2,x3},{x1,x2,x3},
  RegressionReport->PredictedResponseDelta]
{PredictedResponseDelta→{0.575188,0.433307,
  -0.389163,-0.501698,0.586544,0.96858,-0.622104,
  -0.696866,0.420653,-0.556649,0.101065,0.646137,
  -0.152264,0.465906}}
```

We use the RegressionReport option BestFitParametersDelta to obtain the *DFBETAS* values. As in the previous case, none of the values exceeds 1 (in absolute value), so none of the cases have significant influence on the regression function.

```
Regress[dpoints,{1,x1,x2,x3},{x1,x2,x3},
RegressionReport->BestFitParametersDelta]
{BestFitParametersDelta→
  {{-0.0143844,-0.115759,0.382709,-0.136519},
   {-0.0559135,-0.289016,-0.101049,0.209172},
   {-0.123586,0.108709,0.126155,-0.00171137},
   {-0.235483,-0.0653034,-0.0718679,0.257519},
   {-0.0000106592,-0.0497338,0.358846,-0.156987},
   {0.835436,0.62326,-0.674636,-0.59404},
   {-0.226309,-0.307155,-0.220736,0.409338},
   {0.486631,0.446869,-0.124454,-0.530767},
   {-0.355183,-0.301588,0.150272,0.345162},
   {-0.309542,-0.233555,0.47019,0.105511},
   {-0.0680988,-0.059592,-0.00941033,0.0865301},
   {0.25121,0.2822,-0.396119,-0.105235},
   {-0.0263566,-0.0593643,0.0557525,0.0124995},
   {-0.0907156,0.234257,0.199504,-0.10239}}}
```

Random Sampling

For a brief discussion of random sampling refer to Chapter 10.

Independence

When data are collected in a time sequence or in some other sequence, such as geographical, a plot of residuals against the ordering variable is useful for evaluating independence. Under independence a long sequence of positive residuals followed by a long sequence of negative residuals is quite unlikely. Positive and negative residuals which alternate in a regular pattern would also not be expected. Such patterns indicate a correlation between terms that are near each other in the sequence. This condition is called **autocorrelation** or serial correlation.

Durbin–Watson Test for Autocorrelation

The Durbin–Watson test is used to determine whether the data come from a first-order autoregressive process. The model for such a process is the same as that for linear regression, with the additional requirement that the error term at time t be related to the error term at time $t-1$ by the equation $\varepsilon_t = \rho\varepsilon_{t-1} + u_t$, where the parameter ρ, called the autocorrelation parameter, is the correlation coefficient between error terms separated by one time period. The null hypothesis is $H_0: \rho = 0$, which implies that there is no first-order serial correlation. The Durbin–Watson test statistic is

$$d = \frac{\sum_{t=2}^{n} (e_t - e_{t-1})^2}{\sum_{t=2}^{n} e_t^2},$$

where n is the number of cases. Values of d close to zero indicate positive correlation, and values close to 4 indicate negative correlation. Values around two can be taken as evidence of no autocorrelation.

When the alternative hypothesis is $H_a:\rho > 0$, the test statistic is d, and when the alternative hypothesis is $H_a:\rho < 0$, the test statistic is $4 - d$. Durbin and Watson tables have lower and upper bounds d_L and d_U such that if d (respectively $4 - d$) is less than d_L, then reject the null hypothesis, and if it is greater than d_U, then conclude is no first-order autocorrelation. If d (respectively $4 - d$) lies between d_L and d_U, then the test is inclusive.

We use the `RegressionReport` option setting `DurbinWatsonD` to obtain d.

EXAMPLE 4: Calculate the Durbin–Watson d statistic for the data given in Table 11.3.

SOLUTION: Assuming that the **LinearRegression** package is loaded, we use the information defined in `dpoints`, defined in several previous examples, to find that the Durbin–Watson test statistic is approximately 2.09999. Therefore, we we have no reason to conclude that the errors are correlated.

```
Regress[dpoints,{1,x1,x2,x3},{x1,x2,x3},
  RegressionReport->DurbinWatsonD]
```

```
{DurbinWatsonD→2.09999}
```

■

Linearity/Additivity

As we pointed out in the note of caution at the beginning of the chapter, there is a potential for confusion between whether variables are linearly related and whether a particular model can be fit using linear regression methodology. In particular, a model can incorporate independent variables which are related to the response variable in a nonlinear way as long as the model is additive in the predictors and epsilon. Thus, for each independent variable, the challenge is to use predictors that can reflect its relation to the dependent variable in an additive model.

To investigate the form of the relationship between the dependent variable and an independent variable, a scatter plot may be used. To see if a particular model is adequate to represent the relationship between the dependent variable and a particular independent variable, plot the residuals against that independent variable. To evaluate the model for overall appropriateness, plot the residuals against the fitted values. Residual plots that have the points randomly scattered in a band about the zero line can indicate that the model is appropriate for the data. A pattern to the scatter indicates a need to adjust the model. See Example 1 in this section.

EXAMPLE 5: For the data in Table 11.3, create a scatter plot of the studentized residuals against each of the independent variables.

SOLUTION: After making sure that the **LinearRegression** package is loaded and that the data is entered in `dpoints`, we use `Regress` with the option setting `RegressionReport->StandardizedResiduals` to calculate the studentized residuals. Notice that we extract the residuals from the output list and assign the name `multiSTerr`.

```
<<Statistics`LinearRegression`

teamera={3.33,3.51,3.55,3.65,3.80,4.20,4.22,4.27,
  4.31,4.48,4.53,4.55,4.62,5.86};
ownbavg={0.276,0.249,0.249,0.260,0.271,0.241,0.269,
  0.264,0.270,0.240,0.259,0.252,0.258,0.293};
oppbavg={0.240,0.254,0.249,0.245,0.250,0.252,0.254,
  0.270,0.274,0.264,0.280,0.266,0.268,0.286};
winpct={0.625,0.512,0.488,0.524,0.588,0.475,0.513,
  0.463,0.512,0.405,0.450,0.480,0.456,0.506};

Clear[dpoints]
  dpoints=Table[{teamera[[i]],ownbavg[[i]],
  oppbavg[[i]],winpct[[i]]},{i,1,Length[winpct]}];

multiSTerr=Regress[dpoints,{1,x1,x2,x3},{x1,x2,x3},
  RegressionReport->StandardizedResiduals][[1,2]]

{0.659547,0.804585,-0.873353,-1.05465,1.17717,
 1.21243,-1.29964,-1.40762,0.64024,-0.827745,
 0.138544,1.42698,-0.430921,0.268724}
```

Next, using ListPlot, we create a scatter plot of the studentized residuals against the first independent variable (found in teamera). We use eps to denote a small number obtained by dividing the distance between the maximum and minimum values in teamera by the number of elements in teamera. We use this number in selecting the window to display the scatter plot so that all of the points are shown.

```
eps=(Max[teamera]-Min[teamera])/Length[teamera];
ListPlot[Transpose[{teamera,multiSTerr}],
  Prolog->{PointSize[0.02]},
  AxesOrigin->{Min[teamera]-eps,0},
  PlotRange->{{Min[teamera]-eps,Max[teamera]+eps},
   {Min[multiSTerr]-eps,Max[multiSTerr]+eps}}]
```

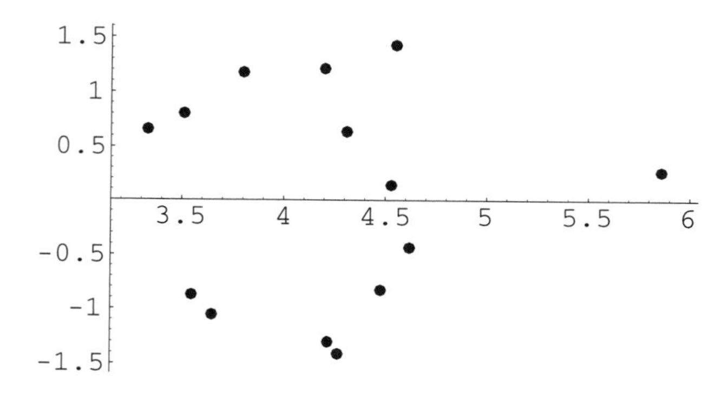

We enter similar commands to generate a scatter plot of the studentized residuals against the second independent variable (found in ownbavg) as well as to obtain the scatter plot of the studentized residuals against the third independent variable (listed in oppbavg). None of these scatter plots strongly suggests a departure from linearity, as the residuals generally appear to be randomly scattered about the zero line. Note that the plot of residuals against teamera (above) may suggest a pattern, although it is difficult to be sure with the small number of points.

```
eps=(Max[ownbavg]-Min[ownbavg])/Length[ownbavg];
ListPlot[Transpose[{ownbavg,multiSTerr}],
  Prolog->{PointSize[0.02]},
  AxesOrigin->{Min[ownbavg]-eps,0},
  PlotRange->{{Min[ownbavg]-eps,Max[ownbavg]+eps},
   {Min[multiSTerr]-eps,Max[multiSTerr]+eps}}]
```

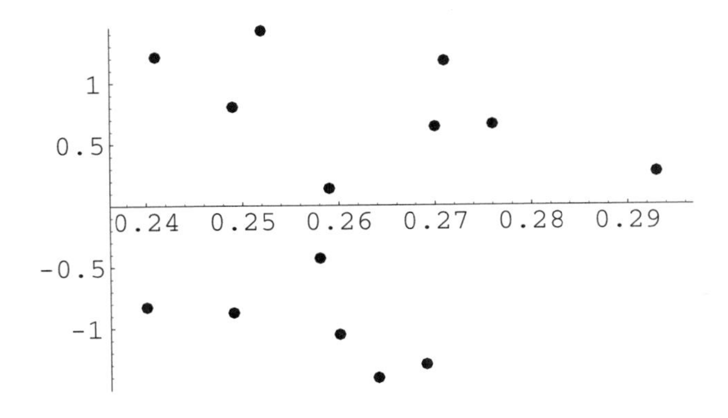


```
eps=(Max[oppbavg]-Min[oppbavg])/Length[oppbavg];
ListPlot[Transpose[{oppbavg,multiSTerr}],
  Prolog->{PointSize[0.02]},
  AxesOrigin->{Min[oppbavg]-eps,0},
  PlotRange->{{Min[oppbavg]-eps,Max[oppbavg]+eps},
   {Min[multiSTerr]-eps,Max[multiSTerr]+eps}}]
```

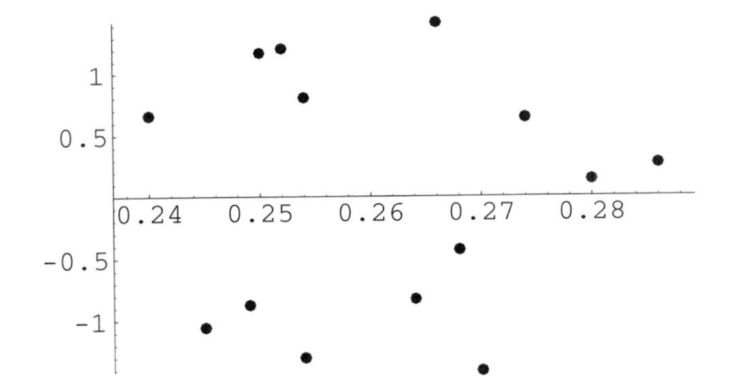

■

Equality of Variance

A plot of the residuals against the predicted values is also useful in evaluating the constant variance assumption. A random scatter of the residuals about zero indicates that the assumption is valid for the data. Nonconstant variance is often demonstrated by a "fanning out" or "funneling in" pattern. "Fanning out" (as in Fig. 11.1) is characteristic of increasing variability with increasing magnitude in the predicted values. On the other hand, "funneling in" (as in Fig. 11.2) is characteristic of decreasing variability with increasing magnitude in the predicted values. In simple regression, if there is nonconstant variance, the "fanning out" or "funneling in" pattern would also be apparent in the plot of residuals against the independent variable. For multiple regression, if nonconstant variance is detected, a plot of residuals against each predictor may identify the variables to which the magnitude of the variability is related. For more information, see the corresponding section in Chapter 10. It is possible to divide the residuals into two or more groups and use the modified Levene's hypothesis test discussed there. In simple regression, the groups are defined by consecutive values of the independent variable. For example, two groups could be formed corresponding to all values of the independent variable less than a particular number and all values greater than that number, respectively.

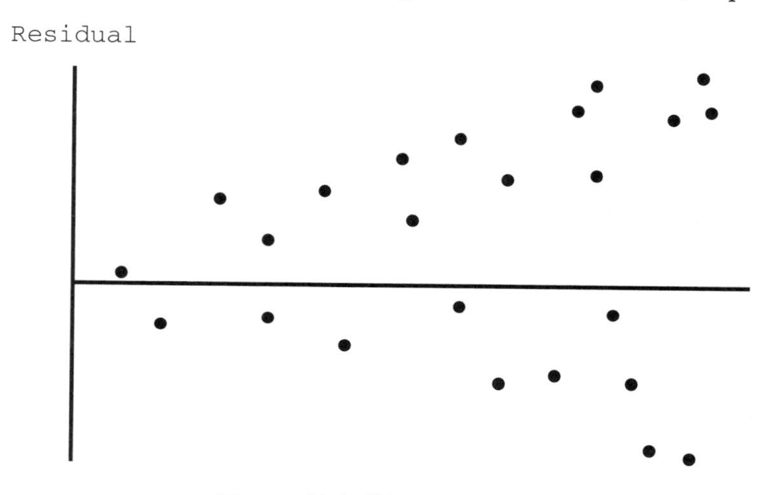

Figure 11.1 "Fanning Out"

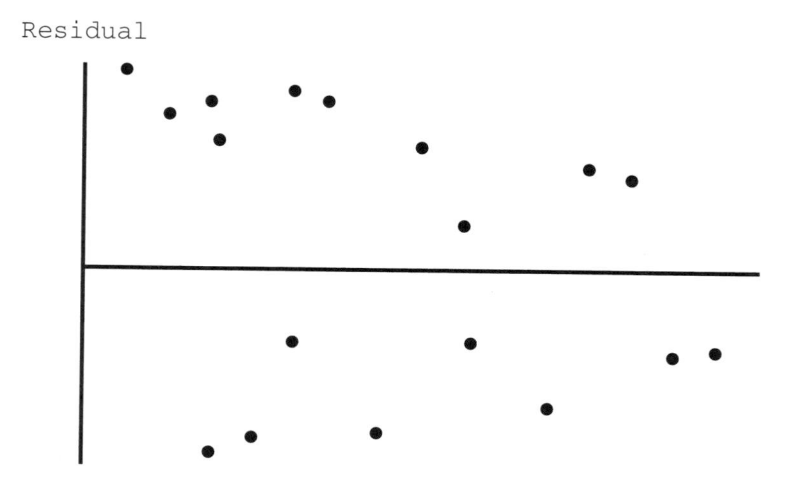

Figure 11.2 "Funneling In"

Normality

Departure from the normality assumption can be studied using the residuals by plotting a histogram or with a normal probability plot. We have written a procedure called **normal-Histogram** to display a histogram with the corresponding normal density curve superimposed on it (see Chapter 7). This can provide a visual appraisal of how closely the residuals reflect a normal distribution. We have also written a procedure called **normal-Probability** to display a normal probability plot (see Chapter 10). Data from a normal distribution form a line on the normal probability plot.

Multicolinearity

Multicolinearity is a term that is used to refer to a high intercorrelation among the predictors in the multiple regression equation. This property is to be avoided. In its most extreme form, the predictors are perfectly related (and the data do not contain any random error component). In this case, many different prediction equations that fit the data exactly are possible, and adding or deleting a predictor can substantially change the estimated regression coefficients of the other predictors. At the other extreme, when predictors are completely uncorrelated, the estimated regression coefficients are the same no matter which predictors are included in the calculations.

In practice, predictors are neither uncorrelated nor perfectly related; however, it is not useful to include highly correlated predictors in the same regression calculation. As intercorrelation increases, the estimated regression coefficients become increasingly imprecise. That is, their standard deviations increase. One common result of multicolinearity is that a set of predictors with a definite relationship to the response variable will individually fail to show the relationship. For example, the *F*-test based on the analysis of variance table will show significance, but none of the regression coefficients will be significant when tested individually using *t*-tests. As an example, see the cubic fit in Example 1 in the section on polynomial regression (Section 11.3). One method of evaluating multicolinearity is to calculate simple correlations between all pairs of predictors. There are, however, no definite guideline values on the correlations above which multicolinearity is considered to be severe. Also, simple correlations do not always indicate the actual nature or extent of the multicolinearity.

A more formal method of detection of multicolinearity is through the use of variance inflation factors. In the **LinearRegression** package, the `RegressionReport` option setting `VarianceInflation` calculates a factor, v_k, corresponding to each estimated regression coefficient, b_k. The variance inflation factor is calculated with

$$v_k = (1 - R_k^2)^{-1},$$

where R_k^2 is the coefficient of multiple determination when the kth predictor is regressed on the other predictor variables. Note that $R_k^2 = 0$ when the kth predictor is not linearly related to the other predictors, and, in this case, $v_k = 1$. When $R_k^2 > 0$, $v_k > 1$. Although no rule of thumb is foolproof, generally any variance inflation factor over 10 (which corresponds to $R_k^2 = 0.9$) indicates a serious problem with multicolinearity. In addition, some authors suggest that values over 4 should raise concern.

A second formal method for detecting multicolinearity involves the eigenvalues and eigenvectors of the **correlation matrix**, a matrix of simple correlation coefficients for all pairs of predictors. The nearness to zero of the smallest eigenvalue is a measure of the magnitude of intercorrelation. Also, the elements of the associated normalized eigenvector represent the weights of the corresponding predictors in the multicolinearity. The `RegressionReport` option setting `EigenstructureTable` calculates eigenvalues (listed from largest to smallest) and, corresponding to each, an index. This option setting also displays for each predictor, the proportion of the variance (in the corresponding regression coefficient) attributable to each eigenvalue. The ith index number is the square root of the largest eigenvalue (λ_{max}) divided by the ith eigenvalue. The ratio

$$\sqrt{\frac{\lambda_{max}}{\lambda_{min}}},$$

where λ_{min} is the smallest eigenvalue, is called the **condition number** of the correlation matrix. Large values indicate multicolinearity. Generally, values of the condition number (or other ratios of the form $\sqrt{\lambda_{max}/\lambda_i}$) over 30 should raise concern about multicolinearity. Note that high-leverage cases can influence the condition number. For example, a high-leverage case can either create multicolinearity or hide it depending on the data set.

EXAMPLE 6: For the data in Table 11.3, calculate: (a) the variance inflation factor for each estimated regression coefficient, b_k; (b) the eigenvalues, the proportion of the variance attributable to each eigenvalue, the index numbers, and the condition number of the correlation matrix.

SOLUTION: After loading **LinearRegression** as well as the data in `dpoints`, we use `Regress` with the option setting `RegressionReport->VarianceInflation` to find the variance inflation factors corresponding to the three independent variables `x1` (team era), `x2` (own batting average), and `x3` (opposing team batting average). We find these values to be $v_1 \approx 4.178558$, $v_2 \approx 1.14993$, and $v_3 \approx 3.95366$. Because none of these values is larger than 10, we do not suspect a serious problem with multicolinearity although the fact that $v_1 > 4$ does raise some concern.

```
<<Statistics`LinearRegression`

teamera={3.33,3.51,3.55,3.65,3.80,4.20,4.22,4.27,
  4.31,4.48,4.53,4.55,4.62,5.86};
ownbavg={0.276,0.249,0.249,0.260,0.271,0.241,0.269,
  0.264,0.270,0.240,0.259,0.252,0.258,0.293};
oppbavg={0.240,0.254,0.249,0.245,0.250,0.252,0.254,
  0.270,0.274,0.264,0.280,0.266,0.268,0.286};
winpct={0.625,0.512,0.488,0.524,0.588,0.475,0.513,
  0.463,0.512,0.405,0.450,0.480,0.456,0.506};

Clear[dpoints]
  dpoints=Table[{teamera[[i]],ownbavg[[i]],
  oppbavg[[i]],winpct[[i]]},{i,1,Length[winpct]}];

Regress[dpoints,{1,x1,x2,x3},{x1,x2,x3},
  RegressionReport->VarianceInflation]

{VarianceInflation→{0.,4.17858,1.14993,3.95366}}
```

Next, we use the option setting `RegressionReport->EigenstructureTable` to generate a table containing the eigenvalues (found in the first column), the index number of each eigenvalue (found in the second column), and for each predictor (`x1`, `x2`, `x3`) the proportion of the variance in the corresponding regression coefficient due to each eigenvalue. In this case, the eigenvalues are $\lambda_1 \approx 2.05464$, $\lambda_2 \approx 0.813183$, and $\lambda_3 \approx 0.132175$. The corresponding index numbers are 1, 1.58955, and 3.9427. Notice that two predictors, `x1` and `x3`, show a large proportion of variance attributable to the smallest eigenvalue, $\lambda_3 \approx 0.132175$. This could lead us to suspect a collinear relationship between these two predictors. However, the condition number, which equals the largest index number, is 3.9427. Because this number is not greater than 30, we are not concerned with multicolinearity.

```
Regress[dpoints,{1,x1,x2,x3},{x1,x2,x3},
  RegressionReport->EigenstructureTable]
```

{EigenstructureTable→

EigenV	Index	x1	x2	x3
2.05464	1.	0.0505137	0.0649444	0.0508254
0.813183	1.58954955	0.0147281	0.899539	0.0338397
0.132175	3.9427	0.934758	0.035517	0.915335

}

■

11.6 Remedial Methods

Introduction

In linear regression, when the data do not meet the assumptions, a method to remedy the situation may be employed. One choice is to change or expand the set of predictors. For example, a predictor based on the square or higher power of an independent variable can address curvature, or an interaction effect can be addressed by a predictor based on the product of the two independent variables. If the measurements exist, a new independent variable included as a predictor might solve the problem. A second choice is to transform the data to bring the adjusted data into agreement with the assumptions. Transformations intended to result in an additive model may be made on the independent variable(s), the dependent variable, or both. Transformations for problems with constant variance and normality are made on the dependent variable. Often, one transformation can solve more than one problem, although it is possible for a transformation to solve one problem and create a new one. With this in mind, if constant variance and normality are satisfied, it would be prudent to transform the independent variable(s) to achieve an additive model because a transformation on the dependent variable may adversely affect one or both of constant variance and normality. On the other hand, a single transformation on the independent variable may solve problems with additivity, constant variance, and normality simultaneously. For additional general information, see Section 10.3, Transformations, in Chapter 10.

A third choice is to develop the prediction equation using weighted least squares, where the sample regression coefficients are chosen to minimize a weighted sum of squared residuals,

$$\sum_{i=1}^{n} w_i e_i^2.$$

The weights, w_i, are chosen according to the problem in the data. We discuss the procedure here and then the choice of the weights in the following subsections as appropriate.

Weighted Least Squares

One of the options of `Regress` called `Weights` allows a vector of weights to be used when calculating the sample regression coefficients. The vector is of the form $\{w_1, w_2, \ldots, w_n\}$ where w_i is a weight corresponding to the ith case. The coefficients are then chosen to minimize the weighted sum of squared residuals,

$$\sum_{i=1}^{n} w_i e_i^2.$$

The weighted least squares criterion generalizes the ordinary least squares criterion by replacing equal weights of one by the w_i's. Because a residual,

$$e_i = y_i - \hat{y}_i,$$

measures the distance between the observed value and the predicted value, its square, e_i^2, can be thought of as a representation of the variability contained in the response, y_i. Therefore, $\sum_{i=1}^{n} w_i e_i^2$. will be the smallest for weights that are inversely proportional to the variance of the corresponding response. For this reason, weighted least squares is most useful for situations where outliers are present or for those in which there are problems with the constant variance assumption. Suggestions for choosing the weights are made in the subsections which follow. The user of weighted least squares should be aware that the confidence intervals, prediction intervals, and hypothesis tests using the resulting estimated regression coefficients are only approximate. Also, the coefficient of determination and, consequently, the adjusted coefficient of determination are not well defined and should not be used for weighted least squares.

Outliers and Influential Cases

Once a case has been determined to be an outlier, the first consideration is whether it is the result of a recording or data processing error, an instrument or machine malfunction, or some other assignable cause. If so, it can be corrected or discarded. If not, the next consideration is whether the outlying case(s) indicates an inadequacy of the model that can be remedied by changing the set of predictors or by some other means. If the model seems to be the correct one, the final consideration is to minimize the impact of the outlier(s) on the calculation of the regression coefficients. This can be accomplished with weighted least squares which is available using the `Regress` option setting `Weights`. The technique to be discussed is **iteratively reweighted least squares** (IRwLS), one of several methods classified as **robust regression**. These are, in general, methods that yield estimated regression coefficients that are nearly as good (efficient) as the least squares estimates when the assumptions are satisfied and better when a particular assumption is not satisfied.

The methodology for IRwLS is as follows:

1. Choose a strategy for calculating the weights.
2. Calculate starting weights for all cases.
3. Use the starting weights in the weighted least squares regression and obtain the residuals.
4. Use these residuals to calculate revised weights.
5. Continue the iteration until it converges.

Convergence is defined as relatively small changes in the calculated weights. It can also be defined as small changes in residuals or estimated regression coefficients or predicted values.

A commonly used strategy for calculating weights is to use the Huber influence function. Weights are determined using

$$w = \begin{cases} 1 & \text{if } |e^*| \le r \\ r/|e^*| & \text{if } |e^*| > r \end{cases},$$

where r is called the **tuning constant** and e^* is a scaled residual (that is, a residual divided by an estimate of variability such as a studentized residual). A reasonable value for r is between 1.0 and 1.5. For example, the value 1.345 will make the IRwLS method 95% efficient when there are no outliers and the other assumptions are satisfied. In IRwLS, a robust scaling factor is used to calculate e^*. That is, the scaling factor should not be sensitive to extreme values as is the mean square error (*MSE*), for example. Possible choices for the scaling factor include 1.5 median $\{|e_i|\}$ and

$$\frac{1}{5745}\text{median}\left\{|e_i - \text{median}\{e_i\}|\right.$$

Then, the scaled residual, e_i^*, is the residual, e_i, divided by the scaling factor. For example, we can choose the weights using

$$w_i = \begin{cases} 1 & \text{if } |e_i^*| \le 1.345 \\ 1.345/|e_i^*| & \text{if } |e_i^*| > 1.345 \end{cases},$$

where $= e_i \Big/ \frac{1}{0.6745}\text{median}\left\{|e_i - \text{median}\{e_i\}\right., i = 1,\ldots,n$. The starting weights will be calculated using the residuals from ordinary least squares regression (i.e., $w_i = 1$, $i = 1,\ldots,n$)

Note that robust regression can help identify outliers in that they are cases with relatively low weights. It can also help justify the use of ordinary least squares results. When robust regression results are similar to those for ordinary least squares, there is reassurance that ordinary least squares is not unduly influenced by the outliers.

For an influential case(s), it is good practice to examine the inferences that would be made with and without the case(s). If inferences are essentially the same, there is no need to take remedial action. If there is a serious change when the case(s) is omitted, then a method to minimize the impact of the case(s) can be considered. Such methods are called **bounded influence regression**. Iteratively reweighted least squares (IRwLS) can be employed as discussed earlier. To make it more sensitive to influential cases, studentized deleted residuals (`RegressionReport` option setting `StudentizedResiduals`) are used in the place of the scaled residuals which were described previously. To provide even stronger control, the weights w_i (calculated using the Huber function and the studentized deleted residuals as e^*) can be modified using,

$$\left(\sqrt{1-h_{ii}}\right)w_i,$$

where h_{ii} is the leverage of the ith case, obtained from the main diagonal of the hat matrix (`RegressionReport` option setting `HatDiagonal`).

EXAMPLE 1: Table 11.4 gives information for 37 states, as well as for the District of Columbia, Guam, and the Virgin Islands, concerning the relationship between a student's home environment and eighth-grade mathematics proficiency in 1990. The dependent variable y represents the average eighth-grade mathematics proficiency, the independent variable x_1 represents the percentage of eighth-grade students with three or more types of reading materials at home, the independent variable x_2 represents the percentage of eighth-grade students that read more than 10 pages per day, and the independent variable x_3 represents the percentage of eighth-grade students that watch television 6 hours or more per day. Use weighted least squares with the Huber weight function to determine a robust fitted regression function.

State	y	x_1	x_2	x_3
Alabama	252	78	34	18
Arizona	259	73	41	12
Arkansas	256	77	28	20
California	256	68	42	11
Colorado	267	85	38	9

Continued

State	y	x_1	x_2	x_3
Connecticut	270	86	43	12
Delaware	261	83	32	18
District of Columbia	231	76	24	33
Florida	255	73	31	19
Georgia	258	80	36	17
Guam	231	64	32	20
Hawaii	251	69	36	23
Idaho	272	84	48	7
Illinois	260	82	43	14
Indiana	267	84	37	11
Iowa	278	88	43	8
Kentucky	256	78	36	14
Louisiana	246	76	36	19
Maryland	260	83	34	19
Michigan	264	84	31	14
Minnesota	276	88	36	7
Montana	280	88	44	6
Nebraska	276	88	42	9
New Hampshire	273	88	40	7
New Jersey	269	84	41	13
New Mexico	256	72	40	11
New York	261	79	35	17
North Carolina	250	78	37	21
North Dakota	281	90	41	6
Ohio	264	84	36	11
Oklahoma	263	78	37	14
Oregon	271	82	41	9
Pennsylvania	266	86	34	10
Rhode Island	260	80	38	12

State	y	x_1	x_2	x_3
Texas	258	70	34	15
Virgin Islands	218	76	23	27
Virginia	264	82	33	16
West Virginia	256	80	36	16
Wisconsin	274	86	38	8
Wyoming	272	86	43	7

Source: ETS Policy Information Center, *America's Smallest School: The Family* (Princeton, New Jersey: Educational Testing Service, 1992)

Table 11.4

SOLUTION: We begin by entering the data in Table 11.4 and naming it `mathtestinfo`. In order to use the average mathematics proficiency (given in the first column) as the independent variable, we must manipulate the columns of `mathtestinfo` so that this information is in the last column. This is accomplished using `Transpose`. Note that `Transpose[mathtestinfo][[2]]` gives the second column in Table 11.4, `Transpose[mathtestinfo][[3]]` the third, `Transpose[mathtestinfo][[4]]` the fourth, and `Transpose[mathtestinfo][[1]]` the first. Placing the columns of Table 11.4 in the appropriate order and then taking the transpose of the result gives us the table `mtestdata`, which is displayed next to make sure that each entry in the list has the average mathematics proficiency in the fourth position.

```
<<Statistics`LinearRegression`

mathtestinfo={{252,78,34,18},{259,73,41,12},
  {256,77,28,20},{256,68,42,11},{267,85,38,9},
  {270,86,43,12},{261,83,32,18},{231,76,24,33},
  {255,73,31,19},{258,80,36,17},{231,64,32,20},
  {251,69,36,23},{272,84,48,7},{260,82,43,14},
  {267,84,37,11},{278,88,43,8},{256,78,36,14},
  {246,76,36,19},{260,83,34,19},{264,84,31,14},
  {276,88,36,7},{280,88,44,6},{276,88,42,9},
  {273,88,40,7},{269,84,41,13},{256,72,40,11},
  {261,79,35,17},{250,78,37,21},{281,90,41,6},
  {264,84,36,11},{263,78,37,14},{271,82,41,9},
  {266,86,34,10},{260,80,38,12},{258,70,34,15},
  {218,76,23,27},{264,82,33,16},{256,80,36,16},
  {274,86,38,8},{272,86,43,7}};
```

```
mtestdata={Transpose[mathtestinfo][[2]],
  Transpose[mathtestinfo][[3]],
  Transpose[mathtestinfo][[4]],
  Transpose[mathtestinfo][[1]]}//
  Transpose
```

```
{{78,34,18,252},{73,41,12,259},{77,28,20,256},
 {68,42,11,256},{85,38,9,267},{86,43,12,270},
 {83,32,18,261},{76,24,33,231},{73,31,19,255},
 {80,36,17,258},{64,32,20,231},{69,36,23,251},
 {84,48,7,272},{82,43,14,260},{84,37,11,267},
 {88,43,8,278},{78,36,14,256},{76,36,19,246},
 {83,34,19,260},{84,31,14,264},{88,36,7,276},
 {88,44,6,280},{88,42,9,276},{88,40,7,273},
 {84,41,13,269},{72,40,11,256},{79,35,17,261},
 {78,37,21,250},{90,41,6,281},{84,36,11,264},
 {78,37,14,263},{82,41,9,271},{86,34,10,266},
 {80,38,12,260},{70,34,15,258},{76,23,27,218},
 {82,33,16,264},{80,36,16,256},{86,38,8,274},
 {86,43,7,272}}
```

Next, we use ordinary least squares to determine the prediction equation. The output indicates that it is $\hat{y} = 119.611 + 0.780427x_1 + 0.401181x_2 - 1.15647x_3$. We also list the residuals with `FitResiduals` and the fitted values with `PredictedResponse`. We name the result `rg` so that we can then extract the residuals, naming the list `res1`.

```
rg=Regress[mtestdata,{1,x1,x2,x3},{x1,x2,x3},
  RegressionReport->
  {BestFit,FitResiduals,PredictedResponse}]
```

```
{BestFit→199.611+0.780427 x1+0.401181 x2-1.15647 x3,
 FitResiduals→{-1.30771,-0.152687,8.19275,-0.808205,
  -3.78369,-0.100605,4.59251,0.612052,7.95444,1.17259,
  -8.26642,9.69613,-4.32802,-4.66595,-0.289135,1.71265,
  -2.73597,-5.39275,3.94662,2.58737,1.36444,0.99852,
  1.2703,-3.24028,2.41909,-3.12755,5.3542,-1.04184,
  1.64121,-2.88795,3.86285,1.35405,-2.80292,-3.41213,
  7.46628,-18.9256,5.65881,-1.98388,1.27941,-3.88297},
 PredictedResponse→{253.308,259.153,247.807,256.808,
  270.784,270.101,256.407,230.388,247.046,256.827,
  239.266,241.304,276.328,264.666,267.289,276.287,
  258.736,251.393,256.053,261.413,274.636,279.001,
  274.73,276.24,266.581,259.128,255.646,251.042,
  279.359,266.888,259.137,269.646,268.803,263.412,
  250.534,236.926,258.341,257.984,272.721,275.883}}
```

```
res1=FitResiduals/.rg
```

```
{-1.30771,-0.152687,8.19275,-0.808205,-3.78369,
 -0.100605,4.59251,0.612052,7.95444,1.17259,-8.26642,
 9.69613,-4.32802,-4.66595,-0.289135,1.71265,-2.73597,
 -5.39275,3.94662,2.58737,1.36444,0.99852, 1.2703,
 -3.24028,2.41909,-3.12755,5.3542,-1.04184, 1.64121,
 -2.88795,3.86285,1.35405,-2.80292,-3.41213, 7.46628,
 -18.9256,5.65881,-1.98388,1.27941,-3.88297}
```

We define the function `scalefactor` so that when given a list of residuals, `elist`, *Mathematica* will compute the value of the scale factor,

$$\frac{1}{5745}\text{median}\left\{\left|e_i - \text{median}\{e_i\}\right|\right.$$

We find that the scale factor for the residuals in `res1` is 4.59772 and name this result `scalefac1`. Next, we define the function `huberW` to find the weights according to the Huber weight function when given a residual `e` and a scale factor `sf`. We create a list of the weights using `Table` and name the result `hwghts1`. Notice that we must load the **DescriptiveStatistics** package in order to use the `Median` command in the definition of `scalefactor`.

```
<<Statistics`DescriptiveStatistics`
```

```
Clear[scalefactor]
  scalefactor[elist_]:=Module[{newlist,med,i},
  med=Median[elist];newlist=Table[Abs[elist[[i]]-med],
  {i,1,Length[elist]}];1/0.6745 Median[newlist]]
```

```
scalefac1=scalefactor[res1]
```

```
4.59772
```

```
huberW[e_,sf_]:=Module[{u},
  u=e/sf;
  If[Abs[u]<=1.345,1,1.345/Abs[u]]]
```

```
hwghts1=Table[huberW[res1[[j]],scalefac1],
  {j,1,Length[res1]}]
```

```
{1,1,0.754805,1,1,1,1,1,0.777419,1,0.748078,
 0.637773,1,1,1,1,1,1,1,1,1,1,1,1,1,1,1,
 1,1,1,1,1,1,0.828247,0.326749,1,1,1,1}
```

We use the weights in hwghts1 to determine a new prediction equation by using Regress with the option setting Weights->hwghts1. In this case, the prediction equation is $\hat{y} = 202.756 + 0.823476x_1 + 0.223465x_2 - 1.14479x_3$. Note the changes in the estimated regression coefficients from those obtained with equal weights earlier. The residuals and predicted values are also different from those obtained with ordinary least squares.

```
iteration1=Regress[mtestdata,{1,x1,x2,x3},{x1,x2,x3},
  Weights->hwghts1,RegressionReport->
  {BestFit,FitResiduals,PredictedResponse}]
```

```
{BestFit→202.756+0.823476 x1+0.223465 x2-1.14479 x3,
 FitResiduals→{-1.97842,0.70596,6.47543,0.455085,
  -3.93974,0.553839,3.35113,-1.92494,6.95415,0.782908,
  -8.71324,9.7099,-2.64049,-3.86267,-0.603212,2.32772,
  -3.00451,-5.6336,3.04899,1.17195,0.747181,1.81467,
  1.69598,-3.14668,2.79251,-2.39189,4.82985,-1.21444,
  1.83811,-3.37975,3.77202,1.8603,-3.72456,-3.38798,
  7.17502,-21.5702,4.66156,-2.36188,1.092,-3.17012},
 PredictedResponse→{253.978,258.294,249.525,255.545,
  270.94,269.446,257.649,232.925,248.046,257.217,
  239.713,241.29,274.64,263.863,267.603,275.672,
  259.005,251.634,256.951,262.828,275.253,278.185,
  274.304,276.147,266.207,258.392,256.17,251.214,
  279.162,267.38,259.228,269.14,269.725,263.388,
  250.825,239.57,259.338,258.362,272.908,275.17}}
```

We follow a similar procedure to obtain the residuals from the first iteration, called res2, and to determine the scale factor for this list, named scalefac2. We also compute the weights and call this result hwghts2. The second iteration of the weighted least squares method gives the prediction equation $\hat{y} = 205.353 + 0.813304x_1 + 0.182479x_2 - 1.15715x_3$. Although to a lesser degree, the estimated regression coefficients, residuals, and predicted values differ from those obtained in the first iteration.

```
res2=FitResiduals/.iteration1;
```

```
scalefac2=scalefactor[res2]
```

```
4.4578
```

```
hwghts2=Table[huberW[res2[[j]],scalefac2],
  {j,1,Length[res2]}]
```

```
{1,1,0.925922,1,1,1,1,1,0.862182,1,0.688119,
 0.617488,1,1,1,1,1,1,1,1,1,1,1,1,1,1,1,
 1,1,1,1,1,1,0.835642,0.277964,1,1,1,1}
```

```
iteration2=Regress[mtestdata,{1,x1,x2,x3},{x1,x2,x3},
  Weights->hwghts2,RegressionReport->
  {BestFit,FitResiduals,PredictedResponse}]
```

```
{BestFit→205.353+0.813304 x1+0.182479 x2-1.15715 x3,
 FitResiduals→{-2.16585,0.680424,6.05663,0.407315,
  -4.00323,0.742517,3.13259,-2.35721,6.60526,0.685438,
  -9.10034,9.57468,-2.32902,-3.68997,-0.693152,2.48731,
  -3.1594,-5.74705,2.92478,0.873168,0.607513,1.99053,
  1.82694,-3.1224,2.89123,-2.48094,4.68122,-1.24183,
  1.91136,-3.51067,3.65812,1.88924,-3.92947,-3.46527,
  6.86913,-22.1176,4.44912,-2.47171,1.02631,-3.04323},
 PredictedResponse→{254.166,258.32,249.943,255.593,
  271.003,269.257,257.867,233.357,248.395,257.315,
  240.1,241.425,274.329,263.69,267.693,275.513,
  259.159,251.747,257.075,263.127,275.392,278.009,
  274.173,276.122,266.109,258.481,256.319,251.242,
  279.089,267.511,259.342,269.111,269.929,263.465,
  251.131,240.118,259.551,258.472,272.974,275.043}}
```

We perform the necessary calculations for a third iteration of the procedure. Notice that the estimated regression coefficients, residuals, and predicted values begin to converge. We can continue this procedure until a desired level of convergence is reached.

```
res3=FitResiduals/.iteration2;
```

```
scalefac3=scalefactor[res3]
```

```
4.4591
```

```
hwghts3=Table[huberW[res3[[j]],scalefac3],
  {j,1,Length[res3]}]
```

```
{1,1,0.990236,1,1,1,1,1,0.907988,1,0.659041,
 0.626391,1,1,1,1,1,1,1,1,1,1,1,1,1,1,1,1,
 1,1,1,1,1,1,0.873107,0.271163,1,1,1,1}
```

```
iteration3=Regress[mtestdata,{1,x1,x2,x3},{x1,x2,x3},
  Weights->hwghts3,RegressionReport->
  {BestFit,FitResiduals,PredictedResponse}]
```

```
{BestFit→206.734+0.803853 x1+0.169497 x2-1.16428 x3,
 FitResiduals→{-2.2401,0.607029,5.90929,0.292519,
  -4.02355,0.81795,3.07963,-2.47327,6.45193,0.648928,
  -9.31861,9.47697,-2.24321,-3.63808,-0.721643,2.55314,
  -3.2362,-5.80711,2.90492,0.788168,0.575334,2.05509,
```

```
 1.88691,-3.10265,2.92893,-2.5839,4.62228,1.25575,
 1.95587,-3.55215,3.59431,1.87952,-3.98514,-3.51145,
 6.69789,-22.2894,4.38543,-2.51535,1.00832,-3.00344},
PredictedResponse→{254.24,258.393,250.091,255.707,
 271.024,269.182,257.92,233.473,248.548,257.351,
 240.319,241.523,274.243,263.638,267.722,275.447,
 259.236,251.807,257.095,263.212,275.425,277.945,
 274.113,276.103,266.071,258.584,256.378,251.256,
 279.044,267.552,259.406,269.12,269.985,263.511,
 251.302,240.289,259.615,258.515,272.992,275.003}}
```

■

Random Sampling and Independence

See the discussion in Chapter 10.

Linearity/Additivity

As discussed in Section 11.5, Diagnostic Procedures, in the subsection on linearity/additivity, linear regression methodology can be used to fit models that involve independent variables that are related to the dependent variable in a nonlinear way as long as the model is additive. The challenge is to develop an additive model that represents the relationships between the dependent variable and the independent variable(s). If this is not possible, an alternative approach would be to use the package **NonLinearFit**, which is found in the **Statistics** folder (or directory) and is discussed later in this chapter. The following discussion will focus on transformations for an independent variable. Transforming the dependent variable can influence constant variance and normality requirements. In multiple regression, the same or different transformations may be required for different independent variables. (Note that in multiple regression, curvature between the dependent variable (y) and an independent variable (x_i) may be difficult to detect using a scatter plot of y against x_i or the residuals against x_i, because multicolinearity can cause such plots to be misleading.)

Remedy 1

If a scatter plot between the dependent variable (y) and an independent variable (x) has the form shown in Fig. 11.3, then consider the following remedies to linearize the regression relationship.

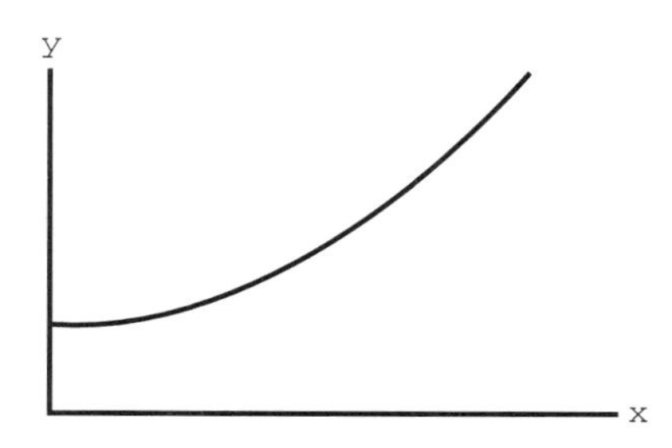

Figure 11.3

1. Include both x and x^2 as predictors in the model.
2. Transform x with one of the following:

$$x' = x^2$$
$$x' = e^x.$$

Several alternatives may be tried and evaluated using scatter plots of residuals against the variable. The remedy that appears most effective is then used. Note that the graph in Fig. 11.3 could reflect a relationship of the form

$$y = \beta_0 x^{\beta_1}.$$

Taking the logarithm on both sides of the equation results in a linear equation in $\log y$ and $\log x$,

$$\log y = \log \beta_0 + \beta_1 \log x.$$

First, if the transformations $y' = \log y$ and $x' = \log x$ are to be tried, we must determine whether the transformation on y will affect the variability or normality assumptions. Second, we must be sure that the model will be additive. If the model for the relationship between x and y is

$$y = \beta_0 x^{\beta_1} \varepsilon,$$

then the transformation will result in an additive model,

$$\log y = \log \beta_0 + \beta_1 \log x + \log \varepsilon.$$

If the effect of other influences on y, which is represented by ε (the error term), is not multiplicative in the first model, it may not be additive in the second model. The additivity of the ultimate regression equation will be reflected by a random scatter of the residuals when plotted, for example, against the predicted values. The third point is that there is usually a way to transform only x as opposed to both x and y. In the preceding example, $x' = x^2$ or some other power may adequately represent the relationship between y and x.

Remedy 2

If the scatter plot between the dependent variable (y) and an independent variable (x) has the form in Fig. 11.4, then consider the transformations

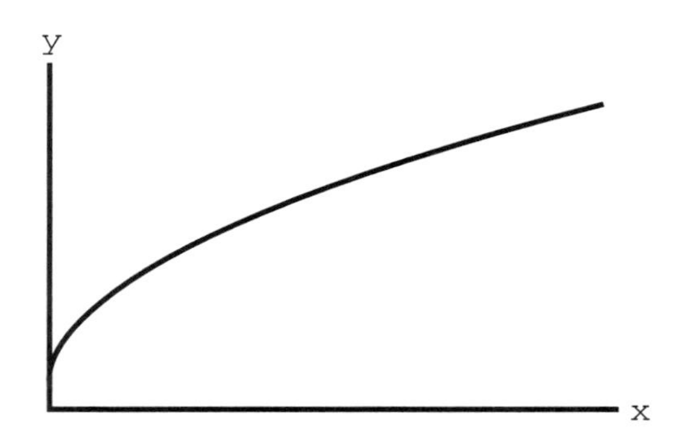

Figure 11.4

$$x' = \sqrt{x}$$
$$x' = e^{-1/x}$$
$$x' = \log x.$$

(Note that logarithms to base e or base 10 work equally well.)

Several alternatives may be tried and evaluated using scatter plots of the residuals against the variable. The remedy that appears most effective is then used.

Remedy 3

If the scatter plot between the dependent variable (y) and an independent variable (x) has the form shown in Fig. 11.5, then consider the transformations

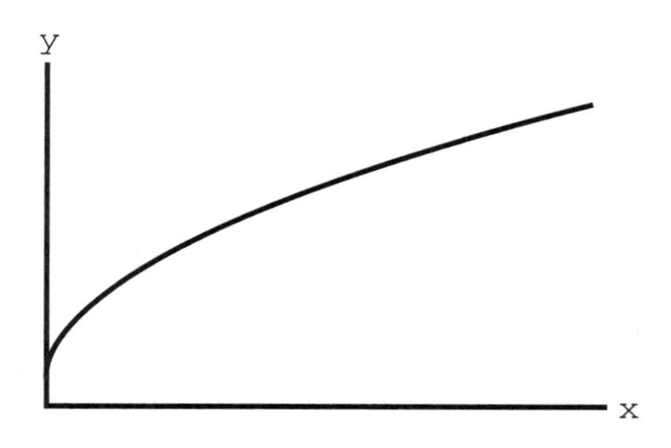

Figure 11.5

$$x' = 1/x$$

$$x' = e^{-x}$$

$$x' = e^{1/x}.$$

Several alternatives may be tried and evaluated using scatter plots of the residuals against the variable. The remedy which appears most effective is then used.

Equality of Variance

Transformations

Unequal variances and nonnormality can occur together, and a transformation on the dependent variable will frequently help solve both problems. The transformation may also linearize a relationship with an independent variable. On the other hand, the transformation on the dependent variable may necessitate a simultaneous transformation on an independent variable to obtain or maintain a linear relationship with the transformed dependent variable. The type of transformation to use to stabilize variance may be suggested by the pattern of variability. To investigate the relationship between variance or standard deviation and the predicted values, use squared residuals to represent variance and the absolute value of residuals to represent standard deviation. See the discussion in the corresponding subsection of Chapter 10 for an idea of how to evaluate the relationships between variance and predicted values (i.e., means) or standard deviation and predicted values (i.e., means) using ratios.

Here are some situations and suggested transformations.

Square Root

Variance is an increasing function of the predicted values characterized by a "fanning out" pattern of the residuals in their scatter plot against predicted values:

$$y' = \sqrt{y},$$

Logarithm

Standard deviation is an increasing function of the predicted values,

$$y' = \log y,$$

where logarithms to base e or base 10 are equally useful.

Reciprocal

Standard deviation is an increasing function of the squared predicted values:

$$y' = 1/y.$$

Square

Standard deviation is a decreasing function of the predicted values:

$$y' = y^2.$$

EXAMPLE 2: Table 11.5 gives the number of hours spent studying outside of class per week as well as the corresponding score on a 200-point final exam for 30 students in a college statistics class. Use a transformation to obtain a regression function to predict the final exam score based on the number of hours spent on the course.

t	*S*	*t*	*S*	*t*	*S*	*t*	*S*	*t*	*S*	*t*	*S*
0.5	40	0.5	50	1.0	75	1.0	80	1.5	80	1.5	95
2.0	100	2.0	90	2.5	114	2.5	103	2.5	101	3.0	116
3.0	120	3.5	123	4.0	138	4.0	133	4.5	146	5.0	152
5.0	147	5.5	157	5.5	164	6.0	167	6.0	162	6.5	164
7.0	173	7.0	179	8.0	186	8.0	193	9.0	190	9.0	180

Table 11.5 Number of Hours Spent on Class and Score on 200 Point Final Exam

SOLUTION: We begin by entering the **LinearRegression** package as well as the data in `gradedata`. Next, using `ListPlot` we graph a scatter plot of the information in `gradedata`, naming the graph `points`.

```
<<Statistics`LinearRegression`

gradedata={{0.5,40},{0.5,50},{1,75},{1,80},{1.5,80},
  {1.5,95},{2,100},{2,90},{2.5,114},{2.5,103},
  {2.5,101},{3,116},{3,120},{3.5,123},{4,138},
  {4,133},{4.5,146},{5,152},{5,147},{5.5,157},
  {5.5,164},{6,167},{6,162},{6.5,164},{7,173},
  {7,179},{8,186},{8,193},{9,190},{9,180}};
```

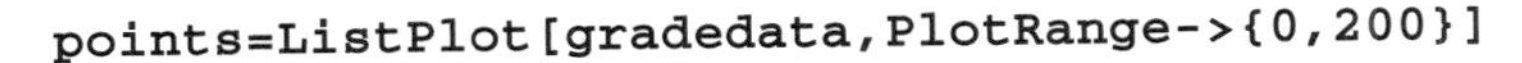

```
points=ListPlot[gradedata,PlotRange->{0,200}]
```

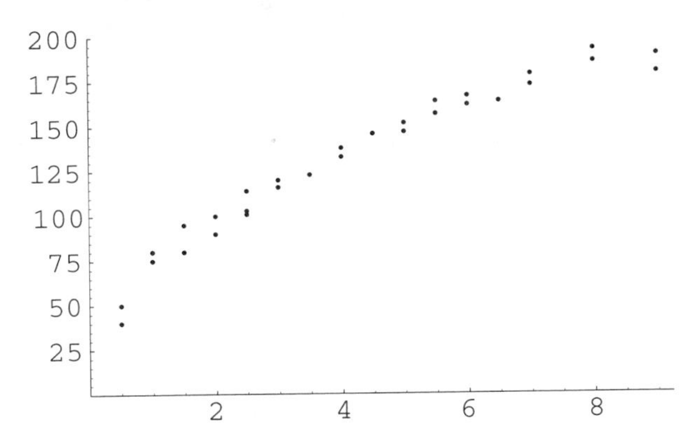

Because the scatter plot of the data is similar to the graph in Fig. 11.4, we select the transformation $x' = \sqrt{x}$. We apply this transformation to the first coordinate of each data point in the list `gradedata` and name this list of points `translateX`. When we graph the scatter plot for the points in `translateX`, we observe that the data appear more linear than that in `points`.

```
translateX=Table[{Sqrt[gradedata[[i,1]]]//N,
  gradedata[[i,2]]},{i,1,Length[gradedata]}]
```

```
{{0.707107,40},{0.707107,50},{1.,75},{1.,80},
 {1.22474,80},{1.22474,95},{1.41421,100},
 {1.41421,90},{1.58114,114},{1.58114,103},
 {1.58114,101},{1.73205,116}, {1.73205,120},
 {1.87083,123},{2.,138},{2.,133},{2.12132,146},
 {2.23607,152},{2.23607,147},{2.34521,157},
 {2.34521,164},{2.44949,167},{2.44949,162},
 {2.54951,164},{2.64575,173},{2.64575,179},
 {2.82843,186}, {2.82843,193},{3.,190},{3.,180}}
```

```
transpoints=ListPlot[translateX,PlotRange->{{0,3},{0,200}}]
```

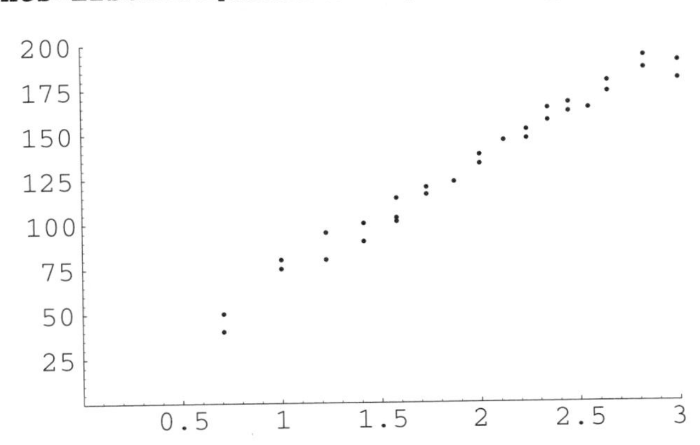

Using the variable name xprime with Regress, we obtain the regression function for the transformed data $\hat{y} = 8.78096 + 62.5235x'$. We graph this function in rgplot and suppress the output initially. We then show the graph with the data in transpoints.

```
rg=Regress[translateX,{1,xprime},xprime,
  RegressionReport->BestFit]
```

```
{BestFit→8.78096+62.5235 xprime}
```

```
rgplot=Plot[rg[[1,2]],{xprime,0,3},
  DisplayFunction->Identity];
```

```
Show[transpoints,rgplot,
  DisplayFunction->$DisplayFunction]
```

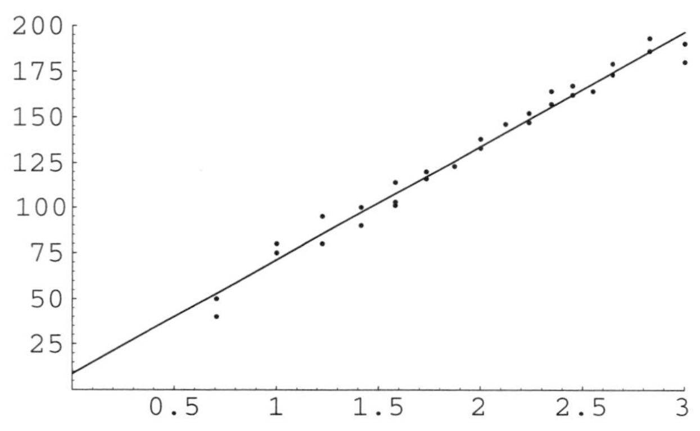

We may convert back to the original (untranslated) variable by replacing xprime in the linear regression equation with $\sqrt{x}$. We do this is the following statement where rg[[1,2]] extracts the expression 8.78096+62.5235 xprime from the output list of rg. Then, /.xprime->Sqrt[x] replaces xprime with $\sqrt{x}$ in the regression equation. In newplot, we graph the equation $y = 8.78096 + 62.5235\sqrt{x}$ over the interval $0 \le x \le 9$. (Note that the graph is suppressed with the option setting DisplayFunction->Identity.) We then use the option setting DisplayFunction->$DisplayFunction to display the graph produced in newplot with the data points (graphed earlier in dots) to observe the accuracy of the fit.

```
newrg=rg[[1,2]]/.xprime->Sqrt[x]
```

$8.78096183169448174 + 62.5234628646058432\sqrt{x}$

```
newplot= Plot[newrg,{x,0,9},DisplayFunction->Identity];
```

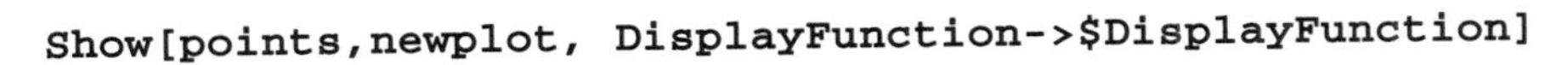

```
Show[points,newplot, DisplayFunction->$DisplayFunction]
```

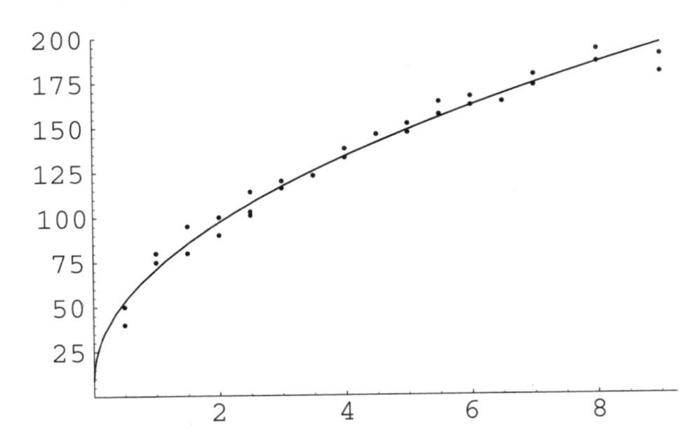

Next, we try to determine if the tranformation affects the normality of the data. We find the residuals of the fit of the transformed data using the option setting `RegressionReport->FitResiduals`, naming the list `res`. After making sure that the our procedure **normalProbability** is loaded, we graph the normal probability plot. The plot does not indicate a departure from normality.

```
res=Regress[translateX,{xprime,1},xprime,
  RegressionReport->FitResiduals][[1,2]];Short[res,2]
```

```
{-12.9917,-2.99173,«27»,-16.3514}
```

```
nplot=normalProbability[res]
```

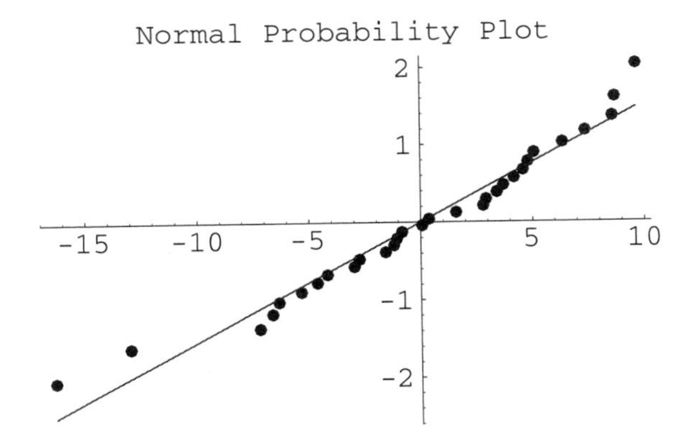


```
Pearson'sCorrelationCoefficient ->0.98064

PValue -> > 0.250"
```

■

Box–Cox Method

The Box–Cox procedure is discussed in Chapter 10. It is a method for selecting a transformation from those of the form $y' = y^{\lambda}$. In addition to the procedure to perform the Box–Cox method described there, we have written a similar procedure for regression called **boxCoxRegression**. The arguments for **boxCoxRegression** are the same as those used with Regress.

> boxCoxRegression[data,pred,indepvars,opts] displays a value for the Box–Cox lambda, a confidence coefficient, and a confidence interval. The options and their default values are limits->{-2,2}, gridsize->20, and confidence->0.95. The limits determine the range of possible values for the Box–Cox lambda. The grid size determines the number of values that are considered at equal intervals along the range of possible values.

To use **boxCoxRegression**, first load the disk that comes with this book and click on its icon. Next, locate the procedure name and click on the rightmost cell bracket. Then press **ENTER** and return to the *Mathematica* work sheet.

EXAMPLE 3: Use the Box–Cox transformation to determine the best transformation for fitting the data in Table 11.5.

SOLUTION: After making sure that the the function **boxCoxRegression** is loaded, we enter the data in gradedata. The result of **boxCoxRegression** indicates that the best transformation is $Y' = Y^2$, where Y is the dependent variable. This is equivalent to taking the square root of the independent variable which was used in the earlier example. Because the value of lambda was at the upper limit of the values specified for consideration, we change the limits and re-run the procedure. The choice of lambda is still seen to be appropriate. Note that the values of the confidence limits are dependent on both the the limits and the grid size.

```
gradedata={{0.5,40},{0.5,50},{1,75},{1,80},{1.5,80},
  {1.5,95},{2,100},{2,90},{2.5,114},{2.5,103},
  {2.5,101},{3,116},{3,120},{3.5,123},{4,138},
  {4,133},{4.5,146},{5,152},{5,147},{5.5,157},
  {5.5,164},{6,167},{6,162},{6.5,164},{7,173},
  {7,179},{8,186},{8,193},{9,190},{9,180}};

boxCoxRegression[gradedata,{1,x},x]

Box-Cox Transformation

{BoxCoxλ→2.,Confidence→95. %,
  ConfidenceInterval→{1.78947,2.}}
```

```
boxCoxRegression[gradedata,{1,x},x,limits->{1.5,3.0}]
Box-Cox Transformation
{BoxCoxλ→2.13158,Confidence→95. %,
  ConfidenceInterval→{1.81579,2.52632}}
```

■

Weighted Least Squares

Weighted least squares can be used as an alternative to transformation when a model has been found that is appropriate except for unequal variance. A transformation under this circumstance could cause departures from other assumptions. To use weighted least squares, a weight must be determined for each case which, in view of unequal variation, should be inversely proportional to the variance of the case.

Method I

One way to determine weights involves estimating variability using repeated observations called replications. If there are, say, six or more cases with the same value(s) for the predictor variable(s), the weight for each such case will be the reciprocal of the sample variance of the dependent variable values, y_i, for these cases. If there are not enough cases with the same value(s) of the predictor variable(s), then groups of cases with values of the predictor variable(s) which are close can be utilized. For each group, the sample variance of the y_i's will be used to weight each case in the group. If we denote the sample variance by $\hat{s}^2$, then the weight $w_i = 1/\hat{s}^2$ will be used for each case in the group.

EXAMPLE 1: Consider the data in Table 11.6 concerning the relationship between age and diastolic blood pressure among 54 healthy women varying between the ages of 20 and 60. Divide the data into eight groups corresponding to the ages 20–24, 25–29, 30–34, 35–39, 40–44, 45–49, 50–54, and 55–59. For each group, calculate a sample variance of the blood pressure and determine the reciprocal of this value to find weights for use in weighted least squares.

Age	BP	Age	BP	Age	BP	Age	BP	Age	BP
27	73	28	67	30	73	44	71	52	86
21	66	26	79	31	80	46	80	53	79
22	63	38	91	37	68	47	96	56	92
24	75	32	76	39	75	45	92	52	85
25	71	33	69	46	89	49	80	50	71
23	70	31	66	49	101	48	70	59	90

Continued

20	65	34	73	40	70	40	90	50	91
20	70	37	78	42	72	42	85	52	100
29	79	38	87	43	80	55	76	58	80
24	72	33	76	46	83	54	71	57	109
25	68	35	79	43	75	57	99		

Table 11.6

SOLUTION: In order to divide the data into groups based on age, we define the function `partitionList` which, when given a data list (`valList`) and a list of upper limits (`cutOffs`) of the division intervals, separates the data into the appropriate groups.

```
partitionList[valList_,cutOffs_]:=Module[{},
  v=Table[{0},{k,1,Length[cutOffs]}];
  iLoop[i_]:=Module[{},
  j=1;
  testQ=False;
  While[And[j<=Length[cutOffs],testQ==False],
  jLoop[i,j];j++]];jLoop[i_,j_]:=
   Module[{},If[valList[[i,1]]<=cutOffs[[j]],
    AppendTo[v[[j]],valList[[i]]]];
    If[valList[[i,1]]<=cutOffs[[j]],testQ=True]];
  Do[iLoop[i],{i,1,Length[valList]}];
  dropZero[j_]:=Drop[v[[j]],1];
  newV=Table[dropZero[k],{k,1,Length[v]}]]
```

We then enter the data in `bpdata` and the upper limits on the intervals 20–24, 25–29, 30–34, 35–39, 40–44, 45–49, 50–54, and 55–59 in `testVals`, and we use partitionList to divide the data in `groups`. (Notice that the output of `partitionList` is a list of sublists where each sublist corresponds to an age interval.)

```
bpdata={{27,73},{21,66},{22,63},{24,75},{25,71},{23,70},
  {20,65},{20,70},{29,79},{24,72},{25,68},{28,67},
  {26,79},{38,91},{32,76},{33,69},{31,66},{34,73},
  {37,78},{38,87},{33,76},{35,79},{30,73},{31,80},
  {37,68},{39,75},{46,89},{49,101},{40,70},{42,72},
  {43,80},{46,83},{43,75},{44,71},{46,80},{47,96},
  {45,92},{49,80},{48,70},{40,90},{42,85},{55,76},
  {54,71},{57,99},{52,86},{53,79},{56,92},{52,85},
  {50,71},{59,90},{50,91},{52,100},{58,80},{57,109}};

testVals={24,29,34,39,44,49,54,59};
```

```
groups=partitionList[bpdata,testVals]

{{{21,66},{22,63},{24,75},{23,70},{20,65},{20,70},{24,72}},
 {{27,73},{25,71},{29,79},{25,68},{28,67},{26,79}},
 {{32,76},{33,69},{31,66},{34,73},{33,76},{30,73},{31,80}},
 {{38,91},{37,78},{38,87},{35,79},{37,68},{39,75}},
 {{40,70},{42,72},{43,80},{43,75},{44,71},{40,90},{42,85}},
 {{46,89},{49,101},{46,83},{46,80},{47,96},{45,92},{49,80},
  {48,70}},
 {{54,71},{52,86},{53,79},{52,85},{50,71},{50,91},{52,100}},
 {{55,76},{57,99},{56,92},{59,90},{58,80},{57,109}}}
```

We load the **DescriptiveStatistics** package so that we may use `Variance` to compute the sample variance of the blood pressure (the second entry of each element of `bpdata`) of each sublist in `groups` in `varTable`. Then in `weights`, we take the reciprocal of each value in `VarTable` to determine the weights. In order to assign the appropriate weight to each element of `bpdata`, we define the function `assignToElement`, which assigns weights to elements of the sorted list of data in `groups`. The list that results in `listOfWeights` is flattened so that we do not see inner parentheses as we did in `groups`. In the next example, we will use these weights to perform a weighted least squares procedure.

```
<<Statistics`DescriptiveStatistics`

varTable=Table[Variance[groups[[j,2]]]//N,
  {j,1,Length[groups]}]

{840.5,1058.,648.,840.5,450.,1352.,578.,882.}

weights=Table[1/varTable[[j]],{j,1,Length[newV]}]

{0.00118977,0.00094518,0.00154321,0.00118977,
 0.00222222,0.000739645,0.0017301,0.00113379}

assignToElement[w_,partList_]:=Module[{},
  subList[j_]:=Table[w[[j]],
  {k,1,Length[partList[[j]]]}];
  weightList=Table[subList[j],{j,1,Length[partList]}];
  weightList]

listOfWeights=assignToElement[weights,groups]//Flatten

{0.00118977,0.00118977,0.00118977,0.00118977,0.00118977,
 0.00118977,0.00118977,0.00094518,0.00094518,0.00094518,
 0.00094518,0.00094518,0.00094518,0.00154321,0.00154321,
 0.00154321,0.00154321,0.00154321,0.00154321,0.00154321,
 0.00118977,0.00118977,0.00118977,0.00118977,0.00118977,
 0.00118977,0.00222222,0.00222222,0.00222222,0.00222222,
```

```
0.00222222,0.00222222,0.00222222,0.000739645,0.000739645,
0.000739645,0.000739645,0.000739645,0.000739645,
0.000739645,0.000739645,0.0017301,0.0017301,0.0017301,
0.0017301,0.0017301,0.0017301,0.0017301,0.00113379,
0.00113379,0.00113379,0.00113379,0.00113379,0.00113379}
```

■

Method II

A second, more sophisticated determination of weights is accomplished using the following steps.

1. Calculate the regression equation using unweighted linear regression, and plot the residuals against each predictor variable as well as against the predicted values. A residual plot against a predictor that exhibits a "fanning out" or "funneling in" pattern identifies that predictor as contributing to unequal variance.
2. Determine which predictors are responsible for departure from constant variance and use them in step 3. If no predictor variables are identified, but the residual plot against the predicted values exhibits a "fanning out" or "funneling in" pattern, then use the predicted values in step 3.
3. Calculate a prediction equation using unweighted linear regression of the absolute residules $|e_i|$ against the predictors (or predicted values) identified in step 2.
4. Use the predicted values (`RegressionReport` option setting `PredictedResponse`) from the equation developed in step 3 to estimate weights for the weighted least squares analysis. The ith predicted value is used to estimate a standard deviation, call it $\hat{s}_i$, corresponding to the ith case, and the corresponding weight is calculated using $w_i = 1/(\hat{s}_i)^2$.

Notes:

1. This method will fail if when calculating the estimated standard deviation $\hat{s}_i$ a negative value is obtained.
2. If the pattern of changing variability appears to be better explained (more extreme) using squared residuals, they should be used in place of absolute residuals in step 3. When evaluating residual plots using squared residuals, e_i^2, the "fanning out" pattern is represented by increasingly larger squared residuals. The "funneling in" pattern is represented by decreasing squared residuals. If the regression in step 3 used squared residuals then the ith predicted value is used to estimate a variance, call it $\hat{s}_i^2$, and the corresponding weight is calculated using $w_i = 1/\hat{s}_i^2$.
3. If there are outliers in the data, it is best to use absolute residuals because the resulting regression is less affected by outliers than when squared residuals are used.
4. Diagnostics and remedial methods can be utilized on the regression equation in step 3 to make it as useful as possible.

5. If the modeling of the standard deviation using absolute residuals or of the variance using squared residuals and steps 1 through 3 is done well, it will be reflected in the weighted least squares regression by a value of the mean square error (*MSE*) close to 1.
6. One or two iterations using the residuals from the weighted least squares for the prior iteration to find weights for the next iteration will improve the final prediction equation.

EXAMPLE 2: Perform a weighted least squares regression using the data in Table 11.6 concerning the relationship between age and diastolic blood pressure among 54 healthy women varying between the ages of 20 and 60.

SOLUTION: We begin by loading **LinearRegression** and by entering the data in `bpdata`. We then generate a scatter plot of the data points. Notice that the scatter plot suggests a linear relationship between age and diastolic blood pressure. We perform linear regression using unweighted least squares and obtain the prediction equation $\hat{y} = 56.1569 + 0.580031x$ where x, represents age and y diastolic blood pressure.

```
<<Statistics`LinearRegression`

bpdata={{27,73},{21,66},{22,63},{24,75},{25,71},
  {23,70},{20,65},{20,70},{29,79},{24,72},{25,68},
  {28,67},{26,79},{38,91},{32,76},{33,69},{31,66},
  {34,73},{37,78},{38,87},{33,76},{35,79},{30,73},
  {31,80},{37,68},{39,75},{46,89},{49,101},{40,70},
  {42,72},{43,80},{46,83},{43,75},{44,71},{46,80},
  {47,96},{45,92},{49,80},{48,70},{40,90},{42,85},
  {55,76},{54,71},{57,99},{52,86},{53,79},{56,92},
  {52,85},{50,71},{59,90},{50,91},{52,100},{58,80},
  {57,109}};

ListPlot[bpdata,PlotRange->{{10,60},{60,110}}]
```

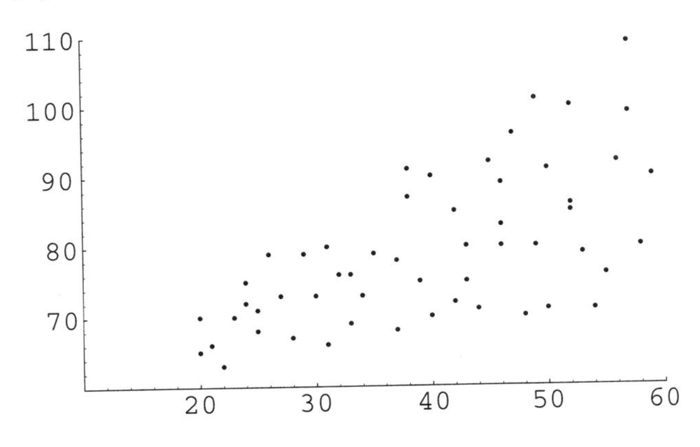

```
rgbp=Regress[bpdata,{1,x},x,RegressionReport->BestFit]
```

```
{BestFit→56.1569+0.580031 x}
```

Note the "fanning out" pattern in the plot. The same pattern will also be apparent in the residual plot. We calculate the residuals, naming the list `rs`. We then extract the first column from `bpdata` and name the result `ages`. In order to plot the residuals against the age data, we form the ordered pairs in `respoints2` and then use `ListPlot` to create the scatter plot.

```
rs=Regress[bpdata,{1,x},x,
  RegressionReport->FitResiduals][[1,2]]
```

```
{1.18224,-2.33758,-5.91761,4.92233,0.342301, 0.502362,
 -2.75755,2.24245,6.02218,1.92233,-2.6577,-5.39779,
 7.76227,12.8019,1.28209,-6.29795,-8.13788,-2.87798,
 0.381931,8.8019,0.702054,2.54199,-0.557853,5.86212,
 -9.61807,-3.77813,6.16165, 16.4216,-9.35816,-8.51822,
 -1.09825,0.161654,-6.09825,-10.6783,-2.83835,12.5816,
 9.74168,-4.57844,-13.9984,10.6418,4.48178,-12.0586,
 -16.4786,9.78132,-0.318531,-7.89856,3.36135,-1.31853,
 -14.1585,-0.378746,5.84153,13.6815,-9.79872,19.7813}
```

```
ages=Transpose[bpdata][[1]]
```

```
{27,21,22,24,25,23,20,20,29,24,25,28,26,38,32,33,31,34,
 37,38,33,35,30,31,37,39,46,49,40,42,43,46,43,44,46,47,
 45,49,48,40,42,55,54,57,52,53,56,52,50,59,50,52,58,57}
```

```
respoints2=Table[{ages[[j]],rs[[j]]},{j,1,Length[ages]}]
```

```
{{27,1.18224},{21,-2.33758},{22,-5.91761},{24,4.92233},
{25,0.342301},{23,0.502362},{20,-2.75755},{20,2.24245},
{29,6.02218},{24,1.92233},{25,-2.6577},{28,-5.39779},
{26,7.76227},{38,12.8019},{32,1.28209},{33,-6.29795},
{31,-8.13788},{34,-2.87798},{37,0.381931},{38,8.8019},
{33,0.702054},{35,2.54199},{30,-0.557853},{31,5.86212},
{37,-9.61807},{39,-3.77813},{46,6.16165},{49,16.4216},
{40,-9.35816},{42,-8.51822},{43,-1.09825},{46,0.161654},
{43,-6.09825},{44,-10.6783},{46,-2.83835},{47,12.5816},
{45,9.74168},{49,-4.57844},{48,-13.9984},{40,10.6418},
{42,4.48178},{55,-12.0586},{54,-16.4786},{57,9.78132},
{52,-0.318531},{53,-7.89856},{56,3.36135},{52,-1.31853},
{50,-14.1585},{59,-0.378746},{50,5.84153},{52,13.6815},
{58,-9.79872},{57,19.7813}}
```

```
ListPlot[respoints2,PlotRange->{{10,60},{-20,20}}]
```

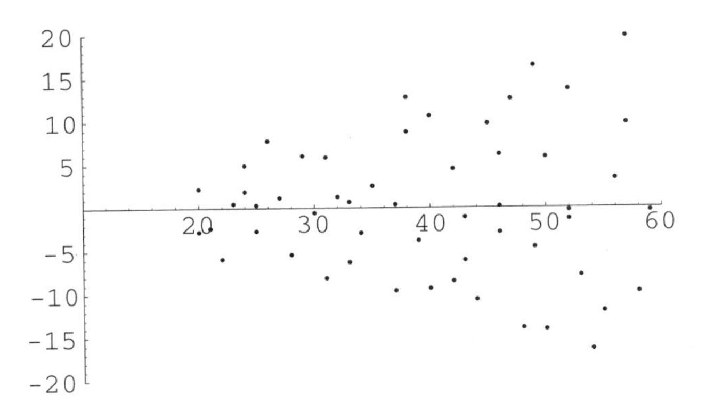

Next, we produce a scatter plot of the squared residuals against the age data. First, we square each residual in `rs` and name the result `sqres`. We continue in `sqrespoints` by forming ordered pairs with the first coordinate indicating age and the second the corresponding squared residual. Finally, we use `ListPlot` to display the scatter plot.

```
sqres=Table[rs[[j]]^2,{j,1,Length[rs]}]

{1.39769,5.46426,35.0181,24.2293,0.11717,0.252368,7.60406,
 5.0286,36.2666,3.69536,7.06337,29.1362,60.2528,163.889,
 1.64374,39.6641,66.2252,8.28275,0.145871,77.4734,0.49288,
 6.46173,0.3112,34.3644,92.5072,14.2743,37.966,269.668,
 87.5752,72.5601,1.20616,0.026132,37.1887,114.026,8.05621,
 158.297,94.9004,20.9621,195.955,113.249,20.0863,145.41,
 271.544,95.6741,0.101462,62.3873,11.2986,1.73852,200.462,
 0.143449,34.1235,187.183,96.0148,391.3}

sqrespoints=Table[{ages[[j]],sqres[[j]]},
 {j,1,Length[ages]}];

ListPlot[sqrespoints,PlotRange->{{10,60},{0,20}}]
```

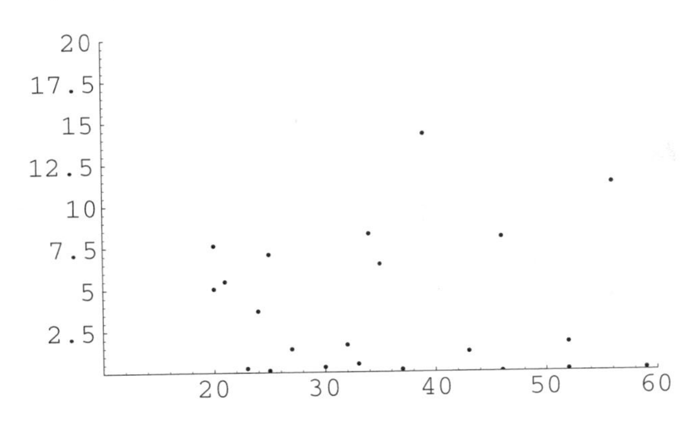

The pattern of changing variability appears to be better explained by the residual plot (as opposed to the plot of squared residuals), so we regress absolute residuals on age to determine a prediction equation to estimate standard deviations. This equation is $\hat{s} = -1.4948 + 0.198172x$, where $\hat{s}$ is the estimated standard deviation.

```
absres=Table[Abs[rs[[j]]],{j,1,Length[rs]}];

absrespoints= Table[{ages[[j]],absres[[j]]},
  {j,1,Length[ages]}];

absregress=Regress[absrespoints,{1,x},x,
  RegressionReport->BestFit]

{BestFit→-1.54948+0.198172 x}
```

We obtain a list of predicted values ($\hat{s}$), named `pred`, using `Regress` with the option setting `RegressionReport->PredictedResponse`. (Note that this procedure is equivalent to evaluating the prediction equation in `absregress` at each age in the list `ages`.) To construct weights of the form $w_i = 1/(\hat{s}_i)^2$, we define the function `f` and then use `Map` to apply `f` to each entry in `pred`. We name the resulting list of weights `wghts1`. Finally, we use `Regress` with the option setting `Weights->wghts1` to obtain the prediction equation $\hat{y} = 55.5658 + 0.596342x$ with the weighted least squares method. Note that the coefficients in this equation differ very little from those obtained with ordinary least squares. Notice that the MSE is approximately 1.47.

```
pred=absregress=Regress[absrespoints,{1,x},x,
  RegressionReport->PredictedResponse]

{PredictedResponse→{3.80117,2.61214,2.81031,3.20666,
  3.40483,3.00849,2.41397,2.41397,4.19752,3.20666,
  3.40483,3.99935,3.603,5.98107,4.79204,4.99021,
  4.59386,5.18838,5.7829,5.98107,4.99021,5.38655,
  4.39569,4.59386,5.7829,6.17924,7.56645,8.16097,
  6.37741,6.77376,6.97193,7.56645,6.97193,7.1701,
  7.56645,7.76462,7.36828,8.16097,7.96279,6.37741,
  6.77376,9.35,9.15183,9.74634,8.75548,8.95365,
  9.54817,8.75548,8.35914,10.1427,8.35914,
  8.75548,9.94452,9.74634}}

f[x_]:=1/x^2

wghts1=Map[f,pred[[1,2]]]

{0.0692093,0.146557,0.126617,0.0972512,0.0862599,
 0.110485,0.171608,0.171608,0.0567564,0.0972512,
 0.0862599,0.0625204,0.077032,0.0279539,0.0435472,
```

```
0.0401571,0.0473853,0.0371481,0.0299026,0.0279539,
0.0401571,0.034465,0.0517542,0.0473853,0.0299026,
0.0261896,0.0174669,0.0150147,0.0245873,0.0217942,
0.0205728,0.0174669,0.0205728,0.0194513,0.0174669,
0.0165867,0.0184191,0.0150147,0.0157714,0.0245873,
0.0217942,0.0114387,0.0119395,0.0105273,0.0130449,
0.0124738,0.0109688,0.0130449,0.0143112 0.00972062,
0.0143112,0.0130449,0.0101119,0.0105273}
```

```
rg1=Regress[bpdata,{1,x},x,Weights->wghts1,
  RegressionReport->{BestFit,ANOVATable}]
```

{BestFit→55.5658+0.596342 x,

ANOVATable→		DF	SumOfSq	MeanSq	FRatio	PValue
	Model	1	83.3408	83.3408	56.64	7.1868×10^{-10} }
	Error	52	76.5135	1.47141		
	Total	53	159.854			

Recall the weights determined when the data in bpdata was sorted in groups in Example 4. We sort the data in bpdata so that the weights in listOfWeights match the appropriate members of bpdata. Then, we use Regress with the weights in listOfWeights. Notice that the MSE is not close to 1.

```
sortList=Sort[bpdata]
{{20,65},{20,70},{21,66},{22,63},{23,70},{24,72},
 {24,75},{25,68},{25,71},{26,79},{27,73},{28,67},
 {29,79},{30,73},{31,66},{31,80},{32,76},{33,69},
 {33,76},{34,73},{35,79},{37,68},{37,78},{38,87},
 {38,91},{39,75},{40,70},{40,90},{42,72},{42,85},
 {43,75},{43,80},{44,71},{45,92},{46,80},{46,83},
 {46,89},{47,96},{48,70},{49,80},{49,101},{50,71},
 {50,91},{52,85},{52,86},{52,100},{53,79},{54,71},
 {55,76},{56,92},{57,99},{57,109},{58,80},{59,90}}
```

```
rg2=Regress[sortList,{1,x},x,Weights->listOfWeights,
  RegressionReport->{BestFit,ANOVATable}]
```

{BestFit→56.6537+0.550678 x,

ANOVATable→		DF	SumOfSq	MeanSq	FRatio	PValue
	Model	1	2.61085	2.61085	29.7756	1.36775×10^{-6} }
	Error	52	4.55958	0.0876842		
	Total	53	7.17043			

■

Normality

In view of the first paragraph under the previous subsection on equality of variance, we will limit our discussion to situations where normality is the only assumption that is not satisfied. The presence of outliers in an otherwise appropriate regression is such a situation, because this can be due to a departure from normality characterized by "heavy tails." That is, the distribution has a bell shape but has more area in the tails of the distribution than does a normal distribution. There is a greater likelihood with such a distribution of values that are distant from the mean (i.e., outliers). In this case, weighted least squares is an appropriate remedy. If normality is the only problem among the assumptions, and the departure is not too severe, it may be safely ignored because of the robustness of the inferential procedures. See the corresponding section in Chapter 10 for other ideas.

Multicolinearity

The simple solution to multicolinearity is to drop one or several predictors from the regression calculation. The idea is to include just one of a group of highly correlated predictors. For polynomial models this is not practical because such models contain one or more independent variables that are present in various powers. For example, the simplest polynomial model is a quadratic model in one independent variable:

$$y_i = \beta_0 + \beta_1 x_i + \beta_2 x_i^2 + \varepsilon_i.$$

This model has two predictors, x and x^2, which may be highly correlated. To avoid the problems associated with multicolinearity, a technique called centering can be used. For each case the value x_i is replaced with $x_i - \bar{x}$, where $\bar{x}$ is the sample mean of the x_i's. The predictors become $x - \bar{x}$ and $(x - \bar{x})^2$ and are called centering variables. In general, centering can reduce the intercorrelation among the powers of an independent variable and thereby avoid multicolinearity. Note that the predicted values and residuals for the regression in terms of the centering variables are the same as for the regression without centering. The estimated standard deviations of the regression coefficients obtained using the centering variables are not the same, however, as the estimated standard deviations of regression coefficients obtained using the original variables.

11.7 Nonlinear Regression

Up to this point we have focused on linear models which can be appropriate even when there is a non linear relationship between an independent variable and the dependent variable. The requirement which must be met is that the model be additive. There are situations in which the model is not additive, and it cannot be transformed into a form which can be fit using linear regression. For example, many problems in biology, thermodynamics, and radioactivity involve exponential growth or decay and require a model of the form $y = \beta_0 e^{\beta_1 x} + \varepsilon$. Because such a model is multiplicative in the β's and additive for the error term ε, it cannot be transformed into an additive model. To fit a set of data to such a model, we must find estimates of the regression parameters which minimize the sum of squared residuals. This necessitates the solution of a nonlinear system of equations using an iterative approach. One available iterative method is called the method of steepest descent or gradient method. It is available in *Mathematica* using the built-in function `FindMinimum`. This method converges slowly. Another method is the Gauss–Newton method which may converge more rapidly, but depending on the form of the model may encounter difficulties. For this reason, several modifications to the Guass–Newton method have been proposed. The method available in the **NonlinearFit** package (found in the **Statistics** folder/directory) is called the `LevenbergMarquardt` method. It combines the method of steepest descent and the Gauss–Newton method.

The **NonlinearFit** package contains two commands that are useful when working with nonlinear models.

`NonlinearFit[data,model,indepVariables,parameters]` fits the data to the model using the independent variables named in the list `indepVariables` and returns the model evaluated at the parameter values found with the least squares method.

`NonlinearRegress[data,model,indepVariables,parameters]` fits the data to the model using the independent variables named in the list `indepVariables` and returns the default list of the nonlinear regression diagnostics.

Options that can be used with both `NonlinearRegress` and `NonlinearFit` with default settings include `MaxIterations->30`, which specifies the maximum number of iterations to use in the search; `Method->LevenbergMarquardt`, which specifies the iterative method to be used; `ShowProgress->False`, which indicates if a report should be given at each step in the iterative process; `Weights->Automatic`, which gives a list of weights for the data points; `WorkingPrecision->$MachinePrecision`, which states the precision of the algorithm used in the solution process; `PrecisionGoal->Automatic`, which indicates the precision for the `ChiSquared` function; `AccuracyGoal->Automatic`, which specifies the accuracy desired for the `ChiSquared` function; `Tolerance->Automatic` which gives the numerical tolerance for certain matrix operations used in the `LevenbergMarquardt` method; and `Gradient->Automatic`, which lists the gradient functions used for the `FindMinimum` method.

EXAMPLE 1: A hot cup of tea is poured and the temperature measured at successive time increments. Obtain a least-squares estimate of the parameters that fit the nonlinear model $T = \beta_0 e^{\beta_1 t} + \varepsilon$. We use the data from Table 11.7.

t	T	t	T	t	T	t	T
0	70.86	1	68.71	2	66.67	3	64.73
4	63.25	5	61.57	6	60.25	7	58.74
8	57.6	9	56.17	10	54.94	11	53.82
12	52.64	13	51.7	14	50.64	15	49.81
16	48.85	17	48.04	18	47.24	19	46.45
20	45.8	21	45.03	22	44.27	23	43.64
24	43.01	25	42.27	26	41.78	27	41.05
28	40.57	29	39.97	30	39.37	31	38.9
32	38.31	33	37.84	34	37.37	35	36.71
36	36.33	37	35.79	38	35.41	39	34.98
40	34.53	41	34.18	42	33.81	43	33.39
44	33.05	45	32.72	46	32.38	47	32.04
48	31.82	49	31.48	50	31.15	51	30.92
52	30.59	53	30.63	54	30.03	55	29.81
56	29.59	57	29.36	58	29.14	59	28.81
60	28.59	61	28.09	62	28.25	63	28.03
64	27.81	65	27.7	66	27.48	67	27.37
68	27.15	69	27.04	70	26.82	71	26.7
72	26.69	73	26.37	74	26.26	75	26.15
76	26.04	77	25.82	78	25.71	79	25.71
80	25.49	81	25.38	82	25.27	83	25.16

84	24.95	85	24.95	86	24.94	87	24.73
88	24.62	89	24.51	90	24.39	91	24.28
92	24.17	93	24.17	94	24.06	95	23.95
96	23.84	97	23.73	98	23.73		

Table 11.7

SOLUTION: We begin by loading the **NonlinearFit** package and then entering the data in `temps`. We then form a new list, called `everyOtherTemps`, which includes every other data point, and we graph these points in `dots`. The scatter plot of the data points suggests an exponential function, so we attempt to fit the data to the nonlinear model $T = \beta_0 e^{\beta_1 t} + \varepsilon$. When we use **NonlinearFit** with only the default option settings, we obtain a less than desirable result, $T \approx 1$. The use of the option `MaxIterations->40` does not improve the approximation.

```
<<Statistics`NonlinearFit`

temps={{0,70.86},{1,68.71},{2,66.67},
  {3,64.73},{4,63.25},{5,61.57},{6,60.25},
  {7,58.74},{8,57.6},{9,56.17},{10,54.94},
  {11,53.82},{12,52.64},{13,51.7},{14,50.64},
  {15,49.81},{16,48.85},{17,48.04},{18,47.24},
  {19,46.45},{20,45.8},{21,45.03},{22,44.27},
  {23,43.64},{24,43.01},{25,42.27},{26,41.78},
  {27,41.05},{28,40.57},{29,39.97},{30,39.37},
  {31,38.9},{32,38.31},{33,37.84},{34,37.37},
  {35,36.71},{36,36.33},{37,35.79},{38,35.41},
  {39,34.98},{40,34.53},{41,34.18},{42,33.81},
  {43,33.39},{44,33.05},{45,32.72},{46,32.38},
  {47,32.04},{48,31.82},{49,31.48},{50,31.15},
  {51,30.92},{52,30.59},{53,30.63},{54,30.03},
  {55,29.81},{56,29.59},{57,29.36},{58,29.14},
  {59,28.81},{60,28.59},{61,28.09},{62,28.25},
  {63,28.03},{64,27.81},{65,27.7},{66,27.48},
  {67,27.37},{68,27.15},{69,27.04},{70,26.82},
  {71,26.7},{72,26.69},{73,26.37},{74,26.26},
  {75,26.15},{76,26.04},{77,25.82},{78,25.71},
  {79,25.71},{80,25.49},{81,25.38},{82,25.27},
  {83,25.16},{84,24.95},{85,24.95},{86,24.94},
  {87,24.73},{88,24.62},{89,24.51},{90,24.39},
  {91,24.28},{92,24.17},{93,24.17},{94,24.06},
  {95,23.95},{96,23.84},{97,23.73},{98,23.73}};
```

```
everyOtherTemps=Table[temps[[j]],{j,1,99,2}];

dots=ListPlot[everyOtherTemps,PlotRange->{0,70},
  PlotStyle→PointSize[0.015]]
```

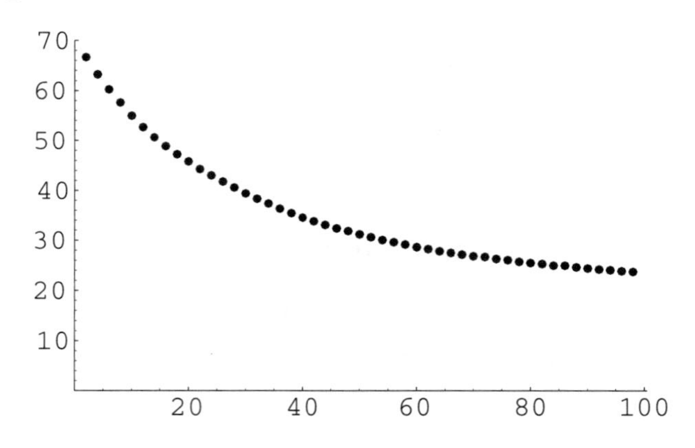


```
NonlinearFit[temps,beta0 Exp[beta1*t]+eps,t,
  {beta0,beta1,eps}]
```

```
NonlinearFit::lmpnocon: Warning: The sum of squares has
  achieved a minimum, but at least one parameter estimate
  fails to satisfy either an accuracy goal of 1 digit(s)
  or a precision goal of 1 digit(s).  These goals are
  less strict than those for the sum of squares,
  specified by AccuracyGoal->6 and PrecisionGoal->6.

NonlinearFit::lmcv: NonlinearFit failed to converge to
  the requested accuracy or precision for the sum of
  squares within 30 iterations.
```

$1. +9.55394\times10^{-23}\ E^{1.t}$

```
NonlinearFit[temps,beta0 Exp[beta1*t]+eps,t,
  {beta0,beta1,eps},MaxIterations->40]
```

```
NonlinearFit::lmpnocon: Warning: The sum of squares has
  achieved a minimum, but at least one parameter estimate
  fails to satisfy either an accuracy goal of 1 digit(s)
  or a precision goal of 1 digit(s).  These goals are
  less strict than those for the sum of squares,
  specified by AccuracyGoal->6 and PrecisionGoal->6.

NonlinearFit::lmcv: NonlinearFit failed to converge to
  the requested accuracy or precision for the sum of
  squares within 40 iterations.
```

$1. +9.33002\times10^{-23}\ E^{1.t}$

Next, we specify the option setting `Method->FindMinimum` in order to employ the method of steepest descent. With this command, we obtain the approximation $\hat{T} = 21.978 + 47.0884e^{-0.0332432t}$, which we define as the function `T` by "cutting-and-pasting" the output of **NonlinearFit**. We graph our approximate solution and name the plot `approx`. We then show `approx` with the data (in `dots`). The approximation appears accurate.

```
NonlinearFit[temps,beta0 Exp[beta1*t]+eps,t,
  {beta0,beta1,eps},Method->FindMinimum]
```

```
FindMinimum::fmmp: Machine precision is insufficient to
  achieve the requested accuracy or precision.
21.978 +47.0884 E^(-0.0332431 t)
```

```
T[t_] :=21.9779835282673641`
  + 47.0883658406797778` E^(-0.0332431321361424636` t)
```

```
approx=Plot[T[t],{t,0,100},PlotRange->{0,70}]
```

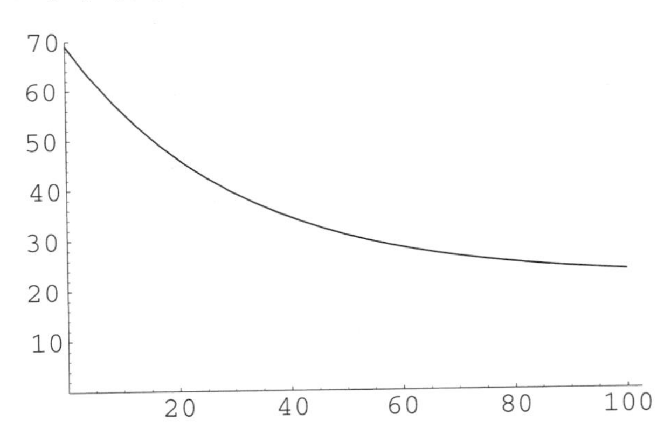

```
Show[approx,dots]
```

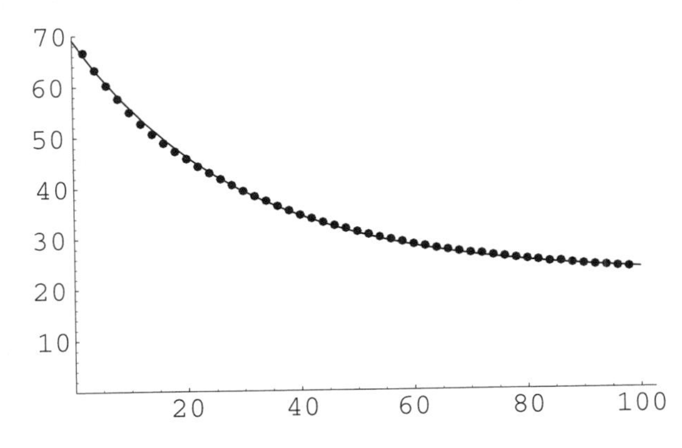

■

Note that the default setting of the option `Method` is `LevenbergMarquardt`. This method gradually shifts the search for the minimum of the error sum of squares function from steepest descent to quadratic minimization (Gauss–Newton).

EXAMPLE 2: The nonlinear model $y = \frac{V_{max}}{k+x}$, the Michaelis–Menten equation (1913), is used to describe the relationship between the concentration x of available dissolved organic matter (μg /liter) and the rate y of uptake (velocity, μg /liter/hr) of that substrate by heterotrophic microbial communities where V_{max} is the maximum velocity and k (μg /l) is the transport constant. This equation has been used for many years to estimate parameters in kinetics. In Table 11.8, we show the data collected when trying to predict the parameters V_{max} and k involving a glucose-type substrate. Use this information to approximate the values of V_{max} and k.

x	y	x	y
0.417	0.0773895	0.417	0.0688714
0.417	0.0819351	0.833	0.0737034
0.833	0.0738753	0.833	0.0712396
1.670	0.0650420	1.670	0.0547667
3.750	0.0497128	3.750	0.0642727
6.250	0.0613005	6.250	0.0643576
6.250	0.0393892		

Source: Department of Biology, Virginia Polytechnic Institute and State University, Blacksburg, VA, 1983

Table 11.8

SOLUTION: After making sure that the **NonlinearFit** package is loaded, we enter the data in `ecology`. Next, we use `NonlinearFit` with only the default settings to obtain the fit $\hat{y} = \frac{0.421171}{4.40374+x}$. However, past experience indicates that reasonable values for V_{max} and k are $V_{max} = 0.5$ and $k = 17$. Using these as starting values with `NonlinearFit`, we find that $\hat{y} = \frac{0.875918}{10.838+x}$. To investigate which approximate solution gives a better fit, we view the results graphically. (Notice that we "cut-and-paste" the output of the two `NonlinearFit` commands to define the functions in `app[x]` and `app2[x]`.) We graph the data points in `ecoplot`, the function $\hat{y} = \frac{0.875918}{10.838 + x}$ in `aplot`, and the function $\hat{y} = \frac{0.421171}{4.40374 + x}$ in `aplot2`. When we show each approximate solution with the data points, we cannot discern a noticeable difference. Therefore, we use another approach based on the residuals.

```
<<Statistics`NonlinearFit`

ecology={{0.417,0.0773895},{0.417,0.0688714},
  {0.417,0.0819351},{0.833,0.0737034},{0.833,0.0738753},
  {0.833,0.0712396},{1.670,0.0650420},{1.670,0.0547667},
  {3.750,0.0497128},{3.750,0.0642727},{6.250,0.0613005},
  {6.250,0.0643576},{6.250,0.0393892}};

NonlinearFit[ecology,v/(k+x),x,{v,k}]
```

$$\frac{0.421171}{4.40374 + x}$$

```
NonlinearFit[ecology,v/(k+x),x,{{v,0.5},{k,17}}]
```

$$\frac{0.875918}{10.838 + x}$$

```
ecoplot=ListPlot[ecology,DisplayFunction->Identity];
```

$$\texttt{app[x\_] :=} \quad \frac{0.87591798`}{10.837971` + x}$$

```
aplot=Plot[app[x],{x,0,7},DisplayFunction->Identity];

Show[ecoplot,aplot,DisplayFunction->$DisplayFunction]
```

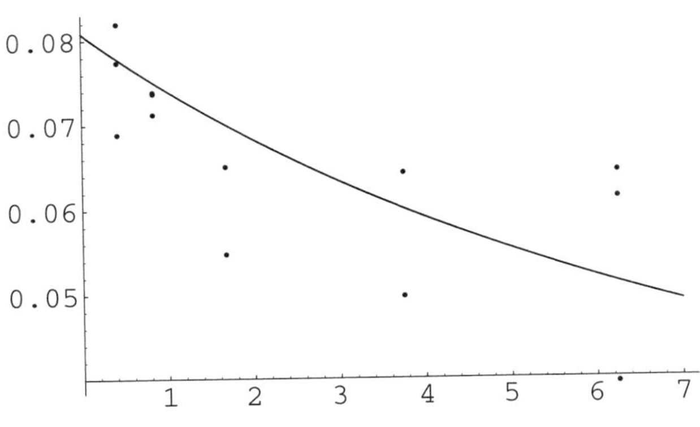

$$\texttt{app2[x\_] :=} \quad \frac{0.42117104`}{4.4037426` + x}$$

```
aplot2=Plot[app2[x],{x,0,7},DisplayFunction->Identity];

Show[ecoplot,aplot2,DisplayFunction->$DisplayFunction]
```

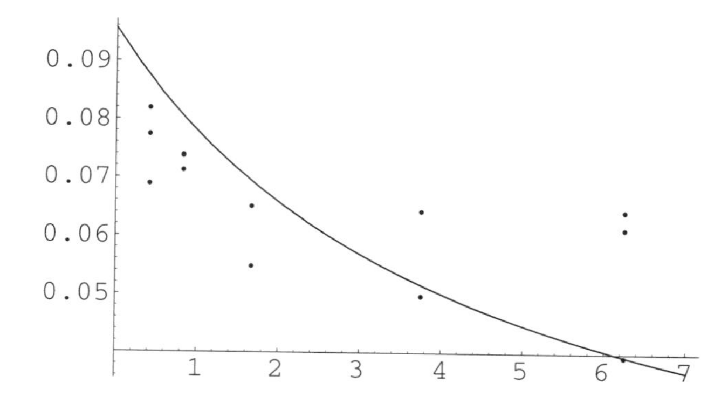

In nrg1, we find the residuals associated with $\hat{y} = \frac{0.421171}{4.40374 + x}$. We extract the list with nrg1[[1,2]] and name the result err1. Finally, we compute the sum of the squared residuals and find this value to be 0.00212771. (Notice that by using the option RegressionReport->StartingParameters, we determine that the starting values of $V_{max} = 1$ and $k = 1$ were used to obtain this approximate solution.) We then follow a similar procedure to find the sum of the squared residuals associated with $\hat{y} = \frac{0.875918}{10.838 + x}$ and find this number to be 0.000910677. Our results indicate that the approximate solution obtained with the starting values of $V_{max} = 0.5$ and $k = 17$ is more accurate.

```
nrg1=NonlinearRegress[ecology,v/(k+x),x,{v,k},
  RegressionReport->FitResiduals]

{FitResiduals→{-0.00997692,-0.018495,-0.00543132,
  -0.00672275,-0.00655085,-0.00918655,-0.00430092,
  -0.0145762,-0.00194091,0.012619,0.0217678,
  0.0248249,-0.000143484}}

err1=nrg1[[1,2]];

sumsq1=Sum[err1[[j]]^2,{j,1,Length[err1]}]

0.00212771

NonlinearRegress[ecology,v/(k+x),x,{v,k},
  RegressionReport->StartingParameters]

{StartingParameters→{v→1,k→1}}
```

```
nrg2=NonlinearRegress[ecology,v/(k+x),x,
  {{v,0.5},{k,17}},RegressionReport->FitResiduals]
```

```
{FitResiduals→{-0.000435491,-0.00895359,0.00411011,
  -0.0013476,-0.0011757,-0.0038114,-0.00498679,
  -0.0152621,-0.0103311,0.00422885,0.0100412,
  0.0130983,-0.0118701}}
```

```
err2=nrg2[[1,2]];
```

```
sumsq2=Sum[err2[[j]]^2,{j,1,Length[err2]}]
```

```
0.000910677
```

One of the `NonlinearRegress` options is `ShowProgress` with default setting `False`. If the setting is `True`, we obtain a step-by-step iteration report. The symbol `ChiSquared` represents the sum of the squared residuals. Notice that the final value of `ChiSquared` in the report matches the value we obtained earlier.

```
NonlinearRegress[ecology,v/(k+x),x,{{v,0.5},{k,17}},
  ShowProgress->True]
```

```
Iteration:1     ChiSquared:0.0211263      Parameters:{0.5, 17.}
Iteration:2     ChiSquared:0.0273833      Parameters:{0.997294, 7.1388}
Iteration:3     ChiSquared:0.00913711     Parameters:{0.924735, 8.25897}
Iteration:4     ChiSquared:0.00340141     Parameters:{0.9117, 9.47141}
Iteration:5     ChiSquared:0.0015738      Parameters:{0.90256, 10.3259}
Iteration:6     ChiSquared:0.00104566     Parameters:{0.885534, 10.6872}
Iteration:7     ChiSquared:0.000927715    Parameters:{0.877484, 10.8141}
Iteration:8     ChiSquared:0.000912423    Parameters:{0.876086, 10.8354}
Iteration:9     ChiSquared:0.000910851    Parameters:{0.875935, 10.8377}
Iteration:10    ChiSquared:0.000910693    Parameters:{0.87592, 10.8379}
Iteration:11    ChiSquared:0.000910677    Parameters:{0.875918, 10.838}
```

$$\frac{0.875918}{10.838 + x}$$

Next, we use `NonlinearRegress` to learn more about the fit. Like `Regress`, `NonlinearRegress` has the option `RegressionReport`. We see from the output that the default setting (`SummaryReport`) for this option with `NonlinearRegress` includes `BestFitParameters`, `ParameterCITable`, `EstimatedVariance`, `ANOVATable`, `AsymptoticCorrelationMatrix`, and `FitCurvatureTable`. Another `RegressionReport` setting is `AsymptoticCovarianceMatrix`, which we illustrate next.

```
NonlinearRegress[ecology,v/(k+x),x,{{v,0.5},{k,17}}]
```

```
{BestFitParameters→{v→0.875918, k→10.838},
```

ParameterCITable→		Estimate	Asymptotic SE	CI
	v	0.875918	0.251463	{0.322452, 1.42938}
	k	10.838	3.55508	{3.01329, 18.6627}

```
EstimatedVariance→0.00008279,
```

```
                                    DF    SumOfSq     MeanSq
               Model                2     0.0557833   0.0278916
ANOVATable→    Error                11    0.0009107   0.00008279,
               Uncorrected Total    13    0.056694
               Corrected Total      12    0.00165764

AsymptoticCorrelationMatrix→( 1.        0.991589 ),
                             ( 0.991589  1.       )

                    Curvature
FitCurvatureTable→  Max Intrinsic                 0.0771249 }
                    Max Parameter-Effects         0.734794
                    95. % Confidence Region       0.50111
```

```
NonlinearRegress[ecology,v/(k+x),x,{{v,0.5},{k,17}},
  RegressionReport->AsymptoticCovarianceMatrix]
{AsymptoticCovarianceMatrix→( 0.0632335 0.886452 )}
                             ( 0.886452  12.6386  )
```

■

Other NonlinearRegress RegressionReport option settings include BestFit, BestFitParameters, ANOVATable, EstimatedVariance, ParameterTable, ParameterCITable, FitResiduals, PredictedResponse, SinglePredictionCITable, MeanPredictionCITable, and AsymptoticCovarianceMatrix. These settings yield the same quantities as those used with Regress, with the exception that standard errors, confidence intervals, test statistics, and matrices are asymptotic.

Statistics`NonlinearFit` contains two RegressionReport options settings, HatDiagonal and StandardizedResiduals, that are useful in identifying influential points and outliers when the nonlinear model is approximately linear. The residuals that result from StandardizedResiduals are those that are scaled by their asymptotic standard errors, computed using the estimated error variance. (According to our earlier notation, these residuals are known as studentized residuals.)

EXAMPLE 3: Use the diagonal elements of the hat matrix and the studentized residuals to determine if the ecology data in Table 11.8 contains any influential points or outliers.

SOLUTION: Assuming that **NonlinearFit** and the data in ecology are loaded, we find the elements along the diagonal of the hat matrix with RegressionReport-> HatDiagonal. Only the last three cases have hat matrix diagonal elements larger than 0.2. We can say that these three cases have moderate leverage, but we would not classify them as influential. The studentized residuals do not indicate any outlying observations.

```
NonlinearRegress[ecology,v/(k+x),x,{{v,0.5},{k,17}},
  RegressionReport->HatDiagonal]
{HatDiagonal→{0.172201,0.172201,0.172201,0.120973,
  0.120973,0.120973,0.0817874,0.0817874,0.133218,
  0.133218,0.230156,0.230156,0.230156}}

p=2;
n=Length[ecology];
2 p/n//N
0.307692

NonlinearRegress[ecology,v/(k+x),x,{{v,0.5},{k,17}},
 RegressionReport->StandardizedResiduals]
{StandardizedResiduals→{-0.0526054,-1.08156,0.496484,
  -0.157969,-0.137819,-0.446784,-0.571957,-1.75048,
  -1.21956,0.499207,1.25776,1.64069,-1.48685}}
```

■

11.8 Correlation

In simple linear regression it is assumed that one variable depends "functionally" on the other, and an equation is determined which can be used to predict a value of the dependent variable, y, using a value of the independent variable, x. In (simple linear) correlation, a linear relationship is again assumed, but there is no requirement that one variable depends "functionally" on the other. The strength of the linear association between the two variables x and y is measured using the **correlation coefficient**, which is also discussed in Chapter 3. Given a set of n points of the form (x_i, y_i), the correlation coefficient is calculated using

$$r = \frac{\sum_{i=1}^{n}(x_i-\bar{x})(y_i-\bar{y})}{\sqrt{\sum_{i=1}^{n}(x_i-\bar{x})^2\sum_{i=1}^{n}(y_i-\bar{y})^2}}.$$

Note that $-1 \le r \le 1$, where $0 \le r \le 1$ indicates positive correlation (an increase in one variable is associated with an increase in the other variable), $-1 \le r \le 0$ indicates negative correlation (an increase in one variable is associated with a decrease in the other variable), and $r = 0$ indicates no correlation. Note that in this case $\sum(x_i-\bar{x})(y_i-\bar{y}) = 0$. The correlation coefficient is also known as the Pearson product-moment correlation coefficient. The square of the correlation coefficient r^2 is called the coefficient of determination in simple linear regression. In the correlation setting, r^2 may be described as the amount of variability in one of the variables accounted for by its linear association with the other variable.

Assumptions

No assumptions are needed in order to calculate a correlation coefficient. But in order to use the correlation coefficient from a set of n points that constitute a random sample from a population of points to make inference about the population, there are underlying assumptions. In correlation, no distinction is made between the two variables. Thus, we assume that for a given value of x, the value of y is randomly chosen from a normal population and for a given value of y, the value of x is randomly chosen from a normal population. A more formal statement of this assumption is that sampling is done from a bivariate normal distribution. When there is only slight population correlation, a deviation from this assumption is inconsequential. However, if the population correlation is high, there may be adverse effects on inferential results due to nonnormality.

Inference for a Single Correlation Coefficient

The sample correlation coefficient r is a point estimate of the population parameter ρ, the correlation coefficient in the population that was sampled. To investigate whether there is a correlation between y and x in the population, test the null hypothesis $H_0{:}\rho = 0$ using Student's t statistic,

$$t = \frac{r}{s_r}, \text{ where } s_r = \sqrt{\frac{1-r^2}{n-2}},$$

and the degrees of freedom $n - 2$. We have written a procedure called **`singleCorrelationTest`** to test the null hypothesis $H_0{:}\rho = \rho_0$. The default is $H_0{:}\rho = 0$, but any appropriate value can be specified for ρ_0. For example, it is possible to test $H_0{:}\rho = 0.75$. The default for the alternative hypothesis is $H_a{:}\rho \neq \rho_0$, although a one-tail alternative may be specified. For values $\rho_0 \neq 0$, we use the transformatio

$$z = 0.5\ln\left(\frac{1+r}{1-r}\right),$$

called **Fisher's** z transformation, where $r = \tanh z$ is the inverse transformation. If $0 < r < 1$, then $0 < z < \infty$; if $-1 < r < 0$, then $-\infty < z < 0$. To test the null hypothesis, we calculate the normal deviate,

$$Z = \frac{z - \zeta_0}{s_r},$$

where z is the transform of r, ζ_0 is the transform of ρ_0, and s_z is the approximate standard deviation of z with $s_z = \sqrt{1/(n-3)}$.

In order to calculate a confidence interval for the population correlation ρ, we first calculate a confidence interval for ζ using the Fisher's z transform of the sample correlation coefficient r. The 100 $1-\alpha$ % confidence limits for ζ are given by $z \pm z_{\alpha/2}\sigma_z$. These limits are both z values, so they can be transformed to corresponding r values using the inverse transformation $r = \tanh z$. These two r values are the confidence limits for ρ. We have written a procedure called **singleCorrelationCI** to calculate confidence limits for ρ. The default confidence coefficient is 95%. However, any confidence level can be specified.

`singleCorrelationTest[data1,data2,opts]` displays the results of a hypothesis test on the population correlation coefficient. The two lists, `data1` and `data2`, must have the same length, and corresponding measurements must be in the same position in their respective lists. The default for the null hypothesis is `rhoZero->0`, but any value between –1 and +1 may be specified as an option. The default for the alternative is `sided->2`, but `sided->1` may be specified as an option.

`singleCorrelationCI[data1,data2,opts]` displays a confidence interval for the population correlation coefficient. The two lists, `data1` and `data2`, must have the same length, and corresponding measurements must be in the same position in their respective lists. The default for the confidence level is `confidence->0.95`, but any value between 0 and 1 may be specified as an option.

To use either of these procedures, first load the disk that comes with this book and click on its icon. Next, locate the procedure name and click on the rightmost cell bracket. Then press **ENTER** and return to the *Mathematica* work sheet.

EXAMPLE 1: Consider the data in Table 11.9. (a) Perform a hypothesis test of $H_0{:}\rho = \rho_0$, $H_a{:}\rho \neq \rho_0$; (b) Perform a hypothesis test of $H_0{:}\rho = -0.5$, $H_a{:}\rho \neq -0.5$; (c) Determine a 95% confidence interval for ρ; (d) Determine a 90% confidence interval for ρ.

Team	Opponent batting average	Team winning percentage
Atlanta	0.240	0.625
Los Angeles	0.254	0.512
Florida	0.249	0.488
San Diego	0.245	0.524
Montreal	0.250	0.588

Continued

Chicago	0.252	0.475
St. Louis	0.254	0.513
New York	0.270	0.463
Houston	0.274	0.512
Philadelphia	0.264	0.405
Pittsburgh	0.280	0.450
Cincinnati	0.266	0.480
San Francisco	0.268	0.456
Colorado	0.286	0.506

Table 11.9 National League: Opponent Batting Average and Winning Percentage (1996 Season)

SOLUTION: (a) First we enter the function **singleCorrelationTest** and the data in Table 11.9, naming the information in the first column `oppbavg` and that in the second `winpct`. Recall that this command has two options. One is used to specify the value of ρ_0, while the other is used to indicate if a one- or two-sided test is to be considered. The default settings for these options are `rhoZero->0` and `sided->2`. When we consider the two-sided test $H_a: \rho \neq \rho_0$, we obtain the p-value 0.0430218. Therefore, we reject H_0 (at the 5% level of significance). The output list for **singleCorrelationTest** also includes the sample correlation coefficient, the test statistic, and the distribution used to calculate the p-value. We also show how a one-sided test is performed with the option setting `sided->1`. As expected, the p-value is one-half of that obtained with the two-sided test.

```
oppbavg={0.240,0.254,0.249,0.245,0.250,0.252,0.254,
  0.270,0.274,0.264,0.280,0.266,0.268,0.286};
winpct={0.625,0.512,0.488,0.524,0.588,0.475,0.513,
  0.463,0.512,0.405,0.450,0.480,0.456,0.506};

singleCorrelationTest[oppbavg,winpct]

{Sample Correlation Coefficient -> -0.546816,
Test Statistic -> -2.26243,
Distribution -> StudentTDistribution[12],
2-sided p-value -> 0.0430218}

singleCorrelationTest[oppbavg,winpct,sided->1]

{Sample Correlation Coefficient -> -0.546816,
Test Statistic -> -2.26243,
Distribution -> StudentTDistribution[12],
1-sided p-value -> 0.0215109}
```

(b) We obtain the p-value 0.830551 when we test $H_0{:}\rho_0 = -0.5$, $\rho_0 \neq -0.5$.

```
singleCorrelationTest[oppbavg,winpct,rhoZero->-0.5]
{Sample Correlation Coefficient -> -0.546816,
Test Statistic -> -0.213995,
Distribution -> NormalDistribution[0,1],
2-sided p-value -> 0.830551}
```

(c) Next we load **singleCorrelationCI** and use it with the default setting. We find the 95% confidence interval to be $-0.835107 \leq \rho \leq -0.0228726$.

```
singleCorrelationCI[oppbavg,winpct]
{-0.835107,-0.0228726}
```

(d) With the option setting `confidence->0.90`, we find that the 90% confidence interval is $-0.803981 \leq \rho \leq -0.117343$.

```
singleCorrelationCI[oppbavg,winpct,confidence->0.90]
{-0.803981,-0.117343}
```

■

Inference for Two Correlation Coefficients

It is possible to compare the correlation coefficients ρ_1 and ρ_2 of two populations. The assumptions are that independent random samples are drawn, one from each population, and that each population has a bivariate normal distribution. Using the sample correlation coefficients, we may test the hypothesis $H_0{:}\rho_1 = \rho_2$ with the test statistic

$$z = \frac{z_1 - z_2}{s_{z_1 - z_2}}, \text{ where } s_{z_1 - z_2} = \sqrt{\frac{1}{n_1 - 3} + \frac{1}{n_2 - 3}},$$

and n_1 and n_2 represent the respective sample sizes. Note that z_1 and z_2 are Fisher's z transforms of the sample correlation coefficients. We have written a procedure called **twoCorrelationTest** to test this null hypothesis. The default for the alternative hypothesis is $H_a{:}\rho_1 \neq \rho_2$, but a one-tail alternative can be specified. Although it would not be correct to conclude that $\rho_1 = \rho_2$ based on this hypothesis test, if we have reason to believe that $\rho_1 = \rho_2$, then both samples came from the same population or from populations with the same correlation coefficients. In this case, we may combine the information from the two samples to calculate a better estimate, call it r_w, of the common population correlation coefficient, ρ. This estimate is found by first calculating $z_w = \dfrac{(n_1 - 3)z_1 + (n_2 - 3)z_2}{(n_1 - 3) + (n_2 - 3)}$ and then using the inverse transformation $r_w = \tanh z_w$. This calculation is available as an option in `twoCorrelationTest`.

`twoCorrelationTest[datapairs1,datapairs2,opts]` displays the results of a hypothesis test on two population correlation coefficients. The default for the alternative is `sided->2`, but `sided->1` may be specified as an option. The other option has default setting `combined->off`, but `combined->on` may be specified.

To use this procedure, first load the disk that comes with this book and click on its icon. Next, locate the procedure name and click on the rightmost cell bracket. Then press **ENTER** and return to the *Mathematica* work sheet.

EXAMPLE 2: Test the null hypothesis that the correlation coefficients relating the team winning percentage to the opposing team batting average of the Central Division (Table 11.10) and the Western Division (Table 11.11) of the National League are equal.

Team	Opponent batting average	Team winning percentage
Chicago	0.252	0.475
St. Louis	0.254	0.513
Houston	0.274	0.512
Pittsburgh	0.280	0.450
Cincinnati	0.266	0.480

Table 11.10 National League Central Division: Opponent Batting Average and Winning Percentage (1996 Season)

Team	Opponent batting average	Team winning percentage
Los Angeles	0.254	0.512
San Diego	0.245	0.524
San Francisco	0.268	0.456
Colorado	0.286	0.506

Table 11.11 National League Western Division: Opponent Batting Average and Winning Percentage (1996 Season)

SOLUTION: First, we load the procedure **twoCorrelationTest**, which has options (with default settings) `sided->2` and `combined->off`. With the setting `combined->on`, **twoCorrelationTest** also lists the value of the combined estimate r_w in the output list. After entering the data corresponding to the Central Division in `central1` and `central2`, we join these two lists in `central`. We enter the data for the Western Division in a similar manner to form the list `west`. We obtain the p-value 0.977231 when we use **twoCorrelationTest** to test the two-sided alternative hypothesis, $H_a:\rho_1 \neq \rho_2$. Using the option setting `combined->on`, we find that the combined correlation is approximately –0.38488.

```
central1={0.252,0.254,0.274,0.280,0.266};
central2={0.475,0.513,0.512,0.450,0.480};
central={central1,central2};

west1={0.254,0.245,0.268,0.286};
west2={0.512,0.524,0.456,0.506};
west={west1,west2};

twoCorrelationTest[central,west]
{Sample Correlation Coefficients -> {-0.356813,-0.386929},
Test Statistic -> 0.0285404,
Distribution -> NormalDistribution[0,1],
2-sided p-value -> 0.977231}

twoCorrelationTest[central,west,combined->on]
{Sample Correlation Coefficients -> {-0.356813,-0.386929},
Test Statistic -> 0.0285404,
Distribution -> NormalDistribution[0,1],
2-sided p-value -> 0.977231}
Combined Sample Correlation Coefficient->-0.366938
```

■

Inference for More Than Two Correlation Coefficients

If we collect k independent, random samples respectively from k populations, each having a bivariate normal distribution, the sample correlation coefficients can be used to test the hypothesis $H_0:\rho_1 = \rho_2 = \cdots = \rho_k$. The alternative hypothesis is that not all of the ρ's are equal. The test statistic is

$$\chi^2 = \sum_{i=1}^{k} (n_i - 3)z_i^2 - \left[\sum_{i=1}^{k} (n_i - 3)z_i\right]^2 \Big/ \sum_{i=1}^{k} (n_i - 3),$$

which we can assume is a value from a chi-square distribution with $k - 1$ degrees of freedom. The n_i's represent the respective sample sizes, and the z_i's are Fisher's z transforms of the sample correlation coefficients. We have written a procedure called **severalCorrelationTest** to test this null hypothesis. Note that if you are testing two correlations, it is best to use **twoCorrelationTest**, which was illustrated in the previous example. If we fail to reject H_0 and have valid reasons to believe that the population correlations are all equal, then we may estimate a common ρ by finding the inverse transformation of the weighted mean z value

$$z_w = \frac{\sum_{i=1}^{k} (n_i - 3) z_i}{\sum_{i=1}^{k} (n_i - 3)}.$$

This calculation is available as an option in **severalCorrelationTest.**

`severalCorrelationTest[popdata,opts]` displays the results of a hypothesis test on three or more population correlation coefficients. The option has default setting `combined->off`, but `combined->on` may be specified.

To use this procedure, first load the disk that comes with this book and click on its icon. Next, locate the procedure name and click on the rightmost cell bracket. Then press **ENTER** and return to the *Mathematica* work sheet.

EXAMPLE 3: Test the hypothesis $H_0: \rho_1 = \rho_2 = \rho_3$ for the data given in Tables 11.10, 11.11, and 11.12 corresponding to the relationship between winning precentage and the opposing team batting average for teams in the East, Central, and Western Divisions of the National League of Major League Baseball.

Team	Opponent batting average	Team winning percentage
Atlanta	0.240	0.625
Florida	0.249	0.488
Montreal	0.250	0.588
New York	0.270	0.463
Philadelphia	0.264	0.405

Table 11.12 National League Eastern Division:
Opponent Batting Average and Winning Percentage (1996 Season)

SOLUTION: First, we the load **severalCorrelationTest**. After entering the data in `central`, `west`, and `east`, we use it to obtain a p-value of approximately 0.696431. Therefore, we fail to find a difference among the locations. In the next command, we calculate a common value for the correlation coefficient with the option setting `combined->on`. In this case, we find this value to be approximately –0.601384.

```
central1={0.252,0.254,0.274,0.280,0.266};
central2={0.475,0.513,0.512,0.450,0.480};
central={central1,central2};

west1={0.254,0.245,0.268,0.286};
west2={0.512,0.524,0.456,0.506};
west={west1,west2};

east1={0.240,0.249,0.250,0.270,0.264};
east2={0.625,0.488,0.588,0.463,0.405};
east={east1,east2};

three={central,west,east};

severalCorrelationTest[three]

{Sample Correlation Coefficients ->
  {-0.356813,-0.386929,-0.821352},
Test Statistic -> 0.723572,
Distribution -> ChiSquareDistribution[2],
2-sided p-value -> 0.696431}

severalCorrelationTest[three,combined->on]

{Sample Correlation Coefficients ->
  {-0.356813,-0.386929,-0.821352},
Test Statistic -> 0.723572,
Distribution -> ChiSquareDistribution[2],
2-sided p-value -> 0.696431}
Combined Sample Correlation Coefficient->-0.601384
```

■

Rank Correlation

When the data we collect come from a bivariate population that is not normal, then the correlation procedures that we have presented earlier may not apply. An alternative approach is to use a rank correlation method. Such methods are also used with ordinal categorical data. One such method is due to Spearman. After ranking the x_i measurements and then separately ranking the y_i measurements, the Spearman rank correlation coefficient is calculated with

$$r_s = 1 - \frac{6}{n^3 - n} \sum_{i=1}^{n} d_i^2,$$

where n is the length of each data list and $d_i = \text{rank}(x_i) - \text{rank}(y_i)$. If there are ties in the data, then the Spearman rank correlation coefficient is calculated with

$$r_s = \frac{(n^3 - n)/6 - \sum d_i^2 - \sum T_x - \sum T_y}{\sqrt{\left[n^3 - n/6 - 2\sum T_x\right]\left[n^3 - n/6 - 2\sum T_y\right]}}.$$

In this formula, $\sum T_x = \frac{1}{12}\sum(t_i^3 - t_i)$, where t_i is the number of tied values of x in the ith group of ties, and $\sum T_y = \frac{1}{12}\sum(t_i^3 - t_i)$, where t_i is the number of tied values of y in the ith group of ties. The package **MultiDescriptiveStatistics** contains the command `SpearmanRankCorrelation[list1,list2]` which calculates r_s for the data in `list1` and `list2`. Note that the exact value of the Spearman rank correlation coefficient can be calculated by using the corresponding ranks with the formula for the Pearson correlation coefficient.

We may test the hypothesis $H_0{:}\rho = 0$ using the Spearman rank correlation coefficient. If $n > 10$, approximate p-values for the test are found using the Student's t distribution with $n - 2$ degrees of freedom, where the test statistic is $t = \dfrac{r_s\sqrt{n-2}}{\sqrt{1 - r_s^2}}$.

EXAMPLE 4: Determine Spearman's rank correlation coefficient and test the hypothesis $H_0{:}\rho_s = 0$ for the data given in Table 11.9 concerning the relationship between winning percentage and opposing team batting average.

SOLUTION: First, we enter the **MultiDescriptiveStatistics** and the **NormalDistribution** packages, and then we enter the data in `oppbavg` and `winpct`. We calculate the Spearman rank correlation coefficient with `SpearmanRankCorrelation`. In this case, $r_s \approx -0.573789$.

```
<<Statistics`MultiDescriptiveStatistics`

<<Statistics`NormalDistribution`

oppbavg={0.240,0.254,0.249,0.245,0.250,0.252,0.254,
  0.270,0.274,0.264,0.280,0.266,0.268,0.286};
winpct={0.625,0.512,0.488,0.524,0.588,0.475,0.513,
  0.463,0.512,0.405,0.450,0.480,0.456,0.506};
```

```
SpearmanRankCorrelation[oppbavg,winpct]//N
-0.573789
```

After defining a function `rank` to assign a rank to the `j`th member of a given list `xlist`, we compute the ranks of the data in `oppbavg` and `winpct`. We name these lists `rank1` and `rank2`, respectively, and use the ranks to compute the value of the rank correlation coefficient with `Correlation`. This value is approximately –0.573789. (Note that this value and that obtained earlier with `SpearmanRankCorrelation` are equal.) Because $n = 12$, we can use the Student's t-distribution to approximate the p-value. We find this value (for a two-tailed test) to be 0.0319112, so we reject the null hypothesis, $H_0:\rho_s = 0$.

```
rank[j_,xlist_]:=Module[{},
  k=1;
  flag=0;
  xsort=Sort[xlist];
  While[xlist[[j]]!=xsort[[k]],k=k+1];
  m=k;
  If[m==Length[xlist],flag=1];
  If[flag<1,While[xsort[[m]]==xsort[[m+1]],m=m+1]];
  num1=m;
  num2=m-k+1;
  Sum[val,{val,k,num1}]/num2//N]

n=Length[oppbavg]
14

rank1=Table[rank[num,oppbavg],{num,1,n}]
{1.,6.5,3.,2.,4.,5.,6.5,11.,12.,8.,13.,9.,10.,14.}

rank2=Table[rank[num,winpct],{num,1,n}]
{14.,9.5,7.,12.,13.,5.,11.,4.,9.5,1.,2.,6.,3.,8.}

rexact=Correlation[rank1,rank2]
-0.573789

tval=Sqrt[n-2] rexact/Sqrt[1-rexact^2]
-2.42693

2(1-CDF[StudentTDistribution[n-2],Abs[tval]])
0.0319112
```

■

CHAPTER 12

Nonparametric Methods

12.1 Introduction

In earlier chapters, we have discussed inferential techniques that apply to population parameters and are based on specific assumptions concerning the distributions of the populations from which the samples are drawn. As a group these are called **parametric methods**. They are generally useful for quantitative data. When the populations do not meet the assumptions and remedial methods are ineffective, or when the data are categorical, alternative inferential techniques are required. The term nonparametric is often used to refer to these methods. As a grouping they are useful under a very general set of assumptions.

As it is conventionally used, the term nonparametric can refer to methods that are not concerned with population parameters, as well as methods that do not depend on the distribution of the population from which the sample is selected. The latter are sometimes referred to as distribution-free methods. In practice, the names nonparametric and distribution-free are not strictly correct, because they refer to methods that are used to make inference about population medians and that require some assumptions about the population(s) being sampled.

It is important to understand that, when the underlying assumptions are satisfied, a parametric method is preferred to its nonparametric counterparts, because more of the information in the data is utilized which results in higher statistical power.

12.2 Methods for Single Samples

Frequently used nonparametric methods that utilize data from a single sample involve inference about central tendency, a population proportion, or randomness.

Central Tendency

The two measures of central tendency that are most frequently used are the mean and the median. Parametric procedures for making inference about a population mean are discussed in Chapter 8. Several nonparametric procedures are available for making inference about the population median. We will discuss two of them. Recall that when the population is symmetric, the mean and median coincide.

Sign Test

The sign test name is used because, as it is traditionally applied, each of the data values in the sample is replaced by a plus or minus sign depending on its value. The test statistic is then the number of plus (or minus) signs.

Assumptions

1. The sample is a simple random sample from a population with unknown median M.
2. The sample values are at least ordinal, categorical data.
3. The population has a continuous distribution.

Hypotheses

The null hypothesis is that the population is centered at M_0. That is, the median of the population is M_0. In symbols, the null hypothesis is $H_0: M = M_0$. It is possible to test both the two-sided alternative $H_a: M \neq M_0$ and one-sided alternatives $H_a: M < M_0$ or $H_a: M > M_0$.

Test Statistic

Given a sample of the form $\{y_1, y_2, \ldots, y_n\}$, for each y_i record a plus (+) if $y_i > M_0$ or a minus (–) if $y_i < M_0$. In case $y_i = M_0$, discard the observed value and reduce n by one (for each value discarded). The test statistic is the number of plus (+) signs. Notice that if the null hypothesis is true, then we expect half of the data values to be greater than M_0 and the other half to be less than M_0. In this case, the value of the test statistic is about $n/2$. A smaller value of the test statistic would imply the median is less than M_0, and a larger value of the test statistic would imply the median is greater than M_0. The p-value is calculated using the binomial probability distribution with parameters n and $p = 1/2$. The value $p = 1/2$ reflects the null hypothesis because, if the median is M_0, then plus (+) and minus (–) will each occur with probability 1/2.

Procedure

We have written a procedure called **`npmSignTest`** to calculate a p-value given a list of data and the value M_0, where the default is two-sided. However, a one-sided alternative can be specified with the option setting `sided->1`. Note that nonnumerical, ordinal observations must be replaced by numeric codes which reflect the order. The value of the median under the null hypothesis, M_0, must be an appropriate number as well. We have also written a procedure called **`npmSignTestFrequencies`** for convenience here and use in Section 12.4. In place of the list and M_0, the input for this procedure consists of two frequencies: the number of plusses and the number of minuses. To use either of these procedures, first load the disk that comes with this book and click on its icon. Next, locate the procedure name and click on the rightmost cell bracket. Then press **ENTER** and return to the *Mathematica* work sheet.

> `npmSignTest[data,m0,opts]` performs the sign test using the values in `data` and the median `m0`. The default option is `sided->2`, but the option setting `sided->1` may be specified. The output includes the sample median, the number of plusses, and the p-value.
>
> `npmSignTestFrequencies[f1,f2,opts]` performs the sign test using the number of plusses `f1` and the number of minuses `f2`. The default option is `sided->2`, but the option setting `sided->1` may be specified. The output includes the number of plusses and the p-value.

Confidence Interval for the Median Based on the Sign Test

Assumptions

1. The sample is a simple random sample from a population with unknown median M.
2. The sample values are at least ordinal, categorical data.
3. The population has a continuous distribution.

Calculations

Given a sample of the form $\{y_1,y_2,\ldots,y_n\}$, the sample median is a point estimate of the median of the population from which the sample was drawn. To calculate a $100(1-\alpha)\%$ confidence interval for the population median, first order the data from smallest to largest. The lower confidence limit is the $(k+1)$th ordered observation and the upper confidence limit is the $(n-k)$th ordered observation, where k is the largest integer such that the cdf at k for the binomial probability distribution with parameters n and $p = 1/2$ is less than or equal to $\alpha/2$.

Procedure

We have written a procedure called **npmSignTestCI** to calculate this confidence interval where the default is a 95% confidence interval, but any confidence coefficient may be specified. Nonnumerical, ordinal observations must be replaced by numeric codes which reflect the order. Note that the binomial distribution is discrete, and it is therefore not possible to calculate a confidence interval with confidence coefficient exactly $1-\alpha$ for every $1-\alpha$. To use this procedure, first load the disk that comes with this book and click on its icon. Next, locate the procedure name and click on the rightmost cell bracket. Then press **ENTER** and return to the *Mathematica* work sheet.

`npmSignTestCI[data,opts]` calculates and displays a confidence interval using the values in `data`. The default option is `confidence->0.95`, but any suitable confidence level may be specified. The output includes an estimate of the actual confidence level corresponding to the displayed interval.

EXAMPLE 1: The median final course average for students taking multivariable calculus courses in the 1980s was 80. Using the sample taken in 1996 (given in Table 12.1), (a) test the hypothesis that the median final course average for students taking this course is the 1990s is 80, $H_0: M = 80$, against $H_a: M \neq 80$. (b) Also test to see whether the median for students taking this course in the 1990s is less than 80. That is, test $H_0: M = 80$ versus $H_a: M < 80$. (c) Determine 95% and 90% confidence intervals for the median based on the sign test.

85.0	94.5	76.7	79.2	83.0	80.2	68.7	89.1	74.1	87.8	44.9
77.6	85.1	75.7	81.5	66.2	83.4	79.8	9.0	91.8	96.3	73.5
82.2	76.1	78.5	69.1	75.4	71.7	78.2	77.7	88.7	79.9	86.1
63.8	78.7	82.6	98.6	81.3	63.4	76.6	84.2	89.7	87.7	54.6

Table 12.1 Final Course Averages

SOLUTION: First, we load **npmSignTest** and **npmSignTestCI**. Then, we enter the data in Table 12.1, naming it `grades`. (a) When we use **npmSignTest**, we obtain a p-value of 0.880396, so we do not reject the null hypothesis $H_0: \mu = 80$. We have also included an example of **npmSignTestFrequencies** using the number of plusses and minuses from this example.

```
grades={85,94.5,76.7,79.2,83,80.2,68.7,89.1,74.1,
  87.8,44.9,77.6,85.1,75.7,81.5,66.2,83.4,79.8,
  94,91.8,96.3,73.5,82.2,76.1,78.5,69.1,75.4,71.7,
  78.2,77.7,88.7,79.9,86.1,63.8,78.7,82.6,98.6,
  81.3,63.4,76.6,84.2,89.7,87.7,54.6};
```

```
npmSignTest[grades,80]
```

```
Title: Sign Test
Estimate: Sample Median -> 79.85
Test Statistic: Number of Pluses is 21
Distribution: BinomialDistribution[44,1/2]
2 - sided p-value -> 0.880396
```

```
npmSignTestFrequencies[21,23]
```

```
Title: Sign Test
Test Statistic: Number of Pluses is 21
Distribution: BinomialDistribution[44,1/2]
2 - sided p-value -> 0.880396
```

(b) The result of the one-sided test, $H_0{:}\mu = 80$ versus $H_a{:}\mu < 80$, also indicates that we do not reject the null hypothesis. (**Note:** The option setting `sided->1` assumes the alternate hypothesis $H_a{:}\mu < 80$, because the median of the sample in `grades` is less than 80. The *p*-value associated with the alternate hypotheses $H_a{:}\mu > 80$ is found with $1 - p$, where *p* is the *p*-value obtained with `sided->1`.)

```
npmSignTest[grades,80,sided->1]
```

```
Title: Sign Test
Estimate: Sample Median -> 79.85
Test Statistic: Number of Pluses is 21
Distribution: BinomialDistribution[44,1/2]
1 - sided p-value -> 0.440198
```

(c) We obtain the 95% confidence interval for the data in `grades` by using **npmSignTestCI** with the default setting for the option `confidence`. This interval is $77.7 \le M \le 82.6$. The output also indicates that the actual confidence coefficient is 0.903858; that is, we are 90.3858% confident that the population median is between 77.7 and 82.6. With the setting `confidence->0.90`, we find that we are 82.5829% confident that the population mean is between 78.2 and 82.2.

```
npmSignTestCI[grades]
```

```
(77.6,83)

Note that the actual confidence level of this interval
  is 95.1233 percent.
```

```
npmSignTestCI[grades,confidence->0.90]
```

```
(77.7,82.6)

Note that the actual confidence level of this interval
  is 90.3858 percent.
```

■

Wilcoxon Signed-Ranks Test

Whereas the sign test utilizes only the sign of the difference between observed values and the hypothesized median, the Wilcoxon signed-ranks test utilizes both the sign and the magnitude of the differences.

Assumptions

1. The sample is a simple random sample from a population with unknown median M.
2. The sample values are at least quantitative having the interval property.
3. The population has a continuous distribution and is symmetric.

Hypotheses

The null hypothesis is that the population is centered at M_0. In symbols, this null hypothesis is $H_0: M = M_0$. It is possible to test the two-sided alternative $H_0: M \neq M_0$ and one-sided alternatives $H_0: M < M_0$ or $H_0: M > M_0$.

Test Statistic

Given a sample of the form $\{y_1, y_2, \ldots, y_n\}$ for each y_i calculate $d_i = y_i - M_0$. If $y_i = M_0$, eliminate it from the calculations and reduce n by 1 (for each value eliminated). Rank the differences from smallest to largest without regard to their signs. In case of a tie, assign each observation in the tie the average of the ranks that would otherwise have been assigned had there been no tie. Assign each rank the sign of the difference of which it is the rank. Calculate the sum of the ranks with positive signs, and call it T_+. The sum of the ranks with negative signs is called T_-, where

$$T_- = \frac{n(n+1)}{2} - T_+.$$

Notice that if the null hypothesis is true, then we expect that $T_+ = T_-$. A small value of T_+ (large value of T_-) implies that the median is less than M_0, and a small value of T_- (large value of T_+) implies that the median is greater than M_0. The test statistic is frequently denoted by $T = \text{minimum}\{T_-, T_+\}$. For small sample sizes $(n \leq 30)$, the p-value is calculated exactly. For $n > 30$, an approximate p-value is calculated using the standard normal distribution.

Procedure

We have written a procedure called **`npmSignedRanksTest`**, where the default is two-sided but a one-sided alternative may be specified. The exact p-value is calculated using a recurrence relationship that is particularly easy to set up in *Mathematica*. For the cases with $n \leq 30$

and T not an integer, **npmSignedRanksTest** gives the average of the exact probabilities found using the smallest integer greater than T and the greatest integer less than T, which are obtained with `Ceiling[T]` and `Floor[T]`, respectively. To use this procedure, first load the disk that comes with this book and click on its icon. Next, locate the procedure name and click on the rightmost cell bracket. Then press **ENTER** and return to the *Mathematica* work sheet.

`npmSignedRanksTest[data,m0,opts]` performs the Wilcoxon signed-ranks test using the values in `data` and the median `m0`. The default option is `sided->2`, but the option setting `sided->1` may be specified. The output includes the sample median, the values of T_+ and T_-, and the p-value. For small sample sizes $(n \le 30)$, the p-value is calculated exactly. For $n > 30$, an approximate p-value is calculated using the standard normal distribution.

Confidence Interval for the Median Based on the Wilcoxon Signed-Ranks Test

Assumptions

1. The sample is a simple random sample from a population with unknown median M.
2. The sample values are at least quantitative having the interval property.
3. The population has a continuous distribution and is symmetric.

Calculations

Given a sample of the form $\{y_1, y_2, \ldots, y_n\}$, calculate all averages of the form

$$\frac{y_i + y_j}{2}.$$

There are $n(n-1)/2$ averages when $i \ne j$ and another n averages when $i = j$. The median of this set of $n(n-1)/2 + n$ averages is the point estimate of the population median. To calculate a $100(1 - \alpha)\%$ confidence interval for the population median, first order the averages from smallest to largest. The lower confidence limit is the $(k + 1)$th ordered average and the upper confidence limit is the $(n(n-1)/2 + n - k)$th ordered average, where k is the largest integer such that $P(T_+ \le k) \le \alpha/2$.

Procedure

We have written a procedure called **npmSignedRanksCI** to calculate this confidence interval where the default is a 95% confidence interval, but any confidence coefficient may be specified. For small sample sizes $(n \le 30)$, the exact probability calculations are made to determine k. For $n > 30$, approximate probabilities are calculated using the standard normal distribution. Note that, when exact calculations are made, it is not necessarily possible to calculate a confidence interval with coefficient exactly as specified. To use this procedure, first load the disk that comes with this book and click on its icon. Next, locate the procedure name and click on the rightmost cell bracket. Then press **ENTER** and return to the *Mathematica* work sheet.

`npmSignedRanksCI[data,opts]` calculates and displays a confidence interval using the values in `data`. The default option is `confidence->0.95`, but any suitable confidence level may be specified. For small sample sizes $(n \leq 30)$, the probability calculations are exact. For $n > 30$, approximate probabilities are calculated using the standard normal distribution. The output includes a point estimate of the population median and, when the exact probability calculations are made, an estimate of the actual confidence level corresponding to the displayed interval.

EXAMPLE 2: (a) Use the Wilcoxon signed ranks test to test $H_0: M = 80$ versus $H_a: M \neq 80$ using the data given in Table 12.1. (b) Construct the 95% confidence interval for the median of the population using the data in Table 12.1.

SOLUTION: (a) After entering the data in `grades`, we use **`npmSignedRanksTest`** to obtain a p-value of 0.97207.

```
grades={85,94.5,76.7,79.2,83,80.2,68.7,89.1,74.1,
  87.8,44.9,77.6,85.1,75.7,81.5,66.2,83.4,79.8,
  94,91.8,96.3,73.5,82.2,76.1,78.5,69.1,75.4,71.7,
  78.2,77.7,88.7,79.9,86.1,63.8,78.7,82.6,98.6,
  81.3,63.4,76.6,84.2,89.7,87.7,54.6};

npmSignedRanksTest[grades,80]
```

```
Title: Wilcoxon Signed-Ranks Test
Sample Median -> 79.85
Test Statistics: T+-> 498.  T_-> 492.  T -> 492.
2 - sided PValue -> 0.97207
```

(b) Using **`npmSignedRanksCI`** (with the default option setting `confidence->0.95`), we find that the 95% confidence interval for the median of the population is $(77.15 \leq M \leq 82.8)$. We also find that the point estimate of the median is 80.05 Notice that because $n = 44 > 30$, the standard normal distribution, $N(0,1)$, is used in the calculation of the upper and lower limits of the confidence interval. To illustrate the results of **`npmSignedRanksCI`** when $n \leq 30$, we form a new data set, named `newgrades`, by selecting every other member of `grades`. In this case, $n = 22 \leq 30$, so that the output of **`npmSignedRankCI`**`[newgrades]` also includes the actual confidence level of the interval.

```
npmSignedRanksCI[grades]
```

```
Point Estimate -> 80.05
Confidence Interval -> {77.15,82.8)
```

```
newgrades=Table[grades[[i]],{i,2,44,2}]

{94.5,79.2,80.2,89.1,87.8,77.6,75.7,66.2,79.8,91.8,73.5,
 76.1,69.1,71.7,77.7,79.9,63.8,82.6,81.3,76.6,89.7,54.6}

npmSignedRanksCI[newgrades]

Point Estimate -> 78.4
Confidence Interval -> {74.15,82.7)
Note that the actual confidence level of this interval
  is 95.384 percent.
```

To demonstrate another feature of the **npmSignedRankCI** command, we construct a small sample by selecting the first five members of `newgrades`, naming this list `smallgroup`. The output of **npmSignedRankCI**`[smallgroup]` indicates that the sample size is too small to construct a 95% confidence interval for the median of the population. (The test fails because $P(T = 0) \geq 0.025(= \alpha/2)$.)

```
smallgroup=Table[newgrades[[i]],{i,1,5}]

{94.5,79.2,80.2,89.1,87.8}

npmSignedRanksCI[smallgroup]

Sample size is too small to calculate a confidence
  interval with confidence coefficient

0.95
```

■

Population Proportion

For a population whose members can be classified into two categories, let p represent the proportion in one of the categories, call it success (S), and $q = 1 - p$ represent the proportion in the other category, call it failure (F). For example, a political pollster may classify as belonging to "success" those voters who favor a proposition on the ballot and as "failure" the remaining voters. The parameter, p, is called the population proportion, and inferences concerning p are made using the number of successes, S, in a sample of size n or using the sample proportion, $\hat{p} = S/n$. An appropriate hypothesis test is called the binomial test.

Binomial Test

The binomial test is based on the binomial probability distribution, which is available in the **DiscreteDistributions** package found in the **Statistics** folder (or directory). For a randomly selected member of the population, $P(\text{success}) = p$ and $P(\text{failure}) = q$, and for a random sample of size n, the probability of s successes is

$$P(s) = \binom{n}{s} p^s q^{n-s}, \; s = 0, 1, \ldots, n.$$

Assumptions

1. The sample is a simple random sample from a population where each member is classified independently of the other members as a success or a failure.
2. It is reasonable that, for each member of the sample, the probability of success is p. (Note that this is true for an infinite population and reasonable for a large finite population.)

Hypotheses

The null hypothesis is that the population proportion has a specific value, which we will denote by p_0. In symbols, the null hypothesis is $H_0: p = p_0$. It is possible to test both the two-sided alternative $H_a: p \neq p_0$ and the one-sided alternatives $H_a: p < p_0$ or $H_a: p > p_0$.

P Value

Given the number in the sample, n, and the number of successes, s, the point estimate of the population proportion, p, is $\hat{p} = s/n$. The p-value that corresponds to the alternative $H_a: p < p_0$ is calculated using `CDF[bdist,s]`, where `bdist` is `BinomialDistribution[n,p0]`. The p-value that corresponds to the alternative $H_a: p > p_0$ is calculated using `1-CDF[bdist,s-1]`. For the special case where the null hypothesis is $H_a: p = 0.5$ (the value of p_0 is 0.5), the two-sided p-value is double the one-sided p-value. For other p_0's, a different strategy has to be used to calculate the two-sided p-value, because the corresponding binomial distribution is not symmetric. One possible strategy is to first calculate the one-sided p-value and then add to that all probabilities from the opposite tail of the distribution that are less than or equal to the probability of the observed number of successes, s. This approach has been called the "principle of minimum likelihood."

Procedure

We have written a procedure called **`npmBinomialPValue`** to calculate p-values as described above. To use this procedure, first load the disk that comes with this book and click on its icon. Next, locate the procedure name and click on the rightmost cell bracket. Then press **ENTER** and return to the *Mathematica* work sheet.

`npmBinomialPValue[n,p0,s]` calculates p-values using the sample size `n`, the value of the population proportion if the null hypothesis is true `p0`, and the number of successes in the sample `s`. The p-values for both a two-sided and a one-sided test are displayed.

EXAMPLE 3: (a) In the sample of 400 consumers, 228 reported that they would not purchase imported products. Use this information to test $H_0{:}p = 0.60$, $H_a{:}p \neq 0.60$ where p is the proportion of the population that does not buy imported products.

SOLUTION: After loading **`npmBinomialPValue`**, we find that the probability is 0.22109 for the two-sided test. Note that in this case, $\hat{p} = \frac{228}{400} = 0.57 < 0.60$, so the one-sided p-value of 0.120488 corresponds to the alternative hypothesis, $H_a{:}p < 0.60$. In either case, we fail to reject the null hypothesis.

```
npmBinomialPValue[400,0.6,228]

One-Sided PValue -> 0.120488
Two-Sided PValue -> 0.22109
```

■

Confidence Interval for the Population Proportion

Assumptions

1. The sample is a simple random sample from a population where each member is classified independently of the other members as a success or a failure.
2. It is reasonable that, for each member of the sample, the probability of success is p. (Note that this is true for an infinite population and reasonable for a large finite population.)

Calculations

To calculate a $100(1 - \alpha)\%$ confidence interval for a population proportion, p, using the number of successes s in a sample of size n, it is possible to use the F (ratio) distribution. We will denote the $(1 - \alpha/2)$th quantile of the F distribution with degrees of freedom v_1 and v_2 by $F_{\alpha/2,v_1,v_2}$. This can be calculated using the **ContinuousDistributions** package with the command `Quantile[fdist,1-α/2]`, where `fdist` is `FRatioDistribution[v1,v2]`. The lower confidence limit for p is

$$\frac{s}{s+(n-s+1)F_{\alpha/2,v_1,v_2}},$$

where $v_1 = 2(n-s+1)$ and $v_2 = 2s$.

The upper confidence limit for p is

$$\frac{(s+1)F_{\alpha/2,v_1,v_2}}{n-s+(s+1)F_{\alpha/2,v_1,v_2}},$$

where $v_1 = 2(s+1)$ and $v_2 = 2(n-s)$.

Procedure

We have written a procedure called **`npmProportionCI`** to calculate this confidence interval. The default is a 95% confidence interval, but any confidence coefficient may be specified. To use this procedure, first load the disk that comes with this book and click on its icon. Next, locate the procedure name and click on the rightmost cell bracket. Then press **ENTER** and return to the *Mathematica* work sheet.

> `npmProportionCI[n,s,opts]` calculates and displays a confidence interval using the number of successes `s` in a sample of size `n`. The default option is `confidence->0.95`, but any suitable confidence level may be specified.

EXAMPLE 4: In the first four games of the 1996 National Basketball Association Finals, Michael Jordan was successful in 41 of 50 free throw attempts. Construct 95% and 90% confidence intervals for the proportion of successful attempts.

SOLUTION: We find that the 95% confidence interval is $(0.685631 \le p \le 0.914238)$. Using the option setting `confidence->0.90`, we see that the 90% confidence interval is $(0.706909 \le p \le 0.902752)$.

```
npmProportionCI[50,41]
(0.685631 , 0.914238)

npmProportionCI[50,41,confidence->0.90]
(0.706909 , 0.902752)
```

■

Randomness

As discussed in Chapters 10 and 11, randomness of sample data is an important requirement of any inferential method. If the randomness of a sample is in question, it is advisable to evaluate this before proceeding with an analysis. There are other uses for methods to investigate randomness as well. Methods for investigating randomness are based on the number of the runs in the data set, where a run is a sequence of like elements that is both preceded and followed by an element of a different type or by no elements. Randomness is questioned when there are either too many or too few runs.

Runs Test

The runs test requires a sequence of observations that are recorded in the order of their appearance and can be categorized into two mutually exclusive types.

Hypotheses

The null hypothesis is of randomness. When there are too few runs, the data are said to be clustered. Too many runs suggests a systematic distribution within the sample. It is possible to test the two-sided alternative, either too few or too many runs, and to test one-sided alternatives.

Test Statistic

Given a sample of the form $\{y_1, y_2, \ldots, y_n\}$ recorded in the order of their occurrence, replace each y_i with a 0 or a 1 depending on the category into which it fits. For example, if the data are ordinal or quantitative, then the median of the list can be calculated and a 0 or 1 assigned depending on whether y_i is less than or greater than the median. There are various strategies for dealing with values which equal the median. One is to discard these values. Let n_1 equal the number of 0's, n_2 equal the number of 1's, and r represent the number of runs. Note that the smallest value of r is 2. This value would occur when all of the 0's are together to form a run of length n_1 and all of the 1's are together to form a run of length n_2. When $n_1 = n_2$, the largest value that can be attained by r is $n_1 + n_2$. This value would occur when the 0's and the 1's alternate so that each run is of length 1. The p-value corresponding to r can be calculated exactly. This is usually done for small $(n_1, n_2 \leq 20)$ samples. For larger sample sizes, the standard normal distribution is usually used.

Procedure

We have written a procedure **npmRunsTest** that accepts a list of 0's and 1's and calculates a p-value corresponding to the number of runs, r. The defaults are two-sided and exact, but a one-sided alternative and an approximate p-value may be specified. To use this procedure, first load the disk that comes with this book and click on its icon. Next, locate the procedure name and click on the rightmost cell bracket. Then press **ENTER** and return to the *Mathematica* work sheet.

`npmRunsTest[zeroOneList,opts]` performs the runs test using the values in `data`. The default options are `sided->2` and `method->exact`, but the option settings `sided->1` and `method->approx` may be specified. The output includes the number of runs and the p-value.

EXAMPLE 5: The total payroll for each Major League Baseball team is given Table 12.2. Use the runs test to test for randomness.

Team	Payroll (million $)	Team	Payroll (million $)
Atlanta (N)	47.93	Milwaukee (A)	20.23
Baltimore (A)	49.36	Minnesota (A)	21.96
Boston (A)	39.42	Montreal (N)	15.41
California (A)	26.89	New York Mets (N)	23.46
Chicago Cubs (N)	31.45	N.Y. Yankees (A)	52.95
Chicago White Sox (A)	41.94	Oakland (A)	19.40
Cincinnati (N)	40.72	Philadelphia (N)	29.72
Cleveland (A)	46.24	Pittsburgh (N)	21.25
Colorado (N)	38.19	San Diego (N)	27.25
Detroit (A)	21.94	San Francisco (N)	34.79
Florida (N)	30.08	Seattle (A)	38.35
Houston (N)	26.89	St. Louis (N)	38.92
Kansas City (A)	18.48	Texas (A)	35.87
Los Angeles (N)	34.65	Toronto (A)	28.49

Table 12.2 Total Payroll for Major League Baseball Teams (1996)
National League, N; American League, A

SOLUTION: First, we define the function `convertToZerosAndOnes` (which uses `dropVals` and `f`) to convert a list of numerical values to a list of zeros and ones where a one is assigned to values greater than the median and a zero is assigned to values less then the median. (Values equal to the median are removed from the list.) This code is part of the code stored with **npmRunsTest** on the disk that accompanies this book.

```
dropVals[x_]:=If[x!=med,AppendTo[newlist,x]]

f[y_,m0_]:=Module[{},
  If[y<m0,0,1]]

convertToZerosAndOnes[datalist_]:=Module[{},
  newlist={0};
  med=Median[datalist];
  addNonMed=Map[dropVals[#]&,datalist];
  total=Length[addNonMed];
  nonMedList=Drop[addNonMed[[total]],1];
  Map[f[#,med]&,datalist]]
```

After entering the data and naming it `totalpayrolls`, we convert the data to zeros and ones and name the result `ZeroOnes`. When we run the two-sided (default) runs test, we obtain the p-value 0.0824696. Notice that this value was computed exactly. When we request an approximate value, we obtain 0.0541266. We also display the result of the one-sided test. There appear to be more runs than would be expected under randomness.

```
totalpayrolls={47.93,49.34,39.43,26.89,
  31.45,41.94,40.72,46.24,38.19,21.94,
  30.08,26.89,18.48,34.65,20.23,21.96,
  15.41,23.46,52.95,19.4,29.72,21.26,
  27.25,34.79,38.35,38.92,35.87,28.49};

ZeroOnes=convertToZerosAndOnes[totalpayrolls]

{1,1,1,0,1,1,1,1,1,0,0,0,0,1,
 0,0,0,0,1,0,0,0,0,1,1,1,1,0}

npmRunsTest[ZeroOnes]

Number of Runs -> 10
Two-Sided PValue -> 0.0824696

npmRunsTest[ZeroOnes,method->approx]

Number of Runs -> 10
Two-Sided PValue -> 0.0541266

npmRunsTest[ZeroOnes,sided->1]

Number of Runs -> 10
One-Sided PValue -> 0.0412348
```

■

12.3 Methods for Two Independent Samples

The methods presented in this section are for data selected randomly from two populations. The samples are assumed to be independent in two respects. First, the elements selected from one population must not depend on the elements selected from the other population. Second, within a sample, each element must be independent of every other element. The frequently used nonparametric methods involve comparison between the two populations with respect to central tendency, variability, or proportionality.

Comparing Central Tendency

Parametric procedures for making inference about the difference between two population means are discussed in Chapter 8. The nonparametric methods discussed here are for comparing population medians although some authors prefer to present the analysis in terms of the location of the distributions without specifying the median.

Mann–Whitney Test

The procedure developed by Mann and Whitney is equivalent to procedures developed at about the same time by other authors, most notably Wilcoxon. Consequently, some references present the Mann–Whitney test and others present the equivalent Wilcoxon rank sum test.

Assumptions

1. The samples are independent and random: one from a population with unknown median M_1 and the other from a population with unknown median M_2.
2. The sample values are at least ordinal, categorical data.
3. The two populations have the same shape. (If the populations differ, it is only in location.)
4. The two populations each have a continuous distribution.

Hypotheses

The null hypothesis is that the populations have equal medians. In symbols, $H_0{:}M_1 = M_2$. It is possible to test the two-sided alternative $H_a{:}M_1 \neq M_2$ and the one-sided alternatives $H_a{:}M_1 < M_2$ or $H_a{:}M_1 > M_2$.

Test Statistic

Given two samples of the form $\{y_{11}, y_{12}, \ldots, y_{1n_1}\}$ and $\{y_{21}, y_{22}, \ldots, y_{2n_2}\}$, combine them and rank all of the observations from smallest to largest. Assign tied observations the average of the ranks that would have been assigned to these observations had there been no tie. When ranking is complete, calculate the rank total corresponding to the sample with the smaller number of observations. We will assume in this discussion that the first sample has the smaller number of observations (i.e., $n_1 \le n_2$). Note that generally (for equal sample sizes) the sample from the population with the smaller median will have the smaller rank total. Denote the sum of the ranks from sample 1 by *W*. This is the Wilcoxon test statistic. The Mann–Whitney test statistic is $U = W - n_1(n_1 + 1)/2$. The *p*-value can be calculated exactly using a recurrence relationship which is particularly easy to set up with *Mathematica*. Usually, the exact calculation is done for small ($n_1 \le 20, n_2 \le 20$) samples and an approximation using the standard normal distribution is done for larger samples.

Procedure

We have written a procedure **`npmMannWhitneyTest`** where the default is two-sided, but a one-sided alternative can be specified. The exact *p*-value should be calculated when sample sizes are small. Otherwise, use the standard normal approximation. The exact calculation (even when sample sizes are small) must be specified with an option. Memory limitations may be a consideration when using the exact procedure. The data set with the smaller size should be entered first. To use this procedure, first load the disk that comes with this book and click on its icon. Next, locate the procedure name and click on the rightmost cell bracket. Then press **ENTER** and return to the *Mathematica* work sheet.

> `npmMannWhitneyTest[list1,list2,opts]` performs the Mann–Whitney test using the values in `list1` and `list2`, where it is assumed that the length of `list1` is less than or equal to the length of `list2`. The default options are `sided->2` and `method->approx`, but the option settings `sided->1` and `method->exact` may be specified. The output includes the sample medians and the *p*-value.

Confidence Interval for the Difference between Two Medians Based on the Mann–Whitney Test

Assumptions

1. The samples are independent and random: one from a population with unknown median M_1 and the other from a population with unknown median M_2.
2. The sample values are at least ordinal, categorical data.
3. The two populations have the same shape. (If the populations differ, it is only in location.)
4. The two populations each have a continuous distribution.

Calculations

Given two independent random samples of the form $\{y_{11}, y_{12}, \ldots, y_{1n_1}\}$ and $\{y_{21}, y_{22}, \ldots, y_{2n_2}\}$, subtract each value in the first sample from each value in the second sample, and order the differences from smallest to largest. The lower confidence limit is the kth smallest difference (counting from the smallest difference) and the upper confidence limit is the kth largest difference (counting from the largest difference) where $P(U \le k) \le \alpha/2$.

Procedure

We have written a procedure called **npmMannWhitneyCI** to calculate the confidence interval where the default is a 95% confidence interval. However, any confidence coefficient may be specified. Note that when the probability calculations are exact, it is not necessarily possible to calculate a confidence interval with confidence coefficient exactly as specified. To use this procedure, first load the disk that comes with this book and click on its icon. Next, locate the procedure name and click on the rightmost cell bracket. Then press **ENTER** and return to the *Mathematica* work sheet.

`npmMannWhitneyCI[list1,list2,opts]` calculates and displays a confidence interval using the values in `list1` and `list2`, where it is assumed that the length of `list1` is less than or equal to the length of `list2`. The default option is `confidence->0.95`, but any suitable confidence level may be specified. For small sample sizes ($n_1 \le 20$, $n_2 \le 30$), the probability calculations are exact. Otherwise, approximate probabilities are calculated using the standard normal distribution. The output includes a point estimate of the difference between the population medians (`med1 - med2`) and, when the exact probability calculations are made, an estimate of the actual confidence level corresponding to the displayed interval.

EXAMPLE 1: (a) Using the data in Table 12.2, test the hypothesis that the total team payrolls of the National League and American League teams have equal medians. (b) Calculate a 95% confidence interval for the difference between the medians.

SOLUTION: (a) After we make sure that we have loaded **npmMannWhitneyTest** and **npmMannWhitneyCI**, we enter the National League team payrolls in `national` and those of the American League teams in `american`. (a) We find that the approximate two-sided p-value is 0.836172. With the option setting `mthd->exact`, we obtain the p-value 0.83879.

```
national={47.93,31.45,40.72,38.19,30.08,26.89,
  34.65,15.41,23.46,29.72,21.26,27.25,34.79,38.9};

american={49.34,39.43,26.89,41.94,46.24,21.94,
  18.48,20.23,21.96,52.95,19.4,38.35,35.87,28.49};
```

```
npmMannWhitneyTest[national,american]
```

```
Title: Mann-Whitney Test
Sample Medians: 30.765, 32.18
Test Statistic: 93.5
Distribution: Normal Approximation
2 - Sided PValue -> 0.836172
```

```
npmMannWhitneyTest[national,american,mthd->exact]
```

```
Title: Mann-Whitney Test
Sample Medians: 30.765, 32.18
Test Statistic: 93.5
Distribution: Exact
2 - Sided PValue -> 0.83879
```

(b) The 95% confidence interval for the difference between the medians is found with **npmMannWhitneyCI** to be $-11.1 \le M_1 - M_2 \le 8.12$, where the actual confidence level is approximately 99.4119 and the difference between the sample medians is –1.415.

```
npmMannWhitneyCI[national,american]
```

```
Difference between sample medians: -1.415
Confidence Interval: (-11.1,8.12)
Note that the actual confidence interval is 99.4119
  percent
```

■

EXAMPLE 2: Test the null hypothesis that the medians of the two populations, from which two randomly generated sets of real numbers are generated, are equal.

SOLUTION: Having loaded **npmMannWhitneyTest**, we use `Random` to generate two sets of four-digit real numbers. The first set named `t1` contains 35 numbers drawn from a uniform population over the interval from 45 to 60, with median 52.5. The second set named `t2` contains 40 numbers drawn from a uniform population over the interval from 40 to 55, with median 47.5. The approximate *p*-value found with **npmMannWhitneyTest** is 0.000444693.

```
t1=Table[Random[Real,{45,60},4],{k,0,35}]
```

```
{54.71,52.,58.31,53.51,51.06,49.84,52.47,56.7,48.68,
 56.49,46.04,47.05,46.42,51.92,50.03,48.93,56.64,49.3,
 50.63,46.83,46.24,57.98,56.11,52.52,51.54,50.98,57.8,
 59.02,45.48,46.14,50.33,47.31,56.8,49.66,49.28,45.26}
```

```
t2=Table[Random[Real,{40,55},4],{k,0,40}]
{43.99,41.27,48.82,54.41,42.15,40.13,43.78,50.41,51.53,
 52.48,41.95,51.62,51.02,49.38,51.3,44.87,47.79,40.69,
 54.31,51.21,40.8,43.65,42.17,45.84,51.81,42.38,48.35,
 46.43,49.65,42.26,44.57,51.02,53.12,44.78,42.62,54.4,
 42.1,50.4,46.32,49.52,49.31}

npmMannWhitneyTest[t1,t2]
Title: Mann-Whitney Test
Sample Medians: 50.8053, 47.7932
Test Statistic: 1082.
Distribution: Normal Approximation
2 - Sided PValue -> 0.000444693
```

■

Comparing Variability

Parametric hypothesis testing procedures for making inference about population variances are discussed in Chapters 8 and 10. Possible nonparametric alternatives to these tests may be called tests of dispersion, spread, or scale, and they all have drawbacks that can render them of little use in practice. For example, a common requirement for the two sample procedures is that the population medians be equal, making such procedures of little use as diagnostic tools to investigate the validity of a parametric test on the population means.

Siegel–Tukey Test

The procedure developed by Siegel and Tukey is similar to several other tests which had been developed earlier. It stands out from the others, because its method of assigning ranks gives the test statistic the same distribution under the null hypothesis as the Mann–Whitney (actually Wilcoxon) test statistic.

Assumptions

1. Samples are independent and random.
2. The sample values are at least ordinal, categorical data.
3. The two populations are identical (including having equal medians) except for a possible difference in variability.
4. The two populations each have a continuous distribution.

Note that in the (unlikely) case that the population medians are known and unequal, in order to use the Siegel–Tukey test the observations in one of the samples must be adjusted. Either subtract the difference between the two population medians from each observation from the population with the higher median, or add the difference to each observation from the population with the lower median. Adjusting the observations using the difference between sample medians is not advisable.

Hypotheses

The null hypothesis is that the populations have the same variability. It is possible to test the two-sided alternative of a difference in variability as well as one-sided alternatives.

Test Statistic

Given two samples of the form $\{y_{11}, y_{12}, \ldots, y_{1n_1}\}$ and $\{y_{21}, y_{22}, \ldots, y_{2n_2}\}$, combine them and order all of the observations from smallest to largest. Ranks are assigned from the extremes to the middle. The lowest observation is given rank 1, the highest rank 2, the second-highest rank 3, the second-lowest rank 4, the third-lowest rank 5, the third-highest rank 6, the fourth-highest rank 7, and so forth. The test statistic is the sum of the ranks of the first sample where it is assumed that $n_1 \leq n_2$.

Procedure

We have written a procedure **npmSiegelTukeyTest** where the default is two-sided, but a one-sided alternative can be specified. The *p*-value is calculated following the same strategy as for the Mann–Whitney test. To use this procedure, first load the disk that comes with this book and click on its icon. Next, locate the procedure name and click on the rightmost cell bracket. Then press **ENTER** and return to the *Mathematica* work sheet.

> `npmSiegelTukeyTest[list1,list2,opts]` performs the Siegel-Tukey test using the values in `list1` and `list2`, where it is assumed that the length of `list1` is less than or equal to the length of `list2`. The default options are `sided->2` and `method->approx`, but the option settings `sided->1` and `method->exact` may be specified. The output includes the sample ranges and the *p*-value.

EXAMPLE 3: Test the null hypothesis that the payrolls of the National League and American League teams have the same variability.

SOLUTION: After loading **npmSiegelTukey**, we find the approximate *p*-value obtained with the standard normal distribution to be 0.0508144. The exact procedure is used to obtain the value of 0.0497355 when the `mthd->exact` option is indicated.

```
national={47.93,31.45,40.72,38.19,30.08,26.89,
  34.65,15.41,23.46,29.72,21.26,27.25,34.79,38.9};

american={49.34,39.43,26.89,41.94,46.24,21.94,
  18.48,20.23,21.96,52.95,19.4,38.35,35.87,28.49};

npmSiegelTukeyTest[national,american]

Title: Siegel-Tukey Test
Sample Ranges: 32.52, 34.47
Test Statistic: 140.5
Distribution: Normal Approximation
2 - Sided PValue -> 0.0508144

npmSiegelTukeyTest[national,american,mthd->exact]

Title: Siegel-Tukey Test
Sample Ranges: 32.52, 34.47
Test Statistic: 140.5
Distribution: Exact
2 - Sided PValue -> 0.0497355
```

■

Wald–Wolfowitz Test

The test developed by Wald and Wolfowitz is a runs test using data from two independent samples. This test differs from the Mann–Whitney test and the Siegel–Tukey test in that the alternative hypothesis is that the populations differ in any respect whatsoever: central tendency, variability, skewness, etc. If the alternative hypothesis of interest is difference in central tendency, then the Mann–Whitney test is preferred because it is more powerful.

Assumptions

1. Samples are independent and random.
2. The sample values are at least ordinal, categorical data.
3. The two populations have a continuous distribution.

Hypotheses

The null hypothesis is the populations have identical distributions. The alternative hypothesis is that the populations are not identically distributed.

Test Statistic

Combine the two samples and order the observations from lowest to highest. Replace the observations in one sample with zeros and the observations in the second sample with ones. The number of runs is the test statistic. The p-value can be calculated using our procedure `npmRunsTest` once the list of zeros and ones has been generated. Note that tied values within either sample do not present any problems in the ordering or calculation of the number of runs. On the other hand, if there are ties across samples, the number of runs is affected by the ordering of the tied observations. One way to handle this situation is to prepare two ordered arrangements, one resulting in the fewest number of runs and the other resulting in the largest number of runs. Calculate the p-value for each list and act accordingly. If both p-values indicate statistical significance (or no significance), then the conclusion is statistical significance (or no significance). If one p-value indicates significance and the other does not, then the experimenter must decide on an appropriate conclusion.

Moses Test

The advantage of the procedure developed by Moses is it does not require the populations to have equal medians. The main disadvantage is that the value of the test statistic is obtained by a random procedure. Thus, it is quite likely that two investigators would obtain different values of the test statistic when analyzing the same data set. Also, it is inappropriate to repeat the procedure until a test statistic is found that leads to a desired conclusion.

Assumptions

1. Samples are independent and random.
2. The sample values are at least interval data.
3. The two populations each have a continuous distribution.

Hypotheses

The null hypothesis is that the populations have the same variability. It is possible to test the two-sided alternative of a difference in variability as well as one-sided alternatives.

Test Statistic

Each sample is randomly divided into subsamples of equal size, call it k, where $2 \le k \le 10$. Each sample should have at least five subsamples. Discard leftover observations. For example, suppose that one sample has 25 observations and the other sample has 29 observations. We might decide to make the subsample size $k = 3$. Then, the first sample would be randomly divided into eight subsamples with one observation discarded. The second sample would have nine randomly chosen subsamples and two observations discarded. For each subsample, calculate the sum of the squared deviations of observations from the subsample mean. Perform the Mann–Whitney test where the samples now consist of the respective sum of squares. The subsample size, k, should be as large as possible but not over 10, but at the same time the number of sums of squares should be large enough to permit meaningful results from the Mann–Whitney test.

We have written a procedure **npmMosesTest** which makes use of a modified version of **npmMannWhitneyTest** to perform the Moses test. Refer to the discussion of the Mann–Whitney test for additional information about the exact option. To use this procedure, first load the disk that comes with this book and click on its icon. Next, locate the procedure name and click on the rightmost cell bracket. Then press **ENTER** and return to the *Mathematica* work sheet.

`npmMosesTest[list1,list2,k,opts]` performs the Moses test using the values in `list1` and `list2`, where it is assumed that the length of `list1` is less than or equal to the length of `list2`, and the subsample size `k`. The default options are `sided->2` and `method->approx`, but the option settings `sided->1` and `method->exact` may be specified. The output includes the sample ranges and the p-value.

EXAMPLE 4: The final point totals (out of a possible 560) for students in two sections of a mathematics course are given in Table 12.3. Use the Moses test to test the null hypothesis that the two classes have equal variability.

Class 1							
	516	523	545	541	516	534	501
	486	537	471	515	558	559	524
	479	542	399	486	456	535	435
	471	555	507	522	545		

Class 2							
	417	522	351	360	461	455	506
	445	543	544	537	440	433	551
	518	335	408	513	437	462	450
	431	508	543	527	523	531	372

Table 12.3 Point Totals for Two Sections of a Mathematics Course

SOLUTION: After loading **npmMosesTest**, we enter the data in `classOne` and `classTwo`. Noting that the lengths of these two lists are 26 and 28, respectively, we select subsamples of size 4 for use with the Moses test. On the first run of the test, we obtain an approximate p-value of 0.010128. However, because the test is based on the random selection of the subsamples, we obtain a different value

(0.153127) on a second run. When we employ the exact procedure for computing the *p*-value with the `mthd->exact` option setting, we find an approximate *p*-value of 0.00815851. Again, if we run the procedure a second time using the exact procedure, we obtain a different value (0.180653).

```
classOne={516,523,545,541,516,534,501,486,
  537,471,515,558,559,524,479,542,399,486,
  456,535,435,471,555,507,522,545};
classTwo={417,522,351,360,461,455,506,445,
  543,544,537,440,433,551,518,335,408,513,
  437,462,450,431,508,543,527,523,531,372};

Length[classTwo]

28

Length[classOne]

26

npmMosesTest[classOne,classTwo,4]

Title: Moses Test
Sample Ranges: 160, 216
Test Statistic: 3.
Distribution: Normal Approximation
2 - Sided PValue -> 0.010128

npmMosesTest[classOne,classTwo,4]

Title: Moses Test
Sample Ranges: 160, 216
Test Statistic: 11.
Distribution: Normal Approximation
2 - Sided PValue -> 0.153127

npmMosesTest[classOne,classTwo,4,mthd->exact]

Title: Moses Test
Sample Ranges: 160, 216
Test Statistic: 3.
Distribution: Exact
2 - Sided PValue -> 0.00815851

npmMosesTest[classOne,classTwo,4,mthd->exact]

Title: Moses Test
Sample Ranges: 160, 216
Test Statistic: 11.
Distribution: Exact
2 - Sided PValue -> 0.180653
```

■

Comparing Proportions

Often the members of a population can be classified into two disjoint categories that we refer to, in general, as success (S) and failure (F). To compare the success rate or proportion of successes between two populations, independent random samples, one from each population can be analyzed. The data is often summarized and displayed in a 2×2 (two by two) contingency table of the form

f_{11}	f_{12}	R_1
f_{21}	f_{22}	R_2
C_1	C_2	n

where f_{ij} represents the frequency observed in row i and column j; R_i is the total of the frequencies in row i; and C_j is the total for column j. Information from one sample is stored in the first column, where f_{11} represents the number of successes and f_{21} the number of failures. The column total C_1 represents the sample size. Information from the other sample is stored similarly in the second column, with C_2 representing the sample size. Some authors distinguish between three types of 2×2 contingency tables and the analytic procedures that are suitable for each type. The type of table depends on whether the totals in the margins (R_1, R_2, C_1, C_2, and n) are fixed in advance of sampling or are an artifact of the sampling procedure. The first type, call it Table Type 1, has been called a double dichotomy and corresponds to a sample for which only n is fixed, and each experimental unit is classified in two ways. For example, for a random sample of 70 children between the ages of 8 and 10 years, each child is classified by gender and whether he or she has access to a personal computer.

Computer	Boys	Girls	Total
Yes	12	6	18
No	23	29	52
Total	35	35	70

The 2×2 table can be analyzed to compare boys versus girls (in the specified age group) for access to a personal computer. The data were collected without specifying the number of boys or girls and without specifying the number with access to a personal computer. Thus, neither column totals nor row totals were fixed in advance. The hypothesis test for this table is often called a test of independence. We prefer the less commonly used terminology (which is a more correct designation) **test of association**.

The second type, call it Table Type 2, results from what has been called a comparative trial and corresponds to sampling where the totals in one of the margins have been set before data collection. Independent random samples of fixed sizes would fit into this setting. For example, independent, random samples of 35 boys and 35 girls between the ages of 8 and 10 years are selected to compare the incidence of access to a personal computer. The previous 2×2 contingency table can be used to display data from this example. The difference from Table Type 1 is the fixing of the sample sizes prior to data collection. The hypothesis test for Table Type 2 is often called a test of homogeneity. Again, we prefer the terminology **test of heterogeneity**.

The third type, call it Table Type 3, has the totals in both margins fixed by the experimenter prior to data collection, or one margin fixed by the experimenter and the other margin fixed by "nature" (i.e., by the influence of an unknown set of factors). The latter table would result from what has been called a fixed margins comparative trial. As an example of the former, the following table was generated for an experiment in which 35 boys and 35 girls between 8 and 10 years old were given a task to perform using a personal computer. The trial was stopped when half of the subjects had completed the task. The data in the table can be analyzed to compare ability level between boys and girls.

Finished	Boys	Girls	Total
Yes	13	22	35
No	22	13	35
Total	35	35	70

In addition to fixing the numbers of boys and girls (i.e., the column totals), the experimenter fixed the row totals by stopping the trial when half of the children had finished the computer task. Thus, totals in both margins were fixed. As an example of a fixed margins comparative trial, the following table contains data from a fictitious trial to compare the incidence of malaria between children who received an experimental vaccine and those who received a placebo.

Malaria	Vaccine	Placebo	Total
No	40	18	58
Yes	10	32	42
Total	50	50	100

The experimenter fixed the column totals; "nature" fixed the row totals.

The three types of 2×2 contingency tables are often not distinguished in statistical textbooks, and the criteria used by experimenters for choosing among the available testing procedures have not usually included table type but often include sample size and computational effort involved. The justification for this lies in the fact that the different procedures generally provide similar results regardless of the type of table. We will discuss the commonly recommended methods for analyzing data in a 2×2 contingency table. First, we discuss the assumptions and hypotheses that can apply for any of the methods.

Assumptions

Observations are randomly and independently selected and frequencies are summarized in a 2×2 contingency table.

Hypotheses

As noted earlier, the wording of the null and alternative hypotheses may change depending on the sampling procedure (and corresponding table type). The following is presented for reference. There is no intention to suggest that one particular set of words or symbols should be used. The experimenter is free to make this choice.

Table Type 1 (Double Dichotomy): The null hypothesis is independence of the two variables, and the alternative hypothesis is association between the two variables. For the example, the null hypothesis may be "access to a personal computer is independent of gender," and the alternative hypothesis may be "access to a personal computer and gender are associated."

Table Type 2: If one considers that two populations which are the same for a characteristic are called homogeneous in that characteristic, then the null hypothesis is homogeneity, and the alternative hypothesis is not homogeneous (i.e., heterogeneity). For the example, the null hypothesis may be "boys and girls are homogeneous in having access to a personal computer," and the alternative may be "boys and girls are not homogeneous in having access to a personal computer."

Table Type 2 and Table Type 3: If one considers the experimental condition as being a comparative trial where independent, random samples of fixed size are selected from respective populations, then the null and alternative hypotheses may be in terms of population proportions. For the example, if we define p_1 to be the proportion of boys to complete the computer task and p_2 the corresponding proportion of girls, then the null hypothesis is $H_0{:}p_1 = p_2$. It is possible to test both the two-sided alternative $H_a{:}p_1 \neq p_2$ and the one-sided alternatives, $H_a{:}p_1 < p_2$ and $H_a{:}p_1 > p_2$.

Fisher Exact Test

The procedure developed by Fisher uses the hypergeometric distribution to calculate the probability of a given 2×2 contingency table. The one-sided p-value is calculated by adding the probabilities for the original table and all tables that are "more extreme." The totals in both margins of the original table are assumed to be fixed when generating the "more extreme" tables. Consequently, some authors suggest that this procedure should only be used with Table Type 3. In practice, it is often used when the total sample size is small, because other commonly used approximate procedures do not apply. In the past, it has been avoided for larger sample sizes because of the lengthy calculations. Of course with *Mathematica*, the amount of calculation is no longer an issue.

P-Value

For a table of the form

f_{11}	f_{12}	R_1
f_{21}	f_{22}	R_2
C_1	C_2	n

the exact probability of the distribution of the f_{ij}'s in the table is calculated using the hypergeometric probability distribution as

$$p = \frac{\binom{C_1}{f_{11}}\binom{C_2}{f_{12}}}{\binom{n}{R_1}}.$$

To find the one-sided p-value, compute the sum of the probabilities of the observed table and all tables more extreme in the same direction. The next more extreme table is found by reducing the smallest f_{ij} by 1 and recalculating the other f_{ij}'s based on the marginal totals being fixed. The most extreme table will have a cell containing zero (that is, $f_{ij} = 0$ for some i and j).

If $R_1 = R_2$ and/or $C_1 = C_2$, the two-sided p-value is twice the one-sided p-value. Otherwise, tables that are as extreme as or more extreme than the observed table but in the opposite direction must be identified and their probabilities added to the one-sided p-value to calculate the two-sided p-value. There is no "best" way to identify these tables. We have chosen to use the procedure that begins by replacing f_{11} in the observed table with $g_{11} = (2R_1C_1 - nf_{11})/n$. Then, calculate the other g_{ij}'s based on the fixed marginal totals of the observed table. Because g_{11} must be an integer, use the following strategy. If $(2R_1C_1 - nf_{11})/n < f_{11}$, truncate at the decimal point so that g_{11} is the integer part of $(2R_1C_1 - nf_{11})/n$. If $(2R_1C_1 - nf_{11})/n > f_{11}$, round off to the next highest integer to calculate g_{11}. The resulting table and any that are more extreme are used for the calculation of the two-sided p-value.

Procedure

We have written a procedure **`npmFisherExactTest`** which calculates both one-sided and two-sided p-values. The user is cautioned to use the two-sided p-value unless a one-sided test is justified prior to collection of the sample data. We assume that the sample proportions are $\hat{p}_1 = f_{11}/C_1$ and $\hat{p}_2 = f_{12}/C_2$. To use this procedure, first load the disk that comes with this book and click on its icon. Next, locate the procedure name and click on the rightmost cell bracket. Then press **ENTER** and return to the *Mathematica* work sheet.

`npmFisherExactTest[fqList]` calculates p-values using the list of frequencies `fqList` where `fqList={{f11,f12},{f21,f22}}`. The p-values for both a two-sided and a one-sided test are displayed.

EXAMPLE 5: Use the Fisher exact test to compute the probability of the 2×2 contingency table given in Table 12.4.

9	1
4	10

Table 12.4 2×2 Contingency Table

SOLUTION: After entering **npmFisherExactTest**, we find that the one-sided p-value is 0.00415601 while the two-sided p-value is 0.00452618.

```
npmFisherExactTest[{{9,1},{4,10}}]
```

```
Title: Fisher Exact Test
Distribution: Hypergeometric
Sample Proportions: 0.692308, 0.0909091
One-Sided P-Value -> 0.00415601
Two-Sided P-Value -> 0.00452618
```

■

Chi-Square Test for 2 × 2 Contingency Tables

The chi-square probability distribution can be used to approximate the two-sided p-value for a 2×2 contingency table. The basic method is presented in the vast majority of elementary statistics textbooks, and there is no controversy associated with applying it to any of the three types. The test statistic involves the use of both observed frequencies (the f_{ij}'s) and expected frequencies where the latter are calculated assuming independence. For Table Type 1, the usual null hypothesis specifies independence, and for Table Types 2 and 3, it is guaranteed by the selection of independent samples. The calculated p-value is not exact because of the approximation of the distribution of the test statistic, which is discrete, by the chi-square probability distribution, which is continuous.

The specific chi-square distribution used in the approximation is determined by its degrees of freedom, which in turn depend on the number of rows and columns in the contingency table. Generally, the number of degrees of freedom is calculated as "the number of rows minus 1 times the number of columns minus 1." In case of a 2×2 contingency table, the degrees of freedom is 1. The accuracy of the approximation is related to sample size and degrees of freedom, with the approximation for a 2×2 table of special concern. To avoid problems, many authors restrict the use of the chi-square test to tables associated with "large sample sizes," where "large" is defined as "all expected frequencies are at least 5." Others suggest the routine use of a correction to the test statistic, called a correction for continuity. The one most often used is known as the Yates correction. Still other authors have suggested that this correction is too severe, leading to inflated p-values. At least one alternative correction for continuity is available which is less conservative than the Yates correction.

In case the column totals and/or the row totals are equal, the one-sided p-value is found by dividing the two-sided p-value from the chi-square approximation by 2. If neither $R_1 = R_2$ nor $C_1 = C_2$, this approach to calculating a one-sided p-value is potentially misleading.

Test Statistic

For a table of the form

f_{11}	f_{12}	R_1
f_{21}	f_{22}	R_2
C_1	C_2	n

the f_{ij}, $i = 1, 2$; $j = 1, 2$ are called observed frequencies. The expected frequency corresponding to f_{ij} is $\hat{f}_{ij} = R_i C_j / n$, and the uncorrected test statistic is denoted χ^2 and calculated as

$$\chi^2 = \sum_{i=1}^{2} \sum_{j=1}^{2} \frac{(f_{ij} - \hat{f}_{ij})^2}{\hat{f}_{ij}} = \frac{n(f_{11}f_{22} - f_{12}f_{21})^2}{R_1 R_2 C_1 C_2}.$$

The two-sided p-value is the probability of a chi-square value which is equal to or greater then the value of the test statistic, χ^2, where the chi-square distribution used has degrees of freedom equal to 1.

The Yates correction for continuity has the effect of decreasing the value of the test statistic and thereby increasing the p-value. The Yates corrected test statistic is denoted by χ^2_Y and calculated as

$$\chi^2_Y = \sum_{i=1}^{2} \sum_{j=1}^{2} \frac{\left(\left|f_{ij} - \hat{f}_{ij}\right| - 0.5\right)^2}{\hat{f}_{ij}} = \frac{n\left(\left|f_{11}f_{22} - f_{12}f_{21}\right| - n/2\right)^2}{R_1 R_2 C_1 C_2}.$$

A less conservative correction, named the Haber method, is done as follows. Let $\hat{f}$ represent the smallest of the four expected frequencies and f represent the corresponding observed frequency. Then

if $f \le 2\hat{f}$ let D = largest multiple of 0.5 that is $< \left|f - \hat{f}\right|$ or

if $f > 2\hat{f}$ let $D = \left|f - \hat{f}\right| - 0.5$

and calculate the test statistic as

$$\chi^2_H = \frac{n^3 D^2}{R_1 R_2 C_1 C_2}.$$

Procedure

We have written a procedure **npmChiSquare2x2Test** which calculates the approximate two-sided p-value. The default is for no correction for continuity, but the user can specify Yates or Haber as an option. If any of the expected frequencies are less than 5, a table of all expected frequencies is displayed. We assume that the sample proportions are $\hat{p}_1 = f_{11}/C_1$ and $\hat{p}_2 = f_{12}/C_2$. To use this procedure, first load the disk that comes with this book and click on its icon. Next, locate the procedure name and click on the rightmost cell bracket. Then press **ENTER** and return to the *Mathematica* work sheet.

> `npmChiSquare2x2Test[fqList,opts]` calculates p-values using the list of frequencies `fqList`, where `fqList={{f11,f12},{f21,f22}}`. The default option is `mthd->uncorrected`. Other options are `mthd->yates` and `mthd->haber`. The sample proportions and the two-sided p-value are displayed. If any of the expected frequencies are under 5, a table of expected frequencies is also displayed.

EXAMPLE 6: Use the chi-square test to determine the two-sided p-value for the 2×2 contingency table given in Table 12.4.

SOLUTION: After entering **npmChiSquare2x2Test**, we find that the two-sided p-value is 0.00290478. Note that, because the minimum f_{ij}, $i = 1, 2; j = 1, 2$, is less than 5, the expected frequencies are printed in the form of a table. We obtain similar results with the option settings `mthd->yates` and `mthd->haber` with two-sided p-values 0.0104025 and 0.00363293, respectively.

```
npmChiSquare2x2Test[{{9,1},{4,10}}]

Title: Chi Square Test
Distribution: Chi Square
Sample Proportions: 0.692308, 0.0909091
Correction: None
Two-Sided P-Value: 0.00290478

                         5.41667   4.58333
Expected Frequencies:
                         7.58333   6.41667
npmChiSquare2x2Test[{{9,1},{4,10}},mthd->yates]

Title: Chi Square Test
Distribution: Chi Square
Sample Proportions: 0.692308, 0.0909091
Correction: Yates
Two-Sided P-Value: 0.0104025

                         5.41667   4.58333
Expected Frequencies:
                         7.58333   6.41667
```

```
npmChiSquare2x2Test[{{9,1},{4,10}},mthd->haber]
Title: Chi Square Test
Distribution: Chi Square
Sample Proportions: 0.692308, 0.0909091
Correction: Haber
Two-Sided P-Value: 0.00363293
                         5.41667   4.58333
Expected Frequencies:
                         7.58333   6.41667
```

■

Log-Likelihood Ratio Test for 2 × 2 Contingency Tables

This test is often called the G test based on the notation for the test statistic. It is used to approximate the two-sided p-value for a 2×2 contingency table based on the chi-square distribution. It was designed to approximate the exact probability for Table Type 1, which is calculated using a multinomial distribution. For Table Type 1, the exact probability for observing cell frequencies f_{11}, f_{12}, f_{21}, and f_{22} in a 2×2 is

$$\frac{n!}{f_{11}!f_{12}!f_{21}!f_{22}!}\left(\frac{f_{11}}{n}\right)^{f_{11}}\left(\frac{f_{12}}{n}\right)^{f_{12}}\left(\frac{f_{21}}{n}\right)^{f_{21}}\left(\frac{f_{22}}{n}\right)^{f_{22}}.$$

Under the null hypothesis of independence, this probability becomes

$$\frac{n!}{f_{11}!f_{12}!f_{21}!f_{22}!}\left(\frac{R_1C_1}{n^2}\right)^{f_{11}}\left(\frac{R_1C_2}{n^2}\right)^{f_{12}}\left(\frac{R_2C_1}{n^2}\right)^{f_{21}}\left(\frac{R_2C_2}{n^2}\right)^{f_{22}}.$$

Taking the natural logarithm of the ratio of these probabilities results in what is known as the log-likelihood ratio. The test statistic, G, is twice this value and is approximately distributed as a chi-square variable with one degree of freedom. The G test is sometimes suggested as a better choice for analysis of any contingency table than the chi-square test. It has been recommended as a better choice specifically when $|f_{ij} - \hat{f}_{ij}| < \hat{f}_{ij}$ for at least one f_{ij}. Other authors prefer the chi-square test.

Test Statistic

For a table of the form

f_{11}	f_{12}	R_1
f_{21}	f_{22}	R_2
C_1	C_2	n

the uncorrected log likelihood ratio test statistic is calculated as

$$G = 2\left[\sum_{i=1}^{2}\sum_{j=1}^{2} f_{ij}\ln f_{ij} - \sum_{i=1}^{2} R_i \ln R_i - \sum_{j=1}^{2} C_j \ln C_j + n\ln n\right].$$

Because the resulting p-value tends to be smaller than it should be, a correction factor is often used. If we let

$$w = 1 + \frac{1}{6n}\left(\frac{n}{C_1} + \frac{n}{C_2} - 1\right)\left(\frac{n}{R_1} + \frac{n}{R_2} - 1\right),$$

then the corrected test statistic is

$$G_w = G/w.$$

The two-sided p-value is the probability of a chi-square value which is equal to or greater than the value of the test statistic, G_w, where the number of degrees of freedom for the chi-square distribution is 1.

Procedure

We have written a procedure, named **npmLogLikelihood2x2Test**, to calculate the approximate two-sided p-value based on the corrected log likelihood ratio, G_w. We assume that the sample proportions are $\hat{p}_1 = f_{11}/C_1$ and $\hat{p}_2 = f_{12}/C_2$. To use this procedure, first load the disk that comes with this book and click on its icon. Next, locate the procedure name and click on the rightmost cell bracket. Then press **ENTER** and return to the *Mathematica* work sheet.

`npmLogLikelihood2x2Test[fqList]` calculates the approximate two-sided p-value based on the corrected log likelihood ratio using the list of frequencies `fqList` where `fqList={{f11,f12},{f21,f22}}`. The sample proportions and the two-sided p-value are displayed.

EXAMPLE 7: Use the log likelihood 2×2 ratio test to find the two-sided p-value based on the corrected log likelihood ratio for the 2×2 contingency table in Table 12.4.

SOLUTION: First, we make sure that we enter **npmLogLikelihood2x2Test**. Then, we find that the two-sided p-value is 0.00236069.

```
npmLogLikelihood2x2Test[{{9,1},{4,10}}]
```

```
Title: LogLikelihood Ratio Test
Distribution: Chi Square
Sample Proportions: 0.692308, 0.0909091
Two-Sided P-Value: 0.00236069
```

■

Hypothesis Test for the Difference between Two Proportions Based on a Normal Approximation

The method that is most often described in elementary statistics textbooks to compare the proportion of successes between two populations based on independent, random samples is based on the normal distribution. This method is, in fact, equivalent to the uncorrected chi-square test when a two-sided test is performed. Because of the approximation there is usually a requirement that the sample sizes be large based on one or another rule of thumb. The advantage of this approach is that one-sided alternative hypotheses are easily tested and a confidence interval on the difference between population proportions is easily calculated.

Test Statistic

For a table of the form

f_{11}	f_{12}	R_1
f_{21}	f_{22}	R_2
C_1	C_2	n

we assume that the sample proportions are, respectively, $\hat{p}_1 = f_{11}/C_1$ and $\hat{p}_2 = f_{12}/C_2$. We also calculate a pooled estimate of the common proportion assuming the null hypothesis, $H_0: p_1 = p_2$, is true as follows:

$$\bar{p} = \frac{f_{11} + f_{12}}{C_1 + C_2}.$$

The test statistic is

$$z = \frac{|\hat{p}_1 - \hat{p}_2|}{\sqrt{\bar{p}(1-\bar{p})(1/C_1 + 1/C_2)}}.$$

A corrected test statistic, which is equivalent to the Yates corrected chi-square, is

$$z_Y = \frac{|\hat{p}_1 - \hat{p}_2| - 1/2(1/C_1 + 1/C_2)}{\sqrt{\bar{p}(1-\bar{p})(1/C_1 + 1/C_2)}}.$$

One- and two-sided p-values are calculated using the standard normal distribution. The one-sided p-value is the probability of a value equal to or greater than z (or z_Y). The two-sided p-value is twice the one-sided p-value.

Procedure

We have written a procedure, called **npmNormalApproximation2x2Test**, to calculate approximate p-values based on the normal approximation. The default is two-sided, but a one-sided p-value may be specified as an option. The default calculation is based on the uncorrected test statistic. The corrected statistic may be specified with an option setting. To use this procedure, first load the disk that comes with this book and click on its icon. Next, locate the procedure name and click on the rightmost cell bracket. Then press **ENTER** and return to the *Mathematica* work sheet.

> `npmNormalApproximation2x2Test[fqList,opts]` calculates p-values using the list of frequencies `fqList` where `fqList={{f11,f12},{f21,f22}}`. The default options are `sided->2` and `mthd->uncorrected`. Other options are `sided->1` and `mthd->yates`. The sample proportions and the p-value are displayed.

Confidence Interval for the Difference between Two Proportions Based on the Normal Approximation

Calculations

A $100(1-\alpha)\%$ confidence interval for $p_1 - p_2$ is calculated using

$$\hat{p}_1 - \hat{p}_2 \pm z_{\alpha/2}\sqrt{\bar{p}(1-\bar{p})(1/C_1 + 1/C_2)},$$

where $z_{\alpha/2}$ is the $(1-\alpha/2)$th quantile of the standard normal distribution. Using the correction term, the interval is calculated as follows:

$$\hat{p}_1 - \hat{p}_2 \pm \left[z_{\alpha/2}\sqrt{\bar{p}(1-\bar{p})(1/C_1 + 1/C_2)} + 1/2(1/C_1 + 1/C_2)\right].$$

Procedure

We have written a procedure **npmNormalApproximationCI** to calculate a confidence interval for $p_1 - p_2$ where the default is a 95% confidence interval, but any confidence coefficient may be specified. The default calculation is based on the uncorrected formula, but the corrected formula may be specified with the option setting, `corrected->true`. To use this procedure, first load the disk that comes with this book and click on its icon. Next, locate the procedure name and click on the rightmost cell bracket. Then press **ENTER** and return to the *Mathematica* work sheet.

> `npmNormalApproximationCI[fqList,opts]` calculates a confidence interval using the list of frequencies `fqList` where `fqList={{f11,f12},{f21,f22}}`. The default

options are `confidence->0.95` and `corrected->false`, but any appropriate confidence coefficient may be specified. The other option is `corrected->true`. The difference between the sample proportions and the confidence interval is displayed.

EXAMPLE 8: Consider the 2×2 contingency table given in Table 12.4. (a) Approximate the two-sided p-value using no correction and that obtained using the Yates corrected test statistic. (b) Determine a 95% confidence interval for $p_1 - p_2$ using no correction and that obtained using the Yates corrected formula. (c) Determine a 90% confidence interval for $p_1 - p_2$ using the Yates corrected formula.

SOLUTION: (a) After loading **npmNormalApproximation2x2Test**, we find the two-sided p-value is 0.000300676 (using no correction). That obtained using the option setting `mthd->yates` is 0.00186874.

```
npmNormalApproximation2x2Test[{{9,1},{4,10}}]

Title: Comparison of Two Proportions
Distribution: Normal
Sample Proportions: 0.692308, 0.0909091
Correction: None
Two-sided P-Value: 0.000300676
```

```
pmNormalApproximation2x2Test[{{9,1},{4,10}},mthd->yates]

Title: Comparison of Two Proportion"
Distribution: Normal
Sample Proportions: 0.692308, 0.0909091
Correction: Yates
Two-sided P-Value: 0.00186874
```

(b) We begin by loading **npmNormalApproximationCI**. Then, we find that the 95% confidence interval for $p_1 - p_2$ is (0.27531, 0.927488). When we use the option setting `corrected->true`, we obtain the confidence interval (0.359226, 1.0114).

```
npmNormalApproximationCI[{{9,1},{4,10}}]

Title: Confidence Interval For p1 - p2
Difference Between Sample Proportions: 0.601399
Confidence Coefficient: 0.95
Correction: None
Confidence Interval: (0.27531,0.927488)
```

```
npmNormalApproximationCI[{{9,1},{4,10}},corrected->true]

Title: Confidence Interval For p1 - p2
Difference Between Sample Proportions: 0.601399
Confidence Coefficient: 0.95
Correction: Yates
Confidence Interval: (0.359226,1.0114)
```

(c) The corrected 90% confidence interval for $p_1 - p_2$ is found to be (0.411652, 0.958977) using the option settings `corrected->true` and `confidence->0.90`.

```
npmNormalApproximationCI[{{9,1},{4,10}},
corrected->true,confidence->0.90]
Title: Confidence Interval For p1 - p2
Difference Between Sample Proportions: 0.601399
Confidence Coefficient: 0.9
Correction: Yates
Confidence Interval: (0.411652,0.958977)
```

■

12.4 Methods for Two Related Samples

The methods presented in this section are for data from two related samples. One of the most common examples of this occurs when observations are taken on the same test subject before and again after some intervening "treatment." Other common situations include one subject receiving two "treatments" or paired subjects where one of the pair receives one "treatment" while the other receives the second "treatment." Methods appropriate for testing central tendency as well as comparing frequencies are included.

Comparing Central Tendency

The single sample parametric procedures discussed in Chapter 8 can be used for making inference about the mean of the differences from related samples. The nonparametric methods focus on a median or the location of the distributions without specifying the median. Because the approach to analysis of two related samples is to take the difference between paired observations and treat the differences as a single sample, the methods presented here are similar to ones presented in Section 12.2.

Sign Test for the Difference between Two Medians (Paired Data)

Given a random sample of pairs of the form $\{(y_{11},y_{12}),(y_{21},y_{22}),\ldots,(y_{n1},y_{n2})\}$, each pair is replaced by a plus or minus sign depending on whether the first value is larger or smaller than the second value. The test statistic is the number of plus signs.

Assumptions

1. The pairs of observations are a random sample, where the first value in each pair is from a population with unknown median M_1 and the second value in each pair is from a related population with unknown median M_2. (An equivalent alternative assumption is that the differences are a random sample from a population of differences with unknown median M.)
2. The observations are at least ordinal, categorical data, so the larger member of the pair can be determined.
3. The populations each have a continuous distribution.

Hypothesis

The null hypothesis is that the populations have equal medians. In symbols, $H_0{:}M_1 = M_2$ or $H_0{:}M_1 - M_2 = 0$. It is possible to test the two-sided alternative $H_a{:}M_1 \neq M_2$ and the one-sided alternatives $H_a{:}M_1 < M_2$ or $H_a{:}M_1 > M_2$.

Test Statistic

Given n pairs of the form (y_{i1}, y_{i2}), assign a plus (+) if $y_{i1} > y_{i2}$ or a minus (–) if $y_{i1} < y_{i2}$. In case $y_{i1} = y_{i2}$, discard the pair and reduce n by 1 (for each discarded pair). The test statistic is the number of plus signs. The calculation of the p-value is discussed in Section 12.2, and the procedure, **npmSignTest**, introduced there may be used for paired observations. If differences of the form $y_{i1} - y_{i2}$ are meaningful (they are not for categorical data but would be for quantitative data), then a list of differences may be input. The value of the median under the null hypothesis is zero (0). Otherwise, the number of pluses followed by the number of minuses can be entered into **npmSignTestFrequencies**.

Confidence Interval for the Median of Differences Based on the Sign Test

Assumptions

1. The pairs of observations are a random sample where the first value in each pair is from a population with unknown median M_1 and the second value in each pair is from a related population with unknown median M_2. (An equivalent alternative assumption is that the differences are a random sample from a population of differences with unknown median M.)
2. The observations are at least ordinal, categorical data, so the larger member of the pair can be determined.
3. The populations each have a continuous distribution.

Calculations

Assume a random sample of pairs of the form $\{(y_{11},y_{12}),(y_{21},y_{22}),\ldots,(y_{n1},y_{n2})\}$ is taken. Provided that differences are meaningful for the type of data collected, a list of differences of the form $d_i = y_{i1} - y_{i2}$ is created and tested as a random sample from a population of differences with unknown median , M. A confidence interval for M may be calculated using the list of differences with the procedure **`npmSignTestCI`** which is presented in Section 12.2. Note that the sign test is most useful for ordinal, categorical data for which differences are not meaningful. Differences are meaningful only for quantitative data, and for this type of data a procedure based on the Wilcoxon test (discussed next) is usually preferred, because more of the information contained in the data is utilized. The Wilcoxon test does require that the population of differences be symmetric. If not, the procedure using the sign test should be used.

Wilcoxon Matched-Pairs Signed Ranks Test

This test is the same as the Wilcoxon signed-ranks test presented in Section 12.2, except each data value is a difference between paired observations. Given a random sample of paired values of the form $\{(y_{11},y_{12}),(y_{21},y_{22}),\ldots,(y_{n1},y_{n2})\}$, a list of differences of the form $d_i = y_{i1} - y_{i2}$ is created and treated as a random sample from a population of differences with unknown median, M.

Assumptions

1. The differences, d_i, are a random sample from a population of differences with unknown median, M.
2. The differences are calculated on quantitative data (having at least the interval property).
3. The population of differences has a continuous distribution and is symmetric.

Hypotheses

The null hypothesis is that the population of differences is centered at zero. This implies equivalence of the (related) distributions from which the paired observations were sampled. In symbols $H_0{:}M = 0$. It is possible to test the two-sided alternative $H_a{:}M \neq 0$ and one-sided alternatives $H_a{:}M < 0$ or $H_a{:}M > 0$.

Test Statistic

The analysis of the sample of differences is the same as that discussed in Section 12.2 for the Wilcoxon signed-ranks test. The procedure **`npmSignedRanksTest`** which is presented there may be used with the list of differences as input, where the value of the median under the null hypothesis is zero (0).

Confidence Interval for the Median of Differences Based on the Wilcoxon Matched-Pairs Signed-Ranks Test

Assumptions

1. The differences, d_i, are a random sample from a population of differences with unknown median, M.
2. The differences are calculated on quantitative data (having at least the interval property).
3. The population of differences has a continuous distribution and is symmetric.

Calculations

As with the Wilcoxon matched-pairs signed-ranks test discussed earlier, we assume a random sample of paired values from which a list of differences is calculated. These differences are treated as a random sample from a population of differences with unknown median, M. A confidence interval for M may be calculated using the list of differences with the procedure called **npmSignedRanksCI**, which is presented in Section 12.2.

EXAMPLE 1: Table 4.2 contains estimates of the number of employed persons 25 years of age and older for each of several occupation categories by gender for the year 1994. (a) Use the sign test to test for a difference between males and females. Also, perform the test using **npmSignTestFrequencies**. (b) Find the 95% confidence interval for the median M of the differences. (c) Use the Wilcoxon matched-pairs signed-ranks test to test for a difference between males and females. (d) Determine the 95% confidence interval for the median M based on the Wilcoxon matched-pairs signed-ranks test.

SOLUTION: (a) First, we make sure that we have entered **npmSignTest**. Then, we enter the data from Table 4.2 in the two lists, women and men. In order to calculate the differences, $y_{i1} - y_{i2}$, we define the function df and find the differences in diffTable. We run the sign test using the numbers in diffTable and median zero. The result is a p-value of 0.790527, so we cannot find a difference between males and females.

```
women={47618 ,15110 ,6555 ,8556 ,2972 ,324 ,19448 ,
  1790 ,5163 ,12495 ,7594 ,1128 ,3724 ,614 };
men={56523 ,16708 ,8883 ,7825 ,1002 ,442 ,11161,
  1661 ,6346,3153,4760,10946,10593,2356};

df[i_,list1_,list2_]:=list1[[i]]-list2[[i]]

diffTable=Table[df[i,women,men],{i,1,Length[women]}]

{-8905,-1598,-2328,731,1970,-118,8287,129,
 -1183,9342,2834,-9818,-6869,-1742}
```

```
npmSignTest[diffTable,0]
```

```
Title: Sign Test
Estimate: Sample Median -> -650.5
Test Statistic: Number of Pluses is 6
Distribution: BinomialDistribution[14,1/2]
2 - sided p-value -> 0.790527
```

We define the function `posOrNeg[x]` to yield a value of +1 if $x > 0$, –1 if $x < 0$, and 0 if $x = 0$. We apply this function to the numbers in `diffTable` to obtain the list `vals` of +1's and –1's. Notice that there are eight –1's found with `Count` in `negCount` and six +1's given in `posCount`. When we perform **`npmSignTestFrequencies`** with these values, we obtain the p-value of 0.790527 as obtained with **`npmSignTest`**.

```
posOrNeg[x_]:=-1/;x<0
posOrNeg[x_]:=0/;x=0
posOrNeg[x_]:=1/;x>0
```

```
vals=Map[posOrNeg,diffTable]
```

```
{-1,-1,-1,1,1,-1,1,1,-1,1,1,-1,-1,-1}
```

```
negCount=Count[vals,-1]
```

```
8
```

```
posCount=Count[vals,1]
```

```
6
```

```
zeroCount=Count[vals,0]
```

```
0
```

```
npmSignTestFrequencies[posCount,negCount]
```

```
Title: Sign Test
Test Statistic: Number of Pluses is 6
Distribution: BinomialDistribution[14,1/2]
2 - sided p-value -> 0.790527
```

(b) After loading **`npmSignTestCI`**, we find that the 95% confidence interval is (–6869,2834) where the actual confidence level is 98.7061%.

```
npmSignTestCI[diffTable]
```

```
(-6869,2834)
```

```
Note that the actual confidence level of this interval
  is 98.7061 percent.
```

(c) Having loaded **npmSignedRanksTest** prior to entering the following command, we obtain $p \approx 0.6698$.

```
npmSignedRanksTest[diffTable,0]
```

```
Title: Wilcoxon Signed-Ranks Test"
Test Statistics: T+ -> 45.  T- -> 60.  T -> 45.
2 - sided PValue -> 0.6698
```

(d) After we load **npmSignedRankCI**, we find that the 95% confidence interval based on the Wilcoxon matched-pairs signed-ranks test is $(-4233.5, 2834)$. Note that the actual confidence level is 95.0562%.

```
npmSignedRankCI[diffTable]
```

```
Point Estimate -> -527.
Confidence Interval -> {-4233.5,2834.)
Note that the actual confidence level of this interval
  is 95.0562 percent.
```

■

McNemar Test

The procedure developed by McNemar is used to analyze frequency data (proportions) from two related samples. For example, a test subject may receive two "treatments" that are each classified as success (S) or failure (F). Alternatively, a pair of matched test subjects may each receive one of the "treatments." The results can be summarized in a 2×2 frequency table of the following form.

	"Treatment" 2		
"Treatment" 1	S	F	Total
S	f_{11}	f_{12}	R_1
F	f_{21}	f_{22}	R_2
Total	C_1	C_2	n

The notation f_{11} represents the frequency or number of times both "treatments" were successful, whereas f_{22} represents the number of times both "treatments" were failures. The frequencies, f_{12} and f_{21}, count the times when the "treatments" had opposite results.

Assumptions

The test subjects (or matched pairs) are chosen independently of other subjects (or matched pairs). The frequencies corresponding to S-S, S-F, F-S, and F-F are calculated.

Hypotheses

Let p_1 be the proportion of successes for "treatment" 1 and p_2 the proportion of successes of "treatment" 2. Then the null hypothesis is $H_0: p_1 = p_2$. It is possible to test the two-sided alternative $H_a: p_1 \neq p_2$ or the one-sided alternatives $H_a: p_1 < p_2$ or $H_a: p_1 > p_2$.

Test Statistic

McNemar has shown that neither f_{11} or f_{22} contributes to the determination of any difference between the "treatments." Thus, any difference will be demonstrated by f_{12} and f_{21}, where $f_{12} > f_{21}$ suggests that "treatment" 1 is superior, and $f_{12} < f_{21}$ suggests that "treatment" 2 is superior. If H_0 is true, meaning the "treatments" are equivalent, then each of the $f_{12} + f_{21}$ observations has a 0.5 chance of contributing to f_{12} and a 0.5 chance of contributing to f_{21}. The p-value for the test is calculated, therefore, using the binomial probability distribution with parameters $n = f_{12} + f_{21}$ and $p = 1/2$, and the calculation is identical to that which was discussed for the sign test. The procedure **npmSignTestFrequencies** can be used to calculate the p-value by entering f_{12} and f_{21} for the required frequencies.

EXAMPLE 2: Two brands of medications were tested with results given in Table 12.5. Use the McNemar test to test the null hypothesis that the proportion of successes is the same for both brands.

	Brand B	
Brand A	*S*	*F*
S	19	11
F	4	16

Table 12.5

SOLUTION: After loading **npmSignTestFrequencies**, we find that $p \approx 0.0351563$, so we reject the null hypothesis. Brand A is superior.

```
npmSignTestFrequencies[11,4]

Title: Sign Test
Test Statistic: Number of Pluses is 11
Distribution: BinomialDistribution[15,1/2]
2 - sided p-value -> 0.0351563
```

■

12.5 Methods for Three or More Independent Samples

The methods presented in this section generalize those presented in Section 12.3. As in that section, independence refers to both within and among the samples under study. The frequently used nonparametric methods involve comparison between three or more populations with respect to central tendency or proportionality.

Comparing Central Tendency

The parametric procedure for making inference about populations means is single factor analysis of variance which is discussed in Chapter 9. The corresponding nonparametric method, due to Kruskal and Wallis, is used to compare population medians, although some authors prefer to present the analysis in terms of the location of the distributions without specifying the median.

Kruskal–Wallis One Way Analysis of Variance by Ranks

The procedure developed by Kruskal and Wallis is equivalent to the parametric analysis of variance being performed on the ranks of the data values instead of the data values themselves. When only two samples are involved it is equivalent to the Mann–Whitney test discussed in Section 12.3.

Assumptions

1. Samples are independent, random samples, one for each of k populations, where the median of population i is denoted by M_i, $i = 1,\ldots,k$.
2. The sample values are at least ordinal, categorical data.
3. The populations all have the same shape. (If the populations differ, this difference is only in location.)
4. The populations each have a continuous distribution.

Hypotheses

The null hypothesis is that the populations have equal medians. In symbols, $H_0{:}M_1 = M_2 = \cdots = M_k$. The alternative hypothesis is that the populations do not all have the same median.

Test Statistic

The k samples are combined and then ranked from smallest to largest. Tied observations are assigned the mean of the ranks that would have been assigned to these observations had there been no tie. When the ranking is complete, the ranks corresponding to each sample are added: Let R_i represent the rank total for sample i. The test statistic is calculated as follows:

$$H = \left(\frac{12}{n(n+1)} \sum_{i=1}^{k} \frac{R_i^2}{n_i} \right) - 3(n+1),$$

where n_i is the size of sample i and $n = \sum_{i=1}^{k} n_i$. The p-value can be calculated by counting all of the sets of possible rank totals which would result in a value of H equal to or greater than the observed value. The result is then divided by the total number of possible sets of rank totals. This divisor is

$$\frac{n!}{n_1! n_2! \cdots n_k}.$$

This form is known as a multinomial coefficient. Most authors suggest using the exact calculation when $k = 3$ and $n_i \le 5$ for $i = 1, 2, 3$. Otherwise, the p-value is approximated using the area to the right of H in the chi-square distribution with degrees of freedom $k - 1$. This approximation is quite good, although it is conservative. That is, the true p-value is less than or equal to the approximate p-value. In case there are ties, a correction factor is used in calculating the approximate p-value to make the result less conservative.

Procedure

We have written a procedure, **`npmKruskalWallisTest`**, to perform this test. The exact p-value may be calculated for $k = 3$ when $n_i \le 5$ for $i = 1, 2, 3$ using the option setting `mthd->exact`. Otherwise, the approximate p-value is calculated (Note that in the exact procedure, we determine all possible arrangements of the n rankings and compute the value of H for each. Therefore, computer memory is a factor in attempting the calculation of an exact p-value for $n_i \ge 6$. In fact, to increase the speed of computation, we have found and stored the frequency tables for the $k = 3$ case with $3 \le n_i \le 5$, $i = 1, 2, 3$. In the 4–5–5 and 5–5–5 cases, we have stored only partial tables. In these two cases, when the test statistic does not appear in the table, we compute the approximate p-value and indicate that the exact value is greater than 0.10.) To use this procedure, first load the disk that comes with this book and click on its icon. Next, locate the procedure name and click on the rightmost cell bracket. Then press **ENTER** and return to the *Mathematica* work sheet.

`npmKruskalWallisTest[samples,opts]` performs the Kruskal–Wallis test using the data in `samples`. The default option is `mthd->approx`. In case there are three groups all with five or less values, specify `mthd->exact`. The sample medians, the value of the test statistic, and the p-value are displayed.

EXAMPLE 1: In the semifinals in the 1996 Women's 400 Meter Hurdles United States Olympic Team Trials, the top two in each heat plus the two best times qualified for the finals. The times in minutes are given in Table 12.6. Use the Kruskal–Wallis test to test for a difference among heats.

Heat 1	Heat 2	Heat 3
54.88	54.67	55.66
54.96	54.87	56.46
55.91	54.95	56.74
55.99	56.27	57.86
56.67	58.33	58.90
57.29	81.90	59.56

Table 12.6 400 Meter Hurdles U.S. Olympic Team Trials

SOLUTION: First, we enter **npmKruskalWallisTest**. Then, we enter the data in `heat1`, `heat2`, and `heat3`, and we place them in one list named `threeheats`. In this case, we have $n_1 = n_2 = n_3 = 6$, so an approximate p-value is found using the Chi Square Distribution with two degrees of freedom. Because $p \approx 0.240052 > 0.05$, we cannot show a difference among heats at the 0.05 significance level.

```
heat1={54.88, 54.96, 55.9,55.99,56.6,57.29};
heat2={ 54.6,54.87, 54.95, 56.27, 58.33,81.90};
heat3 ={55.66,56.4,56.7,57.86, 58.90,59.56};
threeheats={heat1,heat2,heat3};

npmKruskalWallisTest[threeheats,mthd->exact]
```

```
Title: Kruskal Wallis Test
Sample Medians: {55.945,55.61,57.28}
Test Statistic: 2.8538
Distribution: Chi Square
PValue -> 0.240052
The exact PValue cannot be found because n_i>5 for at
  least one i
```

```
npmKruskalWallisTest[threeheats]

Title: Kruskal Wallis Test
Sample Medians: {54.88,54.96,55.9,56.27,58.33,59.56}
Test Statistic: 2.8538
Distribution: Chi Square
PValue -> 0.240052
```

We may consider testing the equality of the heats using only the first three times in each heat. In `firstthree`, we remove the last three entries in each heat (i.e., we drop the last three columns in the matrix `threeheats`). Note that the command `ColumnDrop` is located in the **DataManipulation** package, which was loaded when **npmKruskalWallisTest** was loaded. The results of **npmKruskalWallisTest** indicate that we would not reject the null hypothesis using the *p*-value, $p \approx 0.0581666 > 0.05$, obtained with the approximate method. However, we find that $p \approx 0.0285714 < 0.05$ using the option setting `mthd->exact`, so we would reject the null hypothesis.

```
firstthree=ColumnDrop[threeheats,{4,6}]

{{54.88,54.96,55.9},{54.6,54.87,54.95},
 {55.66,56.4,56.7}}

npmKruskalWallisTest[firstthree]

Title: Kruskal Wallis Test
Sample Medians: {54.96,54.87,56.4}
Test Statistic: 5.68889
Distribution: Chi Square
PValue -> 0.0581666

npmKruskalWallisTest[firstthree,mthd->exact]

Title: Kruskal Wallis Test
Sample Medians: {54.96,54.87,56.4}
Test Statistic: 5.68889
Distribution: Exact
PValue -> 0.0285714
```

■

Friedman Test

The procedure presented here, which was introduced by Friedman, is a nonparametric analogue of a special case of the two-factor analysis of variance when there is one observation per cell. The Friedman test is intended for the analysis of data from a randomized block design, where a block is a homogeneous subgrouping used in order to minimize the influence of other factors on the comparison of the objects under study. In a randomized block design, each of the objects under study, which we will refer to as treatments, is randomly assigned within the block, and the data are collected from several blocks. For example, to compare the effect of four fertilizers on crop yield, an experimenter could divide a field into four quadrants and apply a different fertilizer in each quadrant. However, if soil fertility is different among the quadrants, then the experimenter would not know whether a difference in yields is due to the fertilizers or to the existing soil fertility. The experimenter can avoid this dilemma by dividing the field into several plots which have uniform soil fertility and then randomly assigning the four fertilizers to areas within each plot. The plots are then the blocks in a randomized block design.

Assumptions

1. Each of the k treatments ($k \geq 3$) is randomly assigned within each of b blocks ($b \geq 3$). The data take the form $[\{y_{11}, y_{12}, \ldots, y_{1k}\}, \ldots, \{y_{b1}, y_{b2}, \ldots, y_{bk}\}]$, where y_{ij} represents the observation for the jth treatment in the ith block. The median of the population of values corresponding to the jth treatment is M_j, $j = 1, \ldots, k$.
2. The sample values are at least ordinal, categorical data.
3. The populations each have a continuous distribution.

Hypotheses

The null hypothesis is that the populations have equal medians. In symbols, $H_0\colon M_1 = M_2 = \cdots = M_k$. The alternative hypothesis is that the populations do not all have the same median.

Test Statistic

For each block, the observations are ranked from smallest to largest, so each block contains a separate set of k ranks. Tied observations within a block are assigned the mean of the ranks that would have been assigned to these observations had there been no tie. Let R_j represent the total across all blocks of the ranks corresponding to treatment j. Then the test statistic is calculated as follows:

$$\chi_r^2 = \frac{12}{bk(k+1)} \sum_{j=1}^{k} R_j^2 - 3b(k+1).$$

The p-value can be calculated by counting all of the sets of possible rank totals that would result in a value of χ_r^2 equal to or greater than the observed value. The result is then divided by the total number of possible sets of rank totals, $(k!)^b$. Depending on the tables available to them, authors recommend the use of this exact procedure for values of k up to 6 and values of b up to 15. Otherwise, the p-value is approximated using the area to the right of the observed value of χ_r^2 in the chi-square probability density function with degrees of freedom $k - 1$. This approximation is conservative in that the true p-value is less than or equal to the approximate p-value. In case, there are ties, a correction factor is used in calculating the approximate p-value to make the result less conservative. Another approximation procedure calculates the test statistic as

$$F = \frac{(b-1)(\chi_r^2)}{b(k-1)-(\chi_r^2)}$$

and the p-value using an F (ratio) distribution with degrees of freedom $k-1$ and $(k-1)(b-1)$. This procedure is thought by some authors to be generally superior to that which uses the chi-square distribution.

Note that for a randomized block design with two treatments ($k = 2$), an appropriate nonparametric analysis can be done using the sign test or the Wilcoxon signed-ranks test of Section 12.4. When the number of blocks is two ($b = 2$), an analysis using the Spearman rank correlation discussed in Chapters 3 and 11 is appropriate.

Procedure

We have written a procedure, **npmFriedmanTest**, to perform this test. The exact p-value can be calculated for certain values of k and b. These are $k = 3$ and $2 \le b \le 15$; $k = 4$ and $2 \le b \le 8$; $k = 5$ and $b = 3$. (**Note:** We have stored frequency tables that corresponding to these values of k and b.) The default is to use the chi-square distribution to calculate an approximate p-value. Both the exact calculation and the calculation using the F (ratio) distribution can be specified as options. To use this procedure, first load the disk that comes with this book and click on its icon. Next, locate the procedure name and click on the rightmost cell bracket. Then press **ENTER** and return to the *Mathematica* work sheet.

`npmFriedmanTest[blocks,opts]` performs the Friedman test using the data in `blocks`. The default option is `mthd->chiSquare`. Other options are `mthd->Fdist` and `mthd->exact` The sample medians, the value of the test statistic, and the p-value are displayed.

EXAMPLE 2: The emotions of fear, happiness, depression, and calmness were requested (in random order) from each of eight subjects under hypnosis. Table 12.7 gives these measurements of skin potential (measured in millivolts). Use the Friedman test to test the hypothesis that hypnosis has no significant affect on each emotion. (Damaser, Shore and Orne, "Physiological Effects during Hypnotically Requested Emotions," *Psychosomatic Medicine*, 25: 334–343 (1963), Table 2. Note: Data values were rounded to one decimal place.)

Subject:	1	2	3	4	5	6	7	8
Fear	23.1	57.6	10.5	23.6	11.9	54.6	21.0	20.3
Happiness	22.7	53.2	9.7	19.6	13.8	47.1	13.6	23.6
Depression	22.5	53.7	10.8	21.1	13.7	39.2	13.7	16.3
Calmness	22.6	53.1	8.3	21.6	13.3	37.0	14.8	14.8

Table 12.7 Skin Potential in Millivolts

SOLUTION: After entering **npmFriedmanTest**, we enter the data into the list `emotions`. The result of **npmFriedmanTest** using the default setting `mthd->chiSquare` is a *p*-value slightly larger than 0.05, $p \approx 0.082053$. The exact *p*-value, found using the `mthd->exact`, is $p \approx 0.091$. Finally, with the option setting `mthd->Fdist`, we obtain the *p*-value, $p \approx 0.0708862$. We fail to reject the null hypothesis at the 5% level of significance.

```
emotions={{23.1,22.7,22.5,22.6},{57.6,53.2,53.7,53.1},
  {10.5,9.7,10.8,8.3},{23.6,19.6,21.1,21.6},
  {11.9,13.8,13.7,13.3},{54.6,47.1,39.2,37.0},
  {21.0,13.6,13.7,14.8},{20.3,23.6,16.3,14.8}};
```

```
npmFriedmanTest[emotions]
```

```
Title: Friedman Test
Sample Medians: {22.05,21.15,18.7,18.2}
Test Statistic: 6.45
Distribution: ChiSquare[3]
PValue: .0916554
```

```
npmFriedmanTest[emotions,mthd->exact]
```

```
Title: Friedman Test
Sample Medians: {22.05,21.15,18.7,18.2}
Test Statistic: 6.45
Distribution: Exact
PValue: 0.091
```

```
npmFriedmanTest[emotions,mthd->Fdist]
```

```
Title: Friedman Test
Sample Medians: {22.05,21.15,18.7,18.2}
Test Statistic: 2.57265
Distribution: FRatioDistribution[21]
PValue: 0.0812926
```

■

12.6 Methods for Three or More Samples: Contingency Tables

A table having r rows and c columns where each of the $r \cdot c$ cells contains a count (i.e., frequency) is called an $r \times c$ (r by c) frequency table or an $r \times c$ contingency table. Contingency tables having two rows and two columns are discussed in Section 12.3 under the subheading Comparing Proportions, for two independent samples, and in Section 12.4 (McNemar Test), for two related samples. The methods presented in this section are for tables having more than two rows or more than two columns (or both). In general, an $r \times c$ contingency table will have the form

$$\begin{array}{cccc|c}
f_{11} & f_{12} & \cdots & f_{1c} & R_1 \\
f_{21} & f_{22} & \cdots & f_{2c} & R_2 \\
\vdots & \vdots & & \vdots & \vdots \\
f_{r1} & f_{r2} & \cdots & f_{rc} & R_r \\
\hline
C_1 & C_2 & \cdots & C_c & n
\end{array}$$

where f_{ij} represents the frequency observed in row i and column j; R_i is the total of the frequencies in row i; Cj is the total of the frequencies in column j; and n is the total of all frequencies in the table.

Tests for Association (Tests of Independence)

For a sample of size n taken from a population in which each member can be classified according to two variables such as hair color and eye color, the data can be summarized in an $r \times c$ table, where one of the variables has r categories and the other has c categories. Then, f_{ij} represents the number in the sample with characteristic i for the first variable and characteristic j for the second variable. If the distribution of the frequencies for the categories of the second variable is the same for all categories of the first variable (or vice versa), then the variables are independent. If not, they are said to be associated.

Assumptions

1. The data are a random sample from the population of interest.
2. The variables are categorical, or they may be quantitative with values that can be classified into mutually exclusive categories.

Hypotheses

The null hypothesis is independence of the two variables, and the alternative hypothesis is association of the two variables.

Note that if the null hypothesis is rejected, then the conclusion will be association. If not, then the conclusion will be failure to show an association, *not* independence. Thus, the test to be done is useful to investigate association but *cannot* be used to show independence.

Chi Square Test of Independence

The chi-square probability distribution can be used to approximate the p-value for an $r \times c$ contingency table. The test statistic involves the use of both the observed frequencies (the f_{ij}'s) and expected frequencies which are calculated assuming independence. The calculated p-value is not exact because of the approximation of the distribution of the test statistic, which is discrete, by the chi-square distribution, which is continuous. The specific chi-square distribution used in the approximation is determined by its degrees of freedom, which in turn depends on the number of rows and columns in the contingency table. Generally, the number of degrees of freedom is calculated as "number of rows minus one times number of columns minus one," $(r - 1)(c - 1)$. To be sure that the approximation gives accurate p-values, many authors restrict the use of the chi-square test to contingency tables for which the sample size is large enough that all expected frequencies are at least 5. This may be quite conservative, and other authors have suggested less stringent restrictions. A second widely used restriction permits the minimum expected frequency to as low as 1 if no more than 20% of all expected frequencies are under 5. Still another recommendation, which is even less restrictive than the second, requires the average expected frequency to be 4 or more for testing at the 1% level.

Test Statistic

We denote the observed frequency in cell i, j of the contingency table by f_{ij} and the corresponding expected frequency assuming independence by $\hat{f}_{ij}$, where

$$\hat{f}_{ij} = \frac{R_i C_j}{n}.$$

The test statistic is

$$\chi^2 = \sum_{i=1}^{r} \sum_{j=1}^{c} \left[\frac{(f_{ij} - \hat{f}_{ij})^2}{\hat{f}_{ij}} \right].$$

The p-value is the probability of a chi-square value equal to or greater than the test statistic, χ^2, where the chi-square distribution has degrees of freedom $(r - 1)(c - 1)$.

Procedure

We have written a procedure **npmChiSquareTest** which calculates the approximate p-value for a contingency table. The input lists represent the rows of the table as follows:

$$\left[(f_{11},f_{12},\ldots,f_{1c}),(f_{21},f_{22},\ldots,f_{2c}),\ldots,(f_{r1},f_{r2},\ldots,f_{rc})\right].$$

For data which have not been summarized in contingency table form, see Chapter 4 for a procedure to do this. Although this procedure will work for $r = 2$ and $c = 2$, it is better to use the procedure **npmChiSquare2x2Test**. To use **npmChiSquareTest**, first load the disk that comes with this book and click on its icon. Next, locate the procedure name and click on the rightmost cell bracket. Then press **ENTER** and return to the *Mathematica* work sheet.

`npmChiSquareTest[freqList,opts]` performs the chi-square test of independence using the frequencies in `freqList`. The default option is `mthd->chiSquare`. The other option is `mthd->ExpectedFrequency`. The p-value is displayed.

EXAMPLE 1: Table 12.8 shows the hair color and gender in 100 male subjects and 200 female subjects. Use the chi-square test to test the null hypothesis that hair color is independent of gender in the population sampled. What are the expected frequencies?

Gender	Black	Brown	Blond	Red	Total
Male	32	43	16	9	100
Female	55	65	64	16	200

Table 12.8 Hair Color and Gender Data

SOLUTION: We begin by entering **npmChiSquareTest**. We then find that the p-value is $p \approx 0.0294618$. Therefore, there is an association between gender and hair color. The distribution of hair color for females has a much higher proportion of blond than the distribution for males. The expected frequencies are printed as a 2×4 matrix due to the option setting `mthd->ExpectedFrequencies`.

```
npmChiSquareTest[{{32,43,16,9},{55,65,64,16}},
  mthd->ExpectedFrequencies]
```

```
Title: Chi Square Test
Distribution: ChiSquare[3]
PValue: 0.0294618

29.      36.       26.6667      8.3333
58.      72.       53.3333     16.6667
```

■

Log-Likelihood Ratio Test of Independence

The log-likelihood ratio is often represented by the letter G, so this test has been called the G test. It can be used to approximate the p-value for an $r \times c$ contingency table using the chi-square distribution with degrees of freedom $(r - 1)(c - 1)$. It has been recommended as a better choice than the chi-square test when $|f_{ij} - \hat{f}_{ij}| < \hat{f}_{ij}$ for at least one f_{ij}, although others prefer the chi-square test. The two tests generally have similar results. Because the p-values for the log-likelihood ratio test tend to be too small, a correction factor is often used in calculating the test statistic.

Test Statistic

The uncorrected test statistic is calculated using the formula

$$G = 2\left[\sum_{i=1}^{r}\sum_{j=1}^{c} f_{ij}\ln f_{ij} - \sum_{i=1}^{r} R_i \ln R_i - \sum_{j=1}^{c} C_j \ln C_j + n\ln n\right],$$

where ln stands for the natural logarithm. The correction factor is calculated using

$$w = 1 + \frac{\left(n\sum_{i=1}^{r}\frac{1}{R_i} - 1\right)\left(n\sum_{j=1}^{c}\frac{1}{C_j} - 1\right)}{6n(r-1)(c-1)},$$

and the corrected test statistic is

$$G_w = G/w.$$

The p-value is the probability of a chi-square value equal to or greater than the observed value of the test statistic, where the chi-square distribution has degrees of freedom $(r - 1)(c - 1)$.

Procedure

We have written a procedure **`npmLogLikelihoodTest`** which calculates the approximate p-value for an $r \times c$ contingency table. The input lists represent the rows of the table as follows:

$$\left[(f_{11}, f_{12}, \ldots, f_{1c}), (f_{21}, f_{22}, \ldots, f_{2c}), \ldots, (f_{r1}, f_{r2}, \ldots, f_{rc})\right].$$

For data which have not been summarized in contingency table form, see Chapter 4 for a procedure to do this. Although this procedure will work for $r = 2$ and $c = 2$, it is better to use the procedure **npmLogLikelihood2x2Test**. To use **`npmLogLikelihoodTest`**, first load the disk that comes with this book and click on its icon. Next, locate the procedure name and click on the rightmost cell bracket. Then press **ENTER** and return to the *Mathematica* work sheet.

`npmLogLikelihoodTest[freqList]` performs the G test of independence using the frequencies in `freqList`. The p-value is displayed.

EXAMPLE 2: The data in Table 12.8 show the hair color and gender in 100 male subjects and 200 female subjects. Approximate the p-value for testing the null hypothesis that hair color is independent of gender in the population sampled using the log-likelihood test of independence.

SOLUTION: After we load **npmLogLikelihoodTest**, we obtain the p-value, $p \approx 0.0245917$.

```
npmLogLikelihoodTest[{{32,43,16,9},{55,65,64,16}}]
Title: LogLikelihood Ratio Test
Test Statistic: 9.38456
Distribution: Chi Square Distribution[3]
Two-Sided P-Value: 0.0245917
```

■

Measures of Association

Given a set of data which is summarized in an $r \times c$ contingency table, three main questions arise in the assessment of association: (1) Does an association exist in the population from which the data were taken? (2) How strong is the association? (3) What is its direction? For data from a bivariate normal population, these questions can be answered using Pearson's correlation coefficient. Likewise, if both variables take values that are at least ordinal categorical, then Spearman's rank correlation coefficient can be used to answer the questions. Both of these correlation coefficients are discussed in Chapters 3 and 11. Because the two variables to be evaluated for association using a contingency table need not have the ordinal property, other methods to answer these questions are necessary. The first question is addressed using a hypothesis test as discussed in the previous subsection. If the conclusion is association, the second question is pertinent because any association, no matter how slight, can be statistically significant by using a large enough sample. Note that the third question is only pertinent if there is order inherent in the data.

The following measures of the strength of the association between two variables are used with categorical variables. Generally, a measure of association is interpreted like a correlation coefficient. That is, the value zero indicates no association, and the closer the value is to one or negative one, the stronger the association. Values less than zero indicate an inverse relationship, and values greater then zero indicate a direct relationship. There are no set rules for interpreting the numerical values because strength can be defined differently depending on the experimental situation. A general rule of thumb is to call association weak for values between –0.25 and 0.25, moderate for values between 0.25 and 0.75 in absolute value, and strong for values over 0.75 in absolute value. Finally, different measures of association can give conflicting signals. For example, one measure may indicate strong association while a second, equally appropriate, measure may indicate moderate or weak association for the same data set. This results because different measures of association measure different qualities of the data set all of which are called association.

The first set of measures of association are related to or based on the (uncorrected) value of the test statistic, χ^2, from the chi-square test of independence.

Yule's Q

Yule's measure of association, Q, is for 2×2 contingency tables. It can take values between –1 and 1; however, if the variables are not ordinal, the sign has no real meaning. If the variables are ordinal, then negative values indicate an inverse relationship and positive values a direct relationship:

$$Q = \frac{f_{11}f_{22} - f_{12}f_{21}}{f_{11}f_{22} + f_{12}f_{21}}.$$

Phi Coefficient

The phi coefficient, ϕ, is appropriate for 2×2 as well as $2 \times r$ contingency tables where $r \geq 3$. Its value can be calculated directly from χ^2, and consequently, is between 0 and 1. Generally,

$$\phi = \sqrt{\frac{\chi^2}{n}}.$$

In the case of a 2×2 table, the value of ϕ can be calculated with

$$\phi = \frac{f_{11}f_{22} - f_{12}f_{21}}{\sqrt{R_1 R_2 C_1 C_2}}.$$

Its value is between –1 and 1. Unless the variables are ordinal, the sign is meaningless.

Contingency Coefficient

The contingency coefficient, c, is appropriate for $r \times c$ contingency tables where both r and c are at least 3. Its upper limit is not 1 but is determined by the formula $\sqrt{(r-1)/r}$. It is calculated as follows:

$$c = \sqrt{\frac{\chi^2}{\chi^2 + n}}.$$

Cramer's V

Cramer's measure of association, V, is a modification of the phi coefficient that allows it to be used for tables larger than 2×2:

$$V = \sqrt{\frac{\chi^2}{n(\text{minimum}\{r,c\} - 1)}}.$$

Proportional Reduction in Error (PRE)

The remaining set of measures of the strength of association are based on a concept known as proportional reduction in error (PRE). One variable is assumed to be dependent on the other, and the contingency table is used to calculate two estimates of the error which may be made in predicting a value of the dependent variable. One estimate of error, E_1, is calculated without using the distribution of the independent variable. The second estimate, E_2, is calculated using the distribution of the independent variable. The usefulness of this idea is that if fewer errors are made when using information from the independent variable, then the PRE measure reflects the amount of strength of the implied association. Furthermore, if the result is the same regardless of which variable is assumed to be the dependent variable (the measure is symmetric), then the measure is useful even when there is no dependence relationship. It should be clearly understood that association does not of itself imply that a dependence relationship exists.

Goodman and Kruskal Lambda

The measure lambda (λ) is useful as an illustration of the idea of PRE. Its value depends on which variable is assumed to be the dependent variable. (It is not symmetric.) The possible values range from 0 to 1, and order inherent in the data is not considered. The formula for lambda is

$$\lambda = \frac{E_1 - E_2}{E_1}.$$

Because $E_1 \geq E_2$, the value of λ will be a proportional reduction in error. To illustrate the calculation, we use the following contingency table.

	II_1	II_2	II_3	
I_1	15	14	10	39
I_2	11	35	19	65
I_3	14	17	29	60
	40	66	58	164

Suppose the row variable, I, is the dependent variable, and the column variable, II, is the independent variable. Without any consideration of II, the best estimate of the value of I is I_2, because it corresponds to the largest row total, 65, and is, therefore, the most likely. Because the mode is the most frequently occurring value, I_2 would be called the **modal category**. If I_2 were used as the category for all 164 sample units, then 39 + 60 = 99 errors would be made. Thus, $E_1^{\text{I}} = 99$. To find E_2^{I} we employ the same idea for each column. In this way we include information from II. For column 1, the modal category is I_1, because 15 is the largest frequency in the column. The number of errors is 11 + 14 = 25. For column 2, the modal category is I_2, and the number of errors is 14 + 17 = 31. For column 3, the modal category is I_3, and the number of errors is 10 + 19 = 29. Thus, $E_2^{\text{I}} = 25 + 31 + 29 = 85$, and

$$\lambda_{\text{I}} = \frac{99 - 85}{99} = 0.141.$$

If now we suppose that II is the dependent variable, then $E_1^{\text{II}} = 40 + 58$, and $E_2^{\text{II}} = (14 + 10) + (11 + 19) + (14 + 17)$, and

$$\lambda_{\text{II}} = \frac{98 - 85}{98} = 0.133.$$

In case neither variable is dependent on the other, the two directional calculations are combined to find a value which is called **symmetric lambda**:

$$\text{Sym}\ \lambda = \frac{E_1^{\text{I}} + E_1^{\text{II}} - (E_2^{\text{I}} + E_2^{\text{II}})}{E_1^{\text{I}} + E_1^{\text{II}}}.$$

For our example, $\text{Sym}\ \lambda = \frac{197 - 170}{197} = 0.137$.

Concordant, Discordant, and Tied Pairs

In addition to being PRE measures, the following measures are only valid if both variables are ordinal. To understand them requires knowledge of concepts of concordant pairs, discordant pairs, and tied pairs. A pair of cases is **concordant** if the values of both variables for one case are higher (or both lower) than the corresponding values for the other case. The pair is **discordant** if the value of one variable for a case is larger than the corresponding value of the other case, and the direction is reversed for the second variable. When the two cases have identical values on one or both variables, they are **tied**. A preponderance of concordant pairs suggests positive association; a preponderance of discordant pairs suggests negative association; and similar numbers of concordant and discordant pairs reflect a lack of association. We use the following notation:

P = number of concordant pairs
Q = number of discordant pairs
T_r = number of pairs tied on the row variable
T_c = number of pairs tied on the column variable
T_{rc} = number of pairs tied on both variables.

The preceding 3×3 contingency table represents $n = 164$ cases. We will now assume that $I_1 < I_2 < I_3$ and $II_1 < II_2 < II_3$ (i.e., variables I and II are both ordinal and the categories in the table are ordered from smallest to largest). The following is a partial list of the cases summarized in the table.

Case	I	II
1	I_1	II_1
2	I_2	II_1
3	I_1	II_2
4	I_3	II_1
5	I_3	II_2
6	I_2	II_2
7	I_1	II_2

Note that cases 1 and 5 and cases 1 and 6 are two examples of concordant pairs; cases 2 and 3 and cases 3 and 4 are two examples of discordant pairs; cases 1 and 3 are tied on the row variable (I); cases 5 and 6 are tied on the column variable (II); and cases 3 and 7 are tied on both variables.

For a contingency table where the categories are ordered from smallest to largest down the rows and from left to right across the columns, the following strategy can be used to calculate the numbers of the various kinds of pairs. In particular, cases which form concordant pairs with the 15 cases in the first cell of the table are represented in the four cells which are below and to the right. The number of concordant pairs where one of the pair in the first cell is $15(35 + 19 + 17 + 29) = 1500$. Any cell in the table which has a cell below and to the right must be considered in a similar way when calculating P. Thus,

$$P = 1500 + 14(19 + 29) + 11(17 + 24) + 35(29) = 3693.$$

Cases that form discordant pairs with the 10 cases in the cell in the first row and third column of the table are represented in the four cells that are below and to the left. The number of discordant pairs where one of the pair is in this cell is $10(11 + 35 + 14 + 17) = 770$. Any cell in the table that has a cell below and to the left must be considered in a similar way when calculating Q. Thus,

$$Q = 770 + 14(11 + 14) + 19(14 + 17) + 35(14) = 2199.$$

Cases which form pairs tied on the row variable (I) with the 15 cases in the first cell of the table are represented in the two cells in the same row and to the right. The number of such tied pairs is $15(14+10) = 360$. Any cell in the table which has a cell in the same row to the right must be considered in a similar way when calculating T_r. Thus,

$$T_r = 360 + 14(10) + 11(35+19) + 35(19) + 14(17+29) + 17(29) = 2896.$$

Cases that form pairs tied on the column variable (II) with the 15 cases in the first cell of the table are represented in the two cells in the same column and below. The number of such tied pairs is $15(11+14) = 375$. Any cell in the table that has a cell in the same column and below must be considered in a similar way when calculating T_c. Thus, $T_c = 375 + 11(14) + 14(35+17) + 35(17) + 10(19+29) + 19(29) = 2883$.

Goodman and Kruskal's Gamma

The measure gamma, which ranges from –1 to 1, is often used for ordinal variables. (Note that Yule's Q is a special case of gamma for 2×2 tables.) Although independence implies that gamma = 0, the converse is not always true. That is, gamma = 0 does not necessarily imply that the variables are independent, except for 2×2 tables. A weakness of gamma is that it does not use numbers if they are tied pairs.

$$\text{gamma} = \frac{P-Q}{P+Q}$$

Under the null hypothesis that gamma in the population sampled is zero, an appropriate two-sided p-value is calculated using the standard normal distribution. Thus,

$$z = \text{gamma}\sqrt{\frac{P+Q}{n\left(1-\text{gamma}^2\right)}}$$

and p-value $= 2[1 - \text{CDF}(dist, z)]$ where $dist$ = `NormalDistribution[0,1]`.

Somer's *d*

This measure uses the number of pairs tied (only) on the dependent variable. As such, its value depends on which variable is assumed to be the dependent variable. (It is not symmetric.) Its value can range between –1 and 1. In case neither variable is dependent, there is a symmetric form. If the row variable (I) is dependent, then

$$d_{\text{I}} = \frac{P-Q}{P+Q+T_r}.$$

If the column variable (II) is dependent, then

$$d_{\mathrm{II}} = \frac{P-Q}{P+Q+T_c}.$$

In case neither variable is dependent,

$$\mathrm{sym}\,d = \frac{P-Q}{P+Q+1/2(T_r+T_c)}.$$

Kendall's *tau-b*

This measure considers ties on each variable separately but not on both variables. It takes values between –1 and 1, but it can attain +1 or –1 only for square tables. Consequently, it is often only recommended for tables where $r = c$.

$$\mathit{tau\text{-}b} = \frac{P-Q}{\sqrt{(P+Q+T_r)(P+Q+T_c)}}.$$

Under the null hypothesis that *tau-b* in the population sampled is zero, an appropriate two-sided p-value is calculated using the standard normal distribution. Thus,

$$z = \frac{\mathit{tau\text{-}b}}{\sqrt{\dfrac{4(r+1)(c+1)}{6nrc}}}$$

and p-value $= 2[1 - \mathrm{CDF}(\mathit{dist}, z)]$ where dist = `NormalDistribution[0,1]`.

Kendall's *tau-c*

This measure can attain, or nearly attain, +1 or –1 for any $r \times c$ table, and its value will not differ much from *tau-b* if marginal totals are close in value:

$$\mathit{tau\text{-}c} = \frac{2m(P-Q)}{n^2(m-1)},$$

where $m = \mathrm{minimum}\{r, c\}$.

Procedure

We have written a procedure, **`npmAssociationMeasures2x2`**, to calculate Yule's Q and the Phi Coefficient for contingency tables with two rows and two columns. We have written a procedure **`npmAssociationMeasures`** for any $r \times c$ contingency table to calculate the other measures of association discussed earlier. The user is cautioned to be sure that the categories are in order from smallest to largest down the rows if the row variable is ordinal. Likewise, the categories should be in order from smallest to largest across the columns if the column variable is ordinal. If the variables are not ordinal, then those measures that are intended for ordinal variables should be ignored. To use either of these procedures, first load the disk that comes with this book and click on its icon. Next, locate the procedure name and click on the rightmost cell bracket. Then press **ENTER** and return to the *Mathematica* work sheet.

`npmAssociationMeasures2x2[freqList]` calculates and displays Yule's Q and the Phi Coefficient for the contingency table with two rows and two columns stored as `freqList`, where `freqList={{f11,f12},{f21,f22}}`.

`npmAssociationMeasures[freqList]` calculates and displays measures of association for any $r \times c$ contingency table stored as `freqList`, where `freqList` = $\left[(f_{11},f_{12},\ldots,f_{1c}),(f_{21},f_{22},\ldots,f_{2c}),\ldots,(f_{r1},f_{r2},\ldots,f_{rc})\right]$.

EXAMPLE 3: Find the measures of association for the 2×2 contingency table given in Table 12.4.

SOLUTION: After we load **npmAssociationMeasures2x2**, we find Yule's Q and the phi coefficient for this 2×2 contingency table.

```
npmAssociationMeasures2x2[{{20,14},{14,23}}]
Title:Measures of Association
Yule's Q: 0.402439
Phi Coefficient: 0.209857
```

■

EXAMPLE 4: Find the measures of association for the 3×3 contingency table given in Table 12.9. This table classifies subjects according to one of three levels of social isolation and one three levels of use of a walking aid.

	Use of Walking Aid		
Social Isolation	Never	Occasionally	Always
Low	20	14	8
Medium	14	23	14
High	4	24	14

Table 12.9 Social Isolation versus Use of a Walking Aid

SOLUTION: First, we load **npmAssociationMeasures**. Then, we find the measures of association.

```
npmAssociationMeasures[{{20,14,8},{14,23,14},{4,24,14}}]
Title: Measures of Association
Chi Square: 15.1021
Contingency Coefficient: 0.317194
Cramer's V: 0.236503
Goodman and Kruskal Lambda: (symmetric) 0.0822785
  (row=dependent) 0.0833333 (column=dependent) 0.0810811
Somer's d: (symmetric) 0.24979 (row=dependent) 0.253315
  (column=dependent) 0.246362
Goodman and Kruskal Gamma: 0.372314   PValue: 0.0289511
Kendall's Tau-b: 0.249815   PValue: 0.00109307
Kendall's Tau-c: 0.245267
```

■

Tests for Heterogeneity (Tests of Homogeneity)

For a test of homogeneity, an investigator typically identifies two or more populations ($r \geq 2$) and selects samples from each population. Each sample member is then assigned to one of two or more categories ($c \geq 2$), and the frequencies are summarized in an $r \times c$ contingency table. If the distribution of the frequencies for the categories is the same for each population, then the populations are homogeneous. If not, the populations are called heterogeneous. Although the manner in which the data are collected and the rationale underlying the calculation of expected frequencies are different from the test of independence, the resulting contingency tables are analyzed using the same formulas and procedures based on the chi-square distribution.

Assumptions

1. The data are from independent random samples, one from each of the r populations.
2. The variables are categorical, or they may be quantitative with values that are classified into mutually exclusive categories.

Hypotheses

The null hypothesis is homogeneity of the r populations, and the alternative hypothesis is heterogeneity.

Note that if the null hypothesis is rejected, then the conclusion will be heterogeneity. That is, there is a difference among the populations. If the null hypothesis is not rejected, then the conclusion will be failure to show heterogeneity, *not* homogeneity. Thus, the test is useful to investigate heterogeneity but *cannot* be used to show homogeneity.

Procedures

The procedures **npmChiSquareTest** and **npmLogLikelihoodTest** should be used to test for heterogeneity using an $r \times c$ contingency table.

Comparing Population Proportions (Independent Samples)

If the members of each of c populations can be classified into two disjoint categories, which we may refer to, in general, as success (S) and failure (F), then an $2 \times c$ contingency table can be used to compare the proportions of successes among the populations. This is a special case of a test of homogeneity.

Hypotheses

Let p_i represent the proportion of successes in population i, $i = 1,\ldots,r$. Then the null hypothesis is $H_0{:}p_1 = p_2 = \cdots = p_r$ and the alternative hypothesis is H_a: Not all p_i's, $i = 1,\ldots,r$, are equal.

Procedure

Create a $2 \times c$ contingency table where each column represents a sample. Store the number of successes in the first row and the number of failures in the second row. The procedures **npmChiSquareTest** and **npmLogLikelihoodTest** should be used to test for a difference among the population proportions.

Median Test

The median test is another special case of a test of homogeneity for a $2 \times c$ contingency table. Sample data from c populations are combined, and the combined sample median is calculated. Then each sample member is classified according to its population and whether it is above ($>$) or not above ($\leq$) the combined sample median. The resulting $2 \times c$ contingency table can be used to test for a difference among the population medians. Note that because it uses more of the information in the data, the Kruskal–Wallis test is usually more powerful and is preferred when it can be used.

Assumptions

1. Independent random samples, one for each of c populations, where the median of population i is denoted by M_j, $j = 1,2,\ldots,c$.
2. The sample values are at least ordinal categorical data.
3. If all population medians are equal, each population has the same probability that an observed value will exceed the combined population median.

Hypotheses

The null hypothesis is $H_0{:}M_1 = M_2 = \cdots = M_c$. The alternative hypothesis is is H_a: Not all M_j's, $j = 1,2,\ldots,c$, are equal.

Procedure

Create a $2 \times c$ contingency table as just described. The procedures **npmChiSquareTest** and **npmLogLikelihoodTest** should be used to test for a difference among the population medians.

EXAMPLE 5: In the semifinals in the 1996 Women's 400 Meter Hurdles United States Olympic Team Trials, the top two in each heat plus the two best times qualified for the finals. The times in minutes are given in Table 12.6. Use the chi-square test to test for a difference among heats.

SOLUTION: After loading **npmChiSquareTest**, we enter the data. In order to determine how many members of the list are greater than the median, we first use `Flatten` to create a single list of times in `allheats` and find the median of this list to be 56.365. We also define the function `gOrLtest[x]` that yields `True` if x is greater than the median and `False` otherwise. In `tfTable`, we create a list of `True` and `False` by applying `gOrLtest` to the numbers in `allheats`.

```
heat1={54.88,54.96,55.91,55.99,56.67,57.29};
heat2={54.67,54.87,54.95,56.27,58.33,81.90};
heat3={55.66,56.46,56.74,57.86,58.90,59.56};
allheats={heat1,heat2,heat3};

alltimes=Flatten[allheats];

med=Median[alltimes]

56.365
```

```
gOrLtest[x_]:=If[x>med,True,False]

tfTable=Table[gOrLtest[allheats[[i,j]]],
  {i,1,Length[allheats]},{j,1,Length[allheats[[1]]]}]

{{False,False,False,False,True,True},
 {False,False,False,False,True,True},
 {False,True,True,True,True,True}}
```

In `fqTable`, we count the number of `True` and `False` elements in `heat1`, `heat2`, and `heat3`. The output indicates that there are two members of `heat1` and `heat2` greater than the median and four elements less than or equal to the median. In `heat3`, there are five elements greater than the median and one less than or equal to the median. We take the transpose of `fqTable` to obtain the 2×3 contingency table

$$\begin{pmatrix} 2 & 2 & 5 \\ 4 & 4 & 1 \end{pmatrix}.$$

When we run **npmChiSquareTest**, we find that $p \approx 0.135335$, so we fail to show a difference among heats.

```
fqTable=Table[{Count[tfTable[[i]],True],
   Count[tfTable[[i]],False]},{i,1,Length[tfTable]}]

{{2,4},{2,4},{5,1}}

trans=Transpose[fqTable]

{{2,2,5},{4,4,1}}

npmChiSquareTest[trans]

Title: Chi Square Test
Distribution: ChiSquare[2]
PValue: 0.135335
3.      3.      3.
3.      3.      3.
```

■

Comparing Population Proportions (Related Samples)

In Section 12.4, we discussed the McNemar test, which is used for data from two related samples. In a common application, each test subject receives the two treatments to be compared, and the response to each treatment is classified as either success or failure. Here we will discuss a procedure that is used if more than two "treatments" are received by the test subjects.

Cochran's Q Test

The usual model for the Cochran's Q test involves a set of $c \geq 2$ "treatments" that are applied independently to each of r subjects. In the context of a randomized block design, the "subjects" correspond to the blocks. The result of each treatment is recorded as either 1 for success or 0 for failure. The ones and zeros are then displayed in an $r \times c$ contingency table. Thus, the data for the ith block will be a list $(f_{i1}, f_{i2}, \ldots, f_{ic})$ where $f_{ij} = 1$ or 0 depending on whether the finding of the jth test was success or failure.

Assumptions

1. The r blocks are randomly selected from the population of all possible blocks.
2. The outcomes of the c "treatments" are 1 for success and 0 for failure and are displayed in an $r \times c$ contingency table.

Hypotheses

The null hypothesis is H_0: The "treatments" are equally effective. The alternative hypothesis is H_a: The "treatments" do not have the same effect.

Test Statistic

Recall that the row (block) totals are represented by R_i, $j = 1,\ldots,r$, and the column ("treatment") totals are represented by C_j, $j = 1,\ldots,c$. Cochran's test statistic is then calculated using

$$Q = \frac{c(c-1)\sum_{j=1}^{c} C_j^2 - (c-1)n^2}{cn - \sum_{i=1}^{r} R_i^2},$$

where n is the total number of 1's in the table, and

$$n = \sum_{j=1}^{c} C_j = \sum_{i=1}^{r} R_i \ .$$

Note that the purpose of the test statistic is to determine whether the "treatment" totals (C_j's) differ sufficiently to reject the null hypothesis. Because the value of Q is unaffected by blocks having all zeros or all ones, such rows should not be included in the data. In this case, the usual suggestion is that the p-value be calculated using the exact distribution for small tables. When $rc \geq 24$ and $r \geq 4$, the p-value can be accurately approximated with the chi-square distribution having $c - 1$ degrees of freedom. Note that Cochran's Q test is equivalent to the McNemar test when $c = 2$.

We have written a procedure, **npmCochransQTest**, to perform Cochran's test for data stored in an $r \times c$ contingency table, where the input lists represent the rows of the table. The user should be careful that each list contains no values other than zeros and ones. The procedure will ignore any rows that contain all zeros or all ones. Using the option setting `mtype->exact`, exact p-values can be calculated for $c = 3$ treatments and $c = 4$ treatments when the number of rows is relatively small. (Computer memory leads to this limitation.) For $c \geq 5$ treatments, we assume the user will have well over five rows, which is necessary for a meaningful statistical analysis. Consequently, the chi-square approximation is used to calculate the p-value. For larger numbers of rows, the default calculation uses the chi-square approximation. To use **npmCochransQTest**, first load the disk that comes with this book and click on its icon. Next, locate the procedure name and click on the right-most cell bracket. Then press **ENTER** and return to the *Mathematica* work sheet.

`npmCochransQTest[freqList,opts]` performs Cochran's Q test using the frequencies in `freqList`. Note `freqList` = $\left[(f_{11},f_{12},\ldots,f_{1c}),(f_{21},f_{22},\ldots,f_{2c}),\ldots,(f_{r1},f_{r2},\ldots,f_{rc})\right]$, where each f_{ij} is either zero or one. The default option is `mthd->approx`. The other option is `mthd->exact`. The column totals and the p-value are displayed.

EXAMPLE 6: Each of four basketball fans devised a system for predicting outcomes of collegiate basketball games. Six games were selected at random, and each fan predicted the outcome of each game. The results were tabulated, using 1 for a successful prediction and 0 for an unsuccessful prediction. The results are shown in Table 12.10. Test the null hypothesis that each fan is equally effective in the ability to predict outcomes of the basketball games.

	Fan				
Games	1	2	3	4	Totals
1	1	1	0	0	2
2	1	1	1	0	3
3	1	1	1	0	3
4	0	1	1	0	2
5	0	1	0	0	1
6	1	1	0	1	3
Totals	4	6	3	1	14

Table 12.10

SOLUTION: After loading **npmCochransQTest**, we enter the information by blocks in predictions. With the option setting mthd->approx, we obtain the approximate *p*-value, $p \approx 0.053311$, based on the chi-square distribution with three degrees of freedom. Thus, we would not reject the null hypothesis at the 5% level of significance. On the other hand, based on the *p*-value using the exact calculation procedure, $p \approx 0.0481771$, we can reject the null hypothesis at the 5% level of significance.

```
predictions={{1,1,0,0},{1,1,1,0},{1,1,1,0},
  {0,1,1,0},{0,1,0,0},{1,1,0,1}};
```

```
npmCochransQTest[predictions,mthd->approx]
```

```
Title: Cochran Q Test
Test Statistic: 7.8
Column Totals: {4,6,3,1}
Distribution: Chi Square[3]
PValue: 0.0503311
```

```
npmCochransQTest[predictions,mthd->exact]
```

```
Title: Cochran Q Test
Test Statistic: 7.8
Column Totals: {4,6,3,1}
Distribution: Exact
PValue: 0.0481771
```

■

Selected References

Abell, Martha L. and Braselton, James P., *Differential Equations with Mathematica*, Second Edition, Academic Press, 1996.

Abell, Martha L. and Braselton, James P., *The Mathematica Handbook*, AP Professional, 1992.

Blachman, Nancy, *Mathematica: A Practical Approach*, Prentice-Hall, 1992.

Blachman, Nancy, *Mathematica: Quick Reference, Version 2*, published by Variable Symbols and distributed by Addison-Wesley, 1992.

Bowerman, Bruce L. and O'Connell, Richard T., *Time Series Forecasting: Unified Concepts and Computer Implementation*, Second Edition, Duxbury Press, 1987.

Bradley, James V., *Distribution-Free Statistical Tests*, Prentice-Hall, 1968.

Chatterjee, Samprit and Hadi, Ali S., *Sensitivity Analysis in Linear Regression*, John Wiley & Sons, 1988.

Conover, W. J., *Practical Nonparametric Statistics*, John Wiley & Sons, 1971.

Daniel, Wayne W., *Applied Nonparametric Statistics*, Second Edition, PWS-Kent, 1990.

Draper, N. R. and Smith, H., *Applied Regression Analysis*, Second Edition, John Wiley & Sons, 1981.

Evans, Merran, Hastings, Nicholas, and Peacock, Brian, *Statistical Distributions*, Second Edition, John Wiley & Sons.

Frees, Edward W., *Data Analysis Using Regression Models: The Business Perspective*, Prentice Hall, 1996.

Ghahramani, Saeed, *Fundamentals of Probability*, Prentice Hall, 1996.

Hajek, Jaroslav and Sidak, Zbynek, *Theory of Rank Tests*, Academic Press, 1967.

Hochberg, Yosef and Tamhane, Ajit C., *Multiple Comparison Procedures*, John Wiley & Sons, 1987.

Iman, Ronald L., Quade, Dana, and Alexander, Douglas A., "Exact probability levels for the Kruskal—Wallis test," *Selected Tables in Mathematical Statistics*, Volume 3, Institute of Mathematical Statistics, American Mathematical Society, 1975, pp. 329—384.

Johnson, Norman L. and Kotz, Samuel, *Continuous Univariate Distributions-1 and 2*, Houghton Mifflin, 1970.

Johnson, Norman L. and Kotz, Samuel, *Distributions in Statistics: Continuous Multivariate Distributions*, John Wiley & Sons, 1972.

Johnson, Richard A. and Wichern, Dean W., *Applied Multivariate Statistical Analysis*, Third Edition, Prentice Hall, 1992.

Kelly, Douglas G., *Introduction to Probability*, Macmillan Publishing, 1994.

Kendall, Maurice and Gibbons, Jean Dickenson, *Rank Correlation Methods*, Fifth Edition, Oxford University Press, 1990.

Kreyszig, Erwin, *Advanced Engineering Mathematics*, Seventh Edition, John Wiley & Sons, 1993.

Lehmann, E. L., *Nonparametrics: Statistical Methods Based on Ranks*, Holden-Day, 1975.

Looney, Stephen W. and Gulledge, Thomas R., Jr., "Use of the Correlation Coefficient with Normal Probability Plots," *The American Statistician*, Volume 39, Number 1, February 1985, pp. 75—79.

Maeder, Roman, *The Mathematic Programmer*, AP Professional, 1995.

Maeder, Roman, *Programming in Mathematica*, Second Edition, Addison-Wesley Publishing Co, 1991.

Maxwell, A. E., *Analysing Qualitative Data*, Methuen & Co., Ltd., 1961.

Mathematica in Education (Paul Wellin, Editor, Department of Mathematics, Sonoma State University, 1801 East Cotati Avenue, Rohnert Park, California, 94928, E-mail: wellin@sonoma.edu).

The Mathematica Journal (Academic Press, 525 B Street, Suite 1900, San Diego, CA 92101-4495. E-mail: tmj@acad.com).

Myers, Raymond H., *Classical and Modern Regression with Applications*, Second Edition, PWS-Kent, 1990.

Netter, John, Kutner, Michael H., Nachtsheim, Christopher J., and Wasserman, William, *Applied Linear Statistical Models*, Fourth Edition, Irwin, 1996.

Olkin, Ingram, Gleser, Leon J., and Derman, Cyrus, *Probability Models and Applications*, Second Edition, Macmillan College Publishing, 1994.

Pierce, Albert, *Fundamentals of Nonparametric Statistics*, Dickenson, 1970.
Pratt, John W. and Gibbons, Jean D., *Concepts of Nonparametric Theory*, Springer-Verlag, 1981.
Ross, Sheldon, *A First Course in Probability*, Fourth Edition, Macmillan College Publishing, 1994.
Rossman, Allan J., *Workshop Statistics: Discovery with Data*, Springer, 1996.
Sheskin, David J., *Handbook of Parametric and Nonparametric Statistical Procedures*, CRC, 1997.
Sokal, Robert R. and Rohlf, F. James, *Biometry*, Third Edition, W. H. Freeman, 1995.
Toothaker, Larry E., *Multiple Comparisons for Researchers*, Sage Publications, 1991.
Walsh, Anthony, *Statistics for the Social Sciences with Computer Applications*, Harper & Row, 1990.
Zar, Jerrold H., *Biostatistical Analysis*, Third Edition, Prentice Hall, 1996.

Wolfram Research, Inc. also publishes the following technical reports:

Guide to Standard *Mathematica* Packages
Mathematica Warning Messages
Installation Manual
Release Notes for *Mathematica* Version 3.0
The 3-Script File Format
MathLink Reference Guide
MathSource
PostScript Generated by *Mathematica*
The *Mathematica* Compiler
Upgrading Packages to *Mathematica* 3.0
Major New Features in *Mathematica* Version 3.0

For information, including purchasing information, about *Mathematica*, contact:

Corporate Headquarters:
Wolfram Research, Inc.
100 Trade Center Drive
Champaign, IL 61820
USA
Telephone: 217-398-0700
Fax: 217-398-0747
E-mail: info@wolfram.com
Web: http://www.wolfram.com/

Europe:

Wolfram Research Europe Ltd.
10 Blenheim Office Park
Lower Road, Long Hanborough
Oxfordshire OX8 8LN
UNITED KINGDOM
Telephone: +44-(0) 1993-883400
Fax: +44-(0) 1993-883800
E-mail: info-europe@wolfram.com

Asia:

Wolfram Research Asia Ltd.
Izumi Building 8F
3-2-15 Misaki-cho
Chiyoda-ku, Tokyo 101
JAPAN
Telephone: +81-(0)3-5276-0506
Fax: +81-(0)3-5276-0509
E-mail: info-asia@wolfram.com

For information, including purchasing information, about *The Mathematica Book,* contact:

Wolfram Media, Inc.
100 Trade Center Drive
Champaign, IL 61820, USA
E-mail: info@wolfram-media.com
Web: http://www.wolfram-media.com

Index

Q

R

S

T

U

V

Quick Reference

Frequently Used Abbreviations

Symbol	Meaning
_	Blank
;	CompoundExpression
/;	Condition
'	Derivative
/	Divide
.	Dot
==	Equal
<<	Get
!	Factorial
[...]	(function evaluation)
//	(function evaluation)
>	Greater
>=	GreaterEqual
??	Information
<	Less
<=	LessEqual
\	(line continuation)
{...}	List
/@	Map
-	Minus
%	Out
[[...]]	Part
+	Plus
^	Power
>>	Put
/.	ReplaceAll
->	rule
=	Set
:=	SetDelayed
-	Subtract
*	Times

Frequently Used Commands

Topic	Example	Command
Compute a **determinant**	Compute $\begin{vmatrix} 4 & -3 \\ 2 & 1 \end{vmatrix}$.	`Det[{{4,-3},{2,1}}]`
Compute a **limit**	Calculate $\lim_{x \to 2} \frac{x^2+x-6}{x^2+2x-8}$.	`Limit[(x^2+x-6)/ (x^2+2x-8),x->2]`
Compute a **power series**	Compute the first five terms of the power series for $\sin x$ about $x=0$.	`Series[Sin[x],{x,0,5}]`
Compute the **partial fraction decomposition** of an expression	Find the partial fraction decomposition of $\frac{x}{x^2-3x-4}$.	`Apart[x/(x^2-3x-4)]`
Define a **function**	Define $f(x)=\frac{x^2}{x^2+1}$.	`f[x_]=x^2/(x^2+1)`
Define a **matrix**	Define $\mathbf{A}=\begin{pmatrix} 4 & 3 \\ 5 & 0 \end{pmatrix}$.	`matrixa={{4,3},{5,0}}`
Differentiate an expression	Compute $\frac{d}{dx}(x \sin x)$.	`D[x Sin[x],x]`

(continued on back side)

Display an expression as a single **fraction**.	Write $1+\frac{1}{x}$ as a single fraction.	`Together[1+1/x]`
Factor a polynomial.	Factor $5x^2-8x-4$.	`Factor[5x^2-8x-4]`
Find the **eigenvalues** and corresponding **eigenvectors** of a matrix	Find the eigenvalues and corresponding eigenvectors of $\begin{pmatrix}-17 & -15\\ 20 & 18\end{pmatrix}$.	`Eigensystem[ {{-17,-15},{20,18}}]`
Find the **inverse** of a matrix	Calculate $\begin{pmatrix}-4 & 3\\ 4 & -4\end{pmatrix}^{-1}$.	`Inverse[{{-4,3},{4,-4}}]`
Graph a function	Graph $\frac{x}{x^2+1}$ on the interval $[-6,6]$.	`Plot[x/(x^2+1),{x,-6,6}]`
Graph a **function of two variables** in three dimensions	Graph $\sin x\cos y$ for $0\le x\le 4\pi$ and $0\le y\le 2\pi$.	`Plot3D[Sin[x]Cos[y], {x,0,4Pi},{y,0,2Pi}]`
Graph parametric equations	Graph $\begin{cases} x=\cos t\\ y=4\sin t\end{cases}$ for $0\le t\le 2\pi$.	`ParametricPlot[ {Cos[t],4Sin[t]}, {t,0,2Pi}]`
Graph several functions	Graph $\sin x^2$ and $\sin^2 x$ on the interval $[0,2\pi]$.	`Plot[{Sin[x^2],Sin[x]^2}, {x,0,2Pi},PlotStyle-> {GrayLevel[0], Dashing[{.03}]}]`
Integrate an expression	Compute $\int x\sin x\,dx$.	`Integrate[x Sin[x],x]`
Multiply an algebraic expression.	Compute $(5x+2)(x-2)$.	`Expand[(5x+2)(x-2)]`
Raise a **matrix** to a **power**.	Compute $\begin{pmatrix}-1 & -3\\ -1 & 3\end{pmatrix}^5$.	`MatrixPower[ {{-1,-3},{-1,3}},5]`
Reduce a fraction to lowest terms	Reduce $\frac{x-1}{x^2-1}$ to lowest terms.	`Cancel[(x-1)/(x^2-1)]`
Solve a **differential equation**	Solve $y'=1+y$.	`DSolve[y'[x]==1+y[x], y[x],x]`
Solve a **system of equations**	Solve $\begin{cases} 2x-y=7\\ 4x+2y=2\end{cases}$.	`Solve[{2x-y==7,4x+2y==2}]`
Solve an **equation**	Solve $x^2-4x-5=0$.	`Solve[x^2-4x-5==0]`